Narrative and Cognition in Literature and Science

Literatur- und Naturwissenschaften

Publikationen des
Erlangen Center for Literature and Natural Science/
Erlanger Forschungszentrums für Literatur und
Naturwissenschaften (ELINAS)

Edited by
Aura Heydenreich, Christine Lubkoll and Klaus Mecke

Volume 11

Narrative and Cognition in Literature and Science

Edited by
Michael Sinding, Aura Heydenreich and Klaus Mecke

DE GRUYTER

Funded through ELINAS - Erlangen Center for Literature and Natural Science

ISBN 978-3-11-078279-0
e-ISBN (PDF) 978-3-11-078284-4
e-ISBN (EPUB) 978-3-11-078287-5
ISSN 2365-3434
DOI https://doi.org/10.1515/9783110782844

Library of Congress Control Number: 2025931004

Bibliographic information published by the Deutsche Nationalbibliothek
The Deutsche Nationalbibliothek lists this publication in the Deutsche Nationalbibliografie; detailed bibliographic data are available on the internet at http://dnb.dnb.de.

Cover image: Joseph Reinthaler
Typesetting: Integra Software Services Pvt. Ltd.
Printing and binding: CPI books GmbH, Leck

www.degruyter.com
Questions about General Product Safety Regulation:
productsafety@degruyterbrill.com

Acknowledgements

We would like to offer thanks to all who supported the transformation of the symposium into a publication, including those who helped greatly to develop the concept of the symposium and its design, in particular ELINAS colleagues Klaus Mecke, Aura Heydenreich, Stefan Winter, Maria Wolff, Alex Laska and Miriam Rückelt. Miriam has also been deeply involved in the preparation of the volume, providing invaluable support as a research assistant and editorial assistant. Thanks also to FAU leadership for providing the funding for the conference, as well as the superb setting (the Orangerie) and extensive administrative and technical support.

https://doi.org/10.1515/9783110782844-202

Contents

Michael Sinding

Introduction: Narrative and Cognition in Literature and Science

This volume grew out of a symposium held at the Erlangen Center for Literature and Natural Science (ELINAS). The idea of the symposium was to consider cognition as the link between narrative and science. Let us consider in more detail the main topics and big questions that engaged the collective attention of the authors gathered here.

1 Symposium topics and questions

Part of the conception of ELINAS from its foundation was that certain core literary elements – most importantly *narrative, metaphor, and genre* – have cognitive purchase, in the sciences as well as in literature. Far from being merely ornamental, or formal in a superficial sense, they are fundamental *forms of thought,* and they deeply imprint imagination, discovery and communication in the arts, the sciences, and the echoing between them. The question behind this volume is: What happens when we use cognition to link narrative to literature and science?

1.1 Thumbnail sketch of the entangled big pictures of narrative, cognition, literature & science

An introduction can only gesture towards movements and potential connections in the topics proposed (and the related subfields convened), since they are all so large and vital. But certain artifacts can help to capture how these interests can form conjunctions as they circulate in our culture, like intersecting vectors. For example, a work of popular science journalism, Carolyn Abraham's 2002 book *Possessing Genius: The True Account of the Bizarre Odyssey of Einstein's Brain,* is partly about the dissected brain's travels in mayonnaise jars and pickle jars, and in car trunks, getting passed from hand to hand as people try to study chunks of it. The narrative-genre framing captures the tone: the subtitle calls it an "Odyssey," a reviewer calls it "picaresque." Abraham's book is the only one I know that opens with the story of a developing brain, complete with humanized "characters" of neurons, and metaphors and allegories for their intrepid enterprise. And this choice is tied in to the main neurocognitive story, as we'll see.

https://doi.org/10.1515/9783110782844-001

> [The brain] grew in the belly of a German woman in the summer of 1878. It was barely a speck beneath skirts and petticoats and flesh, drawing forest-scented air from the Swabian Alps northwest of the Danube. [. . .] Cells within the [cellular] tube replicated. Some served the role of nursemaids, catering to those that would grow into mighty neurons. Together, not long after conception, they began to construct the magical scaffolding of human thought. [. . .] By the fourth week, 125,000 cells assembled along the groove in the middle of the neural tube, each an ancestor of the some 100 billion neurons that would eventually stun the world. [. . .] Then they drifted forth, like pioneers into a fluid wilderness, to settle the cerebral frontier. . . . (Abraham 2002, 3–4)

From nursemaid to pioneer, as the neurons turn their hands to sudden expansion to the periphery (suggesting Einstein's own pioneering, of course). And then, a story of metamorphosis, and of hidden treasure.

> It looked like a lima bean by the fall of 1878, just over a centimetre long. The upper curve jutted toward the face, rounding in the lower portion to form the remarkable eyeballs that would float one day in a New Jersey safety deposit box. [. . .] As [the neurons] clumped and shifted, the lima bean transformed into a walnut, its outer shell wrinkling and crumpling to allow its considerable surface area to fit neatly within the confines of the skull. Somewhere amid the valleys and ridges of the grey crust lay the potential to envision and reason, to calculate and write, to translate imagined observations of space and time into profound mathematical equations. (Abraham 2002, 7)

The foreshadowing of the "picaresque" frame of the wild and weird journey (eyeballs floating in the safety deposit box) nuances the pathos with hints of adventure as we await this mysterious-birth-from-buried-treasure.

However, this is also in part a history of neuroscience, as for decades after the autopsy in 1955, brain scientists didn't know enough to study Einstein's brain seriously. No one even wanted it. This becomes a story of reversal after patient waiting, of neuroscience catching up and becoming "hot" in the "Decade of the Brain" (the 1990s) declared by George H. W. Bush and sponsored by large US research organizations, which enabled Sandra Witelson, at the end of the book, finally to discover an unusual feature of the brain, one that corresponds to something unusual in Einstein's thinking. Specifically, "Each hemisphere of his parietal lobes bulged 2.5 centimetres wider than the control group, an average of 15 percent." She suggests that "some chemical instruction had inflated the dimensions of his parietal lobes, effectively damming the normal fault line of the Sylvian fissure [. . .] and leaving a vast, open plane across which his neurons could forge their busy networks" (Witelson in Abraham 2002, 307). And so, 300 pages after the lima bean becomes a walnut, the final revelation has dropped, and we're back again to that pre-natal "speck" – there's where the roots of genius may lie after all:

> Witelson wrote: "Einstein's own description of his scientific thinking was that 'words do not seem to play a role,' but there is 'associative play' of 'more or less clear images' of a visual and muscular type" [. . .]. It was [says Abraham] the kind of thinking that might have sprung from the hind regions of his parietal lobes. She suspected that this wondrous advantage had not emerged from years of obsessively formulating theories and equations, but that it had sprouted long before then, perhaps when the brain was little more than a speck in the belly of a German woman. (Witelson in Abraham 2002, 307)

We've come full circle, and Einstein's thinking is tied in to new developments in cognitive science and narratology: That "visual and muscular" imagination evokes recent "embodied" approaches to mind; *and* this is proposed to have motivated his thought experiments, which are themselves little stories, almost parables in how they set out a schematic situation, then simulate it briefly in order to draw a lesson. For our own lesson, this narrativized trajectory of brain, neuroscience and analysis of scientific thinking can illustrate the increasingly overlapping states of research on narrative, on cognition, and on literature and science, to be explored more fully in what follows.

1.1.1 Narrative research, science and cognitive narratology

Recent decades have seen the expansion of narrative study beyond literary narrative; outwards to many disciplines and research traditions (we'll hear from several), more and more *kinds* of narrative studied, many narratolog*ies*, with the modifier "cognitive" increasingly prominent, and growing interest in narrative in the sciences. However, while the 2005 *Routledge Encyclopedia of Narrative Theory*, still among the essential sources for the subject, illustrates all these other trends in narrative study, it is rather quiet on the sciences. An entry for "Narrative Turn in the Humanities" mentions natural science, but Medicine is the only science really engaging with narrative. There is a "science and narrative" entry by Arkady Plotnitsky, focusing on Jean-François Lyotard's analysis of the *Postmodern Condition*, in which the growth of science is a major factor in the decline of belief in the "grand narratives" of traditional worldviews. Norbert Meuter's 2013 article on "Narration in Various Disciplines" in the *Living Handbook of Narratology* reviews narration in Literary Studies, the Arts, the Historical Sciences, Psychology, Psychoanalysis, Philosophy, Ethics, Sociology, Theology, Pedagogy, Law Studies, Medicine and Philosophy of Science. "Natural Sciences" and "Systems Theory" are appended as "Topics for Further Research" at the end, with the remark "a specifically narratological inquiry in the natural sciences remains a desideratum" (par. 36), with the suggestions that this may require a revised view of nature as a dynamic and creative process (rather than as laws and causal connections), as in

Whitehead's cosmology, and perhaps systems theory. The status of narrative in science as "future research" is beginning to change, as we will find in our Touchstones below, and Whitehead's name will also return as a figure of reconceptualizing both nature and science in a more open and holistic way.

Narratology became cognitive in the 1990s and 2000s, with the publication of two major works of synthesis and several collections. Monika Fludernik's remarkable *Towards a 'Natural' Narratology* (1996) adapted literary linguistics and literary history and theory into a cognition-based reconceptualization of core ideas of narratology. There had been foreshadowings of a mind-focused approach to narrative, but Fludernik's book was expansive and influential enough to mark full fruition. David Herman's monumental *Story Logic* (2002) is an equally important contribution. With its mastery of narrative research, broadened by its well-considered engagement with the wider fields of both literary study and cognitive science, its impact was powerful. The next year produced the ground-breaking *Narrative Theory and the Cognitive Sciences* (2003), edited by Herman, which convened leading thinkers across the fields to produce a multi-perspectival appraisal of their connections. A number of nearly-equally important publications could be mentioned as the field has grown.[1] Today, cognitive narratology is a very well-established, indeed essential, dimension of narrative study.

1.1.2 Cognitive research and literature

Cognitive research too has recently become much deeper and more expansive. Gone are the early days that touted models of mind as computer, and thinking as information-processing. Lakoff and Johnson were already speaking of a post-computational "second-generation" cognitive science in 1999 (*Philosophy in the Flesh*). Many kinds of cognition are explored (multiple modules, multiple intelligences). So-called "4E" cognition is embodied, embedded, enactive, extended ("5E" more accurately, because the emotional is also often specified). We think with our bodies: The abstract and logical remains grounded in the concrete and physical, as we can discern from many kinds of analysis and experimental evidence. Cognitive scientists and neuroscientists propose general theories of mechanisms for this phenomon: for example, the "scaffolding" of higher cognition on early

1 Other major publications in the development of cognitive narratology include Hogan's *The Mind and Its Stories* (2003); Palmer's *Fictional Minds* (2004) and *Social Minds and the Novel* (2010); Zunshine's *Why We Read Fiction* (2006). To see cognitive narratology in the context of other approaches, see the two-part "Histories of Narrative Theory" by David Herman and Monika Fludernik, in Phelan and Rabinowitz's *Companion to Narrative Theory* (2005).

physical experience (Williams, Huang and Bargh 2009); and the "neural reuse" of the same neural systems that support perception and motion for abstract thinking (Anderson 2010).

- Embodied, in that conceptual capacities are shaped by perceptual capacities – the world we think about is the world we experience, and sensory modalities have their own forms of thinking (visual, auditory, haptic, motor, etc.).
- Embedded, in that thought is deeply situated in the physical and sociocultural contexts in which it occurs.
- Enactive, in that thought tends to be geared to action, and preparing for action.
- Extended, in that we think not just with our individual brains, but with tools (e.g. pencil and paper) and instruments (e.g. computers) and other supports – most importantly, other people's minds.
- Emotional, in that the affective systems that are evolutionarily much older than the cerebral cortex tether even the most abstract thinking to the powerful emotional valences that shape our responses to our worlds (pleasure and pain, happiness and sorrow).

This can change our epistemologies and even ontologies; and give new life to cognitive literary studies (CLS), as we'll see below. It can also align with some current studies of Literature and Science (L&S), which emphasize material aspects of science-culture – the body engaged in practices with technologies.[2]

Cognitive literary studies (CLS) began in the late 1980s and early 1990s with far-reaching explorations of how cognitive linguistics and cognitive psychology could recast our understanding of what literature is and how it works, in landmark publications by scholars such as Mark Turner (e.g. *Reading Minds: The Study of English in the Age of Cognitive Science*, 1991; *More Than Cool Reason: A Field Guide to Poetic Metaphor*, co-written with George Lakoff, 1989) and Ellen Spolsky (*Gaps in Nature: Literary Interpretation and the Modular Mind*, 1993). It flowered rapidly, with important introductions and collections appearing – e.g. Peter Stockwell's *Cognitive Poetics* (2002), Patrick Colm Hogan's *Cognitive Science, Literature and the Arts* (2003), Richardson and Spolsky's *The Work of Fiction: Cog-*

2 For recent cognitive studies of literature developing these approaches, see the Special Issue of *Style* on "Second Generation Cognitive Literary Study," edited by Karin Kukkonen and Marco Caracciolo (48.3, 2014); the Special Issue of *Narrative* on "Social Minds in Factual and Fictional Narration," edited by Eva von Contzen and Max Alders (23.2, 2015); *The Cognitive Humanities: Embodied Mind in Literature and Culture*, edited by Peter Garratt (2016); and the Special Issue of *Poetics Today* on "Situated Cognition and the Study of Culture," edited by Ellen Spolsky, Ben Morgan and Sowon Park (38.2, 2017).

nition, Culture, and Complexity (2004). It is now a settled and solid presence in the broader field, if not always and everywhere mainstream.[3] As the next sections will illustrate, it has the potential to interact with L&S in interesting ways.

1.2 Closer encounters: Literature, science, cognition and narrative

To evoke deeper background for this volume, one might ask, how do narrative and cognition figure in the histories of L&S and CLS? To answer this question, I consider three touchstones. The first touchstone examines how narrative and cognition figure in the evolution (i.e. developments over successive editions) of a foundational text of L&S, Gillian Beer's *Darwin's Plots*. The other two touchstones are recent juxtapositions that clarify the actual and potential interconnections between two subfields which do not at first glance have a great deal in common. The second touchstone highlights the movement known as cognitive historicism. The signicance of this movement is revealed in a few rare intersections of L&S and CLS, chiefly in two essays from leading scholars that articulate representative views from each perspective of the other subfield (Joseph Tabbi, Alan Richardson), and one essay by a historian that articulates something in-between (Reviel Netz). The response to the latter essay by the volume editor brings us full circle, in a sense, because it points to an authoritative statement of the importance of CLS to the latest thinking in L&S. The third touchstone highlights the increasing recognition of the potential of the concept of narrative to act as a unifying force between literature and science, via a recent essay on the topic (Daniel Aureliano Newman).

In order to explore these juxtapositions, we will first need a quick bird's-eye-view of L&S as a field.

1.2.1 "Literature and science" as a field: Two cultures reconfigured

The study of L&S can claim a lineage going back at least to the exploration of the "two cultures" by C. P. Snow in 1959, although problems with the "two cultures" view are widely recognized, and lineages tracing the intertwining of the arts and sciences can go back as far as the origins of those phenomena. Perhaps a more

3 For an overview of the field, see Zunshine's *Oxford Handbook of Cognitive Literary Studies* (2015).

meaningful reference point for the conscious creation of a specific subfield is that the Society for Literature, Science and the Arts was inaugurated in 1985, its official publication, the journal *Configurations: A Journal of Literature, Science, and Technology* has been publishing since 1993, and the populations of members, authors and readers has grown steadily. The subfield has always been shaped by studies of science in sociology, history and philosophy. Influential in the early formation of the field were sociologies of science going under names like "science studies" or Science and Technology Studies (STS) or the earlier Sociology of Scientific Knowledge (SSK); Bruno Latour is a major figure. Major figures from the "literature" side tend to hail from the "cultural studies" galaxy (for a handful of leading thinkers, see for example the SLSA "lifetime achievement" award winners, e.g. Barbara Herrnstein Smith, Katherine Hayles, Donna Haraway). But many other thinkers are relevant, because there are so many potentially productive crossover topics and fields (e.g. Charles Bazerman on scientific genres; Mary Hesse, Nancy Nersessian and others on metaphor and analogy in science; many who study the history and philosopy of science).

Some L&S scholars were protagonists in the "science wars" of the 1990s, in which scientists raised the alarm about postmodern/post-structuralist thinkers undermining scientific realism and thereby leading scholars and the public down the dangerous road of relativism about truth and reality. The science wars became quite public with episodes such as the "Sokal Affair" of 1996, in which physicist Alan Sokal published a hoax article ("Transgressing the Boundaries: Towards a Transformative Hermeneutics of Quantum Gravity") in the then-cutting-edge journal *Social Text*, then revealed the hoax simultaneously in the journal *Lingua Franca*, the point being to expose the intellectual vacuity of the "cultural studies" view of science as nothing more than linguistic/social construction. As with C. P. Snow's polemical outline of the "two cultures" decades earlier (celebrating applied scientists and deriding literary intellectuals), responses were swift, fierce and polarizing.

Some respondents formed alliances calling for a "third culture" and a "new humanities" more or less ruled by scientists with humanist aspirations but little expertise in the humanities (Dennett being an exception). These were mostly thinkers who had pushed evolutionary analysis into areas such as philosophy (Daniel Dennett's developed an evolutionary theory of the stream of consciousness, and explored evolution's implications for the "meanings of life"), sociology (Edward O. Wilson's sociobiology), culture (Richard Dawkins's theory of what he called "memes", eventually became a theory of cultural transmission, and has now morphed into a term for bits of internet virality), and psychology (Steven Pinker analyzed everything from language to gender roles to politics to art in terms of evolutionary psychology). These thinkers had congregated on John

Brockman's website Edge.org, with its major project of advancing a *Third Culture* that would remarry the "two cultures" that Snow had declared divorced. For example, Wilson's notion of "consilience" saw humanistic questions studied via scientific methods, an approach that was unsurprsingly generally unwelcome in the humanities.

Part of the problem is the real distance between the humanities and the sciences, with respect to objects of study, theories, methods, etc., and the real difficulty of attaining deep enough knowledge of both (in any of their subfields) to productively combine them. This is a problem of education at the institutional level, but also a simple problem of any individual's time and capacity. A related institutional part of the problem, as many observe, is the apparent discrepancy in power and prestige between literary scholars and scientists. As Peter Dear says, the reason one cannot speak of literature and science as equals is the "special status routinely accorded science" (2017, 505). Scientists are not so sure about this. Scientists may get more attention from politicians and the media than do literary scholars, but that attention is not always positive. Scientists get larger grants than humanists because their research is more expensive; it does not make them rich as tech tycoons. On the other hand, literary scholars can become public intellectuals, while literary writers have their own kind of special status, along with potential fame and wealth far beyond any scientist. And (as Newman observes below), scientists may perceive themselves to be in second place in a subfield whose designation assigns them that place (Literature and Science, not Science and Literature). These two factors result in another real discrepancy, in the relative populations of literary scholars and scientists in L&S. The introduction to a previous ELINAS volume cites Jay Labinger's 2017 question in the *Journal of Literature and Science*, "Where are the scientists in Literature and Science?" (Heydenreich and Mecke 2021, 1).

A more conciliatory response to the science wars was to get scientists and humanists to actually speak to one another on more level footing on questions of common interest, as in Jay Labinger and Harry Collins's collection *The One Culture? A Conversation About Science* (2001). The approach of ELINAS has been in this line, and the results have been satisfying. ELINAS has also sought to redress the balance by bringing more scientists into its organization, and raising the issue in discussion. Dialogue between literature and science has been built into the idea of the organization from the start. ELINAS announces itself as "A Forum for Dialogues" and is "dedicated to the reciprocal transfer of knowledge between physics and literature" (http://elinas.fau.de/index.en.html); it "provides a platform for exchange between physics, literature and literary criticism" (http://elinas.fau.de/about.html). This commitment is central to its "Scientific Vision":

> unconventional avenues of communication between highly specialised expert discourses are necessary to advance research in this field. For scientific fields aware of their social responsibility and therefore interested in a cooperative take on cultural innovations and risks, however, a successful dialogue is essential. (http://elinas.fau.de/research/)

ELINAS aims to create an “interdisciplinary discourse zone” (Heydenreich and Mecke 2021, 3), and with institutional support, its events have been successful in these aims, to varying degrees (some events are clearly weighted towards scientists while others are clearly weighted towards literary scholars).[4] Still the problems of imbalance continue, despite being widely recognized. There remain unresolved issues, and one senses that the embers are still there, waiting to be fanned into flames, given the right occasion. But things have moved on and become more nuanced. Recent handbooks and companions to L&S provide various maps to the field.[5] These often have historical dimensions, with various phases, waves or generations. We will return to the historical interests of the subfield in our second touchstone, as it connects with historicist impulses in CLS. But our first touchstone seeks to gauge the centrality of narrative to the history of L&S, and to register the role of cognition in the ways narrative connects the two, by reconsidering an indispensable foundational text for L&S, Gillian Beer’s superb *Darwin’s Plots*.

1.2.2 First touchstone: Narrative and cognition in the foundations of L&S: Gillian Beer’s *Darwin’s Plots*

Suitably to the topic of evolution in literature, we can foreground certain aspects of Beer’s argument to reveal how later variants of narrative and cognitive analysis descend from ancestral forms – and indeed narrative and cognition seem to have become more pronouncedly entwined over successive editions. The book was originally published in 1983. In her preface to the second edition (1999), Beer elaborates on the rationale for her book’s emphasis on the thought-discourse and argument-narrative interfaces:

---

4 ELINAS hosted the SLSAeu conference in 2023, with the topic of “Models, Metaphors, Simulations: Epistemic Transformations in Science, Literature and the Arts” (https://www.slsaeu23.fau.de/).

5 See, for example, *The Palgrave Handbook of Early Modern Literature and Science*, ed. Marchitello and Tribble (2017); *The Palgrave Handbook of Twentieth and Twenty-First Century Literature and Science*, ed. Ahuja et al. (2020); *The Cambridge Companion to Literature and Science*, ed. Meyer (2018); *The Routledge Handbook of Literature and Science*, ed. Clarke and Rossini (2011). The Society for Literature, Science and the Arts has two book series, *AnthropoScene* and *Proximities*.

> It argues that how Darwin said things was a crucial part of his struggle to think things, not a layer that can be skimmed off without loss. It shows how his non-technical language [. . .] allowed a wide public to read his work and appropriate his terms to a variety of meanings (Nature, race, man, struggle, fit, and family would be examples of story-generating words). My argument demonstrates the degree to which narrative and argument share methods; indeed, it enquires what differences can be maintained between narrative and argument. (Beer 2009, xxv)

She goes on to ask of the present conditions of science and story, "What new tales are being unleashed from scientific work now? and what new forms for storytelling? Must there be new grammars for narrative or is Propp's view of fixed folktale grammars, based on the analogy with morphology, still viable? Are there stories to be told from places and organisms until now unrecognised?" (Beer 2009, xxvii). These are large questions, and contributors to the present collection explore them in various contexts. Yet Beer's reference to Propp is symptomatic of her reliance on mid-twentieth century "classical" narratology, as opposed to today's climate of pluralistic "postclassical" narratology, with its amalgamated wealth of insights from poststructuralist, feminist, postcolonial, rhetorical and cognitive approaches. More strikingly, Beer's third edition (2009) develops her concern with the relation of evolution to cognition by adding a chapter on consciousness – more specifically on what has come to be known as "Theory of Mind" ("Darwin and the consciousness of others"). We will delve into it below.

The main text offers brilliant insights into narrative and cognition in literature, in scientific writing, and in the broader cultures that receive them and imagine the world through them. Yet it also reveals no deep engagement with theories of narrative or cognition. "Narrative" and "plot" are hardly defined; instead there is reliance on common understanding and classical reference points like studies by Propp, Ricoeur and Frye. One of the main cognitive issues of Beer's book is the difficulty of adequately mentally grasping the situation that evolutionary theory/fact describes: The vastness of millions and billions of years, the infinity of profusion of individuals and species, and even more, their interconnection and adaptation. That cognitive challenge is also at once a challenge for language, for metaphor and narrative especially. Thus Beer analyzes how Darwin

> sought to appropriate and to recast inherited mythologies, discourses, and narrative orders. He was telling a new story, against the grain of the language available to tell it in. And as it was told, the story itself proved not to be single or simple. It was, rather, capable of being extended or reclaimed into a number of conflicting systems. [. . .] Reading *The Origin* is an act which involves you in a narrative experience. The experience may seem to diverse readers to be tragic [. . .] or comic [. . .] but it is always subjective and literary. (Beer 2009, 3)

More specifically, evolutionism is:

> rich in contradictory elements which can serve as a metaphorical basis for more than one reading of experience: to give one summary example – the 'ascent' or the 'descent' of man may follow the same route but the terms suggest very diverse evaluations of the experience. The optimistic 'progressive' reading of development can never expunge that other insistence that extinction is more probable than progress, that the individual life span is never a sufficient register for change or for the accomplishment of desire, an insistence which has led one recent critic to characterise Darwinian theory as a myth of death. [. . .] Darwinian theory takes up elements from older orders and particularly from recurrent mythic themes such as transformation and metamorphosis. It retains the idea of natura naturans, or the Great Mother, in its figuring of Nature. It rearranges the elements of creation myths, for example substituting the ocean for the garden but retaining the idea of the 'single progenitor' – though now an uncouth progenitor hard to acknowledge as kin. It foregrounds the concept of kin – and aroused many of the same dreads as fairy-tale in its insistence on the obligations of kinship, and the interdependence between beauty and beast. (Beer 2009, 6–7)

On a theoretical level, Beer explains how a theory of language as anthropomorphic and anthropocentric informs her method, especially in connection with Darwin's struggle against this situation:

> Though the events of the natural world are language-free, language controls our apprehension of knowledge, and is itself determined by current historical conditions and by the order implicit in syntax, grammar, and other rhetorical properties such as metaphor, as well as by the selective intensity of individual experience. Even despite the mathematisation of science, discourse can never be expunged from scientific *enquiry*. And nor, perhaps, can narrative. What particular kinds of story did Darwin's theory tell? What plots did it privilege? Did it invent new ones? And how were they received? (Beer 2009, 46)

Beer sees language as conditioning thought and highlights the importance of narrative in discourse and science. Again, she nearly equates narrative with plot, with *kind* of story. And like many classical narratologists she focuses on narrative's temporality. Narrative thinking is essential to Darwinian thinking because it captures *description* of change over time as essential to scientific innovation, even beyond analysis and exposition:

> As Kuhn says in the Appendix to the second edition of *The Nature of Scientific Revolutions*, 'there are important contexts in which the narrative and the descriptive are inextricably mixed'. The title of Darwin's book signals that this is one of those cases. [. . .] 'On the Origin of Species by means of Natural Selection' is in a very precise sense a narrative, because what it describes cannot be correctly described except through the medium of time. Neither analysis nor exposition would in itself suffice for what was new in Darwin's ideas. Categorisation, classification, description, must all be understood to be implicated in movement, process, and time. (Beer 2009, 58–59)

Postclassical narratology also emphasizes the *spatiality* of narrative, its capacity to build mental models of worlds with spatial structures through which causal forces operate, as several chapters here explore.

Beer also highlights how then-recent analyses of scientific thinking and work emphasized the roles of metaphor and analogy and narrative, as well as culture and society and history (Beer 2009, 83–85, citing Black, Popper, Canguilhem, Hesse, and Schon on metaphor and analogy; and Barnes and Bloor and Young on culture, society and history). George Lakoff and Mark Johnson are not discussed, though their *Metaphors We Live By* (1980) revolutionized the study of metaphor with its argument that linguistic metaphors are based on a rich system of underlying "conceptual mappings" that guide the way we think and act. Similarly important scholars of conceptual structures and processes beneath language are Gilles Fauconnier and Mark Turner (mental spaces, mapping and blending); and Leonard Talmy (spatial concepts, force concepts, fictive motion). Beer's cited thinkers are surely forerunners of Lakoff and Johnson et al., but they do not share the expansive new theoretical frameworks, approaches, methods and topics since developed on the basis of the latter. The cognitive linguistics school puts a different spin on insights that are now well-established (thanks in part to Beer's work): that metaphor and analogy and "heuristic fictions" do real cognitive work in science and are not merely decorative, and that science does not somehow transcend culture, society and ideology.

Beer's new final chapter: Blurring forms of cognition, within and across species

In her new concluding chapter, Beer returns in depth to the topic of human vs. animal cognition in Darwin, examining the questions raised by Darwin's friend and follower George Romanes, especially after Darwin's death:

> Could evolutionary theory provide hope of continuity beyond the colossal losses entailed in the death of the individual – irreplaceable losses, of individual memory, somatic experience, and mind? [. . .] [F]or Romanes, it most offered evidence of 'fellow-feeling and sympathy', from ants to primates. To him, the over-lap of emotional states between other life-forms and humankind was the strongest evidence for the working of intelligence throughout the animate world [. . .] (Beer 2009, 243)

But contemporaries perceived paradox in Darwin's attitudes to such questions:

> Darwin was seen by Samuel Butler, George Bernard Shaw, and others, as the man who banished mind from the universe. Natural Selection gave no place to self-help by the individual organism, nor indeed much survival value to memory, especially across generations, whether human or held within other life-forms. But Darwin himself was fascinated through-

> out his life by the relations between consciousness and instinct, sentience and reason, emotion and reflection, inclination and purposiveness. In plant life he was particularly attracted to movement and activity: plants climbing, reaching out, and devouring raised questions about *intent*. Equally, the issue of animal intelligence was fundamental to his work and he responded with lambent curiosity to human varieties and their very different modes of reasoning and reflection. (Beer 2009, 243–244)

Beer returns to the problem of the anthropocentrism of language in connection with Darwin's investigations of mind, which lead Darwin into speculations that arguably align him with the "post-humanism" that now animates areas of L&S studies:

> 'The half-art, half-instinct of language still bears the stamp of its gradual evolution' he writes in the *Descent*. Agency abuts intent, abuts consciousness. One way beyond the constrictions of human language is to explore the potential for awareness in other life-forms. What shape does intent take elsewhere in organic life and what kinds of consciousness may this imply?
>
> In one of his early notebooks [. . .] Darwin observed, with some tartness:
>
> > People often talk of the wonderful event of intellectual man appearing – the appearance of insects with other senses is more wonderful. Its mind more different probably & introduction of man nothing compared to the first thinking being, although hard to draw the line.
>
> Different minds; other senses; thinking beings: thus he clusters and fans out without hierarchy the properties of all inhabitants, past and present, of the living world. The final demurring phrase is crucial to his mode of thinking too: 'although hard to draw the line', the line between senses and consciousness, between thinking and instinct, sentience and reason. (Beer 2009, 244)

To show Darwin undermining assumed boundaries and hierarchies of mental function within the human species, Beer later cites Darwin's description of a baby's embodied reasoning in moving his own hand in order to get his father's finger in his mouth ("This was not chance, & therefore a kind of reasoning, – It has, perhaps, a little instinct in it"):

> Darwin is fascinated by the shifting boundaries between awareness, intent, and reason on the one hand – which together make up much of what we commonly call consciousness – and instinct, reflex action, and unconsciousness on the other. He is by no means satisfied with the hierarchies implicit in this sorting. (Beer 2009, 251)

Beer then zooms in on "the acuity with which Darwin responded to diverse forms of *being*" (Beer 2009, 246), the variant kinds of reasoning and consciousness across human cultures and across species, which challenge some accepted markers of human uniqueness:

> He explores the kinds of reasoning habitual in other human cultures that he encountered on the Voyage; he seeks out the differing degrees and kinds of consciousness in other life-forms too. His work persistently investigates how far (indeed, whether) it is possible to distinguish *being* from *consciousness*. It is extremely doubtful, he asserts in the *Descent*, that 'the formation of general concepts' and 'self-consciousness' are 'peculiar to man'. (Beer 2009, 246–247)

Beer speaks of the "vertiginous freedom in these private notebook entries, an uninhibited willingness to entertain what might be thought absurd and to cross or tangle categories" (Beer 2009, 247), an "uninhibited yet concentrated play of reverie" that Darwin associates with scientific discovery (Beer 2009, 248). It produces bold speculations on topics central to narrative, such as causation and causal thinking. He considers the roots of causal cognition – on how animals and even plants may develop a "necessary" notion of cause and effect from the experience of simple reactive "willing" and "habitual action" (Beer 2009, 248) – and broadens out to the "question of cause, and of causes" that was "the key issue in Darwin's theory" (Beer 2009, 248).

Beer goes on to note Darwin's proto-psychologist investigations of empathy, inspired by child minds and family dynamics. For Darwin, "one meaning of consciousness was what we would now call empathy" (Beer 2009, 250); Darwin's home life also supplied "materials for brooding on forms of consciousness and their connection with evolutionary history, itself an investigation of cause and effect" (Beer 2009, 250). Soon after the word enters the language, empathy is contrasted with sympathy:

> Empathy (einfuhlung) implies an inhabiting of the thought processes and experiences of others; sympathy (mitgefuhl) implies a dwelling alongside them, and may also move towards the kind of separation implied by 'pity'. (Beer 2009, 250)

Here again current cognitive interests are anticipated – emotion, emotional contagion, simulation of fellow experience and feeling. These concerns have extended into fundamental neuroscience, with the discovery of "mirror neurons" and elaborations of how they underpin mental simulation of observed or represented experience. Emotion and empathy are particularly relevant to literary narrative, and major studies have grounded basic narrative structures and processes in particular emotional structures and processes (Hogan's *Affective Narratology*, 2011), and linked emotion and empathy with the history of major narrative forms (Keen's *Empathy and the Novel*, 2007), and even with the history of human rights (Hunt's *Inventing Human Rights*, 2007).

Beer cites Darwin's observation of his children's curious responses to him, which prompts his claim that "there is no trace of consciousness in very young children – they do not think, whether the person, they are looking at, is thinking of them" (Beer 2009, 250–251). This leads Beer into discussion of another major focus of cognitive narratology, "Theory of Mind":

> He is here suggesting that 'theory of mind' – the capacity to build knowledge by intuiting across different consciousnesses: 'he thought that we thought that she thought' – is a capacity that occurs as part of human maturation and is absent in very young children. [. . .] The capacity for empathy develops later than reasoning [. . .] (Beer 2009, 251)

Scholars such as Lisa Zunshine and Alan Palmer have made Theory of Mind (ToM) central to their cognitive analyses of the novel, and narrative in general, and these studies have been driving forces in the broader development of cognitive approaches to literature and culture. Palmer defines narrative as "the description of fictional mental functioning" and many kinds of such "mind-reading" have been analyzed, across literary genres, periods and cultures. In CLS, ToM is often seen as particularly rich in humans, if not unique to them, and hence as one or the markers of forms of neuro-atypicality such as autism; and scholars draw on now-extensive research on early development of this capacity, and its relation to reasoning. Darwin, on the other hand, along with some science fiction writers and L&S scholars, emphasize possibilities of ToM across species much more strongly.

Specifically, Darwin's cognitive analyses generalize across species by intertwining observations of human children and of other animals:

> by the time he joins in the discussion on child development in the then new psychological journal *Mind* in 1878 he combines the diary he kept of his first child's first year thirty seven years before with a much fuller understanding of how such observations parallel animal behaviour, in the light of his own evolutionary theories.
>
> Indeed, the emphasis on the possibilities of consciousness across kinds was crucial for his argument on the commonality of the human and other animals, whether insects or primates [. . .] (Beer 2009, 253–254)

Hence Beer concludes by celebrating Darwin's "capacity for empathy", which "allowed him to appreciate the aesthetic choices of earthworms, the erotic explorations of orchids, as well as the anger of other primates" (Beer 2009, 254). Despite his sense, late in life, that his aesthetic capacities had dwindled, Beer links Darwin's scientific thinking about mind with his powerful capacity for elementary narrative fiction, for trying out other forms of being in imagination:

> The capacity for double consciousness, for thought experiment, for sceptical belief, and for empathy, all of which fiction provokes, chime with what goes deepest in Darwin's own creativity. His rigorous questioning travels alongside a passionate curiosity and an eagerness to explore – even to try out *being* – whatever is around him, like him and unlike him: children,

> climbing plants, snails, dogs, mountain ranges, baboons, gardens, barnacles, people as animals, and as peculiar human cultures. (Beer 2009, 255)

Hence in a number of ways, Beer's original study, with its strong interest in how fundamental narrative and fiction are to thought, including scientific theorizing and argument, shows the deep roots of the present volume's focus on intersections of CLS and L&S. Her new final chapter shows how the foundations as well as the latest developments in the two subfields have more in common than may at first appear, and this indicates the potential for productive cross-talk and perhaps further collaboration.

1.2.3 Second touchstone: Cognition and history, in CLS and L&S

From around the 1990s (hence in periods later than Beer's first edition, but overlapping with her second and third editions), L&S has occasionally crossed paths with CLS, with mixed results. This section reviews two such crossings, as one leading scholar per subfield surveys and reviews the other subfield. In short, the L&S view of CLS finds much to admire in the way CLS draws on cognitive science to produce new analyses of literature and culture, but ultimately finds it too anthropocentric and over-reaching. The CLS view of L&S finds surprising common ground in the philosophical roots of the two subfields, but ultimately insists on an epistemology that places scientific limits on how far science is "constructed."

L&S Considers CLS

Joseph Tabbi's chapter on cognitive science in the 2011 *Routledge Companion to Literature and Science* proposes that "the literary arts always have been about cognition, consciousness, and their coevolution" because "literary work is cognitive, narrative is an enactment of consciousness, and fictions do a kind of work (consistent with the mind's continual need to fill 'gaps in nature,' the title of a pioneering book by Spolsky)" (Tabbi 2011, 77). He sees broad connections across literary and scientific explorations of cognition as both participating in the "historical development of cognitive knowledge itself" (Tabbi 2011, 79).

Yet he is ambivalent about CLS. He admires Spolsky's sense of the complex relationship between literary studies and cognitive science, where each can be informed but not determined or authorized by the other. He is ready to accept the cognitive critic's emphasis on the interplay of embodied-mind-based universal forms of thought with culturally particular languages, literary conventions and artifacts. Thus he appreciates studies of "the universal cognitive structures that enable a diversity of expression across cultures and languages" by Hogan, Turner,

Zunshine and others (Tabbi 2011, 80), and assesses that what CLS can accomplish is

> a discovery through close reading of the moment-by-moment, word-by-word, and sentence-by-sentence enactment of consciousness in language. Variability [. . .] is of course a primary concern of literary criticism. [. . .] But for variation to be meaningful (and not just "novel"), criticism must also consider, in Turner's words, "the underbrush of unoriginal structures" and largely unconscious operations that do not reach expression in the work, or in thought. (Tabbi 2011, 81)

This ties into an intriguing observation of the potential connections between narrative and metaphor research (another concern of many chapters here):

> cognitive approaches to narrative marry those to metaphor by articulating, in Turner's words, "consciously and systematically the linguistic or conceptual resources" used "in reading and writing" (Turner 1996: 14). [. . .] What cognitive science brings to literary theory, then, is not an entirely new approach but rather a more robust realism grounded in actual, complex, and widely distributed mental processes. When encountering novelty in literary works, the cognitive critic generally calls attention to "the unoriginal aspect of innovation" – a salutary approach in an era when innovation is the rule, not the exception. (Tabbi 2011, 81–82)

Yet Tabbi is suspicious of certain tendencies that seem retrograde against the light of post-humanist thinking. For example, Turner's emphasis on the singularity of human thought – its distinctness from non-human cognition, and its ways of operating by grounding vast networks of conceptualizations in the human scale of the body-brain. Tabbi would prefer those networks be framed by local culture – specifically the vast networks of technocapitalist modernity.

Hence in a historicizing vein, Tabbi outlines some interesting analogies between CLS and post-structuralist thinking, as well as between the development of CLS and modernist, early-postmodernist and late-postmodernist literature, with a nod to Fredric Jameson's analysis of postmodernism as the "cultural logic of late capitalism" in which the globalization of capital and its technologies is central:

> the era of scientific discovery [of the mind] coincided with a rediscovery, by imaginative authors, of the cognition of the work of literature when consciousness is no longer the predominant object of knowledge, "the human" is no longer the primary subject of history, and print itself, predominant during the rise of modernity, is no longer the predominant medium in which knowledge circulates. (Tabbi 2011, 83)

He makes the distinction between Jameson and CLS in terms of consciousness: "Where Jameson's project of cognitive mapping remains centered in consciousness, however, cognitive approaches that incorporate current scientific research tend to diminish the importance of consciousness within the ecology of mind, embodiment, and technics" (Tabbi 2011, 84). Thus he notes the potential links from focusing on "human scale" vs. technologies to broader humanist vs. posthumanist sympathies:

> To a degree, one's decision to focus on "the human scale" or on the emergence of psychic and social systems from other material and symbolic technologies will determine whether one holds on to a humanist or what has come to be described as a "posthumanist" approach to the current cultural situation. (Tabbi 2011, 85)

Hence a critique of the ongoing tendency "among mainstream cognitive critics to look out from the human to cognitive environments, turning the likely universality of brain processes into a dubious cultural unity" (Tabbi 2011, 85).

One might question Jameson's sweeping analysis and object that consciousness and the human cannot so easily be exorcised from the arts, or from anything that can be called thought. Similarly, when Tabbi goes on to position CLS unfavorably vis-à-vis L&S views associated with systems theory, media theory and posthumanism, he makes something of a straw man of Turner's recognition of human "singularity" in the world of cognition:

> Turner's insistence [. . .] that human cognition differs fundamentally from that of animals [. . .] certainly would be contested by critics such as Wolfe (2010) and Clarke (2008), who make the non-human animal and our deep embeddedness in hybrid networks and complex systems starting points for posthumanist explorations of literary and ethical consciousness. (Tabbi 2011, 86)

Such critiques arguably arise from missing the points of the analyses examined, and tend to close off discussion.[6] In the end, having weighed CLS in the balance and found it wanting, Tabbi goes on to propose his own "program for a cognitive literary criticism appropriate to the current media ecology," prescribing more "modest" and "marginal" work (e.g. crowdsourcing for digital literary analysis) to comport better with a "radically new understanding of critical writing" that contrasts higher-level acts of criticism from lower-level networked behavior of communicating about simple tasks (Tabbi 2011, 86–87).

By contrast, it is striking that a more recent authoritative account of L&S precisely identifies CLS as a key development that has allowed a more reciprocal picture of the relations between literature and science to emerge. Marchitello and

6 Contrast Spolsky's 2002 essay on the alignments of cognitive and post-structuralist theory, "Darwin and Derrida: Cognitive Literary Theory as a Species of Post-Structuralism," and compare Crane's 2009 study of how Jameson's large scale models of cultural history rely on schematic spatial concepts, "Surface, Depth and the Spatial Imaginary: A Cognitive Reading of *The Political Unconscious.*"

Tribble's introduction to the 2017 *Palgrave Handbook of Early Modern Literature and Science* recognizes Crane's work in *Losing Touch With Nature* (2014) as exemplary in its insights and approach to current understandings of the relations between literature and science as being mutually constitutive, rather than either being dependent on the other. Her work helped establish the two developments they identify as key to this new approach: First, a "re-turn to history" associated with New Historicism and Cultural Materialism that rejects New Criticism's isolation of the aesthetic object and opens "the possibility of thinking about literature in relation to the fullness of its cultural embeddedness" (Marchitello and Tribble 2017, xxvii). More specifically it mandates "the dissolution of ideas of historical background and literary object occupying the critical foreground" and "enables is a productive blending of the historical and the literary" (Marchitello and Tribble 2017, xxvii), including insight into the "the ways in which both literature and science are systems dedicated to the production of knowledge," and of reciprocity between them. Second, Crane has built on "transformations within the fields of the history and the philosophy of science in recent decades" that have brought to the surface "a new understanding of science" as similarly embedded – that is, "wholly enmeshed within culture and society, and as fully engaged in language – and in discourse – as are any other manifestations of human thought, including literature" (Marchitello and Tribble 2017, xxviii). One should also emphasize how deeply Crane relies on cognitive analyses of metaphor and conceptual structure to make her arguments about the interacting historical trajectories of everyday, literary and scientific concepts.

CLS Considers L&S

Seven years after Tabbi's essay, Alan Richardson's chapter in the 2018 *Cambridge Companion to Literature and Science* notes an initially "tense and unhappy" relationship, as CLS scholars attending SLS meetings in the late 1990s found that their papers and panels were received with "remarkable and quite unexpected hostility" (Richardson 2018, 207–208). Richardson attributes this to the "science wars" debates which "temporarily hardened positions and generated considerable defensiveness on both sides" (Richardson 2018, 208), and cites a colleague explaining at the 1998 convention, "our job isn't to learn from scientists; it's to show them where they're wrong" (Richardson 2018, 208). However, the subsequent criticisms in print of CLS by L&S were relatively naïve, and well-answered long ago.

Richardson does, however, observe some striking common intellectual roots in the kinds of epistemologies developed within the two subfields. He describes how F. Elizabeth Hart's "indispensable essay" "The Epistemology of Cognitive Literary Studies," articulates an epistemology situated between "objective realism and blank-slate style relativism." "Minimal conditions" for such middle ways include "that there is such a thing as species-specific knowledge and that such knowledge must contribute substantially to our philosophical discussions about knowledge;" yet "all knowledge, including species-specific knowledge, is environmentally situated, context-dependent, and culturally indexed, i.e., subject to lesser or greater degrees of constructivity" (Richardson 2018, 210). Richardson points out that Hart adapts her terms from classic essays in L&S: N. Katherine Hayles's "Constrained Constructivism," which builds on Donna Haraway's "Situated Knowledges." Both seek to recognize the "instrumental efficacy" of science without "lapsing into a naïve scientific objectivism" that overlooks the bodily and cultural situatedness of human observers and instruments (Richardson 2018, 211).

Moving from philosophical frameworks to the disciplinary approaches they sanction, Richardson highglights "cognitive historicism" as a vital area in the new field that overlaps with work in L&S. It shares assumptions with new historicism generally, recognizing that literary texts do not merely reflect the science of their time but rather draw on "a common discursive lexicon, sharing in the work of developing and delineating innovative scientific conceptions, and at times even running ahead of science in this common work," while scientific texts are studied not only for their "ideas and methods" but also rhetorically, with special attention to "the ideological environment and implications of literary-scientific discourse" and how the developments they frame may advance or contest "the interests of dominant groups" (Richardson 2018, 212). What makes this vein of historicism "cognitive" is that it recruits "recent developments in cognitive science and neuroscience" in order to "newly perceive the science of the past and to recapture some of the excitement and anxiety it provoked in its own time" (Richardson 2018, 212). Again Mary Crane is a key example – her work "helps her readers appreciate the pre-Cartesian, anti-dualistic tenor of some early modern theories of mind by means of comparison to roughly comparable developments in the 'embodied' psychologies of the present" (Richardson 2018, 212). Richardson stresses that critics recognize that such analogies are approximate and attend as much to the "instructive differences between then and now," the point being "not to naively project the present onto the past but to open up a new perspective on the past that is then developed with scholarly rigor" (Richardson 2018, 212). Moreover, such work can reappraise how current scientific models "can in fact be seen as limited or biased in comparison to some models excavated from the past" (Richardson 2018, 213).

However, a key distinguishing assumption marks a deeper break from much L&S. That is, CLS assumes that "the human brain, along with a range of basic cognitive functions, have not changed fundamentally since the beginnings of recorded culture" (Richardson 2018, 213). Basic mental properties evolved many thousands of years ago, and evolution is generally too slow to have altered over recorded history. For example, "something as basic as the limited amount of available attention," about seven pieces of information in a single gestalt, in the conventional consensus, which is "found currently across cultures, can safely be projected into aesthetic practices and artifacts from the past." This does not imply "that human experiences, representations, and technologies of attention have not changed significantly over historical time" (Richardson 2018, 213). Approaching literary study from these grounds, Richardson emphasizes, "does not entail a flight from history but rather, as Mark Turner has argued, a 'wider notion' of what counts as human history" (Richardson 2018, 213). CLS's engagement with science also leads to critique: critics "speak back to mind and brain research in the present, identifying limitations or biases in scientific research agendas through comparison with past models and formulations," as well as "pointing out the kinds of simplification and distortion that frequently accompany the scientific practice of reducing or segmenting complex behaviors to narrow, neatly defined, and empirically testable bits" (Richardson 2018, 216). Yet, Richardson summarizes, CLS scholars do "subscribe to a basic scientific world view" with an epistemology of "constrained constructivism" in common with "most neuroscientists and cognitive scientists" and adopt a "scientific attitude" with "a willingness to abandon models, assumptions, and theories as soon as these can be shown to be empirically untenable" including those from the cognitive sciences (Richardson 2018, 219).

Richardson's kind of historicizing selectively uses current models of the mind to look back on cruxes of the cultural past with fresh eyes, and reciprocally to frame those current models in the context of their own histories. Tabbi's kind of historicizing uses large-scale post-structuralist models of cultural history to frame the development of cognitive knowledge in terms of the history of literary culture, and this helps him to offer ideological critiques of CLS in light of L&S. One wonders if putting these approaches into dialogue might lead in interesting directions. For example, Tabbi leaves in place the simple contrast of CLS humanism and L&S post-humanism, without considering the possibility that there is no real contradiction between them: human minds are certainly deeply continuous with non-human minds; they are just as certainly highly distinctive from any other species on the planet. We might say they are (paradoxically) both similar in kind and different in kind. One only needs to look around one's everyday world at any time, and consider the implications of such simple questions as, "has any non-human animal ever chatted with tech support because Excel crashed while they were doing their taxes?" The distinction of socio-technological embeddedness is only possible because of the distinction of singular cognitive complexity and creativity. There is at least an opening for a line of discussion that could go much further.

A related promising alignment is that cognitive historicist approaches to literature echo developments in the history and philosophy of science. These fields have long moved beyond "monolithic" stories of steady triumphal Progress by discoveries of individual Genius and unified Rationality. Instead, an emphasis developed on what Lorraine Daston calls a "swarm of microhistories" (809), much inspired by Foucault's "archaeologies of knowledge," and Thomas Kuhn's model of scientific revolutions and paradigm shifts. Here the boundaries between science/non-science are in constant question; and "pseudoscience" can be a contentious term, because it's so flattering to hindsight. Similarly, L&S studies tend to highlight how porous both categories are, and how multiple in reality are the forms of thought, activity and interaction that have been grouped under those grab-bag names in retrospect, before and after disciplinary formations (though Daston's analysis warns of the dangers of histories becoming so "micro" that they lose touch with any larger historical picture or point).

More recently, histories of science have taken on such critiques of scientific objectivity and triumphal history in order to offer more nuanced accounts of the reality of scientific progress. David Wootton's *The Invention of Science* (2015) proposes a new approach to defining the scientific revolution between 1572, when Tycho Brahe saw a nova or new star and sparked the rise of modern astronomy (the "first true science") and 1704, when Newton published his *Opticks*. Wootton resists attacks on the notion of a "scientific revolution" as a legitimate historical category, and devotes considerable time to discussing historical methodology and philosophy, and debating relativist views of science. Wootton's main arguments are cognitive in two ways. First, the scientists involved were *aware* that they were participating in a global shift in the ways in which *knowledge* was established – despite the diversity of activities and terminologies, and even if the outcome was extended and patchy, rather than everywhere-all-at-once. Second, the new science was based on a new form of engagement with sensory reality – the systematic use of the "test of experience": observations and experiments aimed at discovery – a fine example of an emergent form of embodied cognition.

Related to this movement, some *cognitive* histories and philosophies of science map the development of conceptualizations at work, illuminating the richly complex creativity of rational-logical thought, and its engagements with story, metaphor and other forms of mental connection, and the experience that undergirds them. For example, Jürgen Renn proposes an embodied-cognitive answer to "one of the great riddles in the history of science," how to explain both the persistence and the transformation of certain concepts and mental models over "vast periods of time" (Renn mentions number; weight; the relation between force and motion) (Renn 2015, 50). He argues that this points to how scientific theories are complex systems blending scientific with other forms of knowledge – specifically, "the earliest examples of such organizing cognitive structures are rooted in intuitive and practical knowledge" as in mechanics, and "the corresponding cognitive structures continue to this day to shape our experiences" (Renn 2015, 50). The "extraordinary flexibility" of mental models allow details of theories to be changed while the "basic set-up" remains, and the "longevity of some of these mental models" is due to the "layered structure of this architecture of knowledge" where later layers adapt and incorporate rather than replace earlier ones, so that "even the most abstract notions of weight in modern physics can still be related to the operations with a balance from which they originally emerged" (Renn 2015, 50). Thus the "seeming universality" of some concepts is not the mark of "a specific form of rationality" but "the outcome of a specific historical trajectory in the evolution of knowledge" in which human experience continuously grounds and shapes thought (Renn 2015, 50).

L&S Goes Cognitive: Recent developments

Research along these lines is making inroads into the central ground of L&S. The final chapter in *The Cambridge Companion to Literature and Science* (2018) is precisely a cognitive study of the history of literature-science entanglement, although it does not obviously draw on cognitive research. Reviel Netz's "The Long History of Cognitive Practices: Literacy, Numeracy, Aesthetics" focuses on the "Greek inventions" of literature and science, asking a question that frames media and genres as "cognitive tools" for creation and reception: "at a given historical moment, how does one produce literature and how does the literary medium interface with that of science? How is all of that perceived by a mind that forms an aesthetic judgment?" (Netz 2018, 242). Netz concentrates on genres of the exact sciences, because they "would rely essentially on the written tool of the diagram and would thus be a writing-based genre *par excellence*" (Netz 2018, 243); and on those of medicine, because it was "the most important form of science" in most civilizations and "arguably also among the Greeks" (Netz 2018, 244). He examines in detail the poetic and narrative strategies of two key examples of ancient "science-as-literature" – scientists whose writing founds their genres and fields: Archimedes and the exact sciences, and Galen and medicine.

Netz's close reading of the major texts of Archimedes and Galen leads to three key conclusions, all consequent on the main conclusion that literature and science are related because they are "mental, cognitive phenomena" (Netz 2018, 253). First, they "share in a related historical trajectory": "Cognition must take place in the material world and so has to be mediated by material, cognitive tools – media and genres – that are historically formed." Second, because science lives via media and genres, it "can be analyzed [. . .] on aesthetic grounds: the surprise of an Archimedean proof, the bite of a Galenic piece of rhetoric." Third, "cultural practices at any given place and time form a system: if certain practices are common in one part of the system, but not in another, this tends to mark the parts of the system as such and to endow the practices with meaning" (e.g. the opposition between "the performed and the written" informs the later opposition between "literature" and "science" [Netz 2018, 253]). Such oppositions transform through changes in genres, media and other cognitive tools, and evolutions of "literacy and numeracy," leading to new literary forms approaching modern science (Netz 2018, 254).

The editor's concluding essay begins by reflecting on Netz's study of how scientific "cognitive modes" emerged "within concrete practices," and how this registers an "Age of Whitehead" because features of today's "second-wave" L&S were "already stressed by Whitehead nearly a century ago" (Meyer 2018, 256). Not all of these are relevant to our concerns, but a selection helps highlight how some main currents of L&S align with CLS. Meyer lists "criticism of the fact/value dichotomy and the concomitant bifurcation of nature"; "suspicion of the [. . .] two-cultures paradigm"; "expanded and expansive empiricism"; and a "focus on the affective, cognitive, and technological aspects of practice" (Meyer 2018, 256). One might summarize this as an openness about what science is and knows, and an emphasis on the cultural complexity of what science does. Meyer's broader point is to mark the developments that enabled this major shift in the history of the subfield. Most importantly, Meyer's introduction to the volume identifies the "particular disciplinary signature" of the second wave as an idea already suggested by George Rousseau in a 1978 "state of the field" assessment:

> There is no reason to disbelieve on logical or epistemological grounds [. . .] that literature and science affect each other reciprocally. That is, that each influences the other in just about the same degree, although conceivably in different ways. It is also probably valid to assume, although it would be practically impossible to prove, that science shapes literature to the same degree that imaginative literature shapes science. [Yet] only the former has been studied in any depth . . . The latter is an unexplored territory, probably the one in greatest need of cultivation right now and also the one requiring learning so vast that it is hard to imagine it in a single scholar (Rousseau 1978, 587–8, cited in Meyer 2018, 5).

This is a guiding concern of the present volume.

Given the above, one could make the tentative argument that cognition has always been the missing link in this framing of the reciprocal relation between literature and science, and that the maturation of CLS as an interdisciplinary movement can help extend the bridges already built into that half-explored territory.

This brings us to narrative, as a structure and process now recognized as on par with metaphor as a central mode of human cognition, and as such a potential strong bond in those criss-crossing conceptual trajectories.

1.2.4 Third touchstone: Unusual business: Narrative as common ground

The continuing imbalance between literature and science in L&S leads us to wonder what might bridge the gaps, bring the two cultures closer together. What are the deep shared concerns? ELINAS has focused in particular on "the importance of language and metaphors in physical research" and the "discursive and narrative modulations of scientific theories in literary texts" (http://elinas.fau.de/index.en.html). Hence we arrive at one of the key themes of this volume, as ELINAS has proposed *narrative*, along with metaphor and genre, as a key site of potential bridge-building between the fields. ELINAS's core research areas are organized in relation to categories of leading questions that include epistemology, rhetoric, metaphorology, narratology, poetics of genre, and physics methods, concepts and theories (http://elinas.fau.de/about.html). Those middle three subfields stand out, interestingly, as much more specific, and of more recent vintage, than the others, and they have been central to CLS.

A recent article in *Configurations* develops the idea that the concept of narrative could function as a unifying interest for the two fields. Daniel Aureliano Newman (2018) recognizes the importance of narrative not only in science communication but in scientific "theories, models, and procedures," and suggests that it offers "common ground that might help alleviate remnants of suspicion and diverging goals that continue to inhibit truly interdisciplinary work" (Newman 2018, 277). Newman also frames his suggestion as a response to Labinger's "Where are the scientists" question, but sees the recurring status issue differently. Scientists may be put off by the "priority implicitly given to the 'Literature' in 'Literature and Science'" (Newman 2018, 277) and the fact that "much of the research in it is done on humanistic terms," and beyond that, the suspicion that "science is demonized or discredited by 'the arts'" (Newman 2018, 278), so that even the curious may avoid "a field where research may be *about* science but rarely forms *part of* the scientific enterprise" (Newman 2018, 277). In an astute comparison, Newman notes that "literary approaches to science – notably 'suspicious reading' practices – may seem as alienating and invasive to scientists as E. O. Wilson's dream of consilience can seem to humanists" (Newman 2018, 278). Hence "a truly interdisciplinary exchange will require something other than business as usual," and narrative is one important key to that unusual business, with "direct interest and professional benefit" (Newman 2018, 278) for both sides.

Despite the mixed reputation of storytelling in science, Newman cites evidence that "scientists are increasingly receptive to evidence showing that information communicated in narrative formats 'enhances memory, interest, and understanding'" (Newman 2018, 278). As a potential obstacle to this development, Newman singles out the "tendency to understand 'narrative' as a set structure, a preformed story template" so that "[u]sing narrative in science thus becomes a matter of shoehorning scientific facts into preexisting narrative molds" (Newman 2018, 279). Of course, narrative is much more than a seductive Hollywood formula to draw eyeballs and clicks. Instead, Newman proposes reimagining science to focus attention on *parts* of narrative in *parts* of science – that is, "the elements that confer local narrativity to scientific texts – to the generally underappreciated 'narratives *intrinsic* to science'" (Newman 2018, 279).[7] Newman cites Kirk Junker's argument that "scientific explanation shares basic structural features with narrative" especially "'a temporal perception and a causal perception'" (Newman 2018, 280); and he cites other research showing that narratologists have identified "how narrative functions in scientific theories, protocols, and communications: 'as argument,' as an 'explanation' and a 'scaffold' for new hypotheses, as an 'integrative device' linking discrete 'mathematical components of theories,' as the underlying structure of experiments, and as a rhetorical tool" (Newman 2018, 280). Thus scientists could profit from the literary knowledge of the "narrative logic inherent" in science (Newman 2018, 280). Readers will find these and other narrative functions of science elaborated in this volume.

Such an engagement would mean delving into narratology to go beyond fascination with "plot" and "attending to the ways in which scientific discourse 'narrates' rather than 'narrativizes,' since narrative *elements* participate in scientific texts without necessarily determining their global form" (Newman 2018, 281, citing Hayden White). Scientists and literary critics could enter into narratology with shared means despite differing ends (Newman 2018, 281), and collaborations (here citing Anne Dalke and Elizabeth MacCormack) could "maximize serendipity" in a process where "disciplinary presumptions interrogate and unsettle one another to produce novel questions and answers" (Newman 2018, 281–282). They could broaden narrative theory because "Scientific models and texts offer a wealth of narrative structures and devices that could help inform fields such as postclassical narratology and offer new insights into the study of literary history" (Newman 2018, 282) while narrative theory in turn could develop perspectives on

7 At this stage in his argument, to point up how partial narrativity is central to science, Newman cites three of the five keynote speakers who headlined the conference that was the forerunner to this volume: H. Porter Abbott, Marie-Laure Ryan and Hans Fuchs, the last of whom is a contributor.

scientific models and their means of expression. The conjunction of fields could also help critics to counter the too "satisfyingly simple" narratives around scientific and social issues like climate change and nationalism (Newman 2018, 282).

Thus Newman's article is a significant statement at a significant time for narrative and science, articulating a nuanced sense of the promise of linking the two phenomena.

How are the three touchstones examined above reflected in our volume? We can observe at least the following few connections.

A return to plot, in connection with Theory of Mind, and posthumanism

Going beyond Beer's inspirations (Propp et al., noted above), we find a reviving interest in plot as key to narrative in science and science-informed literature, and it is connected with the role of Theory of Mind in narrative, and with posthumanist impulses often associated with climate and ecological concerns. For example, Caracciolo, Bermúdez, Haefs, Brosch, Pennisi, Heydenreich and Vanderbeke all deal with one or more of these topics in their different ways.

Cognitive historicism

As authors take up various kinds of historical questions they develop cognitively inflected approaches to history and historicism. For example, Freiburg, Gambino and Pulvirenti, Sinding and Vanderbeke address historical issues more or less directly, with an eye on the role played by cognitive factors in change over time.

Partial narrative/narratology in science

Naturally all of the chapters address narrative and science, but they do so with varying kinds of cognitive twists, and in ways that Newman only glances at in his reference to postclassical narratology. The interplay of narrative and argument has been a major concern since Beer's book, and it continues in new ways in this volume. Many authors address the idea that plot can be local rather than global, evoked by hints and used for particular sub-purposes, rather than an overall structuring force with a singular goal. Thus we see many connections posited, demonstrated and explored, among narrative, argument, understanding, interest, emotion and persuasion – consider chapters by Heydenreich, Haefs, Fuchs, Dumont and Corni, Lemerle, Gonzalez-Marquez et al. and Miller.

Thus with scholars from multiple disciplines attending to the multiple foci of our title, and merging those forms of attention in various ways, we can hope to help advance both CLS and L&S by bringing to the table fresh looks at: the interplay of narrative and argument; how cognition and epistemology relate to history and historicism (e.g. diachronic perspectives on narrative and literature; the narrative aspects of factual thought and writing); and how narrative depends on and explores minds, and Theory of Mind, within typical human interactions and beyond (into neurodiverse and cross-species interactions, and beyond the linguistic to gesture and other forms of embodied thought and communication).

The volume achieves these ends by emphasizing a few recurring and related themes: the interrelation of narrative and metaphor; the role of embodied simulation and feeling in narrative, and how this may bridge literary and scientific forms of imagination; a renewed consideration of plot and plot types; narrative as an enabler of understanding, via emotion, vividness, memory, and more; narratives and narrativity across media; and the interplay of narrative in chains relating questions or problems to hypotheses, models, solutions, counter-models, revisions, etc.

The volume could have been organized in various ways of course, but the present organization of sections, which identifies "W approaches to XY of Z," seemed felicitous as highlighting both the range of contributing disciplines and approaches, and the emerging alignment of their topics and the ways in which they may be addressed. Zooming in on the narrative-cognition interplay as the core of the literature-science interplay could unearth new visions of the interpenetrating roles of story, minds, literature and science. And with the overlapping turns to narrative and cognition in both literary and science studies, there is good reason to believe that the time is right for this effort. But of course, it will be what we make of it, and with that, I turn to introducing our essay collection for our readers.

2 Chapter introductions

2.1 Cognitive approaches to literary narratives of science

In this first section, authors analyze literary narratives relating to science through various cognitive lenses, in order to better understand how those texts frame scientific topics and convey their thematic significance and emotional impacts to readers.

Rudolf Freiburg's "'Let's hear you apply Heisenberg to ethics': The Problematic Interferences between Science and Literature in Ian McEwan's *Solar*" considers the intellectual distances between the sciences and humanities, apparently two incompatible cultures, focusing on Ian McEwan as predominant among current writers who have attempted to bridge the gap, because of his sense of science as a kind of worldview (vs. religion, mysticism, easy appropriations), and his capacity to "see moral and ethical problems from a scientific point of view and *vice versa*." Freiburg sketches a dialectic of key views of the relations between science and fiction: the Platonic view of fiction as a "lie" that distorts scientific truths, Sidney's countering view of literature as mediating history and philosophy (including natural philosophy or science), and Bacon's reactionary view rejecting all literary imagination and language from science. By contrast, McEwan takes up Werner Heisenberg's enthusiasm for creative imagination of all kinds, including his recognition of the parallels between narrative fiction and scientific hypothesizing, and his development of the "uncertainty principle" in subatomic physics, with its broader implications for epistemology and its analogies with traditional principles of hermeneutics. Freiburg shows how McEwan's novel *Solar* draws on Heisenberg's uncertainy principle to inflect discussions of both science-humanities relations and ethical problems in relation to the climate crisis. McEwan creates characters who embody sciences, humanities and arts, and explores the limits of rationality and science in his portrayal of the arrogant scientist protagonist as both satirizing and satirized, particularly in his rejection of any links between scientific, epistemological, and ethical principles. Narrative ironies and uncertainties interact, leaving the unsettled reader with a role in deciding the reality and extent of those links, especially in light of the climate crisis.

Marco Caracciolo's "The Hard Problem and the End of the World: Narrative and Mind in Postapocalyptic Fiction" examines how certain major postapocalyptic narrative fictions link questions about the nature of mind with the experience of ecological disaster. He considers how a fundamental problem of Western philosophy, the mind-body problem – powerfully restated by David Chalmers as the "hard problem" of how consciousness can arise from material processes – is urgently raised in two recent novels of environmental catastrophe, Bruno Arpaia's *Qualcosa, là fuori* (*Something, Out There*; 2016) and Michel Faber's *The Book of Strange New Things* (2014). Caracciolo situates his study within a broad turn within literary and cultural studies to the "nonhuman," embracing both the natural world that enables human existence, from evolution to ongoing survival, and the omnipresent mediations of technology that enable human exploitation of and pollution of the natural world to the extent of threatening life on Earth. Caracciolo cites Stanisław Lem's novel *Solaris* as a prescient instance of this turn: its planet with an apparently sentient ocean captures the idea of isomorphisms or structural parallels between mind/brain and world, in terms of "patterns" – with complexity of patterns as central to the emergence of mind from matter. The two novels examined similarly use eco-disaster as a "concrete narrative-level analogue of humanity's constitutive interdependency with the material world," despite having very different protagonists and plots. Caracciolo's careful exploration of narrative and linguistic devices and details reveals similar attention to "patterns" as central to the isomorphisms between natural world and mind/brain in both novels; disruptions of which can signal failures of perception and affect. These failures are related to the difficulties in truly grasping and narrating events on the scale of climate change and potential ecological catastrophe – failures of perception in Arpaia's (materialist) protagonist, failures of affect/sympathy in Faber's (religious) protagonist. Thus in different ways, both highlight the the need, in the context of climate crisis, for a "radical rethinking of Western dichotomies" that divide the subjective from the material.

In "Emotion and Matter in *Corps corrosifs* by Lorand Gaspar," Víctor Bermúdez develops an interdisciplinary approach to analyzing abstraction in poetic language. Bermúdez builds on research in CLS to address the emotional basis of conceptualization of matter and space in poetry, focusing on Lorand Gaspar's work *Corps corrosifs*. Bermúdez argues that embodied meaning emerges in this poetic corpus by an "affective abstraction" of space and a scientific treatment of matter. The argument proposes that emotions are involved in our abstract comprehension of the material world, and that literature reveals this kind of emotional comprehension, and enables deeper analysis of how it operates. Bermúdez sketches recent studies of emotion, highlighting how emotion is grounded in the body and its functioning, and precedes and informs behavior, speech and conscious thought, and any form of meaning. He elaborates Mark Johnson's "embodied meaning hypothesis" by showing how it can operate in literature, as a unique set of expressions of emotional experience, conjoined to experience of matter. For Gaspar, verbalizing the matter of the world is modulated by the feeling of space; and emotion is not simply subjective, as the objects, beings and events of a text are also expressions or extensions of emotional experience and response. Gaspar's poems are elusive and surreal, blending perception, fragmented thought and complex feeling in a way reminiscent of synesthesia. Bermúdez unpacks Gaspar's mysterious lines skilfully, allowing fuller comprehension via his close readings of the poems' representations of emotion-inflected landscape and geology as it interacts with climate/weather and multi-perceptual/multi-modal human experience. Thus the chapter underlines the importance of affective cognition as central to the processes of human imagination.

In "On Being Moved: Body Movements as Prompts for Affect and Empathy in Karen Fowler's *We Are All Completely Beside Ourselves*," Renate Brosch argues that fictional renderings of characters can induce experiential and empathetic involvement in readers. Like Bermúdez, she notes the connection between motion and emotion, but draws on empirical studies of links between gestures and empathy to take a new approach to analyzing narrative fiction, focusing on the prompts for embodied simulation and empathy that can be found in narrative depictions of body movement and gesture. While most CLS concentrates on mind-reading and emotion contagion within the story world and among fictional characters, Brosch refocuses on the responses of the reader, proposing that the body language of fictional characters in narratives prompts strong affective involvement, story world immersion, and empathy. Readers of narrative fictions are at some level aware of characters' bodily movements, which express mental states not described in the text. Movement grabs the attention more powerfully than non-moving objects, and it induces enactive immersion in the story. However, different sorts of movements activate different sorts of readerly involvement. Sudden, spontaneous movements can have a surprise effect, but interpersonal gestures are likely associated with the most visceral emotional responses. Gestures are both expressions of emotion and triggers for emotion, so "being moved" by fictions means both imagining story world spaces and perspectives, and also responding affectively and empathetically to them. Brosch identifies narrative prompts of empathy and suggests that body language can function as a trigger for simulation as well as reflection. Brosch analyzes the surprising body language in Karen Fowler's *We Are All Completely Beside Ourselves* (2013) to show that empathy is encouraged in multiple ways by interacting imagery of body movement, embodied simulation, and reversals of expectations. A particular surprsing scene in the novel concentrates these issues and raises new questions about them, including implications for reader identification and empathy with the main character (related to her status as an expert zoologist and psychologist), and reflections on "Theory of Mind" in connection with animal-human cross-species relations (echoing Beer's new chapter on Darwin's study of the topic, and the related post-humanist perspectives mentioned earlier). Brosch elucidates how the novel itself complicates these issues of the nature of theory of mind and empathy, with its tracing of complex histories of highly emotional, morally fraught events and of rememberings of them.

2.2 Literary approaches to narrative cognition in science

The second section turns to the study of scientific texts, in which authors draw on literary and linguistic theories and concepts to discern the narrative qualities of scientific thinking and meaning-making.

ELINAS co-director Aura Heydenreich offers a searching case study of what she calls "interformative" literary-scientific cognitive processes in "The Epistemic Power of Imagination through Interformation between Astrophysics and Literature: Carl Sagan's Novel *Contact* and Kip Thorne's Model of Traversable Wormholes." Heydenreich examines how modeling processes blend literature and physics in the development of the theoretical model of "wormholes," or spacetime tunnels that connect two distant points of the universe. Thorne treated a wormhole as a solution to the field equations of General Relativity Theory, while Sagan's work on his novel *Contact* (1985) posed the question of the potential existence of such a tunnel. Thorne entangles two semio-logical spheres by moving this literary thinking into theoretical astrophysics, contributing to a reorganization of the field. The chapter focuses on reconstructing these interformative processes in Thorne's theoretical modeling. First, Heydenreich describes how Sagan's novel *Contact* develops the concept of traversable wormholes to enable a fictive narrative journey by astronaut Ellie Arroway to the star Vega. The historical background shows that mathematical models of a space-time tunnel have existed since Einstein's publication of the basic field equations, but Sagan and Thorne's collaboration around 1985 revitalized the wormhole metaphor, as Sagan's literary inquiry reconceptualized Thorne's astrophysical studies by attending to the modeling of humanly-traversable wormholes. Heydenreich delves into how interactions between semio-logical spheres engendered this reconceptualization, via a close reading of Thorne and Michael Morris's 1988 article "Wormholes in Spacetime and Their Use for Interstellar Travel."

Elisabeth Haefs's "'Long ago, deep in space': Narrating Gravitational Waves" contributes to studies of rhetoric in scientific thinking and communication, and narrative cognition in not only science communication (news reports of exciting findings) but in core professional scientific expression (journal articles), and in the creative analytical thought informing them. She draws on studies of mythic patterns in literature and culture to illustrate how figurative devices can encode narrativity in science. Narrative is a primal way of structuring knowledge, and even plain factual reports may feature narrative elements. Many scientific papers contain narrative patterns, even plots with literary qualities. Academic interest in the relation between science and literature has generally focused on reflections of scientific ideas in literary texts. Turning this focus in reverse to analyze literary devices in scientific texts poses special challenges but also surprising rewards. Haefs's analysis focuses on the narrative qualities of two texts from different genres reporting on the first discovery of gravitational waves in 2015, Adrian Cho's online news piece "Gravitational waves, Einstein's ripples in spacetime, spotted for first time" from *Science*, and the original paper from *Physical Review Letters*, "Observation of Gravitational Waves from a Binary Black Hole Merger" by B. P. Abbott et al. In these accounts, Haefs identifies an emergent form of quest narrative, in which the scientists, or the gravitational waves themselves, function as heroes. However, the two texts (and their genres) contrast in terms of the "overtness" of their narrativity, with the popular genre using obvious storytelling features deliberately, and the specialized genre using what Haefs calls "subdued narrativity." Yet the presence of the structure in both reveals the pervasiveness of narrative and its fundamental cognitive ordering function. Close readings of the two texts reveals how grammatical and rhetorical features evoke narrative features. Haefs's focus on plot types shows how they can fluidly arise, develop, depart and implicitly combine – quest, detection, exploration, romantic love and others – and interestingly relate to some unconventional action types (e.g. surfing) with metaphorical and narrative functions.

In "Narrative Minds in the Construction and Use of Theories of Forces of Nature: A Model of Experience at Different Scales," the team of Hans U. Fuchs, Elisabeth Dumont and Federico Corni argue for thinking of physics quite differently from the traditional ways – that is, as a product of "imaginative rationality." The consequence of this reframing is an impulse and a capacity to seek out structures of narrative form and conceptual metaphor in formal physical theories, and to regard those structures as "constitutive of the scientific enterprise" rather than merely linguistic features. The authors examine linguistic and methodological evidence that physics is "metaphoric, analogical, and narrative all the way to its conceptual core." They analyze metaphors in language about temperature and heat to discern how those concepts are being defined in terms of underlying schematic structures. They then turn to narratives of macroscopic physical processes, such as how sunlight leads to the creation of fresh water, to examine how forces of nature are framed as agents enacting processes over time. These steps take the authors into reflections on the role of narrative in the process of science, and ultimately to a larger proposal of a model of human-level experience at multiple scales (of time, space and systems), in order to explain how processes of narrative and metaphoric thought create the core concepts of this area of physics. The opening stresses that this study grew out of applied research – working with teachers of physics to very young students (kindergarten and primary school) – that led to narrative pedagogical methods: teaching teachers how to write stories of forces of nature. The authors describe a web of metaphoric conceptual mappings for heat/hotness, in groupings based on linguistic expressions and associated (image-)schematic structures (analogous to Mark Johnson's metaphors for the force of music). They then examine a constructed story of heat as a force of nature to show how narrative allows specific metaphors to be combined into far more complex larger pictures of relations among multiple entities and their dynamics. Building on that, they develop a model of experience at multiple scales, to help explain the roles of metaphor and narrative in science.

2.3 Scientific approaches to narrative cognition in science

This section of the book gathers cognitive analyses from the fields of sociology and psychology of the role of narrative and metaphor in scientific communication and understanding, the role of physical experience in basic narrative structuring, and the role of narrative structure in the understanding of scientific articles.

Sébastien Lemerle's "A Black Box Filled with Highways: A Study of the Uses of Metaphors in French Popular Neuroscience" examines the use of metaphors and analogies within the discourse of the most important event of "live" popularization of neuroscience in France, the "Semaine du Cerveau" (the equivalent of Brain Awareness Week). Lemerle seeks to identify the most frequently occurring metaphors and analogies for (aspects of) the brain, and to compare how they are used by scientists, on the one hand, and the popular narratives of the media who communicate the science to the public, on the other. He describes a kind of struggle over how the brain is framed in public discourse, where scientists accuse the media of distorting and exaggerating their scientific statements for heightened emotional impact and audience attention, yet also rely on the same or similar metaphors in their own professional discourse. Lemerle reviews how popular science is conceived in scholarship, focusing on its role as "an attempt to bridge a gap between specific cognitive worlds" from the ideas and interests of specialized researchers to those of other groups in a broader audience. He sketches the recent history of neuroscience popularization, noting that it has been proceeding for over half a century in all available media, from print to radio, TV and the internet, among a growing body of producers and consumers. Neuroscience has deeply imprinted intellectual and public life, inspiring social scientists and humanists who advocate reconceptualizations of their disciplines, public intellectuals of all stripes articulating views on all manner of public concerns, and politicians crafting policies in education, health and law. The Semaine du Cerveau is meant to be unaffected by the distorting biases of popularization, but here neuroscientists themselves become the popularizers and end up relying on the kinds of frames and methods they elsewhere spurn. Lemerle shows how they seek to balance their specialist purposes with communicational needs as they deploy metaphors and analogies, and what this means for the public understanding of neuroscience. He observes that the range of metaphor is surprisingly narrow, and explains this fact in terms of constraints on the use of metaphor resulting from the ambivalence inherent in popularizing discourses.

In "What Little Red Riding Hood Can Teach Us About Science," Monica Gonzalez-Marquez and her co-authors present a comparative study of ways of teaching science reading that shows that teaching narrative structure supports better comprehension results than the typical pedagogical alternatives. Beginning from a "cognitive narrative" perspective, the study hypothesizes that narrative training for scientific information will be effective because narrative is a primary mental mechanism of information structuring. The experiment tests this hypothesis by examining the ability of three groups of students in different instruction treatment conditions to answer comprehension questions about scientific articles. One group is given no instruction, one group is given conventional instruction in the structure of scientific journal articles, and one group is given instruction in recognizing narratives, and analyzing them in terms how story events characterized by the "universal interrogatives" of who, what, where, when, why and how map to an overall narrative arc. The students then answer questions about what question their scientific article is trying to answer, how the question was asked, and what was the answer. Student answers are coded for correctness, and for the extent to which answers are copied from the text or formulated in students' own words (indicative of deeper understanding). Results show that students with narrative training show more accurate and deeper comprehension in various ways. This has implications for scientific practice as well as science teaching, and the authors propose that the inherently human and processual nature of science as an activity supports a pedagogy and a practice that emphasizes representations of agents and events. This in turn ties in to broader issues of the misleading public stereotype of science as a collection of objective facts; and within the scientific community, how the Open Science movement could help alleviate the "replicability crisis," as replicability hinges on a full and accurate representation of experimental acts and outcomes.

Paola Pennisi's study of "The Contextual, Enabling, and Constitutive Roles of Physical Experience in Narratives" argues that while most narratives are focused on social contents, many narratives (especially in science) are devoid of social content, but all narratives depend on representing physical objects and events. More particularly, she shows how physical experience has three key roles in narrative, from the contextual to the enabling to the constitutive. The contextual role is the development of narrative skills in childhood via physical interaction with the world. The enabling role is illustrated in the creation of both "natural" and "unnatural" narratives, which necessarily draw on physical experience. The constitutive role is illustrated by the necessity of time as a factor in narrative – and the fact that experiencing and conceptualizing time requires experiencing physical space. Pennisi illustrates these roles through different kinds of narratives. The contextual role is shown by comparing narratives from children with Autism Spectrum Disorder and Cerebral Palsy. The enabling role is illustrated in key forms of narrative: first, by the role of the body in "natural" and "unnatural" narratives (in the category of fictional narratives); and second, by the material bases of personal experience and representational media in autobiographical first-person narratives (in the category of factual narratives). The former category is illustrated by Dante's *Divine Comedy*, and Stevenson's *Dr. Jekyll and Mr. Hyde*, while the latter category is illustrated by the neuro-atypical memoir of a synesthete. The constitutive role is demonstrated by examining narrators' interactions with the physical world as a prerequisite for narrativity, and by the ways in which the experience and representation of time necessarily presupposes the experience and conceptualization of space (discussing research in cross-cultural psychology).

2.4 Cognitive Approaches to Narrative Across Literature and Science

The final section expores cognitive angles on narratives that in some way bridge literature and science, from a classic factual text (by a literary writer) representing and analyzing the experience of a fictional visual scene (painted by an artist); to the framing of physical and mental action in terms of energy, comparing semiotic features of literature, science and pseudoscience; to a cognitive approach to key facets of narrative thinking and representation (both literary and scientific); to the cognitive effects of the interplay of novelty and repetition in science fiction narratives of physics, as well as everyday and aesthetic experience.

Renata Gambino and Grazia Pulvirenti's "Embodied Sublime. A Case-Study from German Romanticism: Heinrich von Kleist's ekphrasis *Feelings about a Seascape by Friedrich*" considers Kleist's fascination with Caspar David Friedrich's classic Romantic-era painting *The Monk at the Sea*. The painting was controversial because it disrupted classical principles of landscape, but Kleist rejected the conventional criticism and celebrated the painting's originality and aesthetic impact. The authors argue that cognitive science now gives us new ways to understand and appreciate Kleist's text. They explain Kleist's innovative use of the rhetorical technique of *ekphrasis* (description of an art object) in the context of the contemporary debate over the concept of "the Sublime." Kleist develops an anti-Kantian (implicity pro-Burkean) version of the concept by subverting its abstract intellectual "transcendental" trappings (connected with a view of art as rule-governed), instead linking it to bodily sensory experience. Gambino and Pulvirenti draw on recent neuroscience to argue that Kleist's text goes well beyond the conventional objective description of ekphrasis by *performing* a reconstruction of the complex process of the reader's immersion in the sublime via embodied simulation. This anticipates recent theories in neuroaesthetics, where beauty always balances meaning-making/interpretation and the bodily experience of the viewer.

For Kleist the difference between natural and artistic beauty is that the latter engages the imagination by prompting simulation of situation models associated with experiencing natural scenes, which stretches the cognitive power of the viewer/reader in pushing through habitual imaginative limits.

Michael Sinding's "Energy Models of Causation in Mind, Culture and Science: Electro/Magnetic Thinking in Blake, Galvani and Mesmer" develops an account of how we build mental models of causal events as we experience narratives, and how these models differ across the domans of literature, science and pseudoscience. Cognitive research shows that *sensorimotor schemas* are central to how we construe storyworld persons, objects, settings and events, and to the use of metaphors in narrative. Drawing on "intuitive ontology," the informal models of physics, biology, etc., that we use to navigate our everyday worlds, he proposes that human minds use an *energy principle* that integrates *space, substance, force* and *motion* to model 1. physical structure and causation, and 2. psychosocial structure and causation via metaphor. Basic causation is seen as *transfer of energy between moving objects via contact*; and energy in turn is seen as a form of *force*-bearing *substance* ("umph-stuff"). Sinding illustrates the energy principle in various forms of world-modeling (cosmologies, stories, arguments), then compares forms of energy-thinking across domains via major instances of "electro/magnetic thinking" (preceding the recognition of unified "electromagnetism") in key late eighteenth-century texts by William Blake, Luigi Galvani and Franz Mesmer. A comparative analysis of how literal and metaphorical energy thinking works in these writings leads to proposals on the nature of the distinctions between, and the interactions among, their domains.

Brook Miller's "Reading Narratives" focuses on the remarkable human capacity to have rich aesthetic experiences of simulated participation while reading narratives. He examines how this is possible, arguing that a focus on the prodigious information processing involved is potentially misleading, and that the simulative aspects of the extraordinary experiences of narrative reading are grounded in ordinary perceptual and affective processes. That is, Miller proposes a re-ordering of the conventional hierarchy of cognitive factors involved in narrative experience and effects, arguing that the experiential aspects are integral to "lower level" sense-making processes, rather than subsequent effects of them. Miller's account suggests that we transfer our assumptions about how physical things are built to how narratives are mentally built. Small parts are built into larger parts which are built into still larger wholes, like bricks and boards to walls to houses. Narratives are built from written words that form concepts that are connected to other concepts into a whole story. A reader can derive and synthesize information about the story by drawing on their repositories of many kinds of knowledge (social, natural, aesthetic, etc.). Similarly, a person can navigate through and use the various rooms of a completed house. But this "processing model" says Miller, "puts the cart before the horse" because "qualia are not subsequent to information processing but integral to it." Miller's chapter contributes "a set of gap-filling ideas to further develop models that foreground the phenomenology of reading narrative." It examines isomorphic processes between narrative reading and ordinary conscious experience, and it suggests that while the former adapts the underpinnings of the latter, the phenomenological and the informational intertwine to produce narrative simulations.

Dirk Vanderbeke's "The Comforts of Constancy in Literature and Science" takes on the "cherished assumption that humans love novelty" in experience and variety in art, rather than similarity and repetition. Hence the need for the arts to evolve continuously, to satisfy a public that seeks new aesthetic and cognitive stimuli to fend off boredom. The sciences too nurture the assumption that originality and innovation that challenge status quo ideas, approaches and paradigms are preferred to the familiar. Vanderbeke's chapter challenges these assumptions by considering the pleasures of repetition in both artistic experience and scientific research. He argues that the assumption is rooted in a too-narrow focus in both domains – on the attention-getting "great works" and revolutionary discoveries that are "canonized" and history-defining because of the long shadows they cast. However, the "vast mass" of popular literature and the everyday work of science must be appreciated by those who prefer confirmation of preconceptions to challenge of norms. Vanderbeke goes further, proposing that humans in general prefer the "security of the familiar" in narrative patterns and scientific paradigms, despite the fact that more creative exceptions always trespass boundaries for greater potential recognition and reward, despite greater risk of failure and obscurity.

References

Abraham, Carolyn. *Possessing Genius: The True Account of the Bizarre Odyssey of Einstein's Brain*. Toronto: Penguin, 2002.

Ahuja, Neel, et al., eds. *The Palgrave Handbook of Twentieth and Twenty-First Century Literature and Science*. Cham, Switzerland: Springer Nature-Palgrave Macmillan, 2020.

Anderson, Michael L. "Neural Reuse: A Fundamental Organizational Principle of the Brain." *Behavioral and Brain Sciences* 33.4 (2010): 245–266. doi:10.1017/S0140525X10000853.

Beer, Gillian. *Darwin's Plots: Evolutionary Narrative in Darwin, George Eliot, and Nineteenth-Century Fiction*. 3rd edition. Cambridge: Cambridge UP, 2009 [1983].

Brockman, John, ed. *Edge*. https://www.edge.org/.

Clarke, Bruce, and Manuela Rossini, eds. *The Routledge Companion to Literature and Science*. New York: Routledge, 2011.

Crane, Mary Thomas. *Losing Touch with Nature: Literature and the New Science in Sixteenth-Century England*. Baltimore: Johns Hopkins UP, 2014.

Crane, Mary Thomas. "Surface, Depth and the Spatial Imaginary: A Cognitive Reading of *The Political Unconscious*." *Representations* 108.1 (Fall 2009): 76–97.

Daston, Lorraine. "Science Studies and the History of Science." *Critical Inquiry* 35.4 (2009): 798–813.

Dear, Peter. "Afterword." *The Palgrave Handbook of Early Modern Literature and Science*. London: Springer Nature-Palgrave Macmillan, 2017. 505–514.

Fludernik, Monika. "Histories of Narrative Theory (II): From Structuralism to the Present." *A Companion to Narrative Theory*. Ed. James Phelan and Peter J. Rabinowitz. Blackwell Publishing, 2005. Blackwell Reference Online. 14 November 2012. <http://www.blackwellreference.com/subscriber/tocnode.html?id=g9781405114769_chunk_g97814051147694>

Fludernik, Monika. *Towards a 'Natural' Narratology*. London: Routledge, 1996.

Fludernik, Monika, and Marie-Laure Ryan, eds. *Narrative Factuality: A Handbook*. Berlin and Boston: De Gruyter, 2020.

Garratt, Peter, ed. *The Cognitive Humanities: Embodied Mind in Literature and Culture*. London: Palgrave Macmillan, 2016.

Herman, David. "Histories of Narrative Theory (I): A Genealogy of Early Developments." *A Companion to Narrative Theory*. Ed. James Phelan and Peter J. Rabinowitz. Blackwell Publishing, 2005. Blackwell Reference Online. 14 November 2012. <http://www.blackwellreference.com/subscriber/tocnode.html?id=g9781405114769_chunk_g97814051147693>

Heydenreich, Aura, and Klaus Mecke. "Introduction." *Physics and Literature: Concepts – Transfer – Aestheticization*. Eds. Aura Heydenreich and Klaus Mecke. Berlin and Boston: De Gruyter, 2021. 1–19. https://doi.org/10.1515/9783110481112

Heydenreich, Aura, and Klaus Mecke, eds. *Physics and Literature: Concepts – Transfer – Aestheticization*. Berlin and Boston: De Gruyter, 2021.

Hogan, Patrick Colm. *Affective Narratology: The Emotional Structure of Stories*. Lincoln: University of Nebraska Press, 2011.

Hogan, Patrick Colm. *The Mind and Its Stories: Narrative Universals and Human Emotion*. Cambridge: Cambridge University Press, 2003.

Hunt, Lynn. *Inventing Human Rights: A History*. New York: Norton, 2007.

Keen, Suzanne. *Empathy and the Novel*. Oxford: Oxford University Press, 2007.

Labinger, Jay, and Harry Collins. *The One Culture? A Conversation About Science*. Chicago: University of Chicago Press, 2001.

Marchitello, Howard, and Evelyn Tribble. "Introduction." *The Palgrave Handbook of Early Modern Literature and Science*. London: Springer Nature-Palgrave Macmillan, 2017. xxv–xlvi.

Marchitello, Howard, and Evelyn Tribble, eds. *The Palgrave Handbook of Early Modern Literature and Science*. London: Springer Nature-Palgrave Macmillan, 2017.

Meuter, Norbert. "Narration in Various Disciplines." (2013) *The Living Handbook of Narratology*. Ed. Peter Hühn. Hamburg: Hamburg University Press. https://www.lhn.uni-hamburg.de/node/49.html (12 May 2021).

Meyer, Steven. "Futures Past and Present: Literature and Science in an Age of Whitehead." *The Cambridge Companion to Literature and Science*. Ed. Steven Meyer. Cambridge: Cambridge University Press, 2018. 255–274.

Meyer, Steven. "Introduction." *The Cambridge Companion to Literature and Science*. Ed. Steven Meyer. Cambridge: Cambridge University Press, 2018. 1–21.

Meyer, Steven, ed. *The Cambridge Companion to Literature and Science*. Cambridge Companions to Literature. Cambridge: Cambridge University Press, 2018.

Netz, Reviel. “The Long History of Cognitive Practices: Literacy, Numeracy, Aesthetics.” *The Cambridge Companion to Literature and Science*. Ed. Steven Meyer. Cambridge: Cambridge University Press, 2018. 242–254.

Newman, Daniel Aureliano. “Narrative: Common Ground for Literature and Science?” *Configurations* 26 (2018): 277–282.

Palmer, Alan. *Fictional Minds*. Lincoln: University of Nebraska Press, 2004.

Palmer, Alan. “Intermental Thought in the Novel: The Middlemarch Mind.” *Style* 39.4 (2005): 427–439.

Palmer, Alan. *Social Minds in the Novel*. Columbus: Ohio State UP, 2010.

Renn, Jürgen. “From the History of Science to the History of Knowledge – and Back.” *Centaurus* 57 (2015): 37–53.

Richardson, Alan. “Literary Studies and Cognitive Science.” *The Cambridge Companion to Literature and Science*. Ed. Steven Meyer. Cambridge: Cambridge University Press, 2018. 207–222.

Snow, C. P. *The Two Cultures*. Introd. Stefan Collini. Cambridge: Cambridge University Press, 1998 [1959/1964].

Spolsky, Ellen. *Gaps in Nature: Literary Interpretation and the Modular Mind*. Albany: State University of New York Press, 1993.

Spolsky, Ellen. “Darwin and Derrida: Cognitive Literary Theory as a Species of Post-Structuralism.” *Poetics Today* 23.1 (2002): 43–62.

Spolsky, Ellen, Ben Morgan and Sowon Park, eds. *Poetics Today* Special Issue on “Situated Cognition and the Study of Culture” (38.2 / 2017).

Tabbi, Joseph. “Cognitive Science.” *The Routledge Companion to Literature and Science*. Eds. Bruce Clarke and Manuela Rossini. London and New York: Routledge, 2011. 77–88.

von Contzen, Eva, and Max Alders, eds. *Narrative* Special Issue on “Social Minds in Factual and Fictional Narration” (23.2 / 2015).

Williams, Lawrence E., Julie Y. Huang, and John A. Bargh. “The Scaffolded Mind: Higher Mental Processes Are Grounded in Early Experience of the Physical World.” *European Journal of Social Psychology* 39.7 (2009): 1257–1267.

Wootton, David. *The Invention of Science: A New History of the Scientific Revolution*. New York: HarperCollins, 2015.

Zunshine, Lisa. *Why We Read Fiction: Theory of Mind and the Novel*. Columbus: Ohio State University Press, 2006.

Zunshine, Lisa, ed. *The Oxford Handbook of Cognitive Literary Studies*. Oxford: Oxford University Press, 2015.

Part I: **Cognitive Approaches to Literary Narratives of Science**

Rudolf Freiburg

"Let's hear you apply Heisenberg to ethics"

The Problematic Interferences between Science and Literature in Ian McEwan's *Solar*

Abstract: Science and the Humanities seem to be worlds apart with no *tertium comparationis* they might share. Charles Percy Snow has convincingly described this clash of the two "incompatible" cultures in his famous essay. Some writers, however, have always endeavoured to bridge the gap between these two disparate worlds, and among these, Ian McEwan plays a predominant role in contemporary literature because he manages to see moral and ethical problems from a scientific point of view and *vice versa.*

In this article, I first describe the complex relationship between science and fiction in general, showing that – following Plato's ideas voiced in the parable of the cave – for a long time fiction was considered a "lie," which distorted any expressions of scientific truths. Although Philip Sidney defended poesy (a synonym for literature in his time), since – as an iconography of both history and philosophy (including natural science) – it added a didactic element to the proclaiming of truth, Sir Francis Bacon remained unconvinced and dismissed all acts of (poetic) imagination from science as detrimental to finding the "truth." In his "doctrine of the idols," he categorically recommended a (non-fictional) language exempt from disturbing metaphors, a language which should be adjusted to the clarity of mathematical formulas, a *lingua clara et distincta.* Bacon's ideas intensely influenced the development of scientific thinking in England; the Royal Society heeded Bacon's advice and consistently chose the motto *nullius in verba.* For many centuries, the notion of mathematical plainness remained a linguistic ideal in the communication of science, and it found its twentieth-century expression in the ideas of Ludwig Wittgenstein and the Vienna Circle.

Werner Heisenberg, however, did not follow suit. He esteemed the creative potential, inherent in all acts of imagination, and saw clear parallels between a prose narrative defined as a constructive "fiction" by which meaning is given to reality, and the formulation of a scientific hypothesis conceived to describe an objective correlate in "reality." He stressed the fact that individual factors often enabled – and simultaneously distorted – the results of a scientific analysis in the subatomic world beyond language. In a moment of insight, close to what literary critics of Modernism would probably define as an epiphany, Heisenberg recog-

https://doi.org/10.1515/9783110782844-002

nized the role the "uncertainty principle" plays in acts of epistemology. His description of scientific problems in quantum theory astoundingly reveals many parallels with the tradition of hermeneutics in the Humanities; one might even compare the uncertainty principle to Wolfgang Iser's theory of the gaps.

In *Solar*, McEwan (2010) uses Heisenberg's uncertainty principle as a backdrop for discussions about the relationship between science and the Humanities and for debates on ethical problems. The central theme of the partly satirical novel is the climate crisis. The protagonist Beard, Nobel Prize winner and prestigious scientist, is an arrogant representative of rationality, logic and syllogistic reasoning. Using him simultaneously as a satirical *persona* and as a satirical victim, McEwan shows the limits of both science and rationality. Of course, Beard is right, when he insists that Heisenberg's quantum theory cannot directly be applied to ethics; but on second thought, he is further away from Heisenberg's notions than any of the artists and representatives of the humanities he derides during the conversation that takes place on board a ship in the Arctic. Whereas Heisenberg in his attitude came close to the essayist described by Robert Musil in *A Man without Qualities*, Beard claims the certainty of a scientific tyrant and doctrinaire in *Solar*. Further, most of McEwan's novels, like *Black Dogs*, *Enduring Love*, *Atonement*, *The Children's Act*, and of course *Solar*, prove that ethical problems cannot be solved by apodictic judgments; McEwan makes the reader decide for himself which role he is willing to attribute to the uncertainty principle in ethical and moral contexts. Even at the end of the book, one keeps on wondering if – despite his many moral deficiencies and his questionable character – Beard may have decisively contributed to the solution of the climate crisis.

1 The relationship between poetry and natural philosophy: A brief history

The interferences between natural philosophy and "poetry," as literature was generally called in ancient times, have always been complex. Four centuries before Christ, Plato proposed that all poets should be expelled from his *Republic* (see Destrée and Herrmann 2011), since they distorted truth and produced nothing but lies. Plato's ideas were popular throughout the Middle Ages and the Renaissance, but in the sixteenth century Sir Philip Sidney refused to follow suit and – in his famous work of literary criticism *An Apology of Poetry* or *The Defense of Poesy* (1595), which was published after his death – he justified the role of literature by reanimating Horace's definition of poetry ("*prodesse et delectare*") and by emphasizing its iconographic function (see Horatius 1961):

> Poesy therefore is an art of imitation, for so Aristotle termeth it in the word *μίμησῃ* – that is to say, a representing, counterfeiting, or figuring forth – to speak metaphorically, a speaking picture – with this end, to teach and delight. (Sidney 1978, 25)

Due to poetry's use of images, metaphors, similes and symbols, the poet is able to *illustrate* the abstract truths of both history and philosophy (which included the "humanities" and the "natural sciences" at this time). Two decades later, this positive view of literature was peremptorily rejected by the English philosopher and essayist Sir Francis Bacon, who joined Plato in his fight against poetry. Trying to transcend the epistemological limits Aristotle had fixed in his *Organon* (ca. 340 before Christ), Bacon published *The Advancement of Learning* (1605) and his own *Novum Organum Scientiarum* (1620) as answers to the Greek philosopher. In the *Novum Organum*, he develops the doctrine of the idols as an antidote against the mental diseases of uncritical Aristotelianism (see Bacon 1858, 149–365); he distinguishes between four idols: These, the *idola tribus, idola specus, idola fori* and *idola theatri*, are exposed as irritating prejudices responsible for the falsification of truth and the distortions of scientific results (see Krohn 1987, 93–107). Bacon primarily decried the "idols of the marketplace," that is "words" and "metaphors," and the "idols of the theatre," that is the superstitious notions caused by "tradition," "Aristotelianism" and scholasticism, for obfuscating man's perception of reality – he recommended that all forms of imaginative acts should be avoided in scientific thinking; instead a mathematical language called the *lingua clara et distincta* should be used by mathematicians and physicists alike (see Real and Vienken 1984, 92–94).

In England, this trend towards linguistic scepticism in science was intensified by the scientific program of the influential *Royal Society* and their programmatic motto "*nullius in verba*" ("take nobody's word for it"); the first historian of the *Royal Society*, Thomas Sprat, for example, demanded that all "amplifications," "digressions" and "swellings of style" should be given up to attain the pure status of "mathematical plainness" (see Gamper 2013, 13).

This ideal of mathematical plainness gained momentum throughout the following centuries, and with Ludwig Wittgenstein's mathematical analyses in his *Tractatus Logico-Philosophicus* (1921) and the philosophy of the Vienna Circle (see Wittgenstein 1977), semantic neutrality and purism seemed to be generally accepted as the linguistic norm in all scientific endeavors (see Sullivan and Potter 2013). But even if the dynamics of language were perceived as disturbing factors in all acts of perception and investigation, the scientists of the early twentieth century could not but admit that scientific research beyond the limits of language, or – to put it even more succinctly – *poetic* language, seemed extremely difficult, if not impossible.

2 Werner Heisenberg and Ian McEwan in the interstice between literature and science

Walking up the steep hills of the Hainberg in Göttingen, the intellectual center of mathematics and physics – which, in the early twentieth century, was ruled by Felix Klein, Richard Courant, and Amalie Emmy Noether – Werner Heisenberg and Niels Bohr discussed aspects of the philosophy of language; they were especially interested in the question of whether human language could be applied to the field of subatomic particles completely "free of human taint" (McEwan 2010, 8).[1] Despite all their dissensions, Heisenberg and Bohr were willing to confess that both physicists and poets had more in common than most scholars believed, since in their view the effects of a scientific breakthrough were achieved not by logical calculation but by an act of pure "intuition":

> For the first time, I understood that Bohr was much more skeptical towards his own theory than many other physicists of his time, for example Sommerfeld, and that – for him – his knowledge about the correlations did not result from a mathematical analysis of the underlying assumptions, but from an intense preoccupation with the phenomena, which enabled him to develop an intuitive awareness of the correlations, rather than to deduce them. So this is the way cognition about nature is generated. The second step, then, may lead to a mathematical refinement of what has been found out before, and to subject it to a comprehensive rational analysis. (Heisenberg, quoted in Hermann 1977, 55; trans. by RF)

The scientist, as Bohr and particularly Heisenberg, who blamed Ernst Mach's theories for their lack of poetry (see Hermann 1977, 70), recognized, had to act like a narrator using a kind of "emplotment-strategy" comparable to that described much later in the century by Hayden White as the essential principle of postmodern concepts of metahistory (see White 1978). Seen from this perspective, a scientific theory has to be redefined as a "narrative." Moreover, Heisenberg, Bohr and Carl Friedrich von Weizsäcker were extremely interested in the implications that their scientific discoveries had for all fields of life, especially for philosophy, ethics and religion.[2]

More than eighty years later, in his novel *Solar* (2010), which discusses the role of ethics in the life of a remarkable physicist, Ian McEwan makes his protago-

---

1 References to Ian McEwan's novel *Solar* will from now on be quoted as *S* plus page number.

2 A long section in Heisenberg's book *Physics and Beyond* is dedicated to a discussion about religion with – among others – Paul Dirac, who, differing from the examples of Max Planck and Heisenberg, was firm in his atheistic position, a reason for being made fun of by the others: "Well, our friend Dirac, too, has a religion, and its guiding principle is: 'There is no God and Dirac is His prophet'" (Heisenberg 1971, 87).

nist Beard deride any attempt "to apply Heisenberg to ethics" (*S*, 77), thus denying the structural parallelisms between the sciences and the humanities and pleading for both the detachment and superiority of physics. Beard seems to solidify the notion of the incompatibility of the "two cultures" (Snow 1959), which, however, is permanently belied by Beard's creator, McEwan himself, who in his role as a writer quite often acts as a scientist. McEwan believes that science is a "route to wonder" (Borm 2012, 240); in interviews he confesses "a lifelong interest" in science, and since he has no religion, science "represents [. . .] the only available and credible metaphysics [. . .]" (Cook et al. 2013, 148). McEwan's passion for science has long been confirmed by informed readers who are willing to grant him the status of a scientific expert: "So if any novelist is able to invoke 'quantum harmonic oscillators' or 'Calabi-Yau orbifolds' without sounding as though they have just been looked up on Wikipedia, McEwan is your man" (Gevaux 2010, 236). Thus the worlds of science and art interfere in the person of McEwan, just as they seem to have coalesced in Heisenberg, who was a skilled musician and – from early childhood on – developed a strong interest not only in physics and mathematics but in philosophy and literature, too. McEwan's *Solar* is one of the rare examples in literature which reject the claim that "twentieth-century concepts of physics and modernist or postmodernist literature are worlds apart [. . .]" (Vanderbeke 2011, 193).

3 Heisenberg's "poetical" science

Heisenberg frequently reminded his readers that science is "made by men" (Heisenberg 1971, xvii). His quasi-autobiographical book *Physics and Beyond: Encounters and Conversations* (1971 [1969]) was clearly reminiscent of the historical work of Thucydides (see Monk 2010, 456), he liked to discuss Democritus and Aristotle, and in a well-known anecdote he reports that as a young man he had his first encounter with Platonic philosophy reading on the roof of a priests' seminary in Munich where he enjoyed the warmth of the early morning sun (see Weizsäcker 1977, 319). Although Heisenberg had a certain talent for calculations and a predilection for the universe of numbers, the leading Bavarian mathematician Ferdinand von Lindemann passed a crushing judgment on the young student when he said that he was "completely lost to mathematics" (Heisenberg 1971, 16). Even in his own field of interest, in physics, Heisenberg favoured the theoretical side of his studies, neglecting the empirical part of his education to such a degree that he almost failed his doctoral examination (see Cassidy 1995, 193). This rather wide scope of interests in all fields of human life, this particular passion for philosophical issues, made him congenial to Niels Bohr, whom he had met on the occasion of the famous "Bohr-festival" in Göttingen in 1921, and with whom he remained friends for the rest of his life. Following the traditional ways of scientific research to a certain degree, Heisenberg parted company with Bacon, the *Royal Society*, Mach and the Vienna Circle as far as scientific methods were concerned. He clearly rejected a world picture described in the radically positivistic words of Pierre Simon Laplace, who in his utopian vision was of the opinion that the scientific spirit (Laplace's "demon") would be able to foretell almost everything. Heisenberg was averse to radical positivistic, factual and numeric verities and did not tire of thinking in terms of probabilities reminiscent of the idea of potentiality of Aristotelian provenience. The predictability of future events, the implicit consequence following from positivistic views of a clearly mechanistic world picture, could no longer convince the scientist who had dedicated great parts of his life to the analysis of quantum mechanics. In his personal epiphany, which due to its intensity produced a feeling of giddiness in him,[3] a moment when he remembered Albert Einstein's remark that it is the theory (or the theoretical "narrative" as described above) itself that decides what an observer can see, he seemed to have found the key to locked rooms of quantum mechanics, which led to the development of the famous Uncertainty Principle:

> It must have been one evening after midnight when I suddenly remembered my conversation with Einstein and particularly his statement, "It is the theory which decides what we can observe." I was immediately convinced that the key to the gate that had been closed for so long must be sought right here. I decided to go on a nocturnal walk through Faelled Park and to think further about the matter. We had always said so glibly that the path of the electron in the cloud chamber could be observed. But perhaps what we really observed was something much less. Perhaps we merely saw a series of discrete and ill-defined spots through which the electron had passed. In fact, all we do see in the cloud chamber are individual water droplets which must certainly be much larger than the electron. (Heisenberg 1971, 77–78)

For a moment he knew that he was about to refute the basic procedures of traditional Western thinking, temporal structures and the principle of causality. The emerging paradox became clearer and led to Bohr's famous remark: "Anyone who is not shocked by quantum theory has not understood it" (quoted from Vanderbeke 2011, 194). Heisenberg felt that his "observations" had tremendous philosophical implications, that if they were valid, his own approach and the basis of

---

3 See Heisenberg (1971, 61): "At first, I was deeply alarmed. I had the feeling that, through the surface of atomic phenomena, I was looking at a strangely beautiful interior, and felt almost giddy at the thought that I now had to probe this wealth of mathematical structures nature had so generously spread out before me."

his own scientific thinking, which was bound to take place in a world of clearly structured time and the processes of causality, were infected in an intense way. Whereas traditional scientific theory had accentuated the necessity of objective research, his own observations gave reason to believe that at least some fields demanded the interference of objective and subjective principles, or in his own words:

> In quantum mechanics the departure from this ideal has been even more radical. We can still use the objectifying language of classical physics to make statements about observable facts. For instance, we can say that a photographic plate has been blackened, or that cloud droplets have formed. But we can say nothing about the atoms themselves. And what predictions we base on such findings depend on the way we pose our experimental question, and here the observer has freedom of choice. Naturally, it still makes no difference whether the observer is a man, an animal or a piece of apparatus, but it is no longer possible to make predictions without reference to the observer or the means of observation. To that extent, every physical process may be said to have objective and subjective features. (Heisenberg 1971, 88)

Although in his work he was a stickler for precise calculations, he was certain that significant discoveries were not due to precision and exactness, merely, but that they were made possible by ingenious acts of intuition, imagination and even phantasy (Heisenberg 1971, 186), principles banned from science by Bacon and his disciples. It is true that Heisenberg did not share many attitudes with Mach, but he could not but have subscribed to one of Mach's meditations upon the subject matter of science in general and physics in particular. Mach had compared physicists with "projectors, inventors of pipe dreams, with writers of novels, and the poet of social and technological utopias" (Mach, quoted in Gamper 2013, 114; trans. by RF).

The world of subatomic particles could not but be described in terms of everyday language, let alone the precise logic of mathematical symbols. Although Heisenberg had reanimated the superannuated discipline of "matrix mathematics" in order to make some sense of the arbitrary processes in quantum mechanics, he shared Bohr's opinion that even the alleged precision of formal logic could merely allow an asymptotic approach towards the enigmatic world of the atomic microcosm; the physicist was to become an "essayist."[4] Moreover, he was doomed to use analogies, images, pictures, and fantastic constructs of the imagination and models; eidetic distortions, paradoxically, seemed to be part and parcel of the scientific attempt to make sense of the subatomic world. In his book he quotes from one of his many conversations with Bohr:

4 I discuss "essayism" in more detail below.

> "These models," Bohr replied, "have been deduced, or if you prefer, guessed, from experiments, not from theoretical calculations. I hope that they describe the structure of the atoms as well, but *only* as well, as is possible in the descriptive language of classical physics. We must be clear that, when it comes to atoms, language can be used only as in poetry. The poet, too, is not nearly so concerned with describing facts as with creating images and establishing mental connections." (Heisenberg 1971, 41)

According to Heisenberg, scientists in general, and physicists in particular, had to face an unavoidable dilemma: On the one hand, being dependent on language, they endeavored to make themselves understood by models, analogies, images and especially metaphors taken from everyday life. On the other hand, it became clear that it was exactly these visual concepts that were responsible for so many errors, fallacies and theoretical *cul-de-sacs*. Scientists, paradoxically, needed "eidetic models" only to give them up again in a moment, because precisely these models prevented further success and scientific progress.

One example taken from quantum physics may suffice: Talking to his lifelong friend Wolfgang Pauli, Heisenberg pondered the question of if it were advisable to assume that electrons move on orbits in an atom, because this notion might lead physicists astray, and the change from one orbit to another could then be compared to absurd everyday events such as "high jump" or "long jump" (Heisenberg 1971, 36). Even as a young man, Heisenberg had found fault with the metaphorical description of the atom, a description he had detected in his physics books, where particles were described as having "hooks" and "eyes" (Heisenberg 1971, 2). Launching vitriolic attacks on the *naïveté* of such notions, Heisenberg and his friends started to talk about acts of understanding, remembering what Nicolas Malebranche had found out about parallel structures in the material world and in the universe of understanding. To a representative of the humanities, it must seem that Heisenberg denied the "two-culture-doctrine" made popular by Charles Percy Snow and others (Heisenberg 1971, xvii), and that he became deeply interested in the discipline of hermeneutics.

The German title of his book *Physics and Beyond*, that is *Der Teil und das Ganze (The Part and the Whole)*, can be interpreted as a severe warning against the typical human predilection for "holistic" concepts, and as a plea that the part and the whole – as in any act of understanding – have to be seen as interdependent and interrelational concepts. His – if I may denominate it in these words – "hermeneutics of science" revealed the disaster courted by the inveterate human predilection for holistic structures. Why should whole numbers be more adequate for the understanding of particles in nature than fractions, why should favored numbers like three, seven or ten, with their long symbolic history of meaning be preferred in science, too? His mentor Arnold Sommerfeld and many others had been shocked when Heisenberg had proved to them that the highly problematic

Zeeman-Effect could be made sense of if one used half and not full quantum numbers, thus ignoring the demands of the "holistic prejudice." The stereotypes and clichés that had long since been analyzed in the humanities by scholars such as Mach and Ernst Cassirer, the long tradition of the allegorical meaning of allegedly neutral mathematical numbers – concepts harking back to Pythagoras and the allegorical doctrines of alchemists and Medieval representatives of scholasticism – had not yet ceased completely.

Heisenberg was trying to get rid of any numerical mysticism when he set out to interpret Planck's discoveries in a new light, not shaded by prejudice and erroneous tradition. A scholar of the seventeenth century will be reminded of Sir Thomas Browne's recommendation to forget things in order to get rid of false notions, a recommendation he made in his famous book *Pseudodoxia Epidemica* (1646), a description of the intellectual epidemic of "vulgar errors" by which his contemporaries were "infected" (see Browne 2007, 269–510). After meeting the later eminent physicist Wolfgang Pauli at a lecture, Heisenberg avows that he is afraid of not knowing enough in physics, but – seen from an epistemological point of view – this lack of knowledge is turned into an advantage because it liberates him from the intellectual fashions of the discipline; confessing that atomic theory is much more satisfying aesthetically than Albert Einstein's "theory of relativity,"[5] Pauli warns against quasi-mystic prejudices in physics, when he describes how Sommerfeld and others hope to solve the riddle of subatomic contradictions by the detection of a higher, mystic level of numeric meaning, behind the physical appearances:

> "Sommerfeld hopes that experiments will help us to find some of the new laws. He believes in numerical links, almost in a kind of number mysticism of the kind the Pythagoreans applied to the harmony of vibrating strings. That's why many of us have called this side of his science 'atomysticism,' though, as far as I can tell, no one has been able to suggest anything better. Perhaps it's much easier to find one's way if one isn't too familiar with the magnificent unity of classical physics. You have a decided advantage there," Wolfgang added with a malicious grin, "but then lack of knowledge is no guarantee of success." (Heisenberg 1971, 26)

So the subatomic world puts any language to the test, denying that it makes sense of a world almost beyond human understanding. It is – and Heisenberg thought in a comparable vein – as if one had to invent a new language, not following the structure of the familiar one the moment one starts to enter the *terra incognita* of quantum mechanics. Heisenberg describes this uncanny feeling:

5 Cf. the parallel discussion of the beauty of scientific theories in *Enduring Love* (McEwan 2006, 49–50).

> You see that everyone trying to develop such a theory is really trying the impossible. For we intend to say something about the structure of the atom but lack a language in which we can make ourselves understood. We are in much the same position as a sailor, marooned on a remote island where conditions differ radically from anything he has ever known and where, to make things worse, the natives speak a completely alien tongue. He simply must make himself understood, but has no means of doing so. (Heisenberg 1971, 40)

4 A quantum theory of hermeneutics

Differing from McEwan's protagonist Beard, Heisenberg and his friends were deeply interested in hermeneutics, philosophy and ethics. Facing the weird "textuality" of arbitrarily behaving atoms and particles – reminiscent of the aesthetics of a postmodern narrative characterized by non-linearity, paradoxical presentation of both time and space, and the oscillation between absence and presence of meaning as a result of poststructuralist semantics described by Jacques Derrida – they tried to create a new system of understanding, even more precisely of meta-understanding, that is of understanding the process of understanding. Like a quantum particle, understanding has to be present in the field of the particular, but it can only grasp the particularity of the particular by using the perspective of the whole, and *vice versa*. A kind of "Quantum Theory of Hermeneutics" emerges: Understanding should imitate the erratic movement of the particles, or, as Laurence Sterne had put it centuries before: "Great wits jump" (see Sterne 1986, 179). Logic, rationality, and calculation can pave the way to scientific discoveries, but it is the mental leap, the quantum jump of the intellect by which new levels of understanding can be reached.

Heisenberg's insight into the ontology of the subatomic world, an insight which was to be refined and illustrated later by the well-known experiments arranged by Richard Feynman (see Hawking and Mlodinow 2010, 61–83), could be interpreted as a radical deviation from Newtonian mechanics and also as a deviation from Kantian *a priori* categories of thinking, as Heisenberg recognizes being shocked himself as he tried to clarify in various conversations with his friend the philosopher Grete Hermann, an expert in Kantian philosophy. Heisenberg did not feel that science and the humanities should be separated from each other; he claimed that the discoveries of scientific research would deeply influence the way we perceive the world and the way we feel we should act.

It may be that in his personal life, he considered physics as the self-evident proof that the obscene uniformity and the fascistic unanimity of the National Socialist environment which was starting to emerge were completely absurd. Heisenberg had been shocked when Nazi propaganda vilified Einstein as an "extremely impertinent Jew" and as a "Relativity Jew" (*Relativitätsjude*) (Hermann 1977, 112; 120; trans. by RF). Heisenberg, who had been defamed as the "Ossietsky of Physics" and as a "white Jew" (Hermann 1977, 143; trans by RF), felt emancipated by science; he was certain that there is more than one truth, and that no mode of indoctrination could be justified. Thinking of Einstein's theory of relativity, he said:

> That is why I consider those developments in physics during the last decades which have shown how problematical such concepts as 'objective' and 'subjective' are, a great liberation of thought. The whole thing started with the theory of relativity. In the past, the statement that two events are simultaneous was considered an objective assertion, one that could be communicated quite simply and that was open to verification by any observer. Today we know that 'simultaneity' contains a subjective element, inasmuch as two events that appear simultaneous to an observer at rest are not necessarily simultaneous to an observer in motion. However, the relativistic description is also objective inasmuch as every observer can deduce by calculation what the other observer will perceive or has perceived. For all that, we have come a long way from the classical ideal of objective descriptions. (Heisenberg 1971, 88)

By merely referring to the principle of relativity, fascistic notions of uniformity, consolidation of institutional powers (*Gleichschaltung*) and nationalistic dreams of uniqueness could be criticized or even falsified. And just as relativity implicitly attacked the bigotry of the Hitler-regime, Heisenberg's uncertainty principle could be interpreted as a counterforce to the dark ideology of the Nazis, who based their preposterous world view on the apodictic truths of a self-proclaimed *Führer*.

Whereas parochial minds came to fear the new principles, the intellectuals in both science and the humanities celebrated them. It is small wonder that the famous uncertainty principle was gradually turned into an alluring concept for the humanities. Heisenberg had found out that it is impossible to ascertain the position and the velocity or impulse of a particle at the same time, and that the scientist had to take refuge in the field of stochastics in order to make sense of the subatomic world. Exporting the uncertainty principle from physics into the world of the humanities seemed to have many opportunities, and it is clear that, for example, the hermeneutic doctrines of – let us say – Roman Ingarden and later of Wolfgang Iser reveal at least a certain philosophical parallel with Heisenberg's discovery. The essence of the parallel lies in the recognition of the fact that human minds fill in information although it is not directly given in a text. Talking about a "young pretty woman" in a short story may induce the reader to imagine her having brown eyes, although the color of her eyes is not mentioned anywhere. The proclivity of human minds to complement information, or, as Iser would say, to fill in semantic gaps and fissures, reveals the desire for holistic structures.

5 McEwan's scientific "poesy": The reception of Heisenberg's uncertainty theory in *Solar*

It is precisely this question which is discussed in McEwan's "ecocritical" novel *Solar* (see Garrard 2009, 696). In this book, which can be interpreted as a satirical reaction to the *Fourth Report of the Intergovernmental Panel on Climate Change* (2007) (see Garrard 2013, 123), McEwan tells the story of the amoral, irresponsible and highly egotistical physicist Michael Beard, a kind of postmodern "picaro" (see Kerridge 2010, 157), who cheats on his wife, is left by her and tumbles from one sexual adventure to the next (see also Kirchhofer and Roxburgh 2016, 149–168), filling the intervals with half-hearted attempts to give lectures on quantum mechanics, to bathe in the fame of the significance of the "Beard-Einstein-conflation," a discovery for which he was praised by Paul Dirac and for which he was even awarded the Nobel Prize. Beard spends most of his time drinking vast amounts of alcohol and swallowing enormous portions of food. With his gluttony, lust, avarice and hedonism, Beard, the grotesque incarnation of contemporary consumerism (see Garrard 2013, 127–130), clearly symbolizes "the way his species abuses the planet" (Jones 2010, 6), but his eccentric behavior and his insatiable desires can also be interpreted in terms of psychoanalysis (see Kogan 2012).

Beard becomes the manager of a project by which the energy problems of the world could be solved forever, but the brilliant idea of using artificial photosynthesis for the production of energy, on which the project is based, is not his own. It is a plagiarism, because Beard stole it from his assistant, who is killed by an accident, which is turned into a murder by Beard and blamed on the man for whom his wife left him. Travelling to the arctic, he meets a group of artists who are also interested in solving the energy problem. Whereas he himself doubts that art can pay any tribute to the world's declining resources, he nevertheless participates in their common dinners and the accompanying conversations. But he remains distant and isolated and again and again he reveals his utter contempt for the humanities. The professor of physics cannot stand it, if the expert culture of quantum mechanics is appropriated by the unworthy minds of writers and artists, and he warns them against misinterpreting his own metier as an invitation to mystic and esoteric speculations, a reverberation of what Pauli had sarcastically denominated as "atomysticism." When the self-proclaimed technological geniuses of Great Britain are invited to send in their modest proposals to create machines for the production of ceaseless energy, Beard satirizes both the ignorance and the atomystic desires of dilettantes who "read inexpertly and far too hopefully into quantum field theory to find their esoteric fuel right under their noses, in the voids of the empty air of their sheds or spare bedrooms – zero-point energy" (*S*, 18). He launches vitriolic abuse on those poor projectors who dream of a kind of *perpetuum mobile* constructed according the insights of quantum mechanics:

> Quantum mechanics. What a repository, a dump, of human aspiration it was, the borderland where mathematical rigour defeated common sense, and reason and fantasy irrationally merged. Here the mystically inclined could find whatever they required, and claim science as their proof. And for these ingenious men in their spare time, what ghostly and beautiful music it must be – *spectral asymmetry, resonances, entanglement, quantum harmonic oscillators* – beguiling ancient airs, the harmony of the spheres that might transmute a lead wall into gold, and bring into being the engine that ran on virtually nothing, on virtual particles, that emitted no harm and would power the human enterprise as well as save it. [. . .] (*S*, 19)

Quantum mechanics as atomysticism, Heisenberg's and Pauli's ideas emerge in a new, more cynical and sarcastic shape. The quotation testifies to the fact that in *Solar* McEwan uses Beard as both his mouthpiece and his victim; here Beard partially echoes McEwan's opinions. In many interviews, McEwan has left no doubt that he is an opponent of esoteric thinking and of mysticism; even if he tries to contrast mystic thinking with rationality, as he does in his novels *Black Dogs* (1992) and *Enduring Love* (1997) for instance, it becomes obvious that – despite all criticism concerning the exaggerated arrogance of rational scientists – he clearly sides with rationality and can be considered a "groupie of Enlightenment rationalism" (Ridley 2013, xi), or a "science obsessed writer" (Habibi and Karbalaei 2015, 93). Like his protagonist Beard, McEwan does not think that "science was just one more belief system, nor more or less truthful than religion or astrology" (*S*, 132). Like many of his other novels, *Solar* "interrogates what science may mean in the lives of western subjects at a time of both great material comfort, and multiple threats" (Borm 2012, 240–241), and in general, McEwan approves of the achievements of science (see Cook et al. 2013, 151). For him science is "organized curiosity, which is a fundamentally optimistic state: you might not know it now, but you might know it one day" (see Cook et al. 2013, 151). But then again, McEwan turns his protagonist and *persona* Beard into a satiric victim, exposing his arrogance, his mocking disdain for dilettantes of science, and – here he differs completely from Heisenberg and his coterie – his self-assuredness and his unshakable belief in the righteousness of his own ideas.

To use physical notions as a vantage point for philosophical and ethical discussions appears to be completely absurd in Beard's eyes; in this respect, Beard clearly resembles the positivist and the staunch believer in factuality that Heisenberg and his circle tried to leave behind. Instead of accepting the complementarity of science and art, or science and the humanities – by the way a concept that was of utmost significance for Bohr's scientific and philosophical work – Beard emphasizes the scientist's exclusive authority. He shows off and discloses an arrogant attitude not untypical of some of the Nobel Prize winners. The professor of science does not see any value in literature, he considers the works of John Milton "a monstrous bluff" (*S*, 201), and, as a young man, he only read Milton's *Paradise Lost* (1667) in order to impress his first love, who, predictably, left him soon after. In his study of Milton he "encountered nothing that could remotely be construed as an intellectual challenge" (*S*, 201). He feels nothing but disdain for – as he calls them derisively – "arts people" who fall "out of bed at midday" (*S*, 202), only to spend their time with trivial occupations which will never contribute to the benefit of mankind. Again, *McEwan* uses Beard as his satirical target, blaming him for his vaingloriousness and portraying him as a modern *Alazon* figure bound to be deflated by the stratagems of irony (see Mádrová 2013, 63–77). He clearly shows that this "vaguely unprepossessing, often bald, short, fat, clever" man of "narrowed mental condition, anhedonic, monothematic, stricken" (*S*, 3), is permanently in danger of overestimating himself.

Then again, Beard is turned into McEwan's mouthpiece, for instance, when he characterizes him as the relatively innocent victim of harsh and unfair media campaigns, fundamentally applied *gender studies* and a kind of postmodernism run wild (*S*, 127–144). The story of Beard's supposed sexism is a satirical allusion to the exaggerated media attacks on Harvard's President Larry Summers (because of his allegedly misogynist remarks) and Martin Amis (because of his allegedly Islamophobic sallies) (see Garrard 2013, 131–134). His unwillingness to agree that human genes are culturally constructed and his jovial jokes about the role of women in science induce the mass media to turn him into a "'neo-Nazi' Professor" and a misogynistic patriarch (*S*, 137). This oscillation between using Beard as a satirical *persona* only to turn him into a satirical victim a moment later, reminiscent of Jonathan Swift's satires, is a hallmark of McEwan's *Solar*, the "satire of current debates on climate change" (Borm 2012, 239).

Heisenberg, Planck, Pauli and Bohr came to doubt the infallibility of their own discoveries and the philosophical implications inherent in them; they gradually came to cherish notions, which Robert Musil in his famous novel *The Man without Qualities* (1930–1932) would have described as the "utopia of essayism" ("Utopie des Essayismus") (see Musil 1978, 247–257; trans. by RF). Musil, who like Heisenberg and McEwan, was extremely interested in the interferences between science and the humanities, based his concept on the long tradition of essay writing associated with the revolutionary ideas of Michel de Montaigne (see Freiburg

2006). Montaigne, the "father of essayism" introduced "individuality," "experience" and "subjectivity" into the world of literature by presenting his veritable "book of the self" (Good 1988, 26–42; 27). With Musil, the features of the essay genre are transformed into a mentality, a mindset, a veritable intellectual utopia. Musil's utopian essayist is a *homo curiosus par excellence*, he has internalized Montaigne's Pyrrhonism, and redefined it into a lifelong exercise, "never to take anything for granted." For Montaigne – and Musil – reality is not tantamount to solidity and invariability, it is rather an oscillating phenomenon, characterized by a fluctuation of elements reminiscent of Heraclitus's philosophy of *panta rhei* or of Heisenberg's descriptions of the world of quantum particles. Montaigne's essayism (and Musil's too) defines itself as a counterdiscourse to any forms of apodictic maxims or allegedly eternal truths.

The "essay," etymologically derived from Latin *exagium* (attempt, sample) and French *essai* (attempt, experiment), is innovative and experimental in its core. Instead of walking the beaten paths of tradition, the essayist tries to find new ways, ways sometimes developed by just walking them. The metaphor for this kind of essayism is indeed that of an intellectual promenade which eschews any form of signposts *en route*. The essayist does not deny the power of reason, but he complements it with acts of imagination and justifies radical aspects of individualism and idiosyncrasy as far as the discovery of new thoughts is concerned. Essayism appreciates "intuition" and even "whim" as creative forces. Essayistic thinking defines itself as an alternative to systematic thinking; by defending the beauty of even illogical imaginary acts, it tries to overcome the staleness of any system of thought paralyzed by an obsession with traditional principles and the blind belief in the authority of syllogisms:

> The essay exists outside *any* organization of knowledge, whether medieval or modern. In it, an open mind confronts an open reality. An uncertain, unorganized world enters an unprejudiced awareness, and the essay results as a record and provisional ordering of the encounter. (Good 1988, 4)

Beard, therefore, is the opposite of an essayist; he becomes more and more apodictic, intolerant and aggressive. He lacks Heisenberg's and Bohr's detachment, as well as Einstein's irony and humor, and instead of accepting counterarguments he intends to impose his opinion on his intimidated audience.

Beard is clearly a part of the problem he wants to solve; even his trip to the Pole can hardly be justified if seen from an ecocritical perspective; the "bootroom," where everyone has to change their boots and clothes, is turned into a caustic symbol of human irresponsibility: The individuals who do not tire of discussing the question of how climate catastrophe could be averted (see Habibi and Karbalaei 2015, 98), are unable to keep the room in order; it becomes the synecdoche for exactly that entropy the world suffers from (see Marzec 2015, 80–82). The central scene, in which his intolerance is revealed, takes place on the occasion of one of the evening conversations with the artists. Usually Beard does not take any interest in these conversations, but when one of the artists enters his domain of science he shows a mixture of graciousness and condescension. Asked by the sculptor, who begs Beard to pardon him his "naïvety and ignorance," if "the strange reality described by quantum mechanics [was] a description of the actual world, or [. . .] simply a system that happened to work?" (*S*, 65), Beard reacts with the pretended lenience of a narcissistic scholar:

> He could not have phrased it better himself, for there was no better interrogation of quantum theory than this. It was a matter that had dominated years of Einstein's life and led him to insist that the theory was correct but incomplete. Intuitively, he just could not accept that there was no reality without an observer, or that this reality was defined by the observer, as Bohr and the rest seemed to be saying. In Einstein's memorable phrase, there was out there a "real factual situation." "When a mouse observes," he had once asked, "does that change the state of the universe?" Quantum mechanics seemed to imply that a measurement of the state of one particle could instantaneously determine the state of another, even if it was far away. But this was "spiritualistic" in Einstein's view, it was "spooky action at a distance," for nothing could move faster than the speed of light. Beard the realist was sympathetic to Einstein's extended, failing battle with the brilliant coterie of quantum pioneers, but it had to be faced: the experimental proof suggested that there really could be long-range spooky correlations, and that the texture of reality at the small and large scale really did defy common sense. (*S*, 66)

But his tolerance is not destined to last; when another artist tries to invade his domain, Beard grows furious. His aggressiveness is shown by his bringing his fist down hard on the table:

> Listening, as he usually did, with Jesus at his side from their corner of the mess room, Beard interjected only once, on the last evening when a gangling novelist called Meredith, appearing to forget there was a physicist present, said that Heisenberg's Uncertainty Principle, which stipulated that the more one knew of a particle's position, the less one knew of its velocity, and vice versa, encapsulated for our time the loss of a 'moral compass,' the difficulty of absolute judgments. Beard was peevish in his interruption. It was worthwhile to be correct, he told this crop-haired fellow with rimless glasses. What was at issue was not velocity but momentum, in other words, mass times velocity. At such hair-splitting there were muted groans. Beard said that the principle had no application to the moral sphere. On the contrary, quantum mechanics was a superb predictor of the statistical probability of physical states. The novelist blushed but would not give way. Did he not know who he was talking to? Fine, yes, OK, statistical probability, he insisted, but that was not certainty. And Beard, just finishing his eighth glass of wine and feeling nose and upper lip elevate in contempt for an ignorant trespasser on his field, said loudly that the principle was not incompatible with knowing precisely the state of, say a photon, so long as one could observe it repeatedly. The analogy in the moral sphere might be to re-examine a moral problem a number of times

> before arriving at a conclusion. But this was the point – Heisenberg's Principle would only have application if the sum of right plus wrong divided by the square root of two had any meaning.
>
> The silence in the room was not so much stunned as embarrassed. Meredith stared helplessly as Beard brought his fist down hard on the table. "So come on. Tell me. Let's hear you apply Heisenberg to ethics. Right plus wrong over the square root of two. What the hell does it mean? Nothing!" (*S*, 76–77)

Is this fair? Is the silence in the room an expression of awe or a protest against the pedantry of an expert who is not willing to accept the usual distortions of a concept, once it is transferred from science to ethics? The implicit irony of the situation lies in the fact that although Beard is right, he is wrong too, and that at the same time. The particle of truth, to express this metaphorically, is absent and present simultaneously, is at his command, and it escapes his authority at the same time. The numeric arguments cannot be contradicted – who is the ignorant artist to contradict a Nobel Prize winner – but at the same time he loses his authority because he violates the laws of ethics, and again he breaks more than one law. When Heisenberg and his colleagues chose the term "uncertainty principle" for the physical "behavior" (another metaphor) of particles, they "knew" that they would never be in a situation to do "reality" justice. The importation of a concept from the humanities to physics is accompanied by distortions and by a vagueness typical of these procedures. And the exportation or re-importation of the term from science back to the humanities suffers the same fate, and why not? It is true, "the sum of right plus wrong divided by the square root of two" will not convince any philosopher (*S*, 77); it will never show up in either deontic or teleological ethics (see Quante 2011, 127–142). But the structural parallels between quantum mechanics and the field of ethics are still visible for those willing to use imagination and fantasy, those features claimed by both Bohr and Heisenberg as indispensable preconditions for innovative research.

The story told in *Solar* gives ample evidence for the uncertainty principle in ethics. Perspective and the presence of an observer seem to influence the outcome of an ethical judgment. Although Beard is responsible for the death of Aldous, he is never really pursued by law. In an ironic ethical quantum jump the actual perpetrator even blames the innocent suspect Tarpin for Aldous's death (*S*, 99–100). And the suspect himself, who is doomed to go to prison for a "murder" he has not committed, is guilty in many other respects, especially for his using violence against Patrice and against Beard himself. Beard steals Aldous's scientific ideas, but although this is a clear act of plagiarism, the outcome of his "crime" may be acceptable, because without this transgression the world would not have learned anything about Aldous's plans to use artificial photosynthesis in order to solve the climate crisis. Beard is morally damned as a "Nazi-Professor" and a "misogynist" only because he has opposed the tyranny of an exaggerated form of "political correctness," when he criticizes a female scholar who self-confidently applies the ideas of cultural studies to genetics (*S*, 131). Patrice becomes guilty for believing Beard and for sending Tarpin to prison. More examples could be enumerated; even the episode of the vinegar crisps demonstrates in what degree ethical judgments are based on perceptions (*S*, 122–126), which are themselves not immune to misinterpretation and erroneous "emplotment." Heisenberg's remark, "To that extent, every physical process may be said to have objective and subjective features" (Heisenberg 1971, 88), at least justifies doubt about the alleged self-evidence of things one has seen with one's own eyes, reminding one of the Russian proverb "He lies like an eye-witness," a warning against the omnipresence of uncertainty in everyday life. All McEwan's major novels, from *Black Dogs*, *Enduring Love* (with two central scenes, the initial balloon scene (2006, 3–16) and the narrator's memory of the attack in the restaurant (2006, 174–182)) and *Atonement* (2001) to *The Children's Act* (2014) circumscribe the precariousness, the uncertainty, of ethical judgments.

Solar thus warns against inappropriate certainty in the realms of ethics and morality: Maybe Beard is right in defending Heisenberg against the appropriation of his ideas by the humanities; but by defending him, Beard reveals his own ethos as worlds apart from Heisenberg's. Ethical reflections may be based on vague concepts of imagination; but so may scientific reflections, for example when parts of quantum mechanics are based on calculations with the "root of minus one," a number for which there are no correlates in reality and which is merely the product of "mathematical imagination." The theory, as Einstein had realized, predetermines the outcome of any experiment in quantum physics; the same is true of ethical systems; the preconditions of culture, religion, the implicit value systems inherent in what critical postmodernists would designate as *grand narratives*, shape ethical judgments and axiological systems. The position and the impulse/velocity of a particle cannot be measured or identified at the same time; and the intention and the resulting effect of an intention can never be prognosticated with safety in ethics, as all McEwan's novels prove with such persuasiveness. And due to Georg Cantor's research concerning mathematical infinity, we know that there is not only "one infinity," but "infinite numbers of infinities," and – even more intimidating – there is not one mathematics, but infinite systems of mathematics that could all be "true" at the same time (see Weth 2016, 105–138). The same is true of philosophy and ethics: No one reason rules the world, but there may be infinite forms of reasoning, shaped by culture and history.

McEwan sacrifices Beard in order to plead for the acceptability of Heisenberg's Uncertainty Principle, which is valid in the humanities (see also Habibi and Karbalaei 2015, 91), as well as in science. The way to the uncertain truth can never be found by imperatives and apodictic proclamations. It is only by suggestions and the essayistic spirit of a doubting mind, that the way can be found, even if one has to invent the way in order to find it:

> So science parallels literature as a means by which the world can be understood. There are great, noble and ingenious insights which science has brought us and which literature could never equal. Of course, there are many complex facets of experience for which science has no language and literature does. (Quoted from Cook et al. 2013, 149)

6 Conclusion

The analysis of the reception of the uncertainty principle in *Solar* demonstrates that McEwan has much more in common with Heisenberg, Bohr and Planck than his protagonist Beard. The interference between science and literature is not as problematic as Beard wants us to believe. The novel can be interpreted as a claim for a kind of symbiosis in which philosophers and scientists should live together. In his rigidity Beard reminds the reader of scholastic philosophers too self-assured to see the failures in their authoritative Aristotelian doctrines. Although he once belonged to the elite of the leading physicists in the world, he now reveals a professional bigotry that induces him to transgress ethical borderlines. His logical definition of the uncertainty principle may be correct, but at the same time he offends against the ethics of Heisenberg and his circle. By turning Beard into a satirical victim, McEwan proves that his own ideas differ considerably from those of his protagonist. Like Heisenberg, McEwan advocates for the compatibility of art and science: the humanities can and should learn from science, but scientists should also recognize and accept the charm of imaginative acts which – just like the quantum jumps in the atomic world – might lead them into a new *terra incognita* of physics. Seen in this context, McEwan's essayism focuses on the deficiencies of the two thought systems which claim to be able to explain the world: *mythos* and *logos*. While *mythos* (associated with the humanities) can often be proved wrong, science, too, reveals its blemishes, since it tends to forget that its results are sometimes based on "narrative hypotheses" that do no guarantee any certainty. Transferring this thought to the field of ethics, *Solar*, and probably most of McEwan's novels, could be interpreted as "essayistic comments" on Immanuel Kant's categorical imperative.

Reference list

Bacon, Francis. *Novum Organum* (1620). *The Works of Francis Bacon*. Eds. James Spedding et al. 14 vols. Vol. 1. London: Longman et al., 1858. 149–365.

Bacon, Francis. *The Advancement of Learning* and *New Atlantis* (1605). Ed. Arthur Johnston. Oxford: Clarendon Press, 1974.

Borm, Jan. "Ian McEwan's *Solar*, or Literature and Contemporary Debates on Climate Change." *Literature in Society*. Ed. Regina Rudaitytė. Cambridge: Cambridge Scholars Publishing, 2012. 237–247.

Browne, Sir Thomas. "Pseudodoxia Epidemica [Selections]" (1646). *The Voyce of the World: Selected Writings of Sir Thomas Browne*. Ed. Geoffrey Keynes. London: The Folio Society, 2007 [1968]. 269–510.

Cassidy, David C. *Werner Heisenberg: Leben und Werk*. Trans. Andreas and Gisela Kleinert. Heidelberg et al.: Spektrum Akademischer Verlag, 1995 [1992].

Cook, Jon, Sebastian Groes, and Victor Sage. "Journeys without Maps: An Interview with Ian McEwan." *Ian McEwan: Contemporary Critical Perspectives*. Ed. Sebastian Groes. London et al.: Bloomsbury, 2013. 144–155.

Destrée, Pierre, and Fritz Gregor Herrmann. *Plato and the Poets*. Leiden: Brill, 2011.

Freiburg, Rudolf. "Essayistik." *Einführung in das Studium der englischen Literatur*. Eds. Arno Löffler et al. Tuebingen and Basel: Francke, 2006. 148–172.

Gamper, Michael. "Physik." *Literatur und Wissen: Ein interdisziplinäres Handbuch*. Eds. Roland Borgards, Harald Neumeyer, Nicolas Pethes, and Yvonne Wübben. Stuttgart: Metzler, 2013. 112–118.

Garrard, Greg. "Ian McEwan's Next Novel and the Future of Ecocriticism." *Contemporary Literature* 50.4 (2009): 695–720.

Garrard, Greg. "*Solar*: Apocalypse Not." *Ian McEwan: Contemporary Critical Perspectives*. Ed. Sebastian Groes. London et al.: Bloomsbury, 2013. 123–136.

Gevaux, David. "Physicists in the Glare of the Sun." *Nature Physics* 6 (2010): 236–237.

Good, Graham. *The Observing Self: Rediscovering the Essay*. London and New York: Routledge, 1988.

Habibi, Seyed Javad, and Sara Soleimani Karbalaei. "Incredulity towards Global-Warming Crisis: Ecocriticism in Ian McEwan's *Solar*." *The Southeast Asian Journal of English Language Studies* 21.1 (2015): 91–102.

Hawking, Stephen, and Leonard Mlodinow. *Der Grosse Entwurf: Eine neue Erklärung des Universums*. Trans. Hainer Kober. Reinbek (Hamburg): Rowohlt, 2010.

Heisenberg, Werner. *Physics and Beyond: Encounters and Conversations*. Trans. Arnold J. Pomerans. New York et al.: Harper and Row, 1971.

Heisenberg, Werner. *Der Teil und das Ganze: Gespräche im Umkreis der Atomphysik*. Munich et al.: Piper, 2015 [1969].

Hermann, Armin. *Die Jahrhundertwissenschaft: Werner Heisenberg und die Physik seiner Zeit*. Stuttgart: Deutsche Verlags-Anstalt, 1977.

Horatius Flaccus, Quintus. *Satires, Epistles and Ars Poetica*. Trans. H. Ruston Fairclough. London: Heinemann, 1961 [1926].

Jones, Thomas. "Oh, the Irony." *London Review of Books* 32.6 (2010): 19–20.

Kerridge, Richard. "The Single Source." *Ecozon@: European Journal of Literature, Culture and Environment* 1.1 (2010): 155–161.

Kirchhofer, Anton, and Natalie Roxburgh. "The Scientist as 'Problematic Individual' in Contemporary Anglophone Fiction." *ZAA* 64.2 (2016): 149–168.

Kogan, Ilany. "Ian McEwan's *Solar* through a Psychoanalytic Lens." *Journal of the American Psychoanalytic Association* 60.6 (2012): 1299–1313.

Krohn, Wolfgang. *Francis Bacon*. Munich: C.H. Beck, 1987.

Mádrová, Eva. *Concept of Irony in Ian McEwan's Selected Literary Works*. Master's thesis. Palacký University Olomouc. Olomouc, Czech Republic, 2013. https://theses.cz/id/vvkcwi/00175997-271744137.pdf (3 August 2022).

Marzec, Robert P. "Climate Change and the Evolution of the 9/11 Security State: The Fantasy of Adaptation and Ian McEwan's *Solar*." *Narrating 9/11: Fantasies of State, Security, and Terrorism*. Eds. John N. Duvall and Robert P. Marzec. Baltimore: Johns Hopkins University Press, 2015. 70–97.

McEwan, Ian. *Black Dogs*. London: Vintage Books, 1998 [1992].

McEwan, Ian. *Enduring Love*. London: Vintage Books, 2006 [1997].

McEwan, Ian. *Solar*. London: Jonathan Cape, 2010.

Monk, Ray. "Getting Inside Heisenberg's Head." *A Companion to the Philosophy of Literature*. Eds. Garry L. Hagberg and Walter Jost. Oxford: Wiley-Blackwell, 2010. 453–464.

Musil, Robert. "Auch die Erde, namentlich aber Ulrich, huldigt der Utopie des Essayismus." *Robert Musil: Gesammelte Werke in neun Bänden*. Ed. Adolf Frisé. Vol. I. Reinbek (Hamburg): Rowohlt, 1978. 247–257.

Quante, Michael. *Einführung in die Allgemeine Ethik*. Darmstadt: Wissenschaftliche Buchgesellschaft, 2011.

Real, Hermann Josef, and Heinz J. Vienken. *Jonathan Swift: Gulliver's Travels*. Munich: Fink, 1984.

Ridley, Matt. "Foreword: Ian McEwan and the Rational Mind." *Ian McEwan: Contemporary Critical Perspectives*. Ed. Sebastian Groes. London et al.: Bloomsbury, 2013. ix–xii.

Sidney, Sir Philip. *A Defence of Poetry* (1595). Ed. Jan van Dorsten. Oxford: Oxford University Press, 1978 [1966].

Snow, Charles Percy. *The Two Cultures*. London: Cambridge University Press, 1959.

Sterne, Laurence. *The Life and Opinions of Tristram Shandy, Gentleman*. Ed. Graham Petrie. Harmondsworth: Penguin, 1986 [1759–1767].

Sullivan, Peter, and Michael Potter, eds. *Wittgenstein's Tractatus: History and Interpretation*. Oxford: Oxford University Press, 2013.

Vanderbeke, Dirk. "Physics." *The Routledge Companion to Literature and Science*. Eds. Bruce Clarke and Manuela Rossini. London and New York: Routledge, 2011. 192–202.

Weizsäcker, Carl Friedrich von. *Der Garten des Menschlichen: Beiträge zur geschichtlichen Anthropologie*. Munich: Carl Hanser, 1977.

Weth, Thomas. "Faszination Unendlichkeit: Über die Struktur des Unendlichen." *Unendlichkeit: Erlanger Universitätstage Ansbach 2015*. Ed. Rudolf Freiburg. Erlangen: FAU Press, 2016. 105–138.

White, Hayden. *Tropics of Discourse: Essays in Cultural Criticism*. Baltimore: Johns Hopkins University Press, 1978.

Wittgenstein, Ludwig. *Tractatus Logico-Philosophicus: Logisch-Philosophische Abhandlung*. Frankfurt: Suhrkamp, 1977 [1921].

Marco Caracciolo

The Hard Problem and the End of the World

Narrative and Mind in Postapocalyptic Fiction

Abstract: Anxieties surrounding today's ecological crisis and its consequences for human societies have led to a proliferation of postapocalyptic scenarios in fiction. My chapter focuses on how, in some of these postapocalyptic narratives, questions about the nature of mind become bound up with the experience of environmental disaster. One of the central problems of Western philosophy is how subjectivity – and more specifically the qualitative texture of conscious experience – arises from material processes in our brains and bodies. David Chalmers's discussion of the "hard problem" of consciousness is one of the latest restatements of this fundamental difficulty. In my case studies, Bruno Arpaia's *Qualcosa, là fuori* (2016; Eng: *Something, Out There*) and Michel Faber's *The Book of Strange New Things* (2014), facing environmental cataclysm brings up this hard problem with remarkable clarity and urgency. In this way, these postapocalyptic narratives suggest that the present environmental predicament calls for a radical rethinking of Western dichotomies between subjectivity and materiality.

1 Introduction

> We observe a fragment of the process, the trembling of a single string in a symphonic orchestra of supergiants, and on top of that we know – we only know, without comprehending – that at the same time, above us and beneath us, in the plunging deep, beyond the limits of sight and imagination there are multiple, million fold simultaneous transformations connected to one another like the notes of musical counterpoint. For this reason someone gave them the name of geometric symphony, but if this is the case, we are its unhearing audience. (Lem 2011, Kindle Locations 2179–2183)[1]

Stanisław Lem's influential science-fiction novel *Solaris* – first published in Polish in 1961 – is set on a planet whose surface is covered by an ocean that bears all the hallmarks of intelligent, sentient behavior. The passage above is an excerpt from the narrator's account of the "symmetriads" that erupt on the ocean – weather-

1 I quote from the e-book of *Solaris* because (as of January 2025) no print version exists of Bill Johnston's translation.

https://doi.org/10.1515/9783110782844-003

like events of astounding complexity and spatio-temporal magnitude: no matter what "fragment" the observers focus on, the overall structure eludes their visual and even imaginative range. The closest equivalent to these phenomena is the auditory patterning of music, and in fact the narrator reaches for musical metaphors – "string," "symphonic orchestra," "counterpoint" – to evoke the symmetriads' spectacle. The symmetriads, we learn, are a direct emanation of the sentient ocean's mathematical mind; but the exact meaning they express, Lem's narrator insists, will remain unfathomable to humans.

Yet the *form* of the symmetriads, for all its complexity, is perhaps not entirely alien. Certainly, Lem's musical metaphors make its multilayered patterning more familiar and intelligible. The narrator also draws a suggestive parallel between physical dynamics in the ocean and the neural underpinnings of emotional experience:

> taking as my starting point the innovative research of Bergmann and Reynolds, who from the mosaic of cortical processes had succeeded in identifying and 'filtering out' the components accompanying the most powerful emotions – despair, pain, joy – I went on to juxtapose those recordings with discharges emitted by currents in the ocean, and discovered oscillations and patterns in the curves (in certain parts of the symmetriads' canopy [. . .]) that offered a noteworthy analogy. (Lem 2011, Kindle Locations 3234–3238)

Cortical processes resemble the physical complexity of symmetriads. But for all the resonance of this brain-ocean analogy, the narrator's observation leads to more questions than it solves. First, is this *only* a matter of resemblance, or does it suggest that the symmetriads are, somehow, directly involved in the ocean's thought processes? Second, if emotional experience in humans is accompanied by symmetriad-like complexity in neural terms, why are we not phenomenologically aware of that complexity, but only of seemingly basic – if powerful – emotions such as "despair, pain, joy"?

The questions thus raised by *Solaris* revolve around one of the fundamental issues of philosophy – namely, the bifurcation between physical events (be it electrical patterns of neural activity or the complex configurations of symmetriads) and mental experience. How do states of consciousness – what despair, pain, and joy feel like subjectively – emerge from objectively observable events? Conversely, why are we so intimately aware of whatever is in our experience at any given time, and so oblivious to the physical processes that underlie experience? Philosopher David Chalmers (1995) refers to this issue, famously, as the "hard problem of consciousness."[2] For Chalmers, the behavioral and functional aspects of cognition can be straightforwardly integrated within a neurobiological account of the mind. For instance, pain triggers certain behavioral responses (e.g., removing our hand from a scalding object) and serves a clear function vis-à-vis our survival (protecting our body from harm). Thus, at the behavioral and functional levels mental events such as pain are not particularly mysterious. But why does pain feel the way it does? Why is pain from scalding, for instance, qualitatively different from the pain of a sprained ankle? These questions are the point of departure for Chalmer's discussion of the "hard problem."

Today's scientific worldview assumes that everything, including subjective experience, can be reduced to physical events – a position known as "physicalism" or "materialism" in the philosophy of mind. As Thomas Nagel puts it in *Mind and Cosmos*, "among the scientists and philosophers who do express views about the natural order as a whole, reductive materialism is widely assumed to be the only serious possibility" (2012, 4). Yet there is, currently, no scientific consensus about the exact physical states to which conscious experience can be reduced, and in fact both Chalmers and Nagel take issue with the feasibility of such reduction. Researchers like Terrence Deacon (2006) have embraced complex systems theory as a viable alternative to reductionist accounts of conscious experience. To put this point bluntly: when matter is arranged in sufficiently complex ways, as in living beings, subjective states emerge from physical events. A central concept in the theory of complexity, emergence is directly opposed to reductionist approaches to the hard problem. Deacon writes: "Human consciousness – with its features of autonomous causal locus, self-origination, and implicit 'aboutness' – epitomises the logic of emergence in its very form" (2006, 149). But Deacon himself admits that this is only the beginning of a non-reductionist explanation of consciousness: emergence promises to bridge the gap between physical matter and subjective experience, but – as things stand – the bridge is far from complete.

This, of course, is only the beginning of a formulation of the hard problem. I cannot hope to do justice to philosophical discussions on consciousness in this chapter, and I only take on board the ideas that are directly relevant to my literary analyses.[3] No doubt, philosophers will find this lack of specificity unsatisfac-

2 Chalmers was building on earlier discussions in the philosophy of mind, particularly Joseph Levine's (1983) account of the "explanatory gap" – a phrase that captures many of the same intuitions as Chalmers's hard problem.

3 For more on the philosophical issues surrounding consciousness, see the *Journal of Consciousness Studies* and also Susan Blackmore and Emily Troscianko's *Consciousness: An Introduction* (2018).

tory. However, it should be kept in mind that fiction cannot advance arguments of a philosophical nature or solve scientific problems; it should not be judged by those standards. What fiction *can* do – as argued, among others, by John Gibson (2007) – is demonstrate the relevance of philosophical questions in an ethically engaged and emotionally impactful manner. Thus, Lem's *Solaris* implicates into its plot the vexed link between mind and the physical world and charts its affective consequences, well beyond the passage quoted above. In a preliminary way, though, the symmetriad passage suggests that two analogical strategies are crucial to the narrative exploration of the hard problem: the first is the use of metaphors that – more than direct description – convey the multiscalar nature of the symmetriads, thus serving as an experiential equivalent of their complexity. The second strategy is the explicit foregrounding of patterns that bring together, analogically, the symmetriads and brain activity: *Solaris* highlights – to quote again Deacon – the "logic of emergence in its very form" (2006, 149) by suggesting that emotional experience arises from brain patterns structurally similar to the symmetriads' configurations. Lem's novel thus embraces, and puts to narrative use, what Stephan Besser calls the "isomorphic imagination" – that is, "the general disposition to detect and construct meaningful similarities in terms of form and shape between the human brain and the extracerebral world on the basis of principles of analogy and morphological congruity" (2017, 417). More specifically, the isomorphic imagination involves an analogy between brain structure and patterns in the physical and social world: for instance, James Cameron's science-fiction film *Avatar* (2009) depicts the inhabitants of the planet Pandora as a brain-like assemblage. This kind of isomorphism is a centerpiece of contemporary neuroculture, of which Lem's novel is, arguably, a prefiguration. Brain-world isomorphism privileges the language of patterns: in Besser's primary sources – which range from *Avatar* to contemporary work in the philosophy of mind – the trope of pattern "functions as a privileged interface and suturing device between the domains of the brain/body and the extracerebral world" (Besser 2017, 422). Through this "suturing" effect, the isomorphic imagination advances a possibility of reconciling the physical world and the sphere of subjective experience – if not scientifically, at least rhetorically. Patterns thus serve (although Besser does not raise this point explicitly) as an interim response to the hard problem brought into view by philosophical discussions on consciousness.

As a protracted narrative engagement with the imaginative possibilities of mind-world isomorphism, Lem's novel is hardly an isolated case. This chapter argues that the hard problem of consciousness becomes particularly productive in contemporary fiction that confronts the specter of an ecological disaster.[4] My claim is not that postapocalyptic fiction is single-mindedly obsessed with the hard problem, or that the dilemmas of consciousness have not been explored in other genres. Certainly, literature has contemplated philosophical questions related to the hard problem well before the human overexploitation of fossil fuels started to jeopardize the long-term survival of our species (and countless others). Likewise, I do not want to suggest that *all* postapocalyptic narratives thematize the relationship between consciousness and the material world. However, those that do (including my case studies, discussed below) deserve a closer look. I can offer two preliminary reasons for the relevance of these narratives. First, the evocation of the hard problem shows how radical the challenge presented by the ecological crisis is, how its stakes reach down to the very constitution of our subjective experience. Second, it suggests that the ecological crisis is not a merely "external" event, but one that involves the (quintessentially Western) dichotomy between subjectivity and the physical world: coming to terms with the climate crisis calls for – in addition to urgent political action – a reappraisal of the position of mind vis-à-vis nonhuman materiality. As fiction registers and grapples with the climate crisis and its consequences, it highlights the rift between human subjectivity and the physical world. Literary language responds by way of isomorphic projections that perform an uneasy merger of mind and world. In the strand of postapocalyptic fiction I wish to explore, this integration is made possible by the foregrounding of patterns, at two levels: the sensory patterning of the protagonists' consciousnesses as they strive to make sense of disorienting surroundings; and the linguistic patterns of metaphors and similes that weave together the domains of psychological and material realities, for instance by comparing inner events to physical processes, or by challenging the separation between the characters' bodies and the seemingly external world. My discussion in the next section seeks to ground this complex stylistic-conceptual dynamic in contemporary debates within the so-called "nonhuman turn." I will then examine two examples of this dynamic in recent fiction, Bruno Arpaia's *Qualcosa, là fuori* (2016; Eng. *Something, Out There*) and Michel Faber's *The Book of Strange New Things* (2014). The choice of these novels is inspired by the observation that they take opposite routes to the hard problem, as we'll see; in both texts, however, the evocation of patterns across mind-body-world divides plays a major role as the narrative imagines planetary disaster and traces its repercussions on the protagonists' lives.

---

4 These postapocalyptic novels are a subgenre of what Adam Trexler (2015) calls "Anthropocene fiction," which stages and confronts humanity's entanglement with our planet's climate and geological history. For more on the Anthropocene, see the next section.

2 Facing up to the nonhuman

The "nonhuman" is the term proposed by Richard Grusin (2015) for a wide gamut of realities that are attracting increasing attention in the humanities and social sciences, largely in the wake of posthumanist philosophy. These realities include geological, biological, and climatological phenomena, nonhuman animals, as well as computer technologies. Their common denominator is that they all raise a challenge to the metaphysical primacy and autonomy of the human subject, which have long been fundamental tenets of Western thought. The Anthropocene, a concept introduced by chemist Paul Crutzen (2002) to refer to the current geological epoch, is an egregious example of this challenge: by leaving an indelible mark on the earth's geology (through the burning of fossil fuels or the release of plastic into the environment, for instance), humanity becomes an agent on a par with the natural forces that have shaped our planet's history.[5] In turn, this idea disrupts a view of human societies as fundamentally separate from, and superior to, the nonhuman realities that surround us: we realize that the human species has thrived thanks to environmental conditions that are far from immutable, and that are destabilized by our own collective actions – particularly via the effects of capitalism and large-scale industrialization. The movement known as New Materialism develops these insights in a more explicitly philosophical direction, seeing materiality as an alternative to a long history of dualistic binaries between human subjects and the nonhuman world (see Dolphijn and van der Tuin 2012).

The idea of humanity's "enmeshment" with the nonhuman – to use Timothy Morton's (2010) metaphor – also seeps into human subjectivity and psychology. If humanity is constitutively entangled with nonhuman things and processes, as scholars like Jane Bennett (2010) have shown persuasively, then human subjectivity must be linked to material realities at a fundamental level. Although this point is not typically acknowledged by scholars working within Grusin's nonhuman turn or New Materialism, human-nonhuman entanglement brings up the hard problem of consciousness and its long philosophical history.[6] Mind-matter dualism is certainly not compatible with a (broadly speaking) posthumanist worldview that places an epistemological premium on affective and embodied ways of understanding the world. From that perspective, mind must be embedded and technologically extended into material reality. The difficulty, however, is that a

5 See also Dipesh Chakrabarty's insightful commentary on this convergence of histories in Chakrabarty (2014).

6 An exception is Connolly (2017, 40–44), who discusses Thomas Nagel's *Mind and Cosmos* but also notes that Nagel makes no attempt to link "materialism" in the sense of psychophysical reduction to New Materialism.

formidable combination of cognitive-level predispositions (see Bloom 2005; Slingerland 2008) and, in a Western context, cultural assumptions militate against a view of the mind as fully grounded in physical processes. Put simply: the specter of dualism is hard to exorcise, because dualistic assumptions are part and parcel of our everyday, human-scale experience. The hard problem identified by Chalmers has its roots in these basic dualistic assumptions, even if – of course – scientific or philosophical debates on consciousness cannot be reduced to everyday experience. In his complex systems-inspired discussion of consciousness, Deacon expresses the distance between folk-psychological intuitions about consciousness and scientific accounts as follows: "With so many levels of amplification and causal inversion mediating between brain chemistry, conscious cognition, and symbolic evolution, it is no wonder that we experience symbolically mediated causality as almost completely disconnected from thermodynamic causality, even though its very efficacy is founded upon it" (2006, 149). The loops of complexity that tie together material processes and psychological (conscious) causation cover too much conceptual ground for humans to be able to experience the connection directly, as part of their perceptual world. Even if scientists develop a robust solution to the hard problem, its "hardness" will not be dispelled easily in our prescientific intuitions about the world.

This experiential shortcoming acts as a catalyst for narrative. The reasons for this are straightforward: narrative – a practice whose very name is etymologically linked to knowledge – thrives in epistemic gaps and tensions. Porter Abbott has shown this authoritatively in *Real Mysteries*: certain stories work towards "a form of recalcitrance designed to immerse the reader in a state of unknowing, robbed not only of cognitive mastery but of its resources" (2013, 17). The unreadability of the ocean's mind in Lem's *Solaris* is, of course, a case in point: the mystery the novel foregrounds is how (and if) the symmetriads' spectacular configurations reveal the operations of a nonhuman mind. But the novel also implicates a broader mystery, and a very "real" one at that, in that scientists and philosophers currently don't have an explanation for the emergence of consciousness from material processes.

Equally important to the hard problem's productivity in narrative is the rift it creates between what we experience (a dualistic difference between subjects and objects) and what, if we buy into a materialist worldview, *we think we know* about reality. Narrative, a practice geared towards and shaped by the embodiment of human-scale experience (Fludernik 1996), finds fertile ground in this interstice between subjectivity and material realities. This conceptual dialectic is already present in classic science-fiction works such as *Solaris*; it certainly predates the twentieth century and can be traced back at least to the Romantics' materialist fascination with life (Goldstein 2017). But the mind-matter explanatory gap becomes an even more central concern in contemporary fiction that engages, formally and thematically, with the Anthropocene. In the context of this chapter, I'll focus on two postapocalyptic novels that use environmental catastrophe as a concrete narrative-level analogue of humanity's constitutive interdependency with the material world.[7] Catastrophe is both a possible (and increasingly likely) outcome of the current ecological crisis and a highly tellable event: it brings about dramatic change and deeply affects human characters and communities. My case studies show how the authors of *Qualcosa, là fuori* and *The Book of Strange New Things* graft onto this recognizably Anthropocenic imagination of catastrophe a more circumscribed form of unease about the boundary between the subjective and the material. I choose to focus on these novels because, in addition to being significant examples of contemporary postapocalyptic fiction of the literary variety, they feature protagonists who take widely opposed stances on psychophysical relations: for the protagonist of *Qualcosa, là fuori*, a neuroscientist, materialism is inevitable; for the protagonist of *The Book of Strange New Things*, a Christian minister, consciousness – the soul – is dualistically separate from the physical world. Yet, despite these widely divergent points of departure, the novels come together in the exploration of the fraught boundary between subjectivity and the physical world. This exploration, as we'll see in the next two sections, is supported by stylistic and thematic strategies that lead to the foregrounding of *patterns* across the mind-world divide. This preoccupation with patterns is an extension of the isomorphic imagination we've seen at work in Lem's novel.

3 The brain's failing patterns

The premise of Arpaia's *Qualcosa, là fuori* is that climate change has turned most of the European continent into a scorching desert; only the Scandinavian countries remain habitable. In a clear commentary on the recent refugee crisis, the novel portrays the desperate attempt to escape strife and starvation of thousands of southern Europeans, including the protagonist, Livio, who embark on a perilous journey to Scandinavia. Arpaia's novel interlaces Livio's experiences during that long trek with flashbacks from his past as a Stanford-based neuroscientist. The opening is *in medias res*, with Livio and his travel companions preparing to set up camp for the night during the journey north.

7 For a discussion of postapocalyptic fiction and the narratological challenges it raises, see also Caracciolo (2021, ch. 3).

Arpaia's novel engages with contemporary work in the philosophy of mind from the cover page: as the author points out in an afterword, he drew inspiration for the title from a nonfiction book by Italian philosopher of science Enrico Bellone, *Qualcosa, là fuori: Come il cervello crea la realtà* (2011; *Something, Out There: How the Brain Creates Reality*). Bellone's study centers on the problem of how the human brain, shaped by natural evolution, produces order out of a welter of sense impressions. This order is necessary to make sense of the world and grasp its relevance to our survival: "To orient ourselves in the world, we're always trying to identify the regularities in the tangle of stimuli that shower down on us from the outside" (Bellone 2011, xi).[8] Arpaia's novel repeatedly tries to capture this perceptual process through the stylistic device of internal focalization – i.e., the evocation of the qualitative properties of a character's experience in the context of third-person narrative.[9] We already see this dynamic at work in the opening paragraph: "Perhaps it was a fire, or a lantern, that reddish glimmer on top of the hill. Assuming that the dark mass that he sensed toward the east really was a hill, blacker than the darkness that surrounded it, muddy, dense, without nuances" (Arpaia 2016, 9). The "he" – Livio, as we will soon find out – is here attempting to bring into focus, both visually and conceptually, the "reddish glimmer" and the dark mass next to it. But when Livio looks in that direction through his binoculars, "no trace of the glimmer was left, not even an afterglow. Perhaps he had made a mistake" (Arpaia 2016, 9).

The uncertainty of perception conveys the tight coupling of the fate of the world and the workings of the human mind. The world as we know it, with its social order and technological comforts, is ravaged by anthropogenic climate change. Perception – our means of creating a world – struggles to keep up with the speed at which this familiar world is being undone. Later in the novel, Livio again has the impression of seeing dots of red light on the horizon, "far, very far, beyond the crest of a knot of hills a bit more to the north. Was the quivering of pale and reddish light in the gloom produced by bonfires? Was there something out there, in the middle of the darkness?" (Arpaia 2016, 91). The bonfires may point to a threat, raiders that could swoop down on the travelers and force them to surrender their already meager possessions. But Livio's asking whether there was "something out there," especially when read in light of Bellone's original use of that expression, suggests a more philosophical form of questioning. The following passage, for instance, probes the limitations of human control over (and, per-

8 All translations from Bellone's and Arpaia's books are mine.

9 See Niederhoff (2013) for an overview of narratological work on focalization. See Caracciolo and Kukkonen (2021, ch. 3) for an extended discussion of focalization from the viewpoint of embodied cognition.

haps, cognitive grasp of) a reality marked by environmental cataclysm: "the day would come in which the mechanisms triggered by the rise in temperatures, from the thawing of the tundra to the melting of ice caps, would spiral out of control and recreate the world without humanity's permission" (Arpaia 2016, 17). The Earth's environment is "recreated" by planetary processes in ways that result from the world-creating capacities of the human brain (which is responsible for the technological advances behind the climate crisis); and yet, paradoxically, this anthropogenic crisis may put an end to the human brain's powers.

This struggle is precisely encapsulated by the multiple breakdowns of Livio's perception. The postapocalyptic reality experienced by Livio loses coherence and oscillates, incongruously, between dystopian bleakness and aestheticized beauty:

> Livio watched [Sara and Marta, two fellow travelers] walk away amid the people sprawled on the ground, amid backpacks, camp beds, carts, and dust, under the tenacious light that rained down from the sky, livid like malignant radiation. He stayed where he was, with the muggy heat clenching his lungs, and watched the sun go down and streak the horizon with crimson and greenish flashes. (Arpaia 2016, 48)

Here, as multiple times in the novel, the world is deprived of the stable "regularities" that guide human perception and dissolves instead into chromatic patterns. The anthropomorphized light ("tenacious") creates a striking counterpoint between the dreary quotidian of the migrants' journey and the splendor of the nonhuman world, with "crimson and greenish flashes" embellishing Livio's perception like brushstrokes in an impressionist painting. Elsewhere the light takes on far more dismal qualities: "a nightmarish, morgue-like light, filtered by the dust that floated in the air and made it difficult to breathe" (Arpaia 2016, 70). Arpaia's internally focalized prose is attuned to the strange light that bathes this postapocalyptic world: the predicament faced by the protagonist as a result of environmental catastrophe appears to pare down perception to its constituent elements – vectors of motion, geometric shapes, raw emotional qualities.

As a neuroscientist, Livio is keenly aware of how these patterns without are an extension of the neural patterns within. He knows "that there was something, out there, but that its structure is constructed by our neurons, which elaborate this structure on the basis of perception and then tell us about it in their own way . . ." (Arpaia 2016, 46; ellipsis in the original). Accordingly, in the first pages of the novel Livio leaves no room for transcendent experience as a dualistic alternative to the physical processes of life; dying is simply "becoming dust, like the many others [Livio and his companions] had already left behind" during their journey (Arpaia 2016, 13). Even when describing something as intangible and potentially spiritual as light, the novel insists on the physicality of the environment's impact on the protagonist – the heat or dust filling his lungs. Livio's inner experiences reflect this deeply materialistic worldview; when he recalls the sensation of holding his son in his arms, "it was not a thought, but a sudden electrical discharge that assaulted him point-blank, leaving him in shock" (Arpaia 2016, 88). The metaphor of the "sudden electrical discharge" renders the protagonist's phenomenology in starkly material terms, as do many other metaphorical expressions in Arpaia's novel.[10] As we've seen above in my discussion of Lem's passage, metaphorical language is crucial to bridging the physical and the mental.

But as Livio confronts the inevitability of death, on a personal level, and humanity's lack of futurity, on a species level, a sense of profound unease inflects his scientific materialism. When a friendly hand placed on his shoulder elicits "another shiver, a pleasant thrill," Livio wonders: "How long ago had he learned that all sensations were due to dopamine, serotonin, and oxytocin? That those were only chemical and physical reactions in his brain?" (Arpaia 2016, 19). The protagonist's questions draw their impetus from the hard problem of consciousness: how chemical and physical reactions in the brain can give rise to the comfort of bodily proximity. Suddenly, Livio finds that his scientific belief in the possibility of psychophysical reduction ("only chemical and physical reactions") is shaken by the postapocalyptic surroundings.

Ecological disaster sharpens the edge of experience: it makes personal memories richer and more poignant, bringing out the phenomenological texture of consciousness. As perception breaks down and – simultaneously – the world as we know it collapses, Arpaia's internal focalization reveals the delicate structures of light, color, and emotion that the brain creates in an attempt to come to grips with a dramatically altered reality. In the final scene, Livio is about to breathe his last when "in his brain millions of neurons recomposed the image of Leila and Matías [Livio's wife and son] in the living room of their Redwood home, of him embracing them" (Arpaia 2016, 216). It is reasonable to think that something more than physical will be lost when the neurons in Livio's dead body become unable to conjure up such memories. The specter of this loss further disrupts the equation between physical and mental processes, sustaining the illusion of the autonomy of subjectivity from material biological events (such as death itself). The result – shocking, to Livio – is that the gap between consciousness and the physical world appears to widen. The hard problem of consciousness merges with the paradox that defines humanity's current predicament: while human subjectivity creates the world in the psychological sense, the material effects of human activity (i.e., climate change) and technological creativity have created *a* world that no

10 For more on metaphorical language for mental processes and how it can be used to probe the hard problem of consciousness, see Caracciolo (2018).

longer seems able to sustain human subjectivity. Arpaia's narrative derives its emotional power from the tension between world-making and world-destruction, between appreciation of the brain's creative power and prefiguration of a world without humans.[11]

4 Cosmic distance and dancing rain

Michel Faber's *The Book of Strange New Things* begins with the protagonist, Peter, saying goodbye to his wife before embarking on a space mission to a distant planet known as Oasis. The exact nature of Peter's task remains vague at first, but it gradually transpires that the planet where the USIC corporation has established its outpost is inhabited by an alien species, the Oasans; Peter, a Christian minister, has been recruited as a modern-day missionary to establish contact with the aliens. Going by science-fiction clichés, one may expect the aliens to be aggressive and unreceptive to Peter's teachings, or at least as inscrutable as Lem's ocean; indeed, Peter is initially quite anxious about their response. But his worries turn out to be completely unfounded: the Oasans are extraordinarily meek and well-disposed towards Christianity: "the book of strange new things" is the aliens' way of referring to the Bible, and their devotion quickly earns them the nickname "Jesus Lovers." Yet the aliens seem unable to experience intense emotions; even their language, which Peter decides to learn, does not differentiate between emotional states with the precision of human languages (Faber 2014, 437).

In fact, the dramatic core of Faber's novel has to do less with Peter's encounter with the Oasans than with what he leaves behind on Earth. Communication with our planet is limited to written messages, and Peter's exchanges with his wife, Bea, take on increasingly dark overtones. Peter experiences exhaustion and a deep sense of alienation from Earth, finding himself incapable of maintaining regular contact with Bea. Yet her messages become more and more dramatic over time. Faber is extremely adept at creating a catastrophic crescendo through her emails: in Bea's first message we find only a passing, and easily overlooked, reference to "flooding in some towns in the Midlands, cars floating down the street, etc." (Faber 2014, 92); in the following message she tells Peter about a devastating tsunami in the Maldives; little by little, catastrophic images pile up in Bea's emails, and it becomes increasingly difficult to ignore them (for the reader), but also to formulate an adequate response (for Peter). The world is undergoing a

11 On the imagination of a world without humans, see an insightful article by Greg Garrard (2012), who however focuses on cinematic visions of nonhuman reality, not literary ones.

major, structural crisis; as Bea writes: "Everywhere, things are breaking down. Institutions that have been around forever are going to the wall. We've seen this happening for years, I know, but it's accelerating suddenly. And for once, it's not just the underdogs that suffer while the elites carry on as usual. The elites are being hit just as hard" (Faber 2014, 353). Unexpectedly, the novel enters postapocalyptic territory; the whole purpose of colonizing Oasis, we find out, is recreating human society based on premises other than the exploitative greed and violence that have led to environmental cataclysm.[12] But the apocalypse taking place on Earth is buffered by Peter's cosmic distance: "All these disasters that are befalling the world – the tsunamis and earthquakes and financial meltdowns or whatever – are just so alien to my life here" (Faber 2014, 357), he confesses in a message to Bea. The alienness of Oasis, along with its inhabitants' strange submissiveness and lack of emotions, infiltrate Peter's mind and cause a startling reversal of his allegiances: it is no longer Oasis that seems "alien" to Peter, but the "disasters that are befalling" his native Earth.

The failure of Peter's sympathy for the fate of humanity marks a sharp departure from Livio's postapocalyptic predicament in *Qualcosa, là fuori*. While Livio experiences the full force of environmental catastrophe, being immersed in it with his weakening body and painful memories, Peter is separated from the world's collapse by a seemingly unbridgeable gulf, which Faber's novel equates with the personal rift between Peter and his wife. The differences between Livio and Peter do not end here: Livio's scientific materialism clashes with Peter's Christian worldview, which implies a dualistic understanding of mind and body. In internally focalized prose, we observe Peter rehearsing the doctrine that "bodies [. . .] are just another layer of clothing, an outfit of flesh with an impractically thin leather exterior, in various shades of pink, yellow and brown. The souls alone are real" (Faber 2014, 77). If immaterial souls are the only reality, we couldn't be further away from Livio's neurotransmitters and electrical activity in the brain.

Yet Faber's novel grapples with many of the same questions that trouble Livio. Lamenting the great cost in human life of the disasters she witnesses, Bea asks: "All those people had so many reasons to live and now what are they? Just chemical traces" (Faber 2014, 352). With their implicit materialism, Bea's words contradict the Christian belief in the immortality of the soul, foreshadowing her later, dramatic statement that "there is no God" (Faber 2014, 452) in a one-line

12 See, for instance, this remark by the mission's linguist: "The earth has *had* it. We've mined all the mines, we've exploited all the exploits, we've eaten all the eats" (Faber 2014, 494; italics in the original).

message to Peter. But if Bea's faith is irreversibly shaken by the escalating planetary crisis, Peter's worldview is also significantly – if more subtly – altered by his experiences on Oasis. For the most part, these changes can be imputed to the influence of the Oasan landscape on Peter: "Oasis was the first place he'd ever allowed himself to bond with. The first place he'd ever loved" (Faber 2014, 496). It is via this "bond" that a materialistic account of mind-world relations enters the novel – and makes its way into Peter's consciousness.

As he prepares to leave the Earth's atmosphere, Peter's consciousness becomes fragmented under the influence of a powerful sedative: "This was what happened to the universe when you were no longer able to hold it together. Atoms in clusters, rays of light, forming ephemeral shapes before moving on" (Faber 2014, 36). As the familiar world of terrestrial perception breaks down, shapes and patterns come to the fore, exactly as they did for Livio in Arpaia's novel. But Peter, unlike Earth-bound Livio, only experiences ecological catastrophe through the verbal mediation of Bea's emails. The disruptions in Peter's consciousness are thus not a direct result of the crisis unfolding on Earth, but rather reflect the fact that the character has to adjust to a planet largely untouched by humans, as part of what turns out to be a coordinated attempt to escape from the consequences of environmental exploitation. As soon as he disembarks on Oasis, Peter realizes that there is something unique about its atmosphere; the rain, he notices, "sounded different from rain back home. Its intensity waxed and waned in a rapid cyclical rhythm, three seconds at most between surges. He synchronized the fluctuations with his own breathing, inhaling when the rain fell softer, exhaling when it fell hard" (Faber 2014, 55). The unusual weather patterns of Oasis are reminiscent of Lem's symmetriads in *Solaris*, minus the latter's physical complexity and opaqueness. Instead of Lem's arcane symphony, Faber favors choreographic metaphors throughout the novel to suggest that the Oasan rain speaks to the body directly: "The rain wasn't falling in straight lines, it was . . . dancing! Could one say that about rainfall?" (Faber 2014, 57; ellipsis in the original). These passages demonstrate the embodied nature of Peter's evolving bond with the planet: his breathing immediately becomes attuned to the rain's "rapid cyclical rhythm." The use of the "rhythm" metaphor is even more significant because it had been used a few pages earlier to describe Peter and Bea's relationship.[13] The intimate connection with Oasis rivals, and gradually comes to replace in intensity, Peter's marriage.

13 "Beatrice and Peter got into rhythm, perfectly united in purpose. They'd done this hundreds of times before" (Faber 2014, 21). Alan Palmer discusses these intersubjective rhythms – and their narrative figuration – in *Social Minds in the Novel* (2010).

Once fully synchronized with the rhythms of the alien planet, Peter experiences something akin to what Stacy Alaimo (2010) calls "transcorporeality," an extension of embodied subjectivity into the physical world: "His body felt free and unencumbered, almost a part of the atmosphere, with no division between his skin and the surrounding sky" (Faber 2014, 431). While superficially similar to ecstatic or transcendental experiences common in a religious context, the immediacy of Peter's liberating contact with the nonhuman world and its rhythmic patterns leaves no room for an immaterial soul: Peter's being is defined, without any dualistic residue, by his encounter with Oasis. As the character comes to realize, even the seat of his consciousness, the brain, is not sealed off from the environment: "He'd always imagined his brain as a wholly enclosed thing, safe inside a shell of bone, but maybe, in this strange new environment, the seal was more permeable and his brain was being infiltrated by insidious vapors" (Faber 2014, 113). Brain-world isomorphism is highlighted by the rhythmic permeability of both Oasan weather and Peter's embodiment. Patterns bridge inner experience and the external world, to the point that the former extends into and fuses with the latter, challenging any dualistic separation between the two – and, indirectly, unsettling religious belief in an immaterial soul. The challenge raised by this vision is imaginative and affective: as I discussed above, fiction shouldn't be equated with a philosophical argument against dualism or a religious worldview (although it may certainly inspire readers to look for arguments).

A form of materialism also seems to underlie the Oasans' worldview. When one of the aliens dies, Peter observes with dismay how – in a ritualistic burial ceremony – the dead body is consumed by maggots that are then harvested by the Oasans and transformed into food and other "useful materials" (Faber 2014, 303). The body, for the Oasans, is not a receptacle for an intangible soul – and sacred because of that privileged connection with God – but fully and materially embedded in the world that sustains the Oasan community. This epiphany of embodied materiality, through Peter's double bond with the planet and with its inhabitants, comes close to undermining the character's faith. In the novel's ending, Peter decides to sever his bond with Oasis and travel back to Earth to search for his wife, whose life has become swept up in the postapocalyptic turmoil. The decision is accompanied by a moment of extraordinary disillusionment with the Christian message he had been spreading for many years:

> The holy book he'd spent so much of his life preaching from had one cruel flaw: it was not very good at offering encouragement or hope to those who weren't religious. *With God, nothing shall be impossible*, proclaimed Luke, and that message, which Peter had always thought was the most joyously positive reassurance you could wish for, now turned itself over like a dying insect, and became *Without God, everything shall be impossible*. What use was that to Grainger [one of Peter's fellow colonists, who also decides to travel back to

> Earth]? What use was that to Bea? The way things had turned out, they might need to manage without a savior; they might need to forage and scrabble for whatever future they could get on their own. (Faber 2014, 545; italics in the original)

Remarkable here is the simile comparing Luke's words to a "dying insect" turning over, which evokes the scene with the maggots digesting the Oasan's body and being, in turn, transformed into food. Through the simile, the Gospel becomes physical, like a dying animal, an image that subtly denies reincarnation and undercuts the possibility of spiritual transcendence. The shocking realization articulated in this passage is shaped by Peter's bodily experiences of communion with Oasis. Christian belief falters, the dualistic view of the soul as separable from the physical body is no longer tenable in a world ravaged by catastrophe. If Bea must "forage and scrabble" in God's absence, Peter will join her. It is the insight into the physicality of his embedding in the world, and – crucially – of the postapocalyptic predicament Bea faces on Earth, that prompts Peter to abandon his cosmic distance from human affairs. If Peter comes to question the dualism of Christian doctrine, Livio experiences dismay at the ramifications of his materialist worldview: either way, narrative uses the ecological crisis to stage and put pressure on the subject-object dichotomies that underlie Western views of the conscious mind.

5 Conclusion

Ecocritic Kate Rigby writes that calamities "reveal much about how a given society relates to its physical environment, something that is as much a question of culture, values, and beliefs as it is of economics, politics, and technology" (2015, 15). As a genre that voices and probes our collective anxieties of societal collapse, postapocalyptic fiction is particularly well positioned to expose humankind's reliance on the natural resources of our planet. The retrospective temporality of narrative easily accommodates the post-hoc knowledge created by a catastrophic event: just as stories are generally told after the fact, catastrophe affords retrospective insight into the precariousness of human life and our collective entanglement with nonhuman realities. Narrative, as scholars of science communication have argued (Dahlstrom 2014; Olson 2015), offers a powerful means of translating into human-scale language scientific theories about society's interrelation with the Earth's climate or geological history. By focusing on characters who are experiencing, dramatically, the effects of catastrophe, a fictional story serves as a simulation, a what-if scenario that mediates between abstract scientific models and the affective, embodied experience of everyday interactions with the world.

In that respect, the postapocalyptic novels I have examined in this chapter participate in the broader cultural traffic between (fictional) narrative and science that this volume aims to chart.

While that point applies, in general, to any postapocalyptic narrative, *Qualcosa, là fuori* and *The Book of Strange New Things* approach disaster in ways that resonate strongly with the second conceptual coordinate of this volume – namely, mind. *Qualcosa, là fuori* stages environmental disaster directly, by centering on a group of characters traversing a European continent desertified by global warming. In Faber's novel, human society collapses far from the protagonist, who has been sent to a distant planet to convert its alien inhabitants to Christianity (and, in the process, recreate human community along more sustainable lines). In both novels, the postapocalyptic set-up results in the fine-grained exploration of the protagonists' perceptual and affective responses to catastrophe. Narrative, as narratologists have long acknowledged, is centrally concerned with the representation of how external events affect human or human-like characters – a foregrounding of mental states that Monika Fludernik (1996) captures under the rubric of "experientiality." Furthermore, narrative can *thematize* mind by exploring, and potentially even exposing the limitations of, current scientific knowledge (this is what I call "cognitive thematics" in Caracciolo 2016, 196–202). Both *Qualcosa, là fuori* and *The Book of Strange New Things* probe our cognitive and cultural propensities for thinking dualistically about the world – that is, for seeing mind as qualitatively different from the material stuff that makes up physical reality, including the reality of our bodies. In both novels, patterns are responsible for closing that dualistic divide in imaginative terms. Some of these patterns are sensory configurations of light or weather experienced by the protagonists and rendered through internal focalization as their consciousnesses confront an unfamiliar environment. Other patterns exist at the stylistic level: they are the sophisticated metaphors and similes (such as the "malignant" light or "dancing" rain) through which Arpaia and Faber challenge the separation between subjectivity and physical processes.[14]

Dualism is, certainly in a Western context, deeply bound up with a religious view of the soul's transcendence, as *The Book of Strange New Things* highlights. But the gap between physical phenomena and conscious experience is also one that science has been unable to close so far, and that has attracted significant philosophical attention under headings such as "psychophysical reduction" or "emergence." The "hard problem" of consciousness, in Chalmers's (1995) influen-

14 See also Caracciolo (2021, ch. 6) on how metaphorical language in literary fiction helps bridge the human-nonhuman divide.

tial phrase, lingers experientially because neither psychophysical reduction nor emergence can eliminate the *phenomenological* gap between mind and material processes. Whatever the scientific solution to the hard problem, it is likely to involve highly technical knowledge of processes that play out far below the range of our awareness. This is what Livio, the protagonist of *Qualcosa, là fuori*, realizes: despite his training as a neuroscientist, his physicalist belief that mind is "only" the result of patterns of electrical activity in the brain (and physiological processes in the body) may explain how memory works, but cannot explain away the anguish of remembering his dead wife and son.

By confronting readers with the possibilities and limitations of humans' entanglement with the material world, postapocalyptic fiction can thus query, more or less explicitly, the dualistic separation between mind and world, consciousness and our physical bodies. My case studies in this chapter bring this possibility out in the open: they explore how perception breaks down when the coupling between humans and their environment is destabilized (*Qualcosa, là fuori*), or they show how the body can extend itself into the physical landscape by becoming attuned to the cycles of the weather (*The Book of Strange New Things*). In both cases it is the perceptual and affective patterning of mind-world interactions that comes to the fore, echoing contemporary "enactivist" work in the philosophy of mind that argues for the cognitive primacy of embodied patterns of exploration of the world (Varela et al. 1991; Noë 2004).[15] The two novels' thematic engagement with these philosophical questions goes hand in hand with their formal strategies, particularly their systematic use of internal focalization and metaphorical language that sews together the physical and the mental. In this way, the analysis of postapocalyptic fiction and the formal strategies it deploys to probe characters' consciousnesses can point to a remarkable convergence between philosophical questions on the nature of mind and contemporary discussions on humans' material entanglement with nonhuman forces.

Reference list

Abbott, H. Porter. *Real Mysteries: Narrative and the Unknowable*. Columbus: Ohio State University Press, 2013.

Alaimo, Stacy. *Bodily Natures: Science, Environment, and the Material Self*. Bloomington: Indiana University Press, 2010.

Arpaia, Bruno. *Qualcosa, là fuori*. Milan: Guanda, 2016.

15 See also Caracciolo (2014) for a book-length attempt to cross-fertilize enactivist philosophy and narrative theory.

Bellone, Enrico. *Qualcosa, là fuori: Come il cervello crea la realtà*. Turin: Codice, 2011.

Bennett, Jane. *Vibrant Matter: A Political Ecology of Things*. Durham and London: Duke University Press Books, 2010.

Besser, Stephan. "How Patterns Meet: Tracing the Isomorphic Imagination in Contemporary Neuroculture." *Configurations* 25.4 (2017): 415–445.

Blackmore, Susan, and Emily T. Troscianko. *Consciousness: An Introduction*. Abingdon: Routledge, 2018.

Bloom, Paul. *Descartes' Baby: How the Science of Child Development Explains What Makes Us Human*. New York: Basic Books, 2005.

Cameron, James, dir. *Avatar*. 20th Century Fox, 2009.

Caracciolo, Marco. *The Experientiality of Narrative: An Enactivist Approach*. Berlin: De Gruyter, 2014.

Caracciolo, Marco. "Cognitive Literary Studies and the Status of Interpretation: An Attempt at Conceptual Mapping." *New Literary History* 47.3 (2016): 187–208.

Caracciolo, Marco. "The Nonhuman in Mind: Narrative Challenges to Folk Psychology." *The Edinburgh Companion to Contemporary Narrative Theories*. Eds. Zara Dinnen and Robyn Warhol. Edinburgh: Edinburgh University Press, 2018. 30–42.

Caracciolo, Marco. *Narrating the Mesh: Form and Story in the Anthropocene*. Charlottesville: University of Virginia Press, 2021.

Caracciolo, Marco, and Karin Kukkonen. *With Bodies: Narrative Theory and Embodied Cognition*. Columbus: Ohio State University Press, 2021.

Chakrabarty, Dipesh. "Climate and Capital: On Conjoined Histories." *Critical Inquiry* 41.1 (2014): 1–23.

Chalmers, David J. "Facing Up to the Problem of Consciousness." *Journal of Consciousness Studies* 2.3 (1995): 200–219.

Connolly, William E. *Facing the Planetary: Entangled Humanism and the Politics of Swarming*. Durham: Duke University Press, 2017.

Crutzen, Paul J. "Geology of Mankind." *Nature* 415.6867 (2002): 23.

Dahlstrom, Michael F. "Using Narratives and Storytelling to Communicate Science with Nonexpert Audiences." *Proceedings of the National Academy of Sciences* 111.4 (2014): 13614–13620.

Deacon, Terrence W. "Emergence: The Hole at the Wheel's Hub." *The Re-Emergence of Emergence: The Emergentist Hypothesis from Science to Religion*. Eds. Philip Clayton and Paul Davies. Oxford: Oxford University Press, 2006. 111–150.

Dolphijn, Rick, and Iris van der Tuin, eds. *New Materialism: Interviews & Cartographies*. Ann Arbor: Open Humanities Press, 2012.

Faber, Michel. *The Book of Strange New Things*. London: Hogarth, 2014.

Fludernik, Monika. *Towards a "Natural" Narratology*. London: Routledge, 1996.

Garrard, Greg. "Worlds without Us: Some Types of Disanthropy." *SubStance* 41.1 (2012): 40–60.

Gibson, John. *Fiction and the Weave of Life*. Oxford: Oxford University Press, 2007.

Goldstein, Amanda Jo. *Sweet Science: Romantic Materialism and the New Logics of Life*. Chicago: University of Chicago Press, 2017.

Grusin, Richard, ed. *The Nonhuman Turn*. Minneapolis: University of Minnesota Press, 2015.

Lem, Stanisław. *Solaris*. Trans. Bill Johnston. Northridge, CA: Premier Digital Publishing, 2011.

Levine, Joseph. "Materialism and Qualia: The Explanatory Gap." *Pacific Philosophical Quarterly* 64.4 (1983): 354–361.

Morton, Timothy. *The Ecological Thought*. Cambridge, MA: Harvard University Press, 2010.

Nagel, Thomas. *Mind and Cosmos: Why the Materialist Neo-Darwinian Conception of Nature Is Almost Certainly False*. New York: Oxford University Press, 2012.

Niederhoff, Burkhard. "Focalization" (2013). *The Living Handbook of Narratology*. Ed. Peter Hühn. Hamburg: Hamburg University Press. https://www.lhn.uni-hamburg.de/node/18.html (3 August 2022).

Noë, Alva. *Action in Perception*. Cambridge, MA: MIT Press, 2004.

Olson, Randy. *Houston, We Have a Narrative: Why Science Needs Story*. Chicago: University of Chicago Press, 2015.

Palmer, Alan. *Social Minds in the Novel*. Columbus: Ohio State University Press, 2010.

Rigby, Kate. *Dancing with Disaster: Environmental Histories, Narratives, and Ethics*. Charlottesville: University of Virginia Press, 2015.

Slingerland, Edward. "Who's Afraid of Reductionism? The Study of Religion in the Age of Cognitive Science." *Journal of the American Academy of Religion* 76.2 (2008): 375–411.

Trexler, Adam. *Anthropocene Fictions: The Novel in a Time of Climate Change*. Charlottesville: University of Virginia Press, 2015.

Varela, Francisco J., Evan Thompson, and Eleanor Rosch. *The Embodied Mind: Cognitive Science and Human Experience*. Cambridge, MA: MIT Press, 1991.

Víctor Bermúdez

Emotion and Matter in *Corps corrosifs* by Lorand Gaspar

Abstract: This chapter articulates an interdisciplinary approach to studying abstraction in poetic language. Grounded on the perspective and methodologies of Cognitive Literary Theory, this study addresses the emotional basis of conceptualization of matter and space in poetry, providing an analysis of the work *Corps corrosifs* by French poet Lorand Gaspar. The poetological hypothesis of this essay concerns the emergence of embodied meaning in a poetic corpus characterized by its "affective abstraction" of space and its scientific treatment of matter. The chapter examines the idea that emotions are involved in our abstract comprehension of external materiality, and that literature manifests such an emotional comprehension, allowing us to discern several aspects of the nature of such abstraction. Poems being structures of thought with a profound cognitive value, the theoretical and analytical contribution of this study motivates the claim that abstract thought is ultimately rooted in the perception, movement and feeling of the human body. Additionally, it fosters an insight into literary texts that underlines the relevance of affective cognition as a relevant process of human imagination, which has a potential for further research in cognitive poetics.

1 Introduction

The matter of poetry is the body. One of the procedures of the creative mind consists in abstracting physical reality through an artistic use of language. Hence, embodied understanding of the external world is at the basis of poetic imagination. This theoretical affirmation meets with general approval in the current cognitive sciences concerned with the creative mind, although it requires literary evidence. Therefore, literary analysis should provide evidence for such a statement in order to build an account of the cognition *of* poetics with a hermeneutical methodology. Thus, both philosophical (Robinson 2005) and scientific (De Gelder 2016) research on emotion, along with its outstanding results based on patterns of neural activity in the brain, must be compared and correlated with what can be found in literary texts in more attentive readings.

This chapter addresses the emotional basis of the abstraction and conceptualization of matter in poetry by relying on the methodologies of the discipline of Cognitive Poetics or Cognitive Literary Studies, increasingly widespread in En-

https://doi.org/10.1515/9783110782844-004

glish and German Studies, but still very incipient in Romance literatures. This interdisciplinary field employs theoretical tools from Phenomenology, Semiotics and Philosophy of Mind just as much as from Linguistics, Cognitive Psychology and certain branches of Neurobiology. The term *Cognitive Poetics* (Tsur 2008) refers to a set of significantly diverse approaches to literary phenomena. Understood in a broad way, literary phenomena involve both the writer and the reader of a given literary work; the studies conducted from the perspective of cognitive poetics are deployed in both directions, addressing the experience of reading and textual hermeneutics.[1] As for the reader's experience, several theoretical works and empirical experiments are currently aiming to track the activity of emotional patterns for numerous readers in order to develop hypotheses of literary reception.[2] Specific objects of study and methodologies arise from both perspectives, approaching both the emotional stimuli that are evoked in literary texts and the structure of the emotional experience of the reader, in order to pursue the understanding of the internal mechanisms of poetic language and literary thought. This paper's approach is in the line of the latter, aiming to disclose the procedures of abstract thought in literary texts and how they are linked to emotional structures.[3]

The literary analysis developed in the following essay ponders the perceptual decoding and the emotional responses that are articulated by literary language. Here, the priority is to track the ways in which literary texts inscribe meaning by means of the verbal codification of emotional experience, and to infer what the cognitive correlate of such a codification is. It is therefore necessary to consider the hypothesis that emotions are involved in our abstract comprehension of matter and space, and that literature manifests such an emotional comprehension, allowing us to discern several aspects of such abstraction. Poetic writing configures scenarios that are affectively coherent with the perceptions to which a

1 Hogan offers a synthetic scheme of the standard organization of narrative analysis, which is summarized under the question "where is literary emotion?", as follows: real author – implied author – narrator – plot and verbalization – storyworld and story – narratee – implied reader – real reader (Hogan 2017, 4–5).

2 For detailed accounts of empirical aesthetics, see Kraxenberger and Menninghaus (2017). For the neurocognitive poetics approach, see Arthur Jacobs (2015).

3 In this regard, Hogan underlines that: "Of course, real people are the only vehicles for real emotion. However, we imagine emotions elsewhere and respond emotionally to such imaginations. Such imagination, or "simulation," as it is called in cognitive science, underlies and animates the remaining components. The implied author and implied reader are complex and widely debated constructs. The simplest way of understanding the implied author is as a set of norms for the work. The implied author is related to the real author, but need not be consistent with the real author" (Hogan 2017, 6).

meaning is attributed. Thus, the poem *shows* – that is, it allows perception – just as much as it *moves* – it allows feeling. A cognitive literary perspective is pertinent to the extent that it explores the ways in which literary language is modulated by cognitive mechanisms; such an exploration is pursued by studying perceptual and emotional processes underlying literary reasoning.

Behind this procedure lies the configuration of space on the basis of the nomenclatures describing matter in poetic texts. In this context, it becomes relevant to examine the role that different types of knowledge play in literary texts. The epistemic aspect of the poetic lexicon becomes relevant when portraying nature inasmuch as the representation of matter and space involves different types of knowledge. Therefore, not only is it crucial to observe the conceptual transfers and the semiotic processes that poetry employs by importing several scientific concepts and endowing them with artistic attributes, but the analyses must also examine the procedures of abstraction by considering the cognitive value of literature – that is, to ponder to what extent literature helps us to track the strings of thought by conceiving it as "data" of human mental processes.

As considered in this paper, the cognitive study of literature assumes that some traces of consciousness underlie the hermeneutics of the poetic text.[4] Nevertheless, the way we translate this assumption into the practice of literary scholars implies facing the challenges entailed by the types of evidence that literary theory can provide by means of its analytical resources. This puts the literary theorist in the position of integrating tools from different disciplines in order to propose hypotheses regarding the human mind's processes of imagination. However, every interdisciplinary integration has its own limits and advantages when tracing the phenomenological logic of the sensible, and the production of meaning that operates in poetic expression. Consequently, the versatility of literary theory in incor-

4 By hermeneutics I refer to the generation of interpretation systems of literary texts revealing perceptual, mnemonic, and/or emotional processes of the human mind. Furthermore, I claim that the association mechanism by which a text fosters simultaneous levels of *semiosis* is indeed susceptible of being described and understood by literary theoretical and analytical methods. In other words, the exegesis principle underlying hermeneutics implies that the meaning of a text can be clarified when it is contrasted with external references derived from the text itself. Hermeneutics is at the foundation of literary theory and comparative literature, as the analysis of the form of a text fosters an interpretation that is itself the basis of further theoretical insights, which may be of a literary nature (e.g. poetological) or another (e.g. philosophical). For a historical retrospective of the idea of hermeneutics, see Hösle (2012). However, hermeneutics is not only at the foundation of literary studies, it is also the object of current interdisciplinary approaches. In this regard, a relevant recent notion within cognitive literary studies is that of "Neurohermeneutics," which relates the form of the literary text to its reader's cognitive processes (see Gambino and Pulvirenti 2019).

porating considerations and reasoning from the cognitive sciences must not be detrimental to the type of knowledge that humanists can provide when proposing reliable systems of interpretation of literature. I conclude this introduction with such a statement, following Hogan, for whom:

> [. . .] we may study literature in order to learn something about the human mind, social relations, language, or other worldly phenomena. Most important for our purposes, we may not only study affective science or affective poststructuralism in order to make more sense of literary works and respond to them more fully or more appropriately. We may also study literature in order to learn something about emotion (Hogan 2017, 21).

With Hogan, I believe that literary studies can help not only to provide reliable interpretations of literary works, or historical coherent narratives of movements and artistic influences. When conceived from an interdisciplinary perspective, the relevance of literary research consists of providing other disciplines with ideas and observations about the human mind that can shape their own research topics. This, of course, requires an equally interdisciplinary approach to literature, which underlines both its cultural and biological relevance. Literature then, will not only be enriched by scientific insights but will as well inform the investigation of specific phenomena of the mind taking place in language.

2 Emotion matters

That poetry deals with emotions is not an astonishing statement. Yet, the ways and levels at which this affective language must be examined is not quite as evident. Poetry names, suggests and evokes emotions, just as much as it provides an emotional experience of the spaces that it describes. Following this idea, one can conjecture that poetic language is shaped by the forms of landscape themselves and the enunciation of the specific matter that is the object of this language.[5] Likewise, emotion seems to play a role in the act of speech that poetry entails in its recreation of meaningful affective experiences, and hence, it seems necessary to briefly summarize some theoretical issues regarding the emotional dimension of meaning.

The relation between emotion and knowledge in literary representations requires some elementary theoretical remarks. Firstly, research on literary emotion is mainly focused on two different areas: the emotion that is experienced by the

5 For a theoretical account of the physical grounds of mind, and the material emergence of meaning, feelings and consciousness, see Deacon (2013).

reader of a given literary work, and the emotion that is represented by an author through the verbal codification of a lived or imagined experience. Thus, the first pertinent distinction to establish is that of whose emotions constitute the object of study: the writer's, the reader's or the text's. In his *Thinking about Feeling* (2004) the philosopher Robert C. Solomon claims that "[e]motions are a kind of judgment – or rather, a complex of interlocking judgments, desires and intentions. [. . .] subjective engagements in the world" (2004, 76–77). This idea of emotions as engagements with the world can be relevant for the study of literary emotion because it suggests that traceable "affective data" in textual hermeneutics correlates with both thoughts and feelings. But seeing emotions as conscious evaluations may be restrictive, as a significant part of our affective experience is non-conscious, and, as philosopher Jenefer Robinson points out, "judgement theorists disagree about exactly how to characterize the relation between emotion and judgement. Some think that emotions are identical to judgement, others that judgements are *sufficient* for emotions, and others again that judgements are a necessary condition for emotions but not sufficient" (Robinson 2005, 14). Therefore, for the time being, it may be useful to consider emotions as complex cognitive processes.

The study of literary emotion reveals an intensive debate about the origin, nature and definition of emotions themselves: whether they are an expression of the author's personal experience, from which we could infer biographical data about the writer, or the result of a fictional purpose that functions as a discursive device with no biographical relevance. The relevant remark here is that by investigating the descriptions of emotional processes and their relations to space in poetry, it is possible to better understand our affective experiences in real life, and this is an appropriate task for a literary theorist. However, a brief survey of conceptual distinctions discloses different aspects of affective experience:

> Emotion is one type of affect, other types being *mood, temperament* and *sensation* (for example, pain). Emotions can be understood as either states or as processes. When understood as a state (like being angry or afraid), an emotion is a type of mental state that interacts with other mental states and causes certain behaviors. Understood as a process, it is useful to divide emotion into two parts. The early part of the emotion process is the interval between the perception of the stimulus and the triggering of the bodily response. The later part of the emotion process is a bodily response, for example, changes in heart rate, skin conductance, and facial expression (G. Johnson 2009).

This definition might be sufficiently broad, but Johnson also points out that: "[t]he early part of the process is typically taken to include an evaluation of the stimulus, which means that the occurrence of an emotion depends on how the individual understands or 'sees' the stimulus. Having this evaluative component in the process means that an emotion is not a simple and direct response to a stimulus" (2009). Similarly, from a neurobiological perspective, emotions are specific mechanisms of response with evolutionary advantages, since they function as bodily reactions, and they operate either consciously or unconsciously. In *Looking for Spinoza* (2003), Antonio Damasio defines them as follows:

> Emotions are reactions or movements, many of them public, visible to others as they occur in the face, in the voice, in specific behaviors. To be sure, some components of the emotion process are not visible to the naked eye but can be made "visible" with current scientific probes such as hormonal assays and electrophysiological wave patterns. Feelings, on the other hand, are always hidden, like all mental images necessarily are, unseen to anyone other than their rightful owner, the most private property of the organism in whose brain they occur. Emotions play out in the theater of the body. Feelings play out in the theater of the mind [. . .]. Emotions and related reactions seem to precede feelings in the history of life. Emotions and related phenomena are the foundation for feelings, the mental events that form the bedrock of our minds. (2003, 7)

Likewise, in the field of the philosophy of mind, Mark Johnson highlights the importance of the body in meaning. In contrast with the standard notion that meaning involves the conscious entertaining of concepts and propositions, Johnson focuses on "nonconscious aspects of a person's ability to meaningfully engage their past, present, and future environments" (2007, 10). In *The Meaning of the Body* (2017), Mark Johnson proposes an Embodied Theory of Meaning, by which he argues that it is a mistake to claim that *real* meaning is cognitive meaning of the conceptual/propositional sort. Instead, he underlines the central role of emotion in the way human beings make sense of the world. He also makes decisive statements claiming that "[b]efore there is abstract thinking, before there is reasoning, before there is speech, there is emotion" (2007, 52) and that "[t]here is no cognition without emotion, even though we are often unaware of the emotional aspects of our thinking" (2007, 9). This turns out to be relevant for literary analysis because it extends to another aspect of the study of the acquisition of knowledge and the process of abstraction. For Johnson, "[e]motion and feeling lie at the heart of our capacity to experience meaning" (2007, 53), and this assertion is crucial in literary interpretation. Likewise, within the field of Cognitive Poetics literary scholar Patrick Hogan has an interdisciplinary vision of literary emotion, and his considerations could lead us towards an emotional organization of space in literary representations. In his work *What Literature Teaches Us about Emotion* (2011), he approaches the issue of representation, and points out that:

> The experience of emotions does not in itself constitute an idea of emotions. Experience must be mediated or objectified through an idea or concept. It must be *represented.* Whether we are

> engaging in casual conversation or doing neuroscience, we never think through an experience directly. Rather, we consider some representation of the experience. Indeed, experience itself is never "pure" and "direct." It is mediated by our sensory and cognitive architectures, the innate structures, the acquired processes and contents that shape what occurs in the world into what we think occurs. (Hogan 2011, 14)

Hogan's idea is at the core of our consideration of poetic language as a data source for the emotional representation of space. As this paper will argue, by means of the analysis of the collection of poems *Corps corrosifs* by Lorand Gaspar, literature constitutes a unique set of representations of emotional experience, and such representation is conjoined with that of matter. Similarly, in an analogous perspective to that of Hogan, French phenomenologist and literary theorist Michel Collot proposes the notion of *la matière-émotion*. The fundamental principle underlying this theoretical concept is that the world's image, which is elaborated by a subject through language that "verbalizes the matter," is modulated by his "feeling of space." *La matière-émotion* (2005) is a phenomenological instrument that aims to explore the ways in which poetic expression codifies the physical relation of a subject with their surrounding world. The codification of the "material world" that poetic language undertakes is affected by the "emotional attributes" that permeate the expression of space and that of the subject itself. Although they manifest in the concrete things of the external world, emotional attributes are likely to be examined in the text as extensions or aspects of human consciousness. *La matière-émotion* is applied in a literary context in order to point out that:

> [. . .] l'émotion n'est pas un phénomène purement subjectif. Elle est la réponse affective d'un sujet à la rencontre d'un être ou d'une chose du monde extérieur, qu'il peut tenter d'intérioriser en créant un autre objet, source d'une émotion analogue mais nouvelle: le poème ou l'œuvre d'art. (Collot 2005, 2)[6]

When exploring the transition between the emotional and the poetic experiences, Collot suggests that "l'émotion, loin d'enfermer le poète dans la sphère de la subjectivité, constitue un mode d'ouverture au monde" (2005, 10–11),[7] which is a noteworthy idea if we consider the communicative function of literary discourse, as its characteristic introspection does not prevent it from achieving identification and empathy effects. Solitude and togetherness are both intrinsic aspects of the

6 "Emotion is not a purely subjective phenomenon. It is the emotional response of a subject to the contact with a being or a thing from the outside world, which he can try to internalize by creating another object, source of a similar but new emotion: the poem or the work of art" (Collot 2005, 2; trans. by VB).
7 "far from enclosing the poet in the sphere of subjectivity, emotion is a way of opening to the world" (2005, 10–11; trans. by VB).

poetic act. It is precisely the subjective use of language that makes it a rich source of human experience, and therefore it is this expression of subjectivity that requires analytic strategies that stress the biological and axiological values of poetic discourse. Hence, I do claim that the epistemocritical and the cognitive perspectives are both involved in the reciprocal relation between mind and matter that takes place in poetic language. On the one hand, the mind is somehow materialized, being a higher form of energy that animates all matter. On the other hand, matter itself becomes abstract, being reduced to flows with multiple paths and combinations no longer accessible to practical experience, but only to inspection by the mind. As Mondrian said: "par l'intériorisation de ce qui est connu comme matière et par extériorisation de ce qui est connu comme esprit – jusqu'ici trop séparés! – matière-esprit devient une unité" (in Collot 2005, 55).[8]

Consequently, abstraction operates with both conceptual and experiential content. Matter is conceptualized but it is also experienced bodily, or as cognitive poetics claims, it is conceptualized through its bodily experience. In this context, the notion of *la matière-émotion* is useful to address poetic thought from the relation established between matter and a specific way of perceiving it, somatizing it and representing it. Here, it is possible to distinguish two levels of literary expression. One level conceptually describes matter, its behaviors and its interactions. Another level refers to the body that emotionally experiences it, re-activates it and *moves* it. Mind, body and matter are then seen as entities in interaction and dialogue that will generate new meanings of emotionally experienced space. Emotions matter because they mediate as acts of understanding, helping the subject to remember their previous experiences, to formulate speculations about the immediate present and to develop projections about the future. As Mark Johnson has underlined: "Emotions are [. . .] an integral part of human understanding and meaning. [. . .] Basically, any thing (object, quality, event, person, idea) has meaning just insofar as it points to some experience, either past, present, or future (projected) that is for us connected with it. Things are meaningful because they afford us various possible experiences" (2015, 5). As will be argued, emotions mediate poetic matters by means of associating the subjective and the physical.

Furthermore, one of the tasks of literary theorist is to describe the relation between the formal characteristics of a text and the cognitive functions – whether they are emotional, perceptual, mnemonic – that it fosters in the reader, or which are in the embodied experience of the creative mind itself. A remarkable example of this is the work *Literary Reading, Cognition and Emotion: An Exploration of the*

8 "by internalizing what is known as matter and externalizing what is known as mind – until now too separate! – matter-mind become one" (Collot 2005, 55; trans by VB).

Oceanic Mind (2012), where Michel Burke proposes a "Model of Emotion in Literary Reading" which underlines different inputs such as mood, location, theme, style, leading to what he denominates *affective cognition*, arguing that those literary phenomena can be considered emotive, because of – for instance – "the preparatory excitement and projection of mood; the somatic pleasures of location, the sentimental, cultural universals of themes and the pleasurable balance and rhythm and the breaking thereof in style." Burke concludes that "there are implicit links between cognitive emotion and affective cognition at the reading stage and there will therefore be an ebb and flow of emotive information exchange between them" (Burke 2012, 155). Such flow of emotive information shall be addressed in the following pages.

3 Poetry matters

The purpose of literary analysis is to ponder whether emotions operate as a discursive device that articulates meaning in the text. Since poetic language mediates as an abstraction of the material world, the relation between emotion and other cognitive processes such as perception should be observed more closely. Likewise, it is pertinent to discern *if* and *how* the emotions interfere as mediators of the perceptual content in the configuration of literary meanings. In order to explore abstraction processes, the analysis must then stress the presence of *percepts* and *concepts* that the poem operates with. It is within the percepts-concepts relation that the abstraction and the embodiment of the poem's reality arises. Similarly, observations on the poems should underline the *mental images* that are identifiable in the verses, as if they were embodied in the biological reality of either a reader or an author. The correlation between the poetic representation and the actual lived experience is susceptible of study since, according to Depraz and Varela, the mental image "is not only a recall of a previous perception but can be generated starting from a verbal description. Such language-evoked images are quite comparable to those sensorially induced. In particular, mental scanning and distance comparisons are comparable to those effected on images recalled from previous presentations" (2002, 198).

However, exploring the correlation between the merely perceptual and the conferred emotional meanings within texts allows us to access a reconstruction of space as it is conceived by the lyric subject. In this corpus, the affective processing begins in the very "selection" of a fragment of perceptual reality to which the lyric subject will attribute several poetic values and intensities and new or non-ordinary meanings. Those attributes are rooted in the subject's abstraction of reality, which constitutes a type of introspective subjectivity while it simultaneously appeals to universality. This means that the acts of speech that take place in poetry communicate on the basis of the subject's introspection – that of the author – and thus the poem does not represent the world itself but a specific experience of it.

These elements are deeply rooted in the writing of the French poet Lorand Gaspar. His poetry thrives on references to disciplines such as geology, botany, zoology, physics, and anatomy, among others, extending the conventional nomenclatures of poetic descriptions. Here, all matter becomes writing and the literary language becomes a kind of flesh. A brief overview of some of his work's titles exemplifies the extent to which the organic is at the core of Gaspar's literary concerns. Published in 1966, *Le quatrième état de la matière* (*The Fourth State of Matter*) leaves behind the solid, the liquid and the gaseous, and refers to a metaphorical state of matter when it is illuminated. This work is composed of four sections entitled: "Connaissance de la lumière" ("Knowledge of light"), "Iconostase" ("Iconostasis"), "Écailles" ("Scales") and "Le jardin de pierres" ("The stone garden"), each of which explores the specificity of this metaphorical phenomena.[9] Similarly, *Gisements* (*Deposits*) (1968) is composed of the sections "Expansions," "Pléistocène," "Escarpement" and "Érosions." This work is characterized by a language enfolded with matter, not only through a lexical network but also in the geological processes whose descriptions take place in the poems. This is realized through the specificity of the "hyponymic" features of language, that is, by the use of particular hyponyms instead of generic hypernyms. Such hyponymic features are grounded on rich geological, geographical and botanical descriptions, in which the organic is the fundamental basis of the semantic universe. Therefore, when naming nature, poetic expression involves more than scientific observation, since a key aspect of literary language consists in providing new significance to matter. These new meanings are often built from the embodied relation with space that the lyric subject establishes.

An analogous case can be found in the work *Corps corrosifs* (2006 [1972]), where the idea of the human body as corroding matter is not connotative but denotative.[10] In light of its semiotic processes and its conceptual transfers, *Corrosive Body* constitutes a narrative of subjective apprehension of matter. This eleven-page poem is structured in short descriptions of the landscape and its dynamics

9 For a detailed analysis of *Le quatrième état de la matière* from a cognitive-poetic perspective, providing a neuroaesthetic account along with literary analysis, see Bermúdez (2017).

10 The collection of poems *Corps corrosifs* was originally published by Fata Morgana in 1978. For analytical purposes, this paper quotes several extracts translated by Mary Ann Caws and Nancy Kline (Gaspar 2015).

as they are perceived by the lyric subject. Taking into account the biographical fact that Gaspar himself lived many years in Jerusalem, and travelled along the Judean Desert, which he exhaustively photographed and documented, it is plausible to suggest that the poem's descriptions correspond to Moab, the historical name of a mountainous area of modern Jordan, near the Dead Sea and certain zones of the surrounding Arabic desert.[11] In *Corrosive Body*, the interaction with matter entails a description of nature and of the relation between the subject and its environment. A vision of specific features of the physical world, which reveal the characteristics of the human relation with space, becomes here a source for literary representations. However, at times, the subjective experience of the landscape appears without a subject, as it fades out the first person of the singular and the lyric I along with it. In such episodes, poetic language describes the interaction between matter and matter. *Corrosive Body* starts as follows:

> She would come in the evening's hollow
> dispersing the paths
> would come unclothed of distances
> would come from the non-country
> would come bare tune in her flesh
> never pronounced. (Gaspar 2015, 217)[12]

The text begins naming an ambiguous, fragmented and intriguing presence that approaches from an indeterminate depth, travelling across the space without inherent consistency. This melody appears with no consciousness along to perceive it, as there is no listener to witness its existence ("in her flesh never pronounced"). However, the poem presents a scenario in which a sonorous percept is described by portraying its *movement* across space, as is indicated by the prevailing verbs ("*dispersing* the paths"; "*would come* unclothed of distances"). The observation of the action that takes place within the poems is indeed a pertinent analytical remark, whether the lyric subject is an observer or an active participant in such acts. As Mark Johnson points out, "[. . .] our very ability to understand our world and other people arises from the nature of our bodily existence (including both our physical body and the structures and processes of our brains) plus the embodied interactions we have with our material and cultural environ-

---

11 Surgeon, writer, translator and photographer, Lorand Gaspar cultivated many facets, both in solitary and in collaborations with other artists and scientists, which contributed to his development of both a verbal and a visual language of human perception. For an overview of Gaspar's photography see Sacré and Gaspar (2003).

12 "Au creux du soir elle venait / dispersant les chemins / venait dévêtue de distances / venait du non-pays / venait mélodie nue dans sa chair / imprononcée" (Gaspar 2006, 211).

ments" (M. Johnson 2015, 1). Action, contemplation, and interaction are thus capable of being tracked in the poems.

> In summer when the pomegranates split open and rot
>   the wells
> between the crickets smashed by the bull-dozers
> the earth held up by
> long transfusions of nights.
> Our hands still discover
> under the solar uproar
> the silent snake.
>
> In the almond of the cry that was never opened
> the weight too dense for the name,
> a sowing set down in the estuary of our tongues.
> In the vulva night
> melted in pleasure
> the work of breath and of blood.
> Witnesses surely, but how to come back
> when the shores of the path shatter
> the song shatters in the fire. (Gaspar 2015, 216–217)[13]

To some extent, *Corrosive Body* recreates a progression of the summer by narrating the impact of the season on the space described and its effects on specific elements of flora and fauna ("the pomegranates split open"). The first stanza describes a landscape in which some trucks ("bull-dozers") run over insects ("crickets"), flattening and drying the surface. It is precisely the transformations in the surface that the lyric subject will abstract by its patterns through the entire poem. Within the adversity of summer, then, one could riskily hypothesize that in "the earth held up by / long transfusions of night," "the silent snake" might refer not only to the desert's fauna but also to the water beneath the desert's surface, which is found by the lyric I and his companion ("our hands still discover") "under the solar uproar."[14] As will be

13 "En été quand les grenades se fendent et pourrissent les / puits / entre les cigales qu'écrasent les bull-dozers / la terre maintenue debout par / de longues transfusions de nuits. / Nos mains trouvent encore / sous le vacarme solaire / le serpent silencieux. // Dans l'amande du cri qu'on n'a jamais ouverte / le poids trop dense pour le nom, / semence déposée dans l'estuaire de nos langues. / Dans la nuit vulvaire / fondée en volupté / l'œuvre de souffle et de sang. / Témoins certes, mais comment revenir / lorsque s'écroulent les rives du sentier, / s'écroule le chant dans le feu" (Gaspar 2006, 211–212).

14 For details on soil formation, geology and geomorphology regarding the desert's surface structure, see Schaetzl and Anderson (2012). Such a reference may be of literary interest if it is taken into account that Gaspar, a surgeon by profession in Jerusalem and Tunis, spent several summers traveling across different deserts, and as a result, his work *Sol absolu* (*Absolute Earth*, evident across the different extracts of the poem, the knowledge of landscape will foster different interpretations of the space itself. In this context, the desert will appear as both an austere and a fully-alive space, and as it is conceived by the lyric subject as a "corrosive body," it is an organic and a degradable surface. The lyric I will witness these phenomena and will react to them.

In the second stanza, the first three verses depict a silent mouth ("that was never opened"); here, the almond does not refer to the shape of the eyes, as the reader would expect (the almond-shaped eye is a common image in literary facial descriptions) but to a mouth containing "a sowing set down in the estuary of our tongues," which alludes to the impossibility of verbalizing affective expression ("too dense for the name"). However, the excitement of both breath and blood appears as evidence of pleasure, personifying the night as a female who is capable of experiencing sexual arousal ("In the vulva night"). The night is not an element of landscape itself but a period-of-time phenomenon deriving from astronomical movements. In the desert, the night is a period of activity, as it constitutes a truce of weather hostility, particularly in summer. The night makes possible a less adverse experience of the desert, but one in which it is not possible to properly recognize the way traveled, as "the shores of the path shatter," due to the darkness, and because the wind erases the footprints in the sand. In this context, pleasure appears as a mute "song shatter[ed] in the fire," emphasizing the constraints of its expression. Finally, as the night ends, a soft sunrise displaces it:

> Upon the ardent threshold
> rebellious pilgrim
> I set my insomnia down.
> We precede the day
> with a length of light.
>
> Heaven without beams and without mast
> light maddened by murderous blues
> and beiges, beiges and downy
> bronzes of the slopes where fingers shout
> and you don't understand. (Gaspar 2015, 217)[15]

---

or perhaps more literally *Absolute Soil*) also contains enriched references to the geological composition of those spaces.

15 "Sur le seuil ardent / pèlerin rebelle / je dépose mon insomnie. / Nous devançons le jour / d'une longueur de lumière // Ciel sans poutres et sans mât / lumière foliée de bleus assassins / et des beiges, des beiges et des bronzes / duveteux des pentes où les doigts crient / et tu ne comprends pas" (Gaspar 2006, 212).

The depiction of sunrise is here characterized by a dim glow that is gradually taking the place of the night, as the sun is here a “rebellious pilgrim.” But as the sky clears up (“Heaven without beams and without mast”), the brightness becomes more aggressive and the sun overwhelming and hostile (“light maddened by murderous blues”). The blue sky will then contrast with the tone Gradations of brown in the sand, creating a two-chromatic ambience in which color and temperature (“where fingers shout”) will be the prevailing percepts of the desert.[16] The landscape is characterized by virtue of its hues and oppressive heat as they are subjectively perceived by the lyric I. The enigmatic last verse (“and you don’t understand”) seems to express a feeling of perplexity of the lyric I before this progressively clarifying landscape. Perhaps the notion of biofunctional understanding may be pertinent to reveal the embodied cognition implicit in the poem. As Mark Johnson underlines: “The ‘bio’ component refers to the fact that we are biological organisms evolved both to sustain in our person the conditions of life and to enhance its quality. The ‘functional’ component refers to the fact that our activities as biological organisms give rise to our capacity to perform a wide range of cognitive and affective functions” (2015, 1). Another excerpt from *Corrosive Body* will allow a deeper insight.

> Birds that precede the light in your throats
> prepare the path of the blind.
> Crossing through the anfractuous rock
> with a thin blade of quivering flesh
> you break our mirrors still today.
>
> Word clasped under the teguments of black,
> at the ebb of the springs.
> Path of affluence
> where she never came, never went.
> Under the porch where the barbarian is welcomed
> amid the fire, the arrows and death
> I have laid the steppes on your stomach
> and with burned tongue I listened
> to the deep throb of your waters.
>
> Summer of paths, of dust. Earth without membrane
> or lungs.
> The slight water of the halt
> drunk at the mouths of another burned body.
> Under a great tree’s sleep wandering frees us

16 For further analysis on intermodal perception in Gaspar’s poetry, see the section “Psicología cognitiva en *Patmos et autres poèmes*” in Bermúdez (2017, 101–144).

to a darker flame
without compass and without rod. (Gaspar 2015, 218)[17]

The first stanza suggests that the orientation of the lyric subject within the desert is based on the bird's flight ("prepare the path of the blind"), along with the sound that they produce through bird vocalization, while flying over the rocky surface of the desert ("Crossing through the anfractuous rock"). In the fifth line ("you break our mirrors still today") one can find a reference to mirages, the optical illusions originated by the refraction of sunlight by heated air, producing in the observer the illusory perception of a sheet of water on the desert's surface, and over which the birds' flight and their reflection "break[s] our mirrors." Furthermore, at the beginning of the second stanza, the act of speech appears once again to be muted ("clasped word"), this time by the darkness ("teguments of black"). The desert's night is the "Path of affluence" in which the word "never came, never went." This landscape is "the porch where the barbarian is welcomed / amid the fire, the arrows and death," and even though the lyric subject does not seem able to name it, he can perceive it, as the water reappears once again in the depths: "I listened / to the deep throb of your waters." This tension between nature and speech demonstrates that the distance between matter and the subject generates a void that constitutes a relevant literary concern, as poetic language deals with such interstices, assuming different *affective attitudes* across the entire *Corrosive Body*: fascination, perplexity, anxiety, serenity, fear, angst, nostalgia, etc. Finally, the third stanza stresses the nomadic attitude of the lyric subject, who is shown as a walker of the desert who follows itinerant paths of arid land and uneasy breathing. To this passer-by, the landscape appears as a territory to be perceived and to be felt, and hence it prompts ephemeral actions ("Summer of paths, of dust") of contact with matter ("wandering frees us / to a darker flame / without compass and without rod"), one of which is the attempt to verbalize it through the gesture of poetic writing.[18]

---

17 "Oiseaux qui précédez la lumière dans vos gorges / préparez le chemin de l'aveugle. / Vous qui d'une lame mince de frémissement de chair / traversez la pierre anfractueuse / brisez nos miroirs encore aujourd'hui. // Parole serrée sous les téguments du noir, au reflux des sources. / Chemin d'affluence / où elle ne vint jamais, n'alla jamais. / Sous le porche où l'on accueille le barbare / parmi le feu, les flèches et la mort / j'ai couché la steppe sur ton ventre / et la langue brûlée j'écoutais / le battement profond de tes eaux. // Eté de pistes, de poussières. Terre sans séreuse ni / poumons. / L'eau brève de la halte / bue aux bouches d'un autre corps brûlé. / Sous un grand sommeil d'arbre l'errance nous livre / à une flamme plus sombre / sans boussole et sans barre" (Gaspar 2006, 212–213).

18 For a further insight into the nomadism in Gaspar's writing, see Abdeladhim (2010).

The emotional revitalization of space will gradually increase in the subsequent extracts of *Corrosive Body*:

> Sweetness of the shadows ambushed under the loins
> drawn by the ten pencils of the groping.
> The barbaric rush curved by the hip,
> narrowed by the arches, and our hands slowly
> laid down in the garden floats of the sea.
>
> Blind cry of the white of the eye
> against the white
> of the walls.
> I love you devastating season, matrix broken with heat,
> your fires blued with winds, your liquid distances
> drying upon my skin.
> I love your thirst and your rigorous oblivion
> stabbing the conqueror. (Gaspar 2015, 220–221)[19]

A certain anatomical apprehension of landscape can be found at the beginning of the first stanza as the "sweetness of the shadows," is indeed in the "kidneys" (*les reins*), an organ that is responsible for the excretion of waste substances through urine and also for the balance of homeostasis, regulating the quantity of extracellular fluids, the osmolarity of blood plasma, the balance of electrolytes and the pH of the body. This embodiment of a soft darkness assimilated by the lyric subject seems to precede the gesture of writing itself, as this obscurity is "drawn by the ten pencils of the groping," and in the third verse, the body is slightly inclined for the posture of writing ("The barbaric rush curved by the hip, / narrowed by the arches"). Thus, the poem relates the internal regulation of the organism to the night and the body by intensifying the mediation of the sense of touch, which intercedes between consciousness and nature ("and our hands slowly /laid down in the garden floats of the sea"), as the sense of touching, particularly through the hand in contact with water, conveys a profound notion of intimacy.[20]

Similarly, the second stanza begins by reducing visual perception ("Blind cry of the white of the eye / against the white / of the walls") and favoring a represen-

19 "Douceur des ombres embusquées sous les reins / que dessinent les dix crayons du tâtonnement. / La ruée barbare courbée par la hanche, / fuselée par les arches, et nos mains couchées / lentement dans les nageoires du jardin de mer. // Cri aveugle du blanc de l'œil / contre le blanc / des murs. Je t'aime saison dévastante, matrice rompue de chaleur, / tes feux bleuis par les vents, tes distances liquides / qui sèchent sur ma peau. / J'aime ta soif et ta rigueur d'oubli / qui poignarde le conquérant" (Gaspar 2006, 215–216).

20 For further cognitive and philosophical insights into the sense of touching and its representation, see Radman (2013) and Paterson (2016).

tation of summer ("I love you devastating season") that fosters the relation between the sense of touching and mental consciousness ("your liquid distances drying upon my skin"). Consequently, summer is portrayed here in virtue of its temperature and color ("matrix broken with heat, / your fires blued with winds"), that are perceivable due to thermoreception of heat and cold in relation to the temperature of the perceiving body. An affective appreciation of the elements of nature takes place in the poetic enunciation, as the lyric subject addresses summer itself in order to claim he loves this "devastating season," its "thirst" and its "rigorous oblivion." One more extract from *Corrosive Body* stresses the several percepts of warmth that configure the imaginary of the desert:

> I'm always seeking the same blind evening
> whose earth eaten by solar acids
> ferments in its flesh of such heated browns
> that the eyes tremble in the pulse becalmed.
> Chalk path issuing from the dark hollow
> climbing straight up and rethinking space
> irrigated by your center of redness
> pardon the sky the death of the day and of the god –
> invent ceaselessly, tiring yourself,
> the simple, the paltry, the nothing. (Gaspar 2015, 221–222)[21]

Temperature is perceived through the "solar acids" over the skin and its effects in "flesh," in the trembling eyes, "the redness" in order to create a holistic vision of the emptiness of the landscape ("the simple, the paltry, the nothing"), exemplifying how "emotion and related reactions are aligned with the body, feelings with the mind" (Damasio 2003, 7). By elaborating a representation of the rhythm of the desert's energies, the poem cannot avoid the task of abstracting the characteristics and interactions of space and matter as objects of experience through concepts and percepts, as literary analysis shows.

This procedure is at the basis of the *embodied meaning* hypothesis, which, as Mark Johnson points out, is rooted ultimately in the neural bases of conceptualization itself:

> From a neural perspective, we need to know what concepts are and how they work. The obvious challenge for a theory of embodied meaning is twofold. First, if concepts are not

---

21 "Je cherche toujours le même soir aveugle / dont la terre mangée par les acides solaires / fermente dans sa chair de bruns si chauds / que les yeux tremblent dans le pouls apaisé. / Chemin de craie issu du creux sombre / qui monte vertical et repense l'espace / irrigué par ton centre de rougeur / pardonne au ciel la mort du jour et du dieu – / invente sans répit en t'effritant / le simple, le peu, le rien" (Gaspar 2006, 215–216).

> disembodied symbolic representations, then what are they, from a neural perspective? Second, can *all* concepts be embodied, that is, grounded in sensorimotor aspects of experience? This second question requires an account of both concrete concepts (i.e., concepts of concrete objects, persons, events, and bodily actions) and abstract concepts (e.g., mind, knowledge, ideas, justice, rights, freedom) that are not about concrete, physical entities. First, we must examine evidence that the meaning of concrete concepts is grounded in activations of sensorimotor areas of the brain. Second, we must suggest how abstract concepts (concepts not tied to specific sensory experiences) also rely on sensorimotor areas of the brain and are thus embodied. (2007, 157)

The passage from hermeneutical analysis to the neural basis of cognitive processes such as perceptions and emotions is challenging because (among other reasons) the lived experiences and the verbal literary representation of such experiences differ in intensity. Therefore, the plausible correlation between one and the other can be established in terms of the biological nature of such experiences, and the logic of sensation and the generation of meanings that can be tracked in the prosody of poetic texts, on the assumption that the texts are revealing a coherent narrative of human consciousness. The descriptions that literary language provides about the interactions between space and the subject are evidence of mental functioning as long as this specific type of discourse operates with images and feelings. Therefore, a certain awareness of the lived or imagined experience has preceded verbal encoding. This implied subjectivity is valuable precisely because poetic language provides perceptions and emotions to a subject – that is, an individual pronoun from which all external reality is understood on the basis of its relation to a mind.[22]

> For under the intractable harshness
> when the sun beats us down face to earth
> under our motions of agony and of joy
> we find once more the floats in the sea. (Gaspar 2015, 222)[23]

In this poem, the main effect and purpose of emotion is action. As this episode shows, in the context of the adverse desert ("when the sun beats us down face to earth"), light is associated with both anguish and delight. "The floats in the sea" appear here as part of a body under water, making movement possible ("Our motions of agony and of joy"), but such an action relates to a pronoun ("our") revealing two

---

22 A further analytical attempt would consist in conceptualizing the distinction between oneself and non-self in literary texts. Such a task, however, will not be undertaken here, as the principal focus is on the emotional grounds of abstraction and not on the structures of consciousness. For a philosophical insight into subjectivity research, see Zahavi (2014).

23 "Car sous la dureté intraitable / quand le soleil nous abat face contre terre / sous nos mouvement d'agonie et de joie / nous retrouvons des nageoires de mer" (Gaspar 2006, 216–217).

subjects, and their internal "affective perception" (agony and joy). It is pertinent to remark that the word "emotion" contains an internal "motion," which ultimately allows the adaptation of the subject in the desert. As Antonio Damasio underlines, at a biological level: "[e]motions are complicated collections of chemical and neural responses, forming a pattern; all emotions have some kind of regulatory role to play, leading in one way or another to the creation of circumstances advantageous to the organism exhibiting the phenomenon; emotions are *about* the life of an organism, its body to be precise, and their role is to assist the organism in maintaining life" (1999, 51). Such phenomena are represented in the mood, location, nomenclature, and style of the poem, leading to an actual affective cognition. Once again, the poem can be read from the cognitive notion of biofunctional understanding, in order to underline the way in which its apprehension of matter is grounded in an embodied experience, a "full-bodied, full-blooded, fully passionate process that reaches down into the visceral depths of our incarnate experience and connects us functionally to our physical-cultural world" (M. Johnson 2017, 2), and this is equally an issue of phenomenological relevance, as the above-mentioned notion of *la matière-émotion* has shown. A last extract from *Corrosive Body* exemplifies the materiality of the subjective experience of space:

> Evening, nothing.
> However the black constructs itself, circumspect, careful.
> From near to near, proceeding by gangrene in the massive
> clarity,
> by hemorrhaging.
> Walls and hills beaten down
> daring presences in the distancing.
> Elsewhere
> a chilly underneath of foliage,
> soundless
> rooms placed right at the waiting.
> Patience. Crystals of patience. Lights & shattered gestures.
> Lagunas of air on the loins, under the hand that curves
> them,
> tread of the distance on the familiar threshold, unfolding
> its valves,
> its leavings, its tumults
> and you are seeking how the light breathes
> in so much dark groping. (Gaspar 2015, 226)[24]

24 "Soir, rien. / Pourtant le noir se construit, circonspect, minutieux. / De proche en proche, procédant par gangrène dans la / massive clarté, / par hémorragie. / Murs et collines abattus / présences téméraires dans l'éloignement. / Ailleurs / dessous frileux de feuilles, / chambres, / sans bruit posées à même l'attente. / Patience. Cristaux de patience. Lumières et gestes brisés. / La-

The poem represents the sunset and the progressive beginning of the night, through the gradation of warm maroon tones. The elements of space interact among themselves ("the black constructs itself") and the light of the sun "bleeds" as in anatomical hemorrhage, giving a wide color palette of reds, yellows and purples in order to give birth to the night and space to the muteness ("soundless rooms"). The perception of such phenomena is not indifferent to the lyric subject: the life in the desert is slow ("waiting", "patience"), and such slowness modifies the body ("Lights & shattered gestures") whose physical posture seems affected, along with its internal organism ("Lagunas of air on the loins"). The last beams of the sun appear alive ("the light breathes"), cohabitating with the density of the night ("in so much dark groping").

4 Imagination matters: An open discussion

This chapter has carried out the task of describing the relation between the formal characteristics of a text and the cognitive functions – whether they are emotional, perceptual, mnemonic – that it fosters in the reader, or which are in the embodied experience of the creative mind itself. More specifically, the paper has explained the embodiment implied in poetic language when it describes matter and elements of space, arguing that such subjective perceptions and memories have an affective value. As this chapter has underlined, the extensive poem *Corrosive Body* expresses the relation between perception and action, which is of central relevance for current cognitive neurosciences, arguing that perceptive patterns configurating space as well as the perception of actions are crucial for the affective level of literary enunciation. Moreover, they form part of a major process of association of memories, perceptions and emotions that are at the core of poetic imagination itself. The analysis, along with its theoretical supports, motivates the claim that abstract thought is ultimately rooted in the perception, movement and feeling of the human body. Movement is indeed crucial for the categorization and recognition of spaces, as well as for the attribution of affective values. Thus, following the literary data provided by the analysis, it is possible to argue that movement is the basis of poetic meaning. Furthermore, the study has also underlined that an epistemic dimension of human gesture, as it is expressed by poetic language, and conceptual verbalization might be related. If emotions inter-

gunes d'air sur les reins, sous la main qui les cambre, / foulée du large sur le seuil familier, dépliant ses valves, / ses exodes, ses tumultes / et tu cherches comment la lumière respire / dans tant d'obscur tâtonnement" (Gaspar 2006, 220).

cede in the process of abstracting reality, bodily gestures exemplify a physical interaction with the environment as an attempt to understand external reality. Not only is motion substantial to affective cognition, but understanding emotional dynamics in the poem makes it possible to identify and perceive its actions.

Nevertheless, the possibility of further insights into the cognitive value of poetry remains open for future research, as it is still necessary to explore a diversity of methodological strategies that accentuate what cognitive sciences can learn from literary data, and to what extent interdisciplinary research can be enriched in this process. As an example, given the case study of Lorand Gaspar's poetics, one of its characteristics is its quality of synesthesia, and cross-mapping of senses. At a perceptual level, attributes of one sense are constantly associated with others (see Bermúdez 2017). It would be useful to consider how synesthesia might relate to abstraction, subjectivity and the implications of affective cognition, from an interdisciplinary perspective which can provide evidence on how literary language codifies synesthesia. However, the interdisciplinary dialogue between the humanities and cognitive sciences should be encouraged without neglecting the research objectives of each field.[25] Literary theorists not only analyze literary texts to provide interpretations of specific corpora, they also develop theoretical insights that have the character of statements regarding the relevance of literary texts in society as structures of thought. Such importance stands at both biological and axiological levels and makes human imagination a fruitful field for innumerable interdisciplinary approaches.

Reference list

Abdeladhim, Maha Ben. *Lorand Gaspar, en question de l'errance*. Paris: L'Harmattan, 2010.

Bermúdez, Víctor. *Ciencia y modulación del pensamiento poético: percepción, emoción y metáfora en la escritura de Lorand Gaspar*. Salamanca: University of Salamanca Press, 2017.

Burke, Michael. *Literary Reading, Cognition and Emotion: An Exploration of the Oceanic Mind*. New York and London: Routledge, 2012.

Burke, Michael, and Emily T. Troscianko, eds. *Cognitive Literary Science: Dialogues between Literature and Cognition*. Oxford: Oxford University Press, 2017.

Collot, Michel. *La matière-émotion*. Paris: Presses Universitaires de France, 2005 [1997].

Damasio, Antonio. *The Feeling of What Happens: Body and Emotion in the Making of Consciousness*. New York: Harcourt Brace & Company, 1999.

Damasio, Antonio. *Looking for Spinoza: Joy, Sorrow and the Feeling Brain*. Orlando: Harvest, 2003.

25 For further discussion on current debates and future perspectives in the field of Cognitive Literary Studies, see Zunshine (2015), Hogan (2011) and Burke and Troscianko (2017).

Deacon, Terrence W. *Incomplete Nature: How Mind Emerged from Matter*. New York: W. W. Norton & Company, 2013.

De Gelder, Beatrice. *Emotions and the Body*. Oxford: Oxford University Press, 2016.

Depraz, Natalie, and Francisco J. Varela. "Imagining: Embodiment, Phenomenology, and Transformation." *Buddhism and Science: Breaking New Ground*. Ed. B. Alan Wallace. New York: Columbia University Press, 2002. 195–232.

Gambino, Renata, and Grazia Pulvirenti. "Neurohermeneutics: A Transdisciplinary Approach to Literature." *Gestalt Theory* 41.2 (2019): 185–200.

Gaspar, Lorand. *Sol absolu* précédé de *Le quatrième état de la matière* et suivi de *Corps corrosifs*. Paris: Gallimard, 2006 [1972].

Gaspar, Lorand. *Earth Absolute & Other Texts*. Trans. Mary Ann Caws and Nancy Kline. New York et al.: Contra Mundum Press, 2015.

Hogan, Patrick Colm. *What Literature Teaches Us About Emotion*. Cambridge: Cambridge University Press, 2011.

Hogan, Patrick Colm. *Literature and Emotion*. Abingdon/New York: Routledge, 2017.

Hösle, Vittorio. "Hermeneutics." *The Princeton Encyclopedia of Poetry and Poetics*. Eds. Ronald Greene and Stephen Cushman. Princeton: Princeton University Press, 2012. 618–623.

Jacobs, Arthur M. "Neurocognitive Poetics: Methods and Models for Investigating the Neuronal and Cognitive-Affective Bases of Literature Reception." *Frontiers in Human Neuroscience* 9 (2015): 186.

Johnson, Gregory. "Theories of Emotion." *The Internet Encyclopaedia of Philosophy*. https://web.archive.org/web/20200604163456/https://iep.utm.edu/emotion/ (9 June 2020).

Johnson, Mark. *The Meaning of the Body: Aesthetics of Human Understanding*. Chicago: University of Chicago Press, 2007.

Johnson, Mark. "Embodied Understanding." *Frontiers in Psychology* 6 (2015): 875.

Johnson, Mark. *Embodied Mind, Meaning, and Reason*. Chicago: Chicago University Press, 2017.

Kraxenberger, Maria, and Winfried Menninghaus. "Affinity for Poetry and Aesthetic Appreciation of Joyful and Sad Poems." *Frontiers in Psychology* 7 (2017): 2051.

Paterson, Mark. *Seeing with the Hands: Blindness, Vision and Touch After Descartes*. Edinburgh: Edinburgh University Press, 2016.

Radman, Zdravko, ed. *The Hand, an Organ of the Mind: What the Manual Tells the Mental*. Cambridge: MIT Press, 2013.

Robinson, Jenefer. *Deeper than Reason: Emotion and its Role in Literature, Music and Art*. Oxford: Clarendon Press, 2005.

Sacré, James, and Lorand Gaspar. *Mouvementé de mots et de couleurs*. Mazères: Le Temps qu'il fait, 2003.

Schaetzl, Randall, and Sharon Anderson. *Soils: Genesis and Geomorphology*. Cambridge: Cambridge University Press, 2012.

Solomon, Robert C., ed. *Thinking About Feeling: Contemporary Philosophers on Emotions*. Oxford: Oxford University Press, 2004.

Tsur, Reuven. *Toward a Theory of Cognitive Poetics*. Eastbourne: Sussex Academic Press, 2008.

Zahavi, Dan. *Self and Other: Exploring Subjectivity, Empathy, and Shame*. Oxford: Oxford University Press, 2014.

Zunshine, Lisa, ed. *The Oxford Handbook of Cognitive Literary Studies*. Oxford: Oxford University Press, 2015.

Renate Brosch

On Being Moved

Body Movements as Prompts for Affect and Empathy in Karen Fowler's *We Are All Completely Beside Ourselves*

Abstract: A fictional rendering of characters' aspects and behaviors can induce experiential and empathetic involvement in readers. It is somewhat surprising that most cognitive literary studies concentrate on mindreading and emotion contagion within the story world and among fictional characters, largely neglecting the responses of the reader. In the following chapter, I concentrate on the latter, asking what in particular in narrative fictions produces a high degree of affective involvement, immersion in the story world and empathy with the characters. As in real life, these responses are to a large extent triggered by the body language of the fictional characters. Besides following the actions and events in a narrative fiction, readers are (often subliminally) aware of the characters' bodily movements, which express states and intentions not verbalized in the text. Movement exerts a more immediate grasp on attention than static objects and encourages an enactive immersion. That is not to say, however, that all sorts of motions in narrative fictions activate the readers' involvement in the same way. While sudden, spontaneous movements can have a surprise effect, interpersonal gestures probably elicit the most visceral emotional response. Given that gestures are not only expressive outcomes of emotion but also "eliciting conditions for emotion" (Hogan 2011, 3), "being moved" by fictions does not just entail imagining spaces and mental focus on the story world, but responding with strong affect to something we know is only invented and imagined. By applying the connection between gestures and empathy found in empirical studies to the experience of reading fiction, I look for narrative prompts of empathy, suggesting that body language can function as a trigger for simulation as well as reflection. The case study is Karen Fowler's *We Are All Completely Beside Ourselves* (*WAACBO*), a narrative that provides startlingly unusual examples of body language. The analysis allows me to suggest that empathy is encouraged on several levels by an interaction of vivid imagery of body movement, embodied simulation and surprising reversals of expectations.

https://doi.org/10.1515/9783110782844-005

1 What is readerly empathy and how is it triggered?

Empathy is distinguished from sympathy by its imitative, vicarious character (Keen 2007, 7). The intersubjective sharing it involves is not only, not even primarily, based on reflection (i.e. what some scholars call cognitivist), it is also based on affect and grounded in embodied interaction. In order to understand the mental states of others, we can rely on our innate sensitivity to bodily movements, gestures and facial expressions (Zlatev et al. 2008, 3). Experimental studies show that the human perceptual system is tuned to extract socially relevant information from the actions of the human body (Wachsmuth et al. 2008, 1). Such kinesic intelligence is a normal part of our human capacity to discern and interpret body movements, body postures, gestures, and facial expressions: "Bodily behaviour, expression, and action are essential to (and not merely contingent vehicles of) some basic forms of consciousness. Mental states do not simply serve to explain behaviour; rather, some mental states are directly apprehended in the bodily expressions of people whose mental states they are" (Gallagher and Zahavi 2008, 148).[1] The accurate detection of another's affective state is important for the establishment of successful social interaction.

According to Suzanne Keen, empathy with others can be produced not only by witnessing but also by hearing or reading of another's condition, so that it can be experienced through literature (Keen 2007, 4). Gesturing is productive of empathy in real world encounters, and we can expect gestures in fiction – where they are much rarer – to be powerful tools with a similar effect. The important point for cognitive literary critics is that fictions, like real-life encounters, can trigger what Dan Zahavi defines as "a kind of direct, noninferential, (quasi-) perceptual awareness of other people's emotions, sensations, and other psychological states" (Zahavi and Overgaard 2012, 16). A premise of this argument is the proximity between real-life and fiction-related experiences, as proposed in Monika Fludernik's idea of experientiality. According to Fludernik, readers attempt "to recognize what they find in the text in terms of the natural telling or experiencing or viewing parameters" (Fludernik 1996, 34). Bolens, who is particularly inter-

1 The expression "direct apprehension" comes from recent studies of intermental processes in the cognitive sciences that discard concepts such as theory of mind or simulation theory. Gallagher and Hutto, for instance, take issue with theory of mind and simulation theory on the grounds that both explanatory paradigms start from the assumption that there is "primary separation" between the self and other minds which the individual must somehow bridge employing mental processes either based on a "theory" or "simulation" which are not required and in any case fallible (Zlatev et al. 2008, 2). See below for a discussion of these different approaches.

ested in body language, notes the interconnectedness of language comprehension with the sensori-motor experiences implied by the text (Bolens 2012, 11).

Emotional information can be extracted automatically and without effort from body language. The posterior region of the superior temporal sulcus (STSp for processing human movement) is thought to play a core role in processing the socially relevant information of human movement. Interestingly, the STSp is extensively interconnected with the amygdala, a centre for emotion processing, and therefore responds more strongly to emotional gestures than to instrumental ones (Shiffrar 2008, 102). According to experimental studies, observers even readily identified emotions when looking at point-light movements (Pollick et al. 2001; cf. Shiffrar 2008, 102).

Instead of making a lot of explicit statements about the mental states of characters, fictional narratives can rely on the reader's kinesic intelligence, deploying embodied prompts to trigger empathetic reactions. Body language may be said to be the nexus of psychosomatic and cognitive reader response, the point where memory, semantic interpretation and psycho-somatic simulation intersect, creating a sensory and cognitive bond between novel and reader. The important function of body language in fictions is due to the remarkable fact that gestures are expressive of emotions as well as "eliciting conditions for emotion" (Hogan 2011, 3).

Because of the intersubjective appeal of literature, the overwhelming majority of instances of gesturing in fiction serves the reader's capacities for imaginative perspective taking and matching of affects with characters, in other words the reader's empathy.[2] Neuroscientific studies show that reading includes "experiences of intense identification and emotion-sharing with bodily effects" (Keen 2013, 2). A moment of introspection will confirm this idea: gestures in fiction often trigger a strong impulse to imitate them even though one is sitting on the sofa all by oneself. Hence, there is double experientiality in fictional body language – an embodied and an empathetic response. Because of this dual influence, episodes involving affecting body language tend to remain in memory. In the following, I discuss both reactions, first looking at the revelatory function of movements and then at the eliciting conditions for empathy in communicative gestures.

In an article on reader responses, Naomi Rokotnitz identifies ways by which literary description can intensify experiential involvement in the reading process. On the one hand, what is termed "simulation" is a (quasi-) automatic imitative action by the reader triggered by mirror neurons. On the other, body language can mobilize what neuropsychologists call our "affective consciousness", i.e. a

2 See below for an important distinction between higher and lower empathy levels.

form of pre-reflective emotional reception that arises from bodily experience (Rokotnitz 2017, 274). As Rokotnitz argues, especially powerful embodied prompts in a narrative text work through sensori-motor stimuli that co-opt readers' affective capacities (Rokotnitz 2017, 276). She mentions image clusters around the twins' kinesthetic telepathy in Arundhati Roy's *The God of Small Things* as an example. To develop this idea further, I suggest that the effect is especially powerful when the narrative passage contains certain specific forms of body movements.

It is the experientiality of physical motion that vouchsafes the authenticity of the experience for the reader. The body can rarely fake an emotional outcome and its signs can therefore have a revelatory function, they can expose an inner truth that the narration or the character has kept hidden. Often, the gesturing body is played out as more honest than the character's speech. Intense reading experiences result when such revelations through the body startle the readers into realizing something of which they were previously unaware. These instances in the process of reading "move" the reader in a double sense: in sensori-motor simulation of the action and in an emotional reaction. If, on top of these reactions, an insight into inner motivations or broadened comprehension results, then it is likely that we have a memorable description of body language at hand with empathetic effect, often leading to further interpretive reflection.

2 An example from fiction for embodied and emotive responses

Karen Fowler's novel *We Are All Completely Beside Ourselves (WAACBO)* contains a passage so startling as to remain unforgettable in the reading experience. The I-narrator is Rosemary, a young woman who has severed connections with her family. When she meets her brother again, whom she has not seen for years, he says: "You're all grown up [. . .] I didn't recognize you at all until you climbed onto the table" (Fowler 2013, 171). A young woman, a student, jumps up onto a table in a pub? The remark about her surprising behaviour suddenly captures the essence of the novel's singularity in an iconic moment.

Fowler's novel of 2013 was shortlisted for the Booker Prize 2014 and became an international bestseller. My remarks are unavoidable spoilers, because the extraordinary achievement of the novel rests on a twist so unusual that it is almost unbelievable that it can be pulled off. Until a third of the way into the story (Fowler 2013, 77), the novel makes us believe that the traumatic split-up of Rosemary and her family follows her parents' giving away her sister Fern. But before we can start to wonder at this lack of parental love, it turns out that Fern was not a sister but a chimpanzee. This dramatic twist requires a major adjustment of the primacy effect which the conventional (disrupted) family narrative had previously established in the reader's mind.

The brilliant achievement of hiding the fact that Fern is an ape rests on the construction of the unreliable first-person narrator Rosemary, who immerses us completely in her own experiential world in flashbacks to her childhood and in her current adventures as an unhappy young adult. *WAACBO* resonates with Rosemary's grief for her missing alter ego and "sister," and for her adored brother Lowell; for both of these departures she feels at least partly responsible. She tells the story of her student life and her estrangement from her family, and especially her anger at her parents for deliberately causing the absence of her "sister." The parents, experimental zoologists both, have raised their two children, the narrator and her brother Lowell, together with Fern to study how far a primate could be taught language and social behavior. In the middle of the twentieth century, there were a number of these experiments, aiming to prove the influence of socialization and education. Fowler's novel cites the first such experiment in 1931 by Winthrop and Luella Kellog, who brought up their son Donald and the chimp Gua in exactly the same way. This experiment was terminated in 1932, probably due to Donald's language retardation. While Gua showed no signs of learning human language, her "brother" Donald had begun imitating Gua's chimp noises. In contrast to earlier experiments like this, the novel presents a case for animal rights and respect for creatures, relying on more recent studies, in which direct chimpanzee-child comparisons have been made that delineate a growing set of commonalities between the phenomena of social learning and culture in the lives of chimpanzees and humans, especially the fact that primates can pass on cultural learning to others (Whiten 2017, 129).[3]

In spite of this humanitarian agenda, the story does not put the animal's experience center stage. Of course, the chimpanzee and its adaptation to the human environment is part of the story: we learn, for instance, that Fern masters 200 words in sign language. But since the story's present is years after the family's life with Fern, descriptions of her animal behavior occur only in Rosemary's episodic memories, framed by her narrativizing mental time travel. This focus avoids any sentimental anthropomorphizing of the cute chimpanzee Fern, a strategy that deliberately forfeits an easy road to the reader's affective response. *WAACBO* is not a book for sensitive animal lovers; it is profoundly disturbing in its investigation what it means to be human and to be humane.

---

3 A number of recent experimental projects and studies are mentioned in an Appendix to the novel (Fowler 2013, 321–323).

Starting with the leap onto a table, the novel renders increasingly unsettling depictions of the confusion both Rosemary and Fern experience with regard to their own identity: Tests conducted with Fern in the family home involve making her judge sameness and difference by assigning blue and red poker chips to picture pairs. She can correctly identify most objects, except herself and her "sister" Rosemary; a pair she would consistently sign as "same." "Most home-raised chimps, when asked to sort photographs into piles of chimps and humans, make only the one mistake of putting their own picture into the human pile. This is exactly what Fern did," Rosemary recalls. "What seems not to have been anticipated was my own confusion" (Fowler 2013, 101). So while her father's research measures the effect on the ape of being part of a human family, Rosemary's life becomes for us readers a study of another kind, namely an opportunity to gauge both the haunting lack of distance between humans and primates and their frightening disparity.

Rosemary's confusion is what really draws us into the fictional world, especially the flashbacks to her childhood in which we learn about her rivalry with Fern for parental affection and the competition in skills between them. In the fictional present, the narrator's conversational voice tells the story of her own adventures and misdemeanors as a student, gradually leading us to discover her startling adaptation to and imitation of her "sibling." It is Rosemary's problems as a young adult – informed by her "simian" past – that shape the story world in which we as readers become immersed as it slowly emerges that the girl narrator has been influenced in her corporeality and bodily behavior by her "sister" at least as much as the other way around. This makes her an unreliable narrator, firstly, in so far as she does not fully realize and cannot completely control her spontaneous "animal" behavior. At the same time, secondly, she has been at pains for some time to erase the traces of her early socialization with an ape. The brief remark about her jumping onto a table is later picked up again when Rosemary remembers her mother's admonitions for school: "No[t] jumping on the tables and desks" (Fowler 2013, 102). She even promises the reader, "I'll be on my best behaviour. I won't climb on tables or jump on couches" (Fowler 2013, 298). But the first time the action is mentioned, it has a surprising, relevatory function, instigating a sea-change in our imaginative configuration of the story world. It might be said to constitute a second twist to the narrative, as we now fully realize that it is the similarity of the human to the animal that takes center stage, not the other way round.

3 Perception of body movements and cognition

In real life experience, the world makes itself available to the perceiver through physical movement and interaction; movement is central to perception, cognition, experience and subjectivity (Noë 2006, 1). "Simulation of movement [. . .] is the gateway to human social understanding" (Bolens 2012, 41). According to Goodale and Milner, there are two visual pathways specialized for different modes of vision and mental imagining – the dorsal and the ventral – responsible for vision-for-action and vision-for-perception respectively (Goodale and Milner 2004, 100). The terms refer to two types of visual processing in the brain: vision-for-perception looks to obtain information about objects, vision-for-action is needed to guide and discern movements.[4] While the dorsal stream computes visual input fluently into motor actions, the ventral stream is slower, more enduring and concentrated on object recognition (Clark 2008, 3–4). The ventral stream concerned with object identification may be referred to as the what-pathway, whereas the dorsal stream for movement-vision may be thought of as the where-pathway (Zeman 2002, 195). Both optical systems work together in the production of purposive behavior – the one selecting an object and the other carrying out goal-directed movement (Goodale and Milner 2004, 100). What is important in the present context is that mental imagery respects the distinction made above between spatial and object visualization respectively, as Maria Kozhevnikov's experiments have proved (Kozhevnikov et al. 2010, 29).

Before I go on to analyze the novel, some further clarifications regarding mental imagery while reading are necessary. In another article that addresses the phenomenology of reading, I examine different types of mental processing during an engrossed reading of fictional narratives, in particular what triggers especially vivid mental images (Brosch 2017). During an extended reading process, visualizations reach varying levels of intensity which could be located on a scalar continuum ranging from an automatic, but indistinct "default visualization" to intensive, highlighted and vivid images occurring intermittently. The mental activity accompanying reading that is commonly referred to as "imagining" consists of largely automatic low-level visualizations generated effortlessly when familiar embodied experiences and cultural schemata are evoked. This default mode alternates at significant stages in the narrative with highlighted moments of intensified visualization, directing the reader towards certain elements that command special attention (Brosch 2017, 161). More vivid mental images depend to a signifi-

4 The idea that the processing is done by anatomically different brain networks, the dorsal and the ventral stream, has sometimes been challenged.

cant extent on textual triggers in the narrative, and they tend to remain in memory longer than the rest of the text (Brosch 2018b, 139). The intense effect can be enhanced by a highly subjective, emotional rendering that transmits itself to the reader's empathy. Intensified visualization serves not only to enliven our experience of a narrative but also to bring us emotionally closer to the characters.

It seems that our brains are hard-wired to preferentially notice action, i.e. vision-for-action supersedes vision-for-perception. The enormous and direct impact of motion compared to static objects is not surprising, given the role the visual field plays in warning of potential dangers. When sudden movements occur, alerting mechanisms incontrovertibly notify the perceiver, so that normal cognitive functioning is interrupted (Myin and O'Regan 2009, 196). The capacity of sensory channels to surprise and direct thinking is a significant difference in presence effects between sensations and mental processes like thoughts and imagery (Myin and O'Regan 2009, 196). This is why fictional body movements trigger a more direct, automatic motor resonance than the depiction of thoughts; body movements in fictional characters co-opt the readers' bodies via their simulating minds. Drawing on Maggie Shiffrar's empirical study establishing the crucial role of *movement* in body language, I suggest that "dynamic body language," i.e. whole body actions as opposed to "static body postures," is the most effective appeal to the reader's motor resonance and enactive response (Shiffrar 2008, 96).

There is, however, a special difficulty about this particular scene in the novel. On the one hand, the surprisingly agile jumping action of the narrator possesses a special appeal to visualization in its dynamism. On the other, the image of Rosemary climbing onto a table in an unguarded moment is just too weird in an ordinary social context to encourage enactive mental simulation. I doubt therefore that the passage can induce seamless immersion in the sense of Anezka Kuzmičová's claim that presence effects arise primarily when the reader experiences the phenomenon of motor simulation (Kuzmičová 2012, 29).[5] Shiffrar's research tells us that the perception of movement is optimized when there is a match between what is observed and what the observer can actively produce, i.e. mental imaging

5 Anezka Kuzmičová claims that the most stimulating movements in fiction "ought to be volitional transitive movements, particularly when object directed" (Kuzmičová 2012, 28), as opposed to self or other-directed. She puts forward the argument that volitional motion entails particular attentional focus on the environment interacted with, which is absent from unintended movements (Kuzmičová 2012, 28). This is why Kuzmičová holds that dynamic movements towards objects of the quotidian environment are preferentially simulated by readers from a first-person perspective and thus responsible for enactment (Kuzmičová 2014, 275–276). But that is surely not the case: movements like taking a cup from a table are too ordinary to merit a distinct and vivid imagination.

data show clearer visualizations when, for instance, ballet dancers view dances they can perform themselves (Shiffrar 2008, 100). Mental enactment seems to depend on one's own bodily experiences; there is a greater sensitivity to actions one has performed oneself (Shiffrar 2008, 102).

Since jumping onto tables will not be among the familiar actions performed habitually by readers of Fowler's novel, the readers' sensori-motor resonance will be triggered without yet translating into an enactive emotional understanding. In contrast to empirical research that shows how bodily motion alone in the absence of facial and auditory cues is sufficient for the perception of an emotional state (Oberzaucher and Grammer 2008, 162), this scene does not immediately disclose what feelings and motivations are involved. The strong impact that the reader feels in response to the sentence is largely due to the surprise and shock at the information delivered; it is the first instance in which we realize that our narrator is not quite the same as us.

4 Scenes in the novel prompting vivid imagining

In my own reading experience, the jumping image was nevertheless vivid and lastingly memorable. The reason, I suspect, is that it encapsulates in visual and corporeal form Rosemary's extraordinary socialization and makes powerfully experiential what we have so far only been told in propositional terms (Fowler 2013, 77). The image revises the primacy effect of our assumption of 'normality' for Rosemary by filling in the gaps of her hitherto rather disembodied narration. It contains in compressed ideational form the theme and affective content of the novel. Rosemary's action is the first instance of a series of subtle hints and indications pertaining to the remarkable extent to which the narrator's socialization with a chimp "sibling" has made her imitate chimpanzee behavior. The scene's vividness might even have increased over reading time and in retrospect, given how significant it is in the overall trajectory of the plot and the themes of the novel.

What has impeded our attempts at meaning making is Rosemary's strenuous effort to suppress her spontaneous movements as well as any spontaneous utterance. That she was raised by her scientist parents in company with a "sibling" chimpanzee has seen sensational exposure in the media and caused her a lot of distress. Shame prevents her from acting out a corporeality that is natural to her but beyond weird for others. As the I-narrator of her own story, she is at pains to conceal such impulses, so that we discover only quite late in the novel how far she was influenced by the beloved animal companion of her childhood. Yet, a small number of corporeal signs and incidents hint at her unusual habits. Gradually, after the discussed incident we become more and more aware of the strange behavior and body language performed by the narrator involuntarily, even though mention of them is extremely rare. With comical effect she recalls her childhood antics, "Mom warns me to stay upright. No loping through the snow on my hands and feet" (Fowler 2013, 96). Since Rosemary takes her socialization for granted, she is not fully aware of the extent of deviation in her behavior from that of people raised in normal human environments, but she is painfully conscious of being branded and ridiculed as "monkey girl," a nickname that dogs her school years. Among the rules her mother tries to teach her before she goes to school is: "Not putting my fingers into anyone else's mouth or hair" (Fowler 2013, 102). Even now, in the narrative present when she is at college, she senses that she puts people off with her odd behavior and is keen to shake off her off-kilter body language (Jensen 2014, n.p.). The omissions in her narration are informed by the "deep-rooted identity dysmorphia" setting her apart from her fellow students (Jensen 2014, n.p.). Given her repression, it is cognitively realistic that she herself does not first notice these instances of her own irregular behavior. Rather, she is told by other people that she is behaving strangely: for example, standing too close, intruding on others' personal space (Fowler 2013, 169). In another instance, when she is extremely excited, she notices that the students around her "turned to look" and becomes aware that she had been "wheezing or hissing or panting or something" (Fowler 2013, 148).

These eerie and haunting episodes testify to an expressiveness of the body that is difficult to suppress or hide. In light of the lack of explanatory psychological commentary that is common in novels nowadays, body language has a major function to perform. The genuineness and sincerity of involuntary gestures and the effortless detection of their truth value make this component of fictions a significant factor for eliciting meaning that would otherwise not be available. Especially for an unreliable narrator who cannot be trusted in her evaluation of social interactions, gestures and movements provide a semantic system that ensures a corrective to erroneous views. Contrary to common assessments of first-person narration, a text with an I-character as the deictic centre may also distance the reader. Emmott and Sanford found that extremely personalized focalizers can curb the readers' identification with their perspectives (Emmott and Sanford 2012, 167). They suggest that readers may visualize but at the same not be persuaded by the images generated. In general, unreliable and limited perspectives invite doubt and distrust and prompt readers to counterfocalize. Such counterfocalization is not a mysterious capacity; it is employed in everyday life when we shift perspectives or "read between the lines" of someone's statements, recognizing facets which the person has omitted or downplayed (Emmott and Sanford

2012, 167). Unreliable narration is a standard literary device for provoking the capacity of readers to gather information which the speakers did not intend to divulge. Such participatory strategies are likely to be remembered. When readers become alert to the unreliability of a narrative voice and view, they pay increased attention to other sources of information in the text such as body language and movements. In this case, the oddness of the information provided by the body will cause some resistance to its meaning.

5 Body language and empathy

This reaction is counterbalanced by an extremely knowledgeable discourse by the present-day student Rosemary on zoology and psychology. Indeed, she demonstrates a cutting-edge knowledge in these fields, providing readers with neuropsychological comments on the issues the novel raises. In one of her theory-informed asides, Rosemary states,

> Empathy is natural human behaviour, and natural to chimps as well. When we see someone hurt, our brains respond to some extent as if we'd been hurt ourselves. This response is not located merely in the amygdala, where emotional memories are stored, but also in those regions of the cortex responsible for analysing the behaviour of others. We access our own experiences with pain and extend them to the current sufferer. We're nice that way. (Fowler 2013, 169–170)

These remarks would not be remarkable coming from an authorial narrator. Yet, they are delivered by a personalized narrator whose peculiarities demand of us that we extend our ability to empathize beyond the human, for the effect on Rosemary of spending her first five years with Fern is that she becomes "other," mirroring the chimp's behavior. But she herself is our guide in this since she has learned to see the world at least partly from the point of view of an ape. Many passages depict Rosemary's "mindreading" of Fern, which she translates into words for her parents. For instance, when Fern's finger scratches a picture of a kitten in an illustrated book, she announces, "Fern wants a kitten" (Fowler 2013, 100). As it turns out, however, empathy or mindreading is not as simple or as faultless as Rosemary assumes. In contrast to the narrator as well as some theorists, the novel seems not to endorse (or rather not to favor exclusively) a simulation-based concept of empathy.

Many critics perceive a pre-conscious embodied response in the reader as the high road to empathy. For Marco Caracciolo, who differentiates between "consciousness-attribution" as an external, evaluative, third-person stance towards a character and "consciousness-enactment," the latter is more closely related to empathy (Caracciolo 2014, 113). He defines consciousness-enactment as a complex simulation of a character's state of mind on the basis of emotion contagion, harnessing empathy abilities (Caracciolo 2014, 118).[6] Anezka Kuzmičová, similarly, regards enactment as a vicarious experience in which the reader adopts the character perspective and his/her sensori-motor experience closely (Kuzmičová 2014, 282). Prompted by an imagery that fuses many different sensori-motor modalities, an enactive response is thought to occur spontaneously and seemingly without much cognitive effort. In these cases, the reader-imager comes as "close as one possibly can to forgetting that the experience was in fact mediated by a string of words on a page" (Kuzmičová 2014, 283). Both of these theoretical arguments privilege an automatic and unconscious emotional contagion stemming from embodied simulation as the most intense as well as the most aesthetically rewarding reading experience. I agree, that is one option of responses to narratives, but probably not the most rewarding one.

Like the narrator of *WAACBO* in the quote above, the two scholars appear to subscribe to the so-called perception-action model of empathy, according to which the perception of a target's state activates the observer's own mental representations of that state. In this model, consistent with simulation theory, empathy is viewed as a non-conscious process in which one comes to feel the emotions of another (Rameson and Lieberman 2009, 96). Empathy based on theory of mind (ToM), by contrast, is a more reflective response. It requires the understanding that others have an internal mental landscape that differs from our own (Rameson and Lieberman 2009, 98). Empirical evidence for the less reflective form of empathy is more robust, grounded in the discovery of mirror neurons. When someone observes the facial expressions and bodily movements of someone else, brain scans show that the same cells are activated in the observer as in the target (Coplan and Goldie 2011, xxxvi). A higher-order empathy is an imaginative process that involves emotions but crucially requires conscious reflection, too. According to Alvin Goldman, such "reconstructive empathy" is an effortful process that takes place in stages and depends on shifting to the perspective of another (Goldman 2011, 31).

Some critics credit reading fictional literature with an improvement of empathy and mindreading. Suzanne Keen, for instance, claims that reading produces empathy for fictional characters and can even lead to pro-social behavior or altruism (Keen 2013, 2). Such an educative effect on the ability to compassionately and morally extend ourselves beyond egocentrism is surely a very different pro-

6 Marco Caracciolo and Anezka Kuzmičová define the terms somewhat differently (cf. Caracciolo 2014, 112–113; Kuzmičová 2014, 282–283).

cess from an autonomic somatic mimicry of an observed or imagined experience. Obviously, the low-level empathetic response has little correlation with ethical behavior. An evocation of high-level empathy depends not just on embodied and pre-conscious simulation, but on attentive attunement and reflection. Moral and ethical judgments require more input than mere "emotion contagion" (McFee 2011, 187).

The novel's narrator seems to lean on the side of an overestimation of interpersonal understanding, when she states:

> The flip side of solipsism is called theory of mind. Theory of mind postulates that, even though these cannot be directly observed, we readily impute mental states to others (and also to ourselves, since the bedrock proposal is that we understand our own mental states well enough to generalize from them). And so we constantly infer someone else's intentions, thoughts, knowledge, lack of knowledge, doubts, desires, beliefs, guesses, promises, preferences, purposes, and many, many more things in order to behave as social creatures in the world. (Fowler 2013, 186–187)

This simple summary conveniently conflates diverging views of the capacity to read other minds. According to the adherents of ToM, understanding someone else's state of mind requires a basic understanding that the other person has beliefs, thoughts, knowledge and feelings distinct from our own. Whereas ToM involves a certain amount of cognitive detachment, the kind of mindreading assumed by simulation theorists does not. The latter posit that, in order to understand the experience of another, there is a way in which one can jump into that situation and experience it as though it were happening to oneself, in a very spontaneous way. Spontaneity contradicts the term "theory" in this context, because no theoretical abstraction is needed. "Without an understanding of what people think, what they want and what they believe about the world, it is impossible to operate in any society. [. . .] It is the basic necessity of humanity [. . .]. We interpret people's actions using words that describe their mental states so often that we cease to think what we are actually doing" (O'Connell 1998, 2–3). On this view, the response is automatic, triggered by mirror neurons, and it evolved because it was essential to human survival in the distant past. A third, enactivist view posits that the minds of others are understood primarily through our interactive relations and our ability to directly perceive their intentions and emotions. In this conceptualization of intermental understanding we need neither ToM nor simulation for attributing meaning to other people's actions, since the intended meaning is supposed to be immediately apparent to perception (cf. Zlatev et al. 2008, 2). Such a rapid assessment may be possible in day-to-day encounters, but a further effort, espousing language and cultural learning, is needed to infer all the things mentioned by the novel's narrator: "someone else's intentions, thoughts, knowledge, lack of knowledge, doubts, desires, beliefs, guesses, promises, preferences, purposes" (Fowler 2013, 187).

Of course, readers can experience both types of empathetic reactions, depending on the textual cues for a more identificatory or more detached mode of reading. As I understand it, a simulation-based embodied and enactive reading of fictional characters can lead to empathetic concern and activate lower-order empathy. But concern or sympathy seems a rudimentary response when compared to the complex cognitive processes required for moral considerations in an interpretive empathetic response. When reading fictions, readers alternate between non-conscious, sub-verbal processes and more conscious critical experiences in reading, and between an automatic, affective experience that feels unmediated and a reflective processing that is a more controlled cognitive process. What is interesting in analyzing fictions is the potential of the narrative text to vary or combine these empathetic responses by triggering either experiential or propositional modes of processing.

6 Beyond mental imagining – inducing reflective empathy

In the end, *WAACBO* complicates these alternatives by offering another image of action that severely tests the readers' empathetic dispositions that have been evoked so far. Rosemary's traumatic experience of her family break-up has caused – as in many children of divorced parents – an intense feeling of guilt. She blames herself for what happened to Fern, recalling one particular episode in which she complained to her parents that Fern scared her. As it turns out in the end, this is a self-delusion, her memory is completely at odds with her mother's account. For the reader, who is not treated to this clarification until the end of the novel, the recalled episode is truly scary, because in it the funny little chimp transforms into a dangerous predator. Referring to it as a Freudian "screen memory" (Fowler 2013, 247), Rosemary recounts how she gave one of their cat's newborn kittens to Fern.

> What happened next happened fast. The mother cat was hissing and spitting. The gray kitten in Fern's hand was crying loudly. The mother struck at Fern with her claws. And Fern swung the tiny perfect creature against a tree trunk. He dangled silently from her hand, his mouth loose. She opened him with her fingers like a purse. (Fowler 2013, 250)

The animal's body movements have a revelatory function similar in impact to the earlier table-jumping image. In terms of intensifying response, the scene has the same strong effect of dynamism and body movement that the earlier description of Rosemary's behavior had. Like the earlier scene it causes a reversal of previous assumptions, this time concerning the animal, since Fern suddenly seems to mutate into an alien and dangerous creature. While the earlier scene illustrates our human proximity to animals, this action shocks us with completely senseless, even playful cruelty and hence distances us from the animal. The centrality of movement for the reader's imaginative re-centering and perspective taking that was delineated above is superseded here by a shocked recognition of another side to the story that we have been led to ignore so far. The scene may be said to serve as a warning not to overestimate the power of affective and sensori-motor responses. In contrast to some critics' belief in the beneficial effects of reading literature as a training ground for empathy and altruism (Oatley 2012, 153; Hogan 2011, 30; Keen 2013, 2), the episode seems to illuminate the limits of empathic concern and idealistic sympathy with animals.

In spite of these cautionary notes, however, Fowler's novel as a whole remains ambivalent on the question. *WAACBO's* beautiful achievement is to combine triggers of both spontaneous affect and higher-order empathy by describing body language that appeals to automatic as well as more reflective responses. Among the great number of ways bodies express psychological information, investigators typically distinguish types of gestures according to content and degree of interaction: representational gestures which depict a concrete or abstract concept with the shape or motion of the hands (iconic gestures and metaphoric gestures in McNeill 1992), deictic gestures which point to a referent in physical or imaginary space (concrete or abstract deictic gestures in McNeill 1992), and interactive gestures which connect the speaker and listener (Bavelas et al. 1992).

These categories carry different weight in narrative fictions because instrumental gestures are rarer in fictions than emotional ones, so that the intrapsychological aspect is vastly more important. For response studies, it is notable that all gestures function primarily as decodable signs for the reader, relying on the reader's automatic ability to comprehend body language. I have the (unproven) impression that the majority of instances of gesturing in contemporary fiction consists of spontaneous non-voluntary expressions of emotions, the automatic manifestation of internal states, not the conscious performance of rhetorical or indexical gestures, because the former reveal surplus information. A diachronic survey of gesture description would perhaps show that the frequency and saliency of gestures have increased in fiction as psychological commentary has decreased. Concurring with cognitive science, recent narrative fictions seem to have abandoned the "subjectivity first" model of inner consciousness in favour of externalized character description and inferential emotions via body language. A preference for techniques that gratify their powers of inference may be detected in readers, too, who may like narratives to pay tribute to their adeptness at recognizing affect and cognitive processes.

In a recent empirical study, Mingyuan Chu and her collaborators investigated whether different types of gestures serve different functions, and therefore may be related to different social and cognitive skills. They tested a large, heterogeneous sample of speakers from a wide academic spectrum to discover correlations to the individuals' cognitive abilities and empathy levels (Chu et al. 2014, n.p.).[7] In sum, the experiments proved a positive correlation between gestures and empathy. Chu et al. were able to demonstrate that empathy was the one variable that predicted the saliency of gestures. Speakers with higher levels of empathy produced all types of gestures more saliently and conspicuously than speakers with lower levels of empathy. With more salient gestures, speakers probably wanted to increase the chance that the listener would understand the information expressed in the gesture.[8] This link between empathy and gestures confirms earlier studies showing that individuals with lower levels of empathy produce fewer gestures than controls who gesture more frequently and/or more saliently (Baron-Cohen and Wheelwright 2004). Moreover, this study yielded further specific findings: It also showed a positive relation between empathy and the frequency of *interactive* gestures, gestures used to support communication. This supports the assumption that speakers who care more about the quality of their communication will use more interactive gestures than speakers who care less. Caring about the quality of communication is, of course, related to *empathy*, as empathic individuals are more motivated to create a situation of mutual comprehension.

Similarly, in fictions, interpersonal gestures resonate more with readers' primary interest in characters and their relations. Applying the empirical findings to the reading experience suggests that interpersonal movements and interactive gestures are among the most stimulating elements in narratives. This is not sur-

7 The investigators distinguished representational, conduit gestures and *palm-revealing gestures.* Conduit gestures are similar to representational gestures in that they are used to present the speaker's idea to the listener, but conduit gestures additionally manage interaction by bringing the listener into the conversation. Palm-revealing seems to be a limited gesture indicating uncertainty or indifference (cf. Bavelas et al. 1992) by showing an empty hand.

8 The study also showed a positive relation between empathy and the frequency of *all* except representational gestures, in other words those that depict a concrete or abstract concept and are therefore always speech concordant. This finding supports assumptions about the different function of gestures in fiction. Since literature is able to present maximally successful verbal communication, spontaneous speech-accompanying gestures arising from a cognitive challenge in speech, memory or communication are rarely needed in fictional dialogue, but gestures that replace or contradict speech are a common device in fiction.

prising since interactive gestures express relatedness, and the desire for relatedness or intersubjectivity appears to be innate. The human mind is inherently equipped with motives for sympathy in purposes, interests, and feelings, and in need of dialogic, interpersonal awareness, cooperative action in society and intermental engagement with other minds (Trevarthen 2001, 417). Assuring one another of a shared commitment to interrelation can be a profound occasion for happiness. *WAACBO* goes even further by positing instances of *interspecies* understanding that are able to confound conventional dichotomies of human and animal "consciousness."

An interactive gesture – between the human and the non-human agent – concludes the novel with a powerful and disturbing image that provides an embodied anchor for both affect and empathy. When Rosemary has found the laboratory where Fern is being held in a cage, she plans her visit carefully, taking with her a favourite toy and one of the red chips with which Fern used to mark sameness.

> I signed her name and then our sign for Rosemary. I pressed my palm, poker chip in the middle, against the bullet proof glass. Fern stood heavily and came to me. She placed her own large hand opposite mine, fingers curling slightly, scratching, as if she could reach through and take the poker chip. I signed my name again . . . Then she rested her forehead on the glass. I did the same and we stood that way for a very long time, face-to-face. From that vantage point, I could only see in teary, floating pieces –
>
> her eyes
> the flaring of her nostrils
> the sparse hair on her chin and rimming her ears
> the tiny rise and fall of her rounded shoulders
> the way her breath painted and unpainted the glass. (Fowler 2013, 308)

As Rosemary enumerates her fragmented perceptions of Fern's face when they lean their foreheads on the glass together, we readers are given another instance of body language that speaks to our inherent interpersonal dispositions. We can vividly imagine the closeness and the separation of the two creatures through the vivid image of the breath fogging the glass. The body language is minimal yet significant. Rosemary then concludes: "I didn't know what she was thinking or feeling. Her body had become unfamiliar to me. And yet, at the very same time, I recognized everything about her. My sister, Fern. In the whole wide world, my only red poker chip. As if I were looking in a mirror" (Fowler 2013, 308).

As the narrator denies the reach of empathy into overt disclosure of thoughts and feelings of the other, she simultaneously affirms the proximity of human being and creature in a common inclination to relate to and imitate actions observed in others. Fern's fate has become doubly distressing as we observe that she identifies with her human environment. The scene possesses the appeal of movement and of interactive gestures. Through the body language, the passage triggers our involuntary embodied responses, engages our affective consciousness and evokes empathetic concern. The affect generated by it enhances the readers' immersion in the story world and their pre-conscious and subliminal bodily resonance. What is especially touching in this iconic anchor that concludes the novel is the natural and automatic use of gestures as a means of communicating and as an expression of relatedness. Even more poignant in expressing longing and desire is the futile attempt to achieve something like tactile stimulation ("scratching"). Interpersonal touch is a well-known source of comfort and emotional wellbeing. Its frustration at the moment of reconciliation of the two "sisters" must be felt by readers in the very fibres of their being.

By encouraging imaginative simulation, these interpersonal body movements – between human and animal – render abstract concepts of relatedness physically tangible, and hence induce more intuitive means of understanding. The reader's embodied, affective responses increase the visceral effect and enhance the experience of reading. But as argued above, for an ethical engagement and higher order empathetic understanding, a more conscious effort is required (Brosch 2018a, 183). For reading fictions, the theory that the empathizer matches the emotional state of characters falls short of what is distinctive about higher-level empathy, namely that it is an observer emotion that confronts one with an experience that is precisely not one's own (Zahavi and Michael 2016, 7; Zahavi 2014, 138). It is in literary fictions specifically that empathy preserves interpersonal difference and even highlights other-centeredness. Whereas in real life, empathy is often biased and partial, so that we tend to empathize more with those who are similar and close to ourselves (Zahavi and Michael 2016, 1), in reading a kind of empathy can be trained that is not so common in real life, which reaches towards a sense of justice or moral obligation. This novel follows a progression of the main character from an instinctive and unconscious simulation to a more complex and reflective engagement with another's consciousness. This progression from affective response to rational reflection and moral responsibility is potentially available for the reader to emulate.

As readers of *WAACBO* we may be relatively challenged to empathize with a non-human creature, but we may be equally challenged by the extent of Rosemary's adaptation to animal behavior. It brings home to the reader the need to extend sympathy to someone "who creep[s] you out a little bit" (Fowler 2013, 298) and serves as a reminder of the animal that resides within each of us, however much we might try to forget or dispute it (Seymenliyska 2014, n.p.). As one reviewer puts it, Fowler's novel examines what it means to be human, nonhuman and something in between (Seymenliyska 2014, n.p.). Thus the novel provokes us to think about many aspects of our relationship to animals that most of us would rather ignore. "[T]he novel's real triumph is that the repercussions of the Cooke parents' broken pact with their chimpanzee child extend, by association, to the family of man and its dysfunctional relationship with the entire zoosphere" (Jensen 2014, n.p.). In opening out this wider perspective, the novel belongs to a new kind of narrative that reintegrates the human self into a much larger community than that recognized by dominant scientific, social, legal, and moral norms. These stories in which selfhood extends beyond the human prefigure ways of living on which the continued survival of the earth's entire biotic community arguably depends (Herman 2014, n.p.).

Reference list

Baron-Cohen, Simon, and Sally Wheelwright. "The Empathy Quotient: An Investigation of Adults with Asperger Syndrome or High Functioning Autism, and Normal Sex Differences." *Journal of Autism and Developmental Disorders* 34 (2004): 163–175. doi:10.1023/B:JADD.0000022607.19833.00.

Bavelas, Janet Beavin, Nicole Chovil, Douglas Lawrie, and Allan Wade. "Interactive Gestures." *Discourse Processes* 15.4 (1992): 469–489.

Bolens, Guillemette. *The Style of Gestures: Embodiment and Cognition in Literary Narrative*. Baltimore: Johns Hopkins, 2012.

Brosch, Renate. "Experiencing Narratives: Default and Vivid Modes of Visualization." *Situated Cognition and the Study of Culture*. Eds. Ben Morgan, Sowon Park, and Ellen Spolsky. Special Issue *Poetics Today* 38.2 (2017): 255–272.

Brosch, Renate. "Reading Through the Body: The Damaged Mind in Tom McCarthy's *Remainder*." *The Wounded Hero in Contemporary Fiction: A Paradoxical Quest*. Eds. Susana Onega and Jean-Michel Ganteau. London: Routledge, 2018a. 178–200.

Brosch, Renate. "What We 'See' When We Read: Visualization and Vividness in Reading Fictional Narratives." *The Eye's Mind: Visual Imagination, Neuroscience and the Humanities*. Eds. Adam Zeman, Matthew MacKisack, and John Onians. Special Issue *Cortex* 105 (2018b): 135–143. doi:10.1016/j.cortex.2017.08.020.

Caracciolo, Marco. *The Experientiality of Narrative: An Enactivist Approach*. Berlin: De Gruyter, 2014.

Chu, Mingyuan, Antje Meyer, Lucy Foulkes, and Sotaro Kita. "Individual Differences in Frequency and Saliency of Speech-Accompanying Gestures: The Role of Cognitive Abilities and Empathy." *Journal of Experimental Psychology: General* 143.2 (2014): 694–709. doi:10.1037/a0033861.

Clark, Andy. "Perception, Action, and Experience: Unravelling the Golden Braid." *Perception and Action*. Eds. Robert McIntosh and Thomas Schenk. Special Issue *Neuropsychologia* 47.6 (2008): 1460–1468, online 1–54. doi:10.1016/j.neuropsychologia.2008.10.020.

Coplan, Amy, and Peter Goldie. "Introduction." *Empathy: Philosophical and Psychological Perspectives*. Eds. Amy Coplan and Peter Goldie. Oxford: Oxford University Press, 2011. ix–xlvii.

Emmott, Catherine, and Anthony J. Sanford. *Mind, Brain and Narrative*. Cambridge: Cambridge University Press, 2012.

Fludernik, Monika. *Towards a "Natural" Narratology*. London and New York: Routledge, 1996.

Fowler, Karen. *We Are All Completely Beside Ourselves*. London: Profile Books, 2013.

Gallagher, Shaun, and Dan Zahavi. *The Phenomenological Mind: An Introduction to Philosophy of Mind and Cognitive Science*. London and New York: Routledge, 2008.
Goldman, Alvin I. "Two Routes to Empathy: Insights from Cognitive Neuroscience." *Empathy: Philosophical and Psychological Perspectives*. Eds. Amy Coplan and Peter Goldie. Oxford: Oxford University Press, 2011. 31–44.
Goodale, Melvyn A., and A. David Milner. *Sight Unseen: An Exploration of Conscious and Unconscious Vision*. Oxford: Oxford University Press, 2004.
Herman, David. "Narratology beyond the Human." *Diegesis* 3.2 (2014): 131–143. https://www.diegesis.uni-wuppertal.de/index.php/diegesis/article/view/165/229.
Hogan, Patrick Colm. *Affective Narratology: The Emotional Structure of Stories*. Lincoln: University of Nebraska Press, 2011.
Jensen, Liz. "We Are All Completely Beside Ourselves review – 'A provocative take on family love'." *The Guardian* (20 March 2014). https://www.theguardian.com/books/2014/mar/20/completely-beside-ourselves-family-love-review (3 August 2022).
Keen, Suzanne. *Empathy and the Novel*. Oxford: Oxford University Press, 2007.
Keen, Suzanne. "Empathy in Reading: Considerations of Gender and Ethnicity." *Anglistik: International Journal of English Studies* 24.2 (2013): 49–65.
Kozhevnikov, Maria, Olesya Blazhenkova, and Michael Becker. "Trade-off in Object versus Spatial Visualization Abilities: Restriction in the Development of Visual-Processing Resources." *Psychonomic Bulletin & Review* 17.1 (2010): 29–35. doi:10.3758/PBR.17.1.29.
Kuzmičová, Anezka. "Presence in the Reading of Literary Narrative: A Case for Motor Enactment." *Semiotica* 189.1/4 (2012): 23–48.
Kuzmičová, Anezka. "Literary Narrative and Mental Imagery: A View from Embodied Cognition." *Style* 48.3 (2014): 275–293.
McFee, Graham. "Empathy: Interpersonal vs Artistic?" *Empathy: Philosophical and Psychological Perspectives*. Eds. Amy Coplan and Peter Goldie. Oxford: Oxford University Press, 2011. 185–210.
McNeill, David. *Hand and Mind*. Chicago: Chicago University Press, 1992.
Myin, Erik, and J. Kevin O'Regan. "Situated Perception and Sensation in Vision and Other Modalities: A Sensorimotor Approach." *The Cambridge Handbook of Situated Cognition*. Eds. Philip Robbins and Murat Ayede. Cambridge: Cambridge University Press, 2009. 185–200.
Noë, Alva. *Action in Perception*. Cambridge, MA: MIT Press, 2006.
Oatley, Keith. *The Passionate Muse: Exploring Emotion in Stories*. Oxford: Oxford University Press, 2012.
O'Connell, Sanjida. *Mindreading: An Investigation into How We Learn to Love and Lie*. New York: Doubleday, 1998.
Oberzaucher, Elisabeth, and Karl Grammer. "Everything is Movement: On the Nature of Embodied Communication." *Embodied Communication in Humans and Machines*. Eds. Ipke Wachsmuth, Manuela Lenzen, and Günter Knoblich. Oxford: Oxford University Press, 2008. 151–177.
Pollick, Frank E., Helena M. Paterson, Armin Bruderlin, and Anthony J Sanford. "Perceiving Affect from Arm Movement." *Cognition* 82.2 (2001): B51–B61. doi:10.1016/S0010-0277(01)00147-0.
Rameson, Liam T., and Matthew D. Liebermann. "Empathy: A Social Cognitive Neuroscience Approach." *Social and Personality Psychology Compass* 3.1 (2009): 94–110. doi:10.1111/j.1751-9004.2008.00154.x.
Rokotnitz, Naomi. "Goosebumps, Shivers, Visualization, and Embodied Resonance in the Reading Experience: *The God of Small Things*." *Situated Cognition and the Study of Culture*. Eds. Ben Morgan, Sowon Park, and Ellen Spolsky. Special Issue *Poetics Today* 38:2 (2017): 273–293.
Seymenliyska, Elena. "*We Are All Completely Beside Ourselves* by Karen Joy Fowler, review: 'deeply moving'." *The Telegraph* (13 October 2014). https://www.telegraph.co.uk/culture/books/bookre

views/11160844/We-Are-All-Completely-Beside-Ourselves-Karen-Joy-Fowler-Man-Booker-2014.html (3 August 2022).

Shiffrar, Maggie. "The Visual Perception of Dynamic Body Language." *Embodied Communication in Humans and Machines*. Eds. Ipke Wachsmuth, Manuela Lenzen, and Günter Knoblich. Oxford: Oxford University Press, 2008. 95–109.

Trevarthen, Colwyn. "Intersubjectivity." *The MIT Encyclopedia of the Cognitive Sciences*. Eds. Robert A. Wilson and Frank C. Keil. Cambridge: MIT Press, 2001. 415–419.

Wachsmuth, Ipke, Manuela Lenzen, and Günter Knoblich. "Introduction to Embodied Communication: Why Communication Needs the Body." *Embodied Communication in Humans and Machines*. Eds. Ipke Wachsmuth, Manuela Lenzen, and Günter Knoblich. Oxford: Oxford University Press, 2008. 1–27.

Whiten, Andrew. "Social Learning and Culture in Child and Chimpanzee." *Annual Review of Psychology* 68 (2017): 129–154. doi:10.1146/annurev-psych-010416-044108.

Zahavi, Dan. "Empathy and Other-Directed Intentionality." *Topoi* 33.1 (2014): 129–142.

Zahavi, Dan, and John Michael. "Beyond Mirroring: 4E Perspectives on Empathy." *The Oxford Handbook of 4E Cognition*. Eds. Albert Newen, Leon de Bruin, and Shaun Gallagher. Oxford: Oxford University Press. 2016. 589–606.

Zahavi, Dan, and Soren Overgaard. "Empathy without Isomorphism: A Phenomenological Account." *Empathy: From Bench to Bedside*. Ed. Jean Decety. Cambridge, MA: MIT Press, 2012. 3–20.

Zeman, Adam. *Consciousness: A User's Guide*. New Haven and London: Yale University Press, 2002.

Zlatev, Jordan, Timothy P. Racine, Chris Sinha, and Esa Itkonen. "Intersubjectivity: What Makes Us Human?" *The Shared Mind: Perspectives on Intersubjectivity*. Eds. Jordan Zlatev, Timothy P. Racine, Chris Sinha, and Esa Itkonen. Amsterdam: John Benjamins, 2008. 1–16.

Part II: **Literary Approaches to Narrative Cognition in Science**

Aura Heydenreich

The Epistemic Power of Imagination through Interformation between Astrophysics and Literature: Carl Sagan's Novel *Contact* and Kip Thorne's Model of Traversable Wormholes

Abstract: This paper is centered on modes of exchange between literature and astrophysics and addresses the new concept of *interformation* (Heydenreich 2024) through which discourses and practices of knowledge production, representation and simulation are intersected, leading to epistemic transformations. The chapter addresses the interformative exchanges between Carl Sagan's novel *Contact* (1985) and Kip Thorne and Michael Morris's research on traversable wormholes, which resulted in the scientific paper "Wormholes in Spacetime and Their Use for Interstellar Travel", published in the *American Journal of Physics* in 1988. The chapter analyzes how the intersection between literature and physics can lead to new epistemic approaches within scientific discourse through explorative modeling and simulation. Additionally, it shows how Morris and Thorne employ narrative techniques like focalization, cognitive immersion, experientiality, and metalepsis to explore the new theoretical model of traversable wormholes. Thorne and Morris plead in their scientific paper for the epistemic power of imagination evolving from narration in order to reconceptualize existing research questions on traversable wormholes and to reorganize research discourses, as the scientific impact of their paper demonstrates.

Metaphors, models and simulations are epistemic tools (Gelfert 2017) for mathematics, physics, astronomy, climatology, earth system sciences, chemistry, biology (Godfrey-Smith 2006, 2009), life sciences, medicine and robotics. Model-based reasoning (Magnani, Nersessian and Thagard 1999) is employed in the social sciences, cognitive sciences, computer sciences, archaeology and architecture. While scientific research usually employs experiments, theories and their falsifications, mainly based on technical devices and mathematical formalism, recent approaches in philosophy of science investigate how scientific models and simulations are embedded in cultural processes (see Gelfert 2016, 2019), and ask how they are formed or epistemically transformed as parts of material cultures. Moreover, they ask how metaphors, models and simulations are ascribed a certain agency and autonomy due to their artifactual character (Knuuttila 2005) and their mediation (Morgan and Morrison 1999) and exemplification functions (Frigg

https://doi.org/10.1515/9783110782844-006

2023). The philosophy of science goes so far as to ask about the epistemic functions of fiction (Frigg 2010, Woods 2010) in the process of scientific modeling and simulation (Suarez 2009). In *The Fiction View of Models Reloaded*, Roman Frigg and James Nguyen draw parallels between scientific models and fictional characters, objects or places in literary fiction (Frigg and Nguyen 2016).

The starting point of my research question[1] is based on a twofold observation: firstly, that the philosophy of science (Frigg 2006), as already mentioned, has recently been conducting a critical debate (Falkenburg and Muschik 1998) about forms of modeling (Suarez 1999 and 2003) and representation (Van Fraassen 2013) in the natural sciences (Frigg and Hartmann 2018, Morrison 1998) and in doing so has also problematized the function of fiction (Morrison 2009, Frigg 2010, see also the critical view of Giere 2009) in scientific modeling (Bokulich 2009). On the other hand, the question of a poetics of modeling is also being discussed in contemporary literary studies (see Wendler 2013; Balke, Siegert and Vogl 2014; Erdbeer/Kläger/Stiersdorfer 2018; Erdbeer 2015). The debate is fruitful, raising the question of whether and how models can be examined as mediators between different spheres of discourse, as proposed by Morgan and Morrison (1999). According to Axel Gelfert, models are understood here as "*functional entities*" (Gelfert 2017, 5–6), which can be configured symbolically, semiotically, mathematically, diagrammatically or aesthetically. For Morgan and Morrison, they exhibit an explorative dimension in the construction of theory and thus function as mediators between denotation and representation through experimental simulation and the exemplification of new symbolic correlations (Morgan and Morrison 1999). This would then lead to the reorganization of sign functions and thus, after experimental tests, also to the reorganization of what had previously been denoted. For Knuutttila, models can also be understood as *epistemic tools*, as epistemic arti-

1 This chapter is based on research within ELINAS, the Erlangen Center for Literature and Natural Sciences, in whose interdisciplinary research context this work was written. I would like to thank Klaus Mecke from the Institute for Theoretical Physics at Friedrich-Alexander-Universität Erlangen-Nürnberg for numerous valuable suggestions from a theoretical physics perspective. See Klaus Mecke: "Zahl und Erzählung. Metaphern in Erkenntnisprozessen der Physik". *Quarks and Letters. Natural Sciences in Contemporary Literature and Culture*. Eds. Aura Heydenreich and Klaus Mecke. Berlin and Boston: De Gruyter, 2015. 31–85. doi:10.1515/9783110406542-002. One version of this chapter was published in German in Aura Heydenreich: *Literatur und Naturwissenschaft: Interformation und epistemische Transformation zwischen zwischen Literatur und Astronomie, Mikrobiologie, Relativitätstheorie und Quantentheorie 1600- 2020*. Berlin and Boston: De Gruyter, 2024. 441–481. A modified version was published in: Michael Gamper and Lukas Wolff (eds.): *Physiker lesen, Physiker schreiben. Wissen der Literatur und humanistische Bildung in der modernen Physik*. Göttingen: Wallstein, 2024. 169–226. The chapter has been revised and updated for this book and translated by Aura Heydenreich and Michael Sinding.

facts (Knuuttila 2005). They make it possible to formally, medially, symbolically or materially configure the knowledge relevant to understanding a certain area in order to recorrelate and reinterpret it accordingly.

In the context of analytical philosophy of art, Nelson Goodman introduced the distinction between the understanding of the process of "modeling of" or "*representation of*", which alludes to the earlier theory of representation but has to be reframed, and "modeling as" or "*representation as*" whose determinant is not denotation but exemplification (Goodman 2008, 27–31), in the case of art, aesthetic principles (Elgin 2010). Bas van Fraassen takes up this distinction in the philosophy of science in *Scientific Representation: Paradoxes of Perspective* and spells it out for the entire process of scientific representation (Van Fraassen 2013). These approaches to the philosophy of modeling in art and science is important for my present chapter, which will raise the following questions: How does science recur to literary techniques in order to open new research avenues to scientific questions through models and simulations? What kind of functions do literary techniques have in providing new epistemic approaches in astrophysical research? What kind of problems arise during the transfer of research questions due to the fact that different textual, diagrammatic, and discursive practices come into play in the two spheres of discourse?

Analysis of the interactions between the two fields presupposes recognition that mathematical physics has its own practice of using signs and symbols in its theoretical modeling; and on the other hand, that literature also has its own way of using signs and symbols in its aesthetic modeling. Only when these differences have been worked out from the perspectives of cultural semiotics, narratology, and theory of physics, can the possible interactions between the two fields be investigated. This is a question of semio-ethics (Petrilli 2014). It implies recognizing differences in the use of sign systems and taking them into account when investigating the interactions between the two fields. In my broader monograph on the epistemic interactions between physics and literature, the above-mentioned connections are examined historically and systematically (see Heydenreich 2024). In order to describe the processes of modeling in literature and science, it is necessary to define a space of signs in which both operate and to ask: What is that space of signs that allows us to describe phases in which the modeling practices of the two fields differ from each other *globally*, as well as those specific phases in which both fields interact with each other *locally* in order to exemplify[2] how epistemic routines can be reframed?

2 The concept of 'exemplification' goes back to Goodman 2008 and Elgin 2010.

In line with his concept of literature as a secondary modeling system, Lotman proposes the concept of the semiosphere (Lotman 2005):[3] "The semiosphere is that semiotic space outside of which the existence of semioses is impossible" (Lotman 2005, 205). Etymologically, the concept of the semiosphere can be traced back to the Greek words 'σημεῖον', 'sêmeion' = 'sign' and 'σφαῖρα', 'globe', 'sphere' = 'space'. Lotman's concept of the semiosphere should be understood as the space of signs that allows both the specialization of fields according to their own sign- and symbol-theoretical rules, which are centrally metasemiologically reflected and systematically codified, as well as the selective interaction of fields at their borders, where the exact opposite metasemiological debates are conducted – about the possibilities and consequences of crossing borders and the entanglement with fields of semiosis that use different sign logics.[4] Albrecht Koschorke has adopted the concept of the semiosphere for the design of a culturally oriented narratology in his study *Fact and Fiction. Elements of a General Theory of Narrative* and locates it within the framework of current cultural theories, demonstrating its merits as a cultural theoretical paradigm in contrast to Luhmann's systems theory and in confrontation with poststructuralist theories. (Koschorke 2018, 90–100; see also Koschorke 1999). Koschorke suggests that Lotman's theory should be taken into account because it points to a third way of overcoming the gap between systems theory and discourse analysis approaches – and also between structuralist and deconstructivist approaches (see Koschorke 2012). The dialectic between system and process dynamics could be observed through the lens of the dialectic between "codes and narratives" (Koschorke 2004, 174). When it comes to the sedimentation and consolidation of systems, these rely on metasemiotic communication on the consolidation of their own systems of rules, on their conventionalization and habitualization. This happens through a relatively porous demarcation from another cultural semiotic environment. But it is precisely this porous boundary, which serves to functionally stabilize the identity of the system, that registers the difference between the two semiological fields that separates them from one another, and unfolds a metasemiotic dynamic in the opposite direction. The boundary activates practices of both semiological fields and plays them against each other.

Lotman and Koschorke relate the concept to linguistically constituted cultural spaces. I would like to adopt this concept and develop it further for the investiga-

---

3 For a detailed presentation of the concept and its embedding in the cultural semiotic theory of the Tartu School see Lotman 2000, 123–171.

4 Cf. for a more detailed description of the system and process dynamics of the semiosphere Heydenreich 2024, 59–84.

tion of the interactions between physics and literature. In the following, the semiosphere is therefore defined as a field continuum of signs, sign functions and sign correlations, which comprises the following: linguistic signs and mathematical, graphic and diagrammatic symbols; the sign relations and sign functions existing between them; the codes that regulate the meaning of sign conventions and sign functions; finally, the actors as sign users who use the sign functions and sign relations in processes of modeling, communication, information and *interformation.*

My study refers to those fields of the semiosphere whose sign processes are regulated by different sign-logics (such as those of physics and literature) as *semio-logical fields.* The hyphen underlines that different 'logics of the use of signs' are employed in the physics and literary fields. It is therefore important to consider different approaches to sign and symbol theory, such as those of Lotman, Cassirer and Goodman, in order to show how the two spheres can be differentiated in terms of sign logic.[5] I propose the concept of *interformation* for the process of crossing boundaries between the fields.[6] The concept of interformation is to be understood as distinct from that of information. In the case of information, many communication models assume that the sender and receiver have the same code, so that the transmission of a message can function without any problems. In the case of *interformation*, however, the transfer of the message does not work according to this familiar model – because physics and literature have different logics of sign usage. However, there are possible interpolations. Interformation will have to be described as the processual modeling correlation between semio-logic fields that use different codes in order to transgress discursive boundaries, to simulate possible reframings of epistemic practices and to exemplify exploratory modeling practices. The interformation process can be traced in both

5 This comprehensive analysis cannot be carried out within this chapter. For a broader view on the interactions between cultural semiotics, narrative theory and philosophy of science see Heydenreich 2024, 59–200.

6 I also offer a more detailed presentation of the concept of interformation in the context of the analysis and reconstruction of the semiological, narratological and physics-theoretical foundations of Einstein's Special Theory of Relativity in: Aura Heydenreich: "Epistemic Narrativity in Albert Einstein's Treatise on Special Relativity. A Narratological Approach to *The Electrodynamics of Moving Bodies*. The Process of Interformation (Part I)". *Physics and Literature. Concepts – Transfer – Aestheticization.* Eds. Aura Heydenreich, Klaus Mecke. Berlin and Boston: De Gruyter, 2021. 49–103. Cf. also: "Albert Einstein's *Physics and Reality* and *The Electrodynamics of Moving Bodies.* The Process of Interformation, Semiologic Foundations and Epistemic Transformations (Part II)". *Physics and Literature. Concepts – Transfer – Aestheticization.* Eds. Aura Heydenreich, Klaus Mecke. Berlin and Boston: De Gruyter 2021. 105–138.

literary and in scientific texts (see Heydenreich 2024), where it works differently, but has the same functions: to question given boundaries/rules of discursive sign-correlation, to simulate changes in knowledge structures, to exemplify the changes and possibly to reorganize discourses.

The questions of my above-mentioned monograph are: Under which sign- and symbol-theoretical conditions can the interactions between the two secondary modeling systems of physics and literature be observed and described? To what extent, in what way and within what framework can the narrative modeling of literature draw on formal aspects of physics? To what extent – on the other hand – can the theoretical modeling of physics, under certain conditions, incorporate secondary modeling techniques of literature? An important thesis here is that the two fields of literature and physics can be conceptualized on a theoretical-systematic level due to their functional difference, but certain areas of their semio-logical practices intersect locally – in certain historical contexts, under certain conditions, in concrete individual texts. This induces a boundary process and opens up an enabling space for epistemic interactions. As will be shown, interformation opens up the possibility of explorative modeling in the sense of Axel Gelfert through reciprocal *constraints* of both discursive orders,[7] in order to reorient conventional research questions, to reframe approaches to solutions and thus to reorganize research discourses. This approach also allows us to observe how theoretical physics draws on narrative techniques and makes them heuristically fruitful for its own modeling.

In the following, I examine interformative processes of epistemic and aesthetic modeling between literature and general relativity in the context of theoretical astrophysics. I will focus on the theoretical model of the traversable wormhole[8] (Morris and Thorne 1988, 395; see also Morris, Thorne and Yurtserver 1988), an astrophysical object based on Einstein's field equations of general relativity (see Einstein 1914; Einstein 1915; Einstein 1916). The question of the potential existence of such an astrophysical object, which could connect two remote places in the universe, specifically arose for the author Carl Sagan in 1984/85 when he was working on his novel *Contact* (Sagan 1985). Sagan was working at Cornell Univer-

7 See the chapter "Exploratory Uses of Scientific Models". Axel Gelfert: *How to Do Science with Models. A Philosophical Primer.* Dordrecht and Cham: Springer, 2016. 71–101. See also Grant Fisher, Axel Gelfert, Friedrich Steinle: "Exploratory Models and Exploratory Modeling in Science: Introduction". *Perspectives on Science* 29/4 (2021): 355–358.

8 In the following, I will use the terms 'wormhole' and 'spacetime tunnel' as synonyms, as is done in the specialist literature, for example in the relevant monograph *Lorentzian Wormholes* by Visser 1996, 97–201.

sity as Director of the *Center for Space Research* and approached Kip Thorne, one of the leading experts in the field of general relativity and astrophysics,[9] who, together with Rainer Weiss and Barry C. Barish, was awarded the Nobel Prize in Physics in 2017 for the detection of gravitational waves predicted by Einstein's theory. Thorne's book *Black Holes and Time Warps: Einstein's Outrageous Legacy* (Thorne 1994) sheds light on the collaboration between Kip Thorne and Carl Sagan. Thorne reports how he took up Sagan's literary question, declared it to be his own research question and worked with his doctoral student Michael Morris for three years on its theoretical modeling. In 1988, two scientific publications appeared in the renowned journals *Physical Review Letters* (Morris, Thorne and Yurtserver 1988) and *American Journal of Physics* (Morris and Thorne 1988). In the following, I will analyze this exchange, which I propose to call an "interformative process", which transfers a question from literature to theoretical physics, where it is reframed and transformed into a theoretical problem that reorganizes the research discourses for this particular field of astrophysics, because it leads to a change of approach to solving Einstein's field equations.

I examine the exemplary case study in three steps: In the first section, I will briefly introduce Carl Sagan's *Contact*. The second section is devoted to the scientific background of the solutions to Einstein's field equations (Einstein 1915) that could not provide a solution for the problem of traversable wormholes before Sagan's question was formulated 1985 and Thorne and Morris's papers were written 1988. The third part of the paper shows how Thorne and Morris use mathematical models, physics intuition and narrative techniques of focalization inspired by the novel in order to reframe the approach to solving Einstein's equations. The third section also offers a detailed reading of the paper *Wormholes in Spacetime and Their Use for Interstellar Travel* by Thorne and Morris from the perspective of science narratology, in which I analyze the epistemic functions of the interformative techniques used for the reconceptualization of the research question.

In the narrative world of the novel, the concept of traversable wormholes was necessary to enable the astronaut Ellie Arroway to make a fictive journey to Vega, a star 25 light years away from the sun. Model proposals for wormholes had already emerged in 1916, immediately after Einstein published the basic equations of the general theory of relativity (Einstein 1916). But all existing models were conceptualized for elementary particles. The collaboration between Sagan and Thorne in the context of *Contact* is regarded as the starting point of

---

9 Together with John Archibald Wheeler and Charles Misner, Thorne published the book *Gravitation* in 1973, which is still regarded as a reference work on Einstein's general theory of relativity.

the renaissance of the wormhole concept that could be traversable for humans, as Mat Visser states in *Lorentzian Wormholes: From Einstein to Hawking* (see Visser 1996, 97–201).

1 The search for extraterrestrial intelligence in *Contact*: Communication codes of an information-theoretical palimpsest

The novel *Contact*[10] depicts the passionate commitment of astronomer Ellen Arroway to the implementation of a project called *Argus*, which aims to search for messages from extraterrestrial intelligence in interplanetary space using radio telescopes. Arroway has dedicated almost the entire capacity of NASA's *Argus* Observatory to this task, but the *National Scientific Agency* has a critical view on this project.

Although the novel was a great success in American literature, also thanks to the film adaptation[11] by Robert Zemeckis with Jodie Foster in the leading role, no paper has so far been devoted to its scientific-historical context or the links between literature and the general theory of relativity. So far, literary studies have examined conceptualizations of transcendence at the interface between theology, science and literature (see Stone 1998, Douglas 2016), and the search for extraterrestrial forms of intelligence (see Morrissey 2002, Coleman 2009).

In the novel, Ellen Arroway and her research team one day receive a signal from space that appears to have been sent from the vicinity of the star Vega. The signal is checked, but deciphering the message requires cracking the code in which the message was encoded. Cultural semiotics, information theory and physics expertise are required for this, as it is not clear on what basis this communication is taking place. Do the extraterrestrial beings use the same modulation frequencies as the Argus station? How can you even communicate with another community without knowing whether you have a common code with them? Without having clarified this, one can speak of neither "message" nor "communication".

The difficulties are numerous: the sender of the message is unknown, the channel through which the message was transmitted is unexplored, the code is

10 In the following, quotations from the novel are referenced directly in the text using the abbreviation *C* and the page number.

11 See Robert Zemeckis (director): *Contact*, Warner Home Video, 1997.

considered indecipherable for a long time and the message is complex. It comprises a three-level communication structure and requires the combined expertise of various disciplines to decipher it. Ellie Arroway compares the structure of the message to a palimpsest: "In classical time, thousands of years ago, when parchment was in short supply, people would write over an old parchment, making what's called a palimpsest. There was writing under writing under writing." (*C*, 100). But this is the definition of a palimpsest, which merely materially synthesizes layers of writing and cultural techniques of the past. The message in the novel, however, has a structure that superimposes layers of media history and information-theoretical cultural techniques. The first layer requires the deciphering of algorithmic formulas. This coding level is that of mathematical thinking, and so the research team initially discovers a series of prime numbers. The code of prime numbers differs from the rest of the sounds and noises in the universe in that it suggests the existence of a symbolic language. A sequence of prime numbers would belong to the symbolic sphere and thus to the semiosphere.

This leads to a second layer, a media image message, a film recording. This second signal stage uses modulation of the polarization of light waves as a transmission code. To decipher the code, it is important whether the polarization is left- or right-handed and at what frequency it was transmitted. According to Ellie, statistical analyses show that the different frequencies of polarization modulation can be translated into a code that contains an encrypted data collection of diagrams: "When a wave of light comes at you – visible light, radio light, any kind of light – it's vibrating at right angles to your line of sight. If that vibration rotates, it is said to be elliptically polarized. If it rotates clock-wise, the polarization is called right-handed; counterclockwise, it's left-handed. [. . .] Anyway, by varying between the two kinds of polarization, you could transmit information" (*C*, 83–84). Ultimately, it turns out that the third level of communication provides the instructions for the construction of a dodecahedron, which is to function as a space probe to enable a journey to Vega.

The question that Carl Sagan asked himself for his novel in 1988 went beyond the limits of human knowledge: was it possible for a superior extraterrestrial civilization to create a bridge through space-time so that an astronaut could make an interstellar journey to Vega? As an astronomer, Carl Sagan did not have enough theoretical physics expertise and therefore turned to Kip Thorne with his question. Thorne took the question seriously and devoted no less than three years of research to it (from 1985 to 1988) – with the aim of creating a mathematical model for the space-time tunnel in question that could be considered a solution to Einstein's equations of general relativity. The conditions that had to be met for a human being to survive interstellar time travel and not violate the laws of physics were: The traversable wormhole through interstellar space-time should be represented by a mathematically correct model; the journey should be able to be covered within a manageable period of time, so the average duration of a human life should not be exceeded; and the observers who remained on Earth and registered the departure of the journey should still be able to observe the return of the traveler despite the relativity of simultaneity, as postulated by the special theory of relativity. Although it was known that the model was not yet technically feasible, it should nevertheless be possible to calculate its feasibility mathematically so that it would also be plausible for the fictional world of the novel. A mathematical model was therefore sought that would be consistent with the principles of relativity, quantum theory and astrophysics.

2 Kip Thorne's scientific paper "Wormholes in Spacetime"

Sagan's novel poses the questions that need to be clarified within the framework of theoretical physics. He does this by imagining the technical possibilities of an advanced alien civilization on Vega that would make it possible to build a space-time tunnel. What conditions would be imposed on this civilization by the laws of physics? This is the starting point of the argumentation of the scientific publication of Michael Morris and Kip Thorne. I quote the initial question of the article "Wormholes, Time Machines, and the Weak Energy Condition", which appeared in *Physical Review Letters* in 1988, because it follows the same line of argumentation as the article in the *American Journal of Physics:*

> Normally theoretical physicists ask, "What are the laws of physics?" and/or "What do those laws predict about the Universe?" In this Letter we ask, instead, "What constraints do the laws of physics place on the activities of an arbitrarily advanced civilization?" This will lead to some intriguing queries about the laws themselves. We begin by asking whether the laws of physics permit an arbitrarily advanced civilization to construct and maintain wormholes for interstellar travel. (Morris, Thorne and Yourtserver, 1446)

Thorne and Morris intend to fundamentally reconceptualize the theoretical question: no longer to ask what the laws of nature prohibit, but what could be imagined through theoretical modeling? If one were to assume that the Copernican principle is correct and that the laws of nature that apply to humans would also apply to the alien civilization – which constraints would have to be considered in the theoretical modeling of such a traversable space-time tunnel?

First, a few basic concepts of general relativity should be explained very briefly. For this, I refer to Matt Visser's standard work *Lorentzian Wormholes.*

From Einstein to Hawking (see Visser 1996). The historical genealogy of spacetime tunnels can be traced back to the beginnings of Einstein's general theory of relativity. All attempts made in physics to theoretically postulate these spacetime tunnels are mathematical objects that represent solutions to Einstein's field equations of general relativity:

$$G_{\alpha\beta} + \Lambda g_{\alpha\beta} = KT_{\alpha\beta}$$

After Einstein published the general theory of relativity in 1915 (Einstein 1915), Karl Schwarzschild presented the first solution in 1916 (Schwarzschild 1916). He proposed the so-called Schwarzschild metric, which described the gravitational field of a non-rotating and electrically non-charged (spherically symmetrical) point of mass or gravitational field. The general theory of relativity is a field theory; the mathematics with which theoretical physics operates here is differential geometry. Space and time are united in a common spacetime continuum that exhibits curvature. In the general theory of relativity, gravity is represented by the curvature of spacetime. Spacetime is a four-dimensional differentiable manifold with a Lorentz metric that represents its signature. The Lorentz signature has four dimensions, which are represented as (–, +, +, +). The three "+" signatures represent the three dimensions of space, the "–" signature represents the time dimension. Together they form the four-dimensional spacetime continuum. The curvature of spacetime in the vicinity of the object corresponds to the gravitational field of this object. To describe this gravitational field, the metric of spacetime is important (see Visser 1996, 9).

A wormhole would be a tunnel through space-time that could directly connect two distant space-time points in the universe. As fantastic as this idea sounds, Einstein himself took it up after Hermann Weyl had also speculated about it in his article "Field and Matter" (see Weyl 1921). In 1935, Einstein and Rosen presented a solution to Einstein's field equations and thus also the concept of the Einstein-Rosen bridge, which represented an important stage in the historical genealogy of wormholes. In the summary that precedes the paper, Einstein and Rosen state:

> The writers investigate the possibility of an atomistic theory of matter and electricity which, while excluding singularities of the field, makes use of no other variables than the $g_{\mu\nu}$ of the general relativity theory and the ϕ_μ of the Maxwell theory. By the consideration of a simple example they are led to modify slightly the gravitational equations which then admit regular solutions for the static spherically symmetric case. These solutions involve the mathematical representation of physical space by a space of two identical sheets, a particle being represented by a "bridge" connecting these sheets. (Einstein and Rosen 1935)

The Einstein-Rosen bridge was described by John Wheeler in 1957 with the metaphorical term "wormhole".

> This analysis forces one to consider situations [. . .] where there is a net flux of lines of force, through what topologists would call "a handle" of the multiply-connected space, and what physicists might perhaps be excused for more vividly terming a "wormhole". (Wheeler 1955)

The metaphorical term, for which Wheeler apologized in his original article, ultimately prevailed (see Wheeler 1955, Wheeler 1957). Thorne and Morris explain the object of wormholes using a simple definition: "Such a wormhole is a short 'handle' in the topology of space, which links widely separated regions of the universe" (Morris, Thorne and Yurtserver 1988, 1446). To illustrate this, they provide the following diagram (Fig. 1), which they use to introduce a typology of astrophysical objects. They distinguish between the *inter-universe wormhole* (a), which could connect our universe with another universe, and the *intra-universe wormhole* (b). This could connect two regions that are very far apart within our universe (see Morris and Thorne 1988, 396).[12]

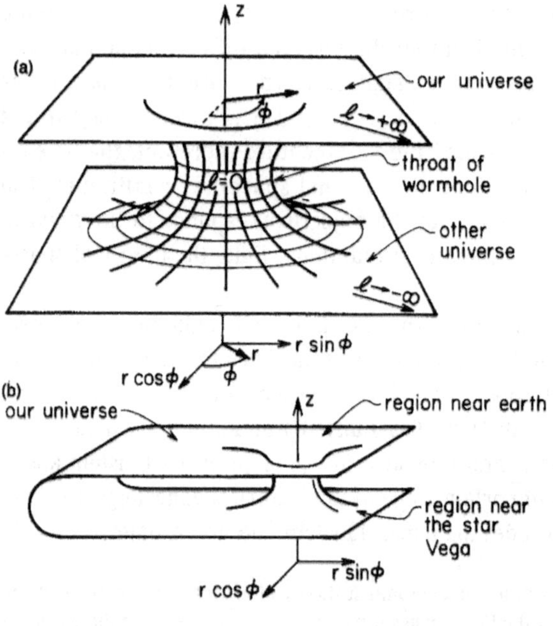

Fig. 1: Wormhole typology according to Morris and Thorne (WiS, 396).

12 In the following, quotations from the paper *Wormholes in Spacetime* (Morris and Thorne 1988) are cited directly in the text with the abbreviation "WiS" and the page reference.

The so-called Schwarzschild radius defines the event horizon of this object. If an observer is in the inner sphere of the object, a connection with the outer sphere is no longer possible. Due to the structure of space-time and the finite nature of the speed of light the outer and inner spheres are separated by the "event horizon", which forms the boundary for the exchange of information and for causal relationships. However, an event horizon only forms when the mass of an astronomical object compresses to such an extent that it reaches a value below the Schwarzschild radius. Any object whose mass compresses to below the Schwarzschild radius becomes a "black hole."

Wormholes are conceptualized primarily through their bridging function. The distinction between black holes and wormholes as astrophysical objects is difficult because a wormhole can transform into a black hole under certain conditions. The term "black hole" was introduced in 1962 by John Wheeler, Kip Thorne's doctoral supervisor, to designate an object that was identified with the *technical term* "gravitationally completely collapsed object" (Wheeler 1962). This refers to an astrophysical object, a region of space that exerts such strong gravitational forces on its surroundings that everything that comes close to its event horizon is pulled into its interior. Within the event horizon, the object exhibits a curvature singularity. This would be the place where the gravitational forces are so strong that the curvature of space-time diverges. Black holes are known to completely absorb matter and electromagnetic energy (e.g. light) in their vicinity.

The mathematical object of wormholes already existed before the theoretical modeling by Morris and Thorne. However, earlier modeling did not envisage that these objects could be traversed by humans. Thorne summarizes the state of research before he began his work on wormholes in *Black Holes and Time Warps* as follows:

> Wormholes are not mere figments of a science fiction writer's imagination. They were discovered mathematically, as a solution to Einstein's field equation, in 1916, just a few months after Einstein formulated his field equation; and John Wheeler and his research group studied them extensively, by a variety of mathematical calculations, in the 1950s. However, none of the wormholes that had been found as solutions of Einstein's equation, prior to [. . .] 1985, was suitable for Carl Sagan's novel, because none of them could be traversed safely. (Thorne 1994, 486)

All earlier models were microscopic astrophysical objects – until 1984/1985, when Sagan took up the model in his novel *Contact* for the above-described journey of Ellie and her fellow astronauts to Vega. He sent a version of this novel draft to Kip Thorne, an expert on the general theory of relativity, with the question of how this wormhole should be modeled so that it would correspond to current astrophysical theories. The decisive condition demanded by the novel – that the space-time tunnel should be traversable by humans – had not been fulfilled by any of the existing models: The question of it had simply not arisen. Now, however, the requirement of the fictional storyworld became a challenge for theoretical modeling in astrophysics.

The research-historical details had to be explained here in order to show that the starting point of this research process is actually motivated by both astrophysics and literature. Thorne and Morris themselves describe the starting point of their research activity in their paper:

> We were stimulated to find them in the summer of 1985, when Carl Sagan sent one of us a prepublication draft of his novel Contact and requested assistance in making the gravitational physics in it as accurate as possible. Sagan, in response to our preliminary description of these solutions' properties, incorporated them into his novel at the galley proof stage. (WiS, 397)

Here it becomes clear that Sagan had even literarily reframed a preliminary version of these solutions and integrated them into his text. Sagan's novel is quoted extensively in the scientific paper. Morris and Thorne place the quotations in a box. I call this marked citation of the discursive practice of the literary sphere in the scientific article "transdiscursive interpolation". In editorial philology, the term "interpolation" refers to an insertion or change made in a text by a third party. Here, "different hand" is to be understood as a different semio-logical discursive practice, which in this case is clearly marked in its own (here: scholarly) context. The quotations from the novel appear on the second page of the scientific text and are typographically framed and entitled "Box 1: Excerpts from *Contact* by Carl Sagan" (WiS, 397). This in turn marks the starting point for the process of interformation, as I will show. The detailed quotations from the novel are set off graphically in the scientific paper (see Fig. 2).

The authors are aware of the fact that they are crossing the boundaries between two semio-logical spheres, and they discuss this (cf. WiS, 397). Kip Thorne himself reflects on this in his book *Black Holes and Time Warps*, published six years later – which could be described as an epitext to the scientific publication. The literary question, Thorne admits, was unique for the constitution of theory because in theoretical physics[13] one did not dare to ask such questions as those posed by Carl Sagan:

13 Seven years later, Thorne reported in *Black Holes and Time Warps* that, as a doctoral supervisor, he had great reservations about leaving this topic to his doctoral student Michael Morris and publishing on it together with him. The question arose as to whether this would 'jeopardize' his doctoral student's career. But Morris himself did not see this as a problem. (See Thorne 1995, 508).

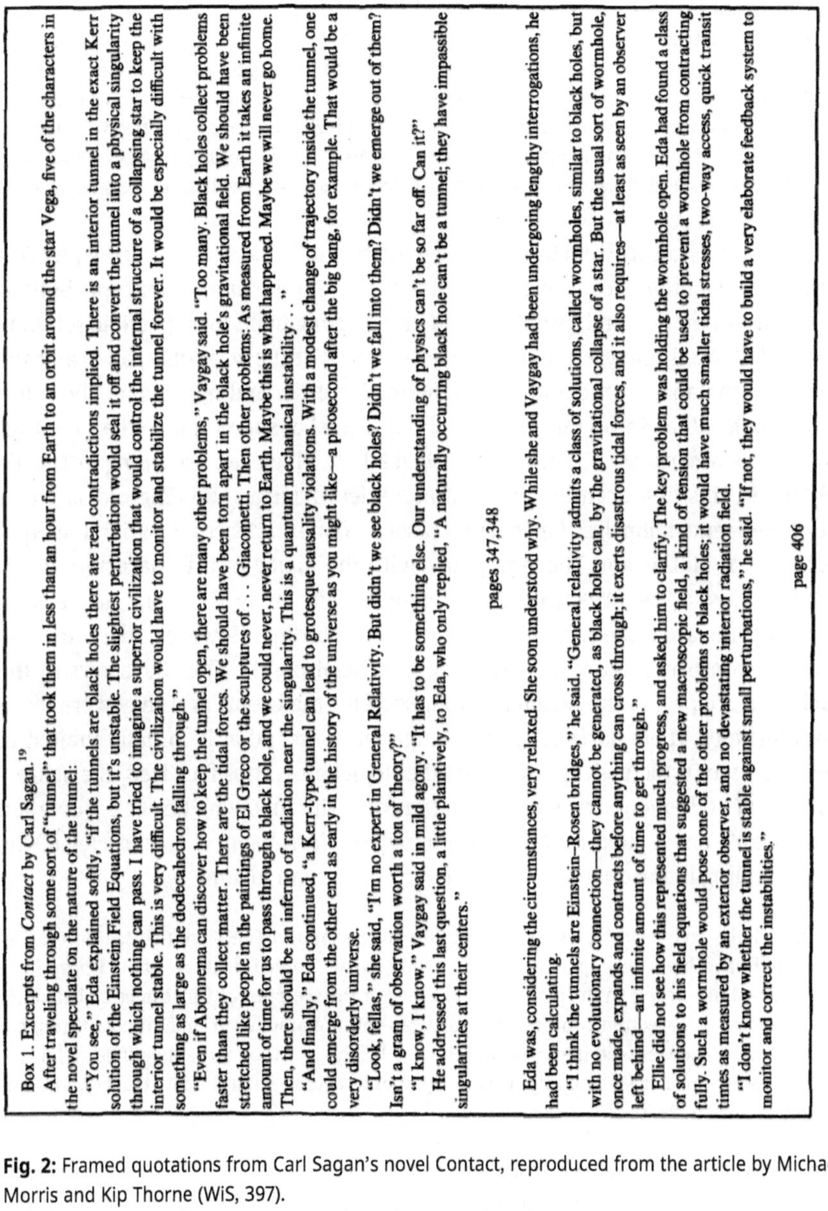

Box 1. Excerpts from *Contact* by Carl Sagan.[19]

After traveling through some sort of "tunnel" that took them in less than an hour from Earth to an orbit around the star Vega, five of the characters in the novel speculate on the nature of the tunnel:

"You see," Eda explained softly, "if the tunnels are black holes there are real contradictions implied. There is an interior tunnel in the exact Kerr solution of the Einstein Field Equations, but it's unstable. The slightest perturbation would seal it off and convert the tunnel into a physical singularity through which nothing can pass. I have tried to imagine a superior civilization that would control the internal structure of a collapsing star to keep the interior tunnel stable. This is very difficult. The civilization would have to monitor and stabilize the tunnel forever. It would be especially difficult with something as large as the dodecahedron falling through."

"Even if Abonnema can discover how to keep the tunnel open, there are many other problems," Vaygay said. "Too many. Black holes collect problems faster than they collect matter. There are the tidal forces. We should have been torn apart in the black hole's gravitational field. We should have been stretched like people in the paintings of El Greco or the sculptures of . . . Giacometti. Then other problems: As measured from Earth it takes an infinite amount of time for us to pass through a black hole, and we could never, never return to Earth. Maybe this is what happened. Maybe we will never go home. Then, there should be an inferno of radiation near the singularity. This is a quantum mechanical instability. . . "

"And finally," Eda continued, "a Kerr-type tunnel can lead to grotesque causality violations. With a modest change of trajectory inside the tunnel, one could emerge from the other end as early in the history of the universe as you might like—a picosecond after the big bang, for example. That would be a very disorderly universe.

"Look, fellas," she said, "I'm no expert in General Relativity. But didn't we see black holes? Didn't we fall into them? Didn't we emerge out of them? Isn't a gram of observation worth a ton of theory?"

"I know, I know," Vaygay said in mild agony. "It has to be something else. Our understanding of physics can't be so far off. Can it?"

He addressed this last question, a little plaintively, to Eda, who only replied, "A naturally occurring black hole can't be a tunnel; they have impassible singularities at their centers."

pages 347,348

Eda was, considering the circumstances, very relaxed. She soon understood why. While she and Vaygay had been undergoing lengthy interrogations, he had been calculating.

"I think the tunnels are Einstein–Rosen bridges," he said. "General relativity admits a class of solutions, called wormholes, similar to black holes, but with no evolutionary connection—they cannot be generated, as black holes can, by the gravitational collapse of a star. But the usual sort of wormhole, once made, expands and contracts before anything can cross through; it exerts disastrous tidal forces, and it also requires—at least as seen by an observer left behind—an infinite amount of time to get through."

Ellie did not see how this represented much progress, and asked him to clarify. The key problem was holding the wormhole open. Eda had found a class of solutions to his field equations that suggested a new macroscopic field, a kind of tension that could be used to prevent a wormhole from contracting fully. Such a wormhole would pose none of the other problems of black holes; it would have much smaller tidal stresses, two-way access, quick transit times as measured by an exterior observer, and no devastating interior radiation field.

"I don't know whether the tunnel is stable against small perturbations," he said. "If not, they would have to build a very elaborate feedback system to monitor and correct the instabilities."

page 406

397 Am. J. Phys., Vol. 56, No. 5, May 1988 M. S. Morris and K. S. Thorne 397

Fig. 2: Framed quotations from Carl Sagan's novel Contact, reproduced from the article by Michael Morris and Kip Thorne (WiS, 397).

> We physicists, I believe, have tended to avoid such questions because they are so close to science fiction. While many of us may enjoy reading science fiction or may even write some, we fear ridicule from our colleagues for working on research close to the science fiction fringe. We therefore have tended to focus on [. . .] other, less radical, types of questions: "What kinds of things *occur* naturally in the Universe?" [. . .] By 1988 it seemed clear to me that we physicists had been much too conservative in our questions. Already, one *Sagan-type question* (as I shall call them) was beginning to bring a payoff. (see Thorne 1995, 493, emphasis in the original text).

In the literary excerpts from Sagan's novel the scientific paper outlines the physical research problem *in nuce*. Thorne and Morris present the literary excerpts both as a summary of the research problems and as a starting point for a possible solution. In what follows, I analyze the discursive practices of the scientific paper and the novel, first juxtaposing them to show how they overlap and intertwine. The astrophysical object that Sagan imagined for the traversal of space-time in the first draft of his novel, according to Thorne, was a black hole. However, according to Thorne and Morris, this first draft led to a plot and scientific dead end. The crossing of a black hole seemed hopeless for several reasons. If one compares the scientific argumentation of Morris and Thorne's paper with the dialog of the characters in Sagan's novel, who are already on the interstellar journey to Vega, one notices that the arguments are quite similar – even in their wording. However, there are also differences in presentation and spelling at crucial points. The characters in the novel, who are part of the fictional world, perceive the events as if they were focalized internally. Their horizon of knowledge and perception is initially limited to the space-time in the dodecahedron. As the depiction is thus focalized, the characters' affects and emotions also come to the fore. The following text is a quotation from Sagan's novel cited in the scientific paper. Here Vaygay, a character of the novel traveling in the dodecahedron explains their conundrum:

> Black holes collect problems faster than they collect matter. There are the tidal forces. We should have been torn apart in the black hole's gravitational field. We should have been stretched like people in the painting of El Greco or the sculptures of Giacometti. Then other problems: as measured from Earth it takes an infinite amount of time to pass through a black hole, and we could never, never return to Earth [. . .] Then there should be an inferno of radiation near the singularity. This is a quantum mechanical instability [. . .] (*C*, 347–348; quoted in WiS, 397)

With Marie-Laure Ryan, one could say that the text-strategic effect of the literary discourse that is cited in the scientific paper is the cognitive immersion of the reader (see Ryan 2001, 2003). With Fludernik and Caracciolo, one could also use the category of narrative experientiality (see Fludernik 2018, Caracciolo 2014).

To summarize, Vaygay, the fictional character in the novel, addresses the following problems: Einstein's field equations stipulate that the tidal forces inside the black hole become infinitely strong. Therefore, the entrance to a black hole is a one-way street – if an object falls into it, it will be destroyed. In addition, space-time ceases to exist in the curvature singularity of the black hole. It is therefore impossible for the figures to travel back and forth to Earth. This is postulated by the general theory of relativity. But now the laws of quantum gravity come into play, which change the character of space-time. In the vicinity of the curvature singularity of the black hole, space-time no longer has a metric. It can only be calculated probabilistically. In this sphere, space no longer has a certain curvature or a certain topology, but only a certain probability for a curvature or topology.

Exactly the same arguments for the inadequacy of the black hole model are discussed both by the fictional characters in the novel *Contact* and by the authors of the scientific text. At the beginning of their article, Thorne and Morris name the same reasons in order to make the reconceptualization of the question plausible: The boundary regions of the black hole do not allow the passage of matter, light and energy between inside and outside (see WiS, 395). At the horizon of the black hole, the tidal forces are extremely high, so that no astronaut could pass this threshold alive: "[T]he adventurer will be killed by tidal gravity before even reaching the horizon" (WiS, 395). It follows that the entrance to the black hole is practically a one-way street: "Things can fall in, but nothing can emerge. Thus, two-way travel (which is soften invoked) is forbidden" (WiS, 395). Thorne and Morris also consider rotating black holes that exhibit the so-called Kerr metric. The rapid contractions and expansions of black holes generate probabilistic space curvatures that lead to spacetime singularities. However, Thorne and Morris also assume for rotating black holes in the sense of Kerr that they cannot form tunnels to other spacetime regions of the universe in their interior, but rather curvature singularities with high tidal forces that would tear apart any astronaut. Interestingly, this solution is also discussed and rejected by the characters in the novel:

> "You see," Eda explained softly, "if the tunnels are black holes, there are real contradictions implied. There is an interior tunnel in the exact Kerr solution of the Einstein Field Equations, but it's unstable. The slightest perturbation would seal it off and convert the tunnel into a physical singularity through which nothing can pass." (*C*, 347)

Morris and Thorne then move on to another class of solutions to Einstein's equations, the so-called Schwarzschild wormholes, which, however, bear the same problems as the black holes that exhibit a Schwarzschild metric. These are the gravitational forces, the rapid expansions and contractions that would tear the astronauts apart (see WiS, 396). After these reasons have been given in the scientific text and in the novel, Morris and Thorne point out that the second passage of the quotation from Carl Sagan's novel contains a possible solution to the problem:

> Two relevant passages from Sagan's novel are reproduced in Box 1. The first passage describes the objections to black holes as means of interstellar travel, and a portion of the second passage describes the objections to Schwarzschild wormholes. [. . .] The remainder of the second passage describes some key features of the traversible wormhole solutions, including the *wormholes' ability to remain always open, very small tidal forces,* [. . .] (WiS, 397; emphasis A. H.)

This sentence is interrupted by "Box 1" in Morris and Thorne's paper. The reader can find therein an excerpt from the novel, which in turn is framed by the scientific text (see Fig. 3).

The literary text presents the explorative model within a dialogue between the characters in the novel:

> "I think the tunnels are Einstein-Rosen bridges," he said. "General relativity admits a class of solutions, called wormholes, similar to black holes, but with no evolutionary connection – they cannot be generated, as black holes can, by the gravitational collapse of a star. But the usual sort of wormhole, once made, expands and contracts before anything can cross through; it exerts disastrous tidal forces, and it also requires – at least as seen by an observer left behind – an infinite amount of time to get through." [. . .] *The key problem was holding the wormhole open.* (C, 406; quoted in WiS, 397; emphasis A. H.)

A possible solution to this problem can be found in the novel quoted in the scientific text:

> Eda had found a class of solutions to his field equations that suggested a new macroscopic field, a kind of tension that could be used to prevent a wormhole from contracting fully. Such a wormhole would pose none of the other problems of black holes; it would have *much smaller tidal stresses, two-way access, quick transit times as measured by an exterior observer, and no devastating interior radiation field.* (*C*, 406; quoted in WiS, 397; emphasis A. H.)

The box that frames the literary quotation is then closed and the scientific text is continued – but with a new proposal of how to reframe the scientific approach to the problem. Thorne and Morris present their own model, the theoretical conception of which is inspired by Carl Sagan's novel. If one compares the literary and the scientific text, one notices that the arguments are similar right down to the choice of words. This is why I propose to call the epistemic conceptualization of a new research approach that incorporates elements of both discursive practices "interformation." At this point, the scientific text is taken up again and the boundary conditions for the scientific model are formulated:

> [. . .] *two-way travel (no horizons), rapid transit times as seen by both travelers and external observers,* lack of intense, singularity-produced *radiation fluxes,* and also the requirement

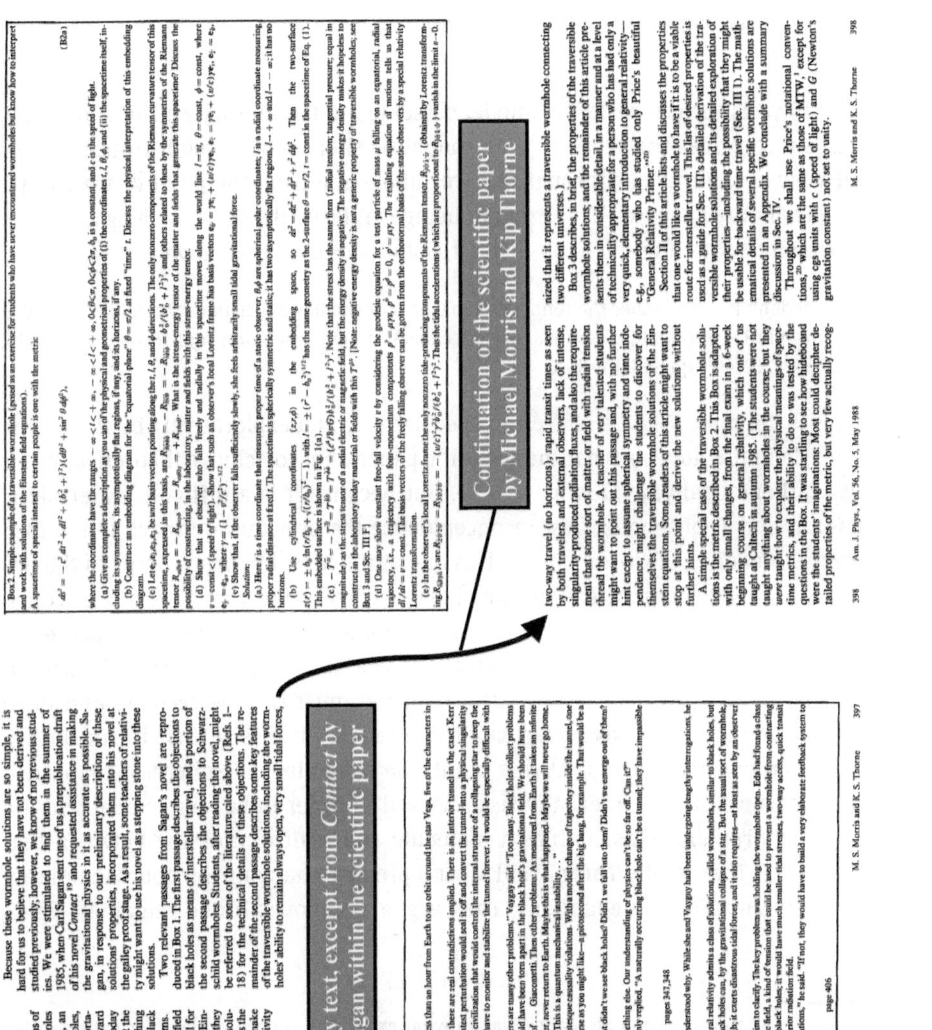

Fig. 3: Excerpt from the scientific paper by Thorne/Morris with the quotation from the novel Contact, framed here on the left-hand side (WiS, 397–398).

> that some sort of *matter or field with radial tension* thread the wormhole. (WiS, 398; emphasis A. H.)

The physical properties of the model presented in the literary and scientific texts largely coincide. The crossing time should be short enough, both from the point of view of the traveler through the wormhole and from the point of view of the external observers, so that a human being can leave the wormhole again during his lifetime. Eda, the relativity theorist, points this out in the novel:

> "I don't know whether the tunnel is stable against small perturbations" he said. "If not, they would have to build a very elaborate feedback system to monitor and correct the instabilities" (C, 406; quoted in WiS, 397)

This is the crux. What is addressed speculatively and exploratively by the character Eda in the novel is what the scientific paper aims to achieve theoretically, that is mathematically, in further course. But because it is a difficult undertaking, Morris and Thorne must first reconceptualize the previous approach to the theoretical modeling of wormholes. This is one of the most important steps in the process of interformation. The literary idea of traversable wormholes is of course not sufficient in itself. Here we touch on another crucial point in the process of interformation: the necessity of reframing the new problem within the specialized field of knowledge. The central question must first be recoded and introduced into the discursive practice of theoretical physics so that it can then be modeled mathematically. At this point, the semio-logical sphere of theoretical physics must also be internally reconfigured in order to be able to deal with the literary problem. This enables a plausible and fruitful crossing between both spheres.[14] That is why Thorne and Morris propose changing the approach to solving the Einstein field equations. They argue as follows: If one remained rigidly

14 At this point, it should be made clear to what extent my concept of interformation differs from Jürgen Link's interdiscourse theory, although it certainly owes important impulses to it. According to Link, the dimension of interdiscursivity is based on elementary everyday language, "popular religion, conversations (talk shows), art and literature, popular science, mediopolitics and media entertainment" (Link 2006, 412–413 my translation). My concept of interformation, on the other hand, is narrower and refers to the nexus between literature and the scientific discourses exemplified by theoretical physics in this study. With the approach of interformation I propose to consider not only the "elementary literary forms of perception" but also those operative concepts such as "mathematical formalization, classification schemes, measuring procedures" which Jürgen Link excludes: "Now it is obvious that [. . .] literature does not take up and process all interdiscursive elements, procedures and substructures at will. *Operative-interdiscursive elements such as mathematical formalization, classification schemes, measurement procedures, etc.* largely or completely recede behind imaginary elements such as pictorial analo-

within the boundaries of the expert discourse of physics and mathematics, one would not find a solution for traversable wormholes. Nevertheless, the safe ground of the discourse of mathematics and theoretical physics cannot be left lightly. Therefore, Thorne and Morris differentiate three dimensions of modeling: firstly, the rigorous mathematical-theoretical dimension: secondly, a more flexible physics dimension, which could certainly be modeled with narrative procedures in the form of thought experiments: and a third "aesthetic" dimension. Thorne and Morris admit that in moving from the first to the second and third stages of modeling, they must also necessarily change their style of argumentation, which I will analyze in the next section of this chapter.

3 Interformation and analysis of scientific argumentation

3.1 Criteria for interformative modeling

Thorne and Morris present seven criteria for traversable wormholes. The criteria themselves derive from the superimposition of the discursive spheres. The first four criteria are mathematical constraints and non-negotiable:

(1) The model of the traversable wormhole is supposed to be a solution to Einstein's field equations.
(2) The metric that defines the signature of spacetime should be spherical and static.
(3) Spacetime should have no curvature singularities.
(4) The object is said to have two openings that create a connection between two different points in the universe.

The first four demands are those of mathematical consistency: Thorne and Morris call them "basic wormhole criteria" (WiS, 400). Requirements (5) and (6) result from the novel's storyworld. Thorne and Morris call them "the wormhole's parameters for human physiological comfort, so we shall call them 'usability crite-

gies, metaphors, symbols, figurations of human subjects, etc. The totality of the *(non-operative) interdiscursive elements* could be understood as *elementary literary forms of perception*, which are generated from the tendency to reintegrate the special discourses" (Link 1988, 286, my translation and emphasis).

ria'." (WiS, 400) This opens the way to explorative modeling in the sense of Axel Gelfert.[15]

(5) The object must not have an event horizon, as this would prevent the astronaut from entering or exiting and thus traveling back and forth through the space-time tunnel. The gravitational forces experienced by the astronaut must be limited. They must be low enough for the astronaut to survive the journey.

(6) The traveler must be able to cross the wormhole in a suitable time: Thorne and Morris state approximately one year, measured both from the perspective of the traveler and from the perspective of the external observers on Earth.

Finally, the seventh requirement is about aesthetic criteria, as Thorne and Morris themselves state.

> By means of Property 7, we shall tune the parameters *for our own aesthetic comfort* – i.e., we shall tune them to make the wormhole's construction material as compatible as possible with our present prejudices about the forms of matter allowed by the laws of physics. (WiS, 400; emphasis A. H.)

The parameters that are to be modeled according to aesthetic criteria are the central crux of the research problem.

(7) The matter and fields that influence the geometry of the wormhole, and which determine the curvature of space-time, should be appropriate from a physics perspective.

Thorne and Morris use the term "physically reasonable". This means that the spacetime tunnels should have the structural stability to ensure that a space probe can traverse them (cf. WiS, 399–400).

The quintessence of the process of interformation can be summarized here in the fact that criteria (1) to (4) must be adhered to from a system-dynamic perspective. At the same time, requirements (5) to (7) should allow a process-dynamic explorative modeling and the reconceptualization of certain categories.

How does the flexibilization of epistemic routines of modeling take place? Interformation is the decisive point of innovation, as I would like to show. Theoreti-

---

15 Axel Gelfert coined this concept of explorative modeling (Gelfert 2016) based on Friedrich Steinle's studies on explorative experimentation in the history of science (Steinle 2005). See for a recent overview Fisher, Gelfert, and Steinle 2021.

cal modelings before 1988 took as their starting point the energy-momentum tensor ($KT_{\alpha\beta}$), the right-hand side of Einstein's field equations:

$$G_{\alpha\beta} + \Lambda g_{\alpha\beta} = KT_{\alpha\beta}$$

with the aim of calculating the geometry of the spacetime tunnel, which is represented by the left-hand side of the equation. After the energy-momentum tensor ($KT_{\alpha\beta}$) was calculated, Einstein's field equations were solved, and it was then investigated whether curvature singularities were present in the solutions. If curvature singularities were present, then these wormhole solutions could not be traversed by humans (see Visser 1996, 101).

The theoretical framework of traversable wormholes is situated at the interface of the two fundamental physical theories of the 20th and 21st centuries, between general relativity and quantum field theory. General relativity is the theory represented by the left-hand side of Einstein's field equations ($G_{\alpha\beta} + \Lambda g_{\alpha\beta}$), which represents the geometry of space-time.

$$G_{\alpha\beta} + \Lambda g_{\alpha\beta} = KT_{\alpha\beta}$$

Quantum field theory is the theory represented by the right-hand side of the equation; it represents the energy-momentum tensor $KT_{\alpha\beta}$. The basic problem, which must also be addressed through the interformative process, is that the left-hand side of the equation would allow solutions to represent the world of traversible wormholes, while the right-hand side of the equation imposes very tight constraints.

3.2 New approach: reconceptualization of the research question

The two theories are based on different principles. Thorne and Morris make use of this. Challenged by the novel idea, they argue that general relativity would allow a spacetime geometry that could be modeled as a spacetime tunnel. Consequently, the right-hand side of the equation is no longer understood as a restriction, but as an epistemic challenge for new explorations. It becomes a key problem for future research discourses in physics. Thorne and Morris open up a new scientific approach with this reconceptualization of the question.

> It turns out that there are very simple, exact solutions of the Einstein field equations, which describe wormholes that have none of the above problems. [. . .] As we shall see below, it is not clear today whether the laws of physics prohibit or actually permit the construction of such "traversible wormholes"; but nothing we know makes them seem nearly as impossible as black hole or Schwarzschild wormhole transportation systems. (WiS, 397)

Morris and Thorne state that they have decided to deviate from the conventional way of theoretical modeling and take a different path. They speak of "changing the philosophy of solving field equations" (WiS, 402). As a starting point, they no longer choose the quantum field theory approach, but the relativity theory approach. They therefore no longer start from the energy-momentum tensor $KT_{\alpha\beta}$, as was always conventionally calculated before them, but from the necessary spacetime geometry $g_{\alpha\beta}$. They declare this to be relevant because it can fulfill the requirements of human traversability. They first calculate what the spacetime geometry of the tunnel would have to look like for it to be traversable by humans, i. e. characters in the novel. Once they have determined this spacetime geometry, they look for a suitable energy-momentum tensor. The energy-momentum tensor is therefore no longer the starting point but becomes the target of future explorative theoretical modeling.

The epistemic effect of the literary question is not that a suitable solution to Einstein's field equations for the wormholes is found, but that the question is reconceptualized. This reconceptualization of the question is also accompanied by a new approach to solving the equations, which can be described as process-dynamic modeling: The conventional solution, which I would like to call "*intraformative*" because it remains within the semio-logical sphere of physics, proves to be the wrong path. However, the question is not abandoned, but realigned from a new perspective: A new *interformative* solution path is thus taken. Since Thorne and Morris demanded a specific theoretical model of wormholes, which – as first evolved in the novel – also had to meet certain human and not just mathematical criteria, they also adopt this explorative perspective.

> In our study of traversible wormholes the philosophy of solving the field equations [. . .] must be altered somewhat from the usual one. We desire solutions with certain properties (enumerated in Sec. II), and to achieve them we must be willing to let the builders of a wormhole synthesize, or search throughout the universe for materials or fields with whatever stress-energy tensor might be required. (WiS, 402)

They then come to the requirements resulting from the criterion of "human usability": the absence of an event horizon, fast traversal times, low gravitational tidal forces. The new approach to solving the equations stipulates that the right-hand side of the equation, the energy-field tensor conditioned by quantum field theory, should not be the starting point of theoretical modeling, but the goal: "Stated mathematically, we wish to control the functions b(r) and Φ (r) so as to shape the wormhole to our specifications; [. . .]" (WiS, 402).

Thorne and Morris now start from the left-hand side of Einstein's field equations, from the geometry of space-time. They calculate the spacetime curvature for this geometry and derive from Einstein's field equations what matter-energy distribution is necessary to keep the wormholes stably traversable. They then establish that the required energy density violates current quantum field theory, such as the law of conservation of energy. They postulate that a certain physical entity, yet to be determined, is needed to stabilize the walls of the wormholes. For this task, they introduce an epistemic object,[16] the so-called *exotic matter* – which initially serves as a placeholder for an epistemic void in order to organize future astrophysical research discourses that search for the solution of this problem. It turned out that, according to the new modeling approach, the interior of the spacetime tunnel had a negative energy density. Around 1988, nobody in physics had any idea what this would look like. The following quote is the opening sentence of the Morris and Thorne paper in the *American Journal of Physics.*

> Science fiction stories [. . .] and films often use black holes for rapid interstellar travel: Intrepid adventurers plunge into a black hole and find themselves almost immediately emerging at some distant location in our universe – much to the annoyance of afficionados who can marshal a long list of objections [. . .]. (WiS, 395)

Here, Thorne and Morris approach literary science fiction discourse and simultaneously distance themselves from it. Their paper is not about what is already technically feasible, but about exploring the scope of theoretical modeling in physics. What is conceptually possible should therefore not be reduced to what is technically feasible. A narratively configured idea could inspire the epistemic power of imagination (see Badura and Kind 2021, Kind and Kung 2016) to find out what the solutions to Einstein's equations would theoretically yield. The scope is in turn limited by what mathematics and logic allow and by what previous models have already shown.

3.3 Epistemic functions of interformation

The process of interformation has several epistemic functions, which I would like to call the *explorative*, the *performative*, the *immersive* and the *persuasive* function. For it is not only the idea of the novel that is continued in Thorne and Morris – one can also show interformative narrative techniques. These are used scientifically to demonstrate various aspects of the world of wormhole models. The narration opens up the possibility of epistemic exploration. The first part of the paper, the introduction, had presented the difficulties of existing models. The second part pre-

16 I use this term in the sense of Hans-Jörg Rheinberger 1992 and 2019.

sented the necessary characteristics of the astrophysical object that make human traversal conceivable. The third part of the paper is devoted to the concrete implementation of the theoretical model. Here it becomes clear that the argumentation is divided into three different sections. The first section provides the consistent mathematical solution to Einstein's field equations. The mathematical form of the spacetime metric (cf. WiS, 400), which is static and spherical, and the Einstein tensors, which determine the geometry of spacetime, are taken into account.

This is followed by exploratory modeling, which refers to criteria (5) to (7), which are constraints of the novel's storyworld. Thorne and Morris return in their scientific argumentation to the situation of the characters of the novel and so we can see how the narrative discourse is functionalized epistemically, immersively, exploratively and persuasively.

A journey through a traversable wormhole is difficult to imagine as long as it is not clear what this mathematical, physical or cosmological object is supposed to be like. It is therefore worth introducing the perspective of an observer for whom the theoretically modeled spacetime tunnel should be traversable.

The characters in the novel are already on their way to Vega and act in this setting. For them, the scientific argumentation has an existential dimension. The narrative text can thus unfold the pull of cognitive immersion (Ryan 2001, 2003). The emplotment suggests the existential threat to the characters in this situation on their way to Wega. It is a "liminal" situation: old, known laws of nature *no longer* apply, and new models of new laws of nature, such as the astrophysical model of the space-time tunnel, *do not yet* apply. Perhaps this is the background for the interpolation of decisive dialogue quotes from the novel *Contact* into the scientific paper. Ellie, the main character of the novel, is quoted in the scientific paper during her exchange with her astronaut colleagues:

> "Look, fellas," she said, "I'm no expert in General Relativity. But didn't we see black holes? Didn't we fall into them? Didn't we emerge out of them? Isn't a gram of observation worth a ton of theory?"
>
> "I know, I know," Vaygay said in mild agony. "It has to be something else. Our understanding of physics can't be so far off. Can it?"
>
> He addressed this last question, a little plaintively, to Eda, who only replied, "A naturally occurring black hole can't be a tunnel; they have impassible singularities at their centers." (*C*, 347–348; quoted in WiS, 397)

Thorne and Morris refer back to the novel and introduce the perspective of the novel character Ellie Arroway – as a so-called vector for cognitive immersion. They suggest adopting the following perspective in order to imagine what has been scientifically explained:

> If somehow an adventurer were to pass unscathed through a Kerr tunnel and emerge into some distant region of our own universe, by adjusting slightly her trajectory through the tunnel *(we adopt the spirit of Contact in which the protagonist and wormhole traveler is Dr. Ellie Arroway)* she could emerge whenever she wishes [. . .] (WiS, 396; emphasis A. H.)

The reader of the scientific paper can witness how the boundaries of the fictional world of the novel are transgressed when Thorne and Morris suggest that Ellie, a fictional character of the novel, should be also adopted as a focalizer within the factual, scientific discourse. They retain Ellie's pronouns "she/her" for the whole argumentation of the scientific paper. A fictional character becomes the focalizer of a thought experiment in a scientific paper – is this a special form of scientific metalepsis? Why do the authors of the scientific publication decide to adopt the perspective of the fictional Ellie? Because the narrative part of the argumentation is not just a rhetorical-persuasive one. The narrative has its own part to play in the creation of this explorative world model. It seems paradoxical to use a fictional character to make a scientific argument plausible. However, this can be motivated by cognitive immersion. In Jean-Marie Schaeffer's terminology, the figure of Ellie would in this case be the vector of immersion: "An immersion vector is in some way the key of access thanks to which we can enter into that universe."[17]

Ellie as a character is the key to exemplifying the new world of the wormhole. Thorne and Morris also consistently cite Ellie's perspective at other points in their paper. The argumentation of the technical text thus switches to a thought experiment, which can be read as an interformative strategy between the novel and the scientific paper, as the quotes below show:

> Turn attention now to a thought experiment in which a traveler journeys radially through a wormhole, beginning at rest in a space station in the lower universe, [. . .] and ending at rest in a space station in the upper universe, [. . .]. Denote by $v(r)$ the radial *velocity of the traveler as she passes radius* r, as measured by a static observer there [. . .] (WiS, 403, emphasis A. H.)

> [. . .] if the observer falls sufficiently slowly, *she feels* arbitrarily small tidal gravitational force. [. . .] Traveler feels tidal-gravity accelerations *between different parts of her body* with magnitude [. . .]. (*C*, 398, emphasis A. H.)

> *Her* velocity is also constrained by the demand that *she not feels* too large an acceleration. (*C*, 409, emphasis A. H.)

17 "Immersion postures [. . .] determine the aspectuality or the particular modality with which the universe manifests itself to us from the fact that we enter it thanks to a key of access, an immersion vector." (Schaeffer 2010, 218–219).

> For concreteness, ask *her* to accelerate away from the lower station [. . .] until *she* is halfway to the throat, then decelerate [. . .] until *she* comes to rest at the throat, then accelerate away from the throat [. . .] until *she* is halfway to the upper station, then finally decelerate [. . .] until *she* comes to rest at the upper station. (*C*, 409, emphasis A. H.)

3.4 Introduction of a new epistemic object

The most delicate area of the wormhole construction is that metaphorically referred to as "the throat." Thorne and Morris envisage a certain spacetime geometry to ensure that it remains stable for humans to pass through. In order to stabilize the walls of the wormhole, they have to set constraints for the energy-momentum tensor that are problematic. To satisfy these conditions, they introduce an epistemic object that has never been observed before. They call it "exotic matter":

> The constraint $\tau_0 > \rho_0 c^2$ is deeply troublesome; it says that in the throat the tension must be so large as to exceed the total density of mass-energy $\rho_0 c^2$. We shall call material with this property, $\tau > \rho c^2 > 0$, "exotic". (WiS, 405)

Such a metaphor is not without precedent in contemporary cosmology – think of the term "dark energy." The exotic material is needed to ensure that the walls of the wormhole remain stably navigable. Thorne and Morris continue the argument from the perspective of Ellie, the novel's main character. In the summary of the theoretical model that Thorne and Morris provide in Box 3 on page 399, the function of this narrative technique of *internal focalization* can be clearly seen: it can be contrasted with the *external focalization* of the observers that are left on the earth and has to do with the concept of "Eigenzeit" or "proper time" in the special theory of relativity.[18] The scientists are concerned with the focalization with regard to the proper reference system of an observer. The solutions of the field equations and the requirement of the absence of a curvature singularity result:

> (Field equations) + (absence of horizon at throat) $\Rightarrow \tau > \rho c^2$ in throat $\Rightarrow$ traveler moving through throat at very high speed sees negative mass-energy density $\Rightarrow$ violation of weak, strong, and dominant energy conditions in throat ("exotic" matter that *might* – but it is not known for sure – be forbidden by laws of physics). (WiS, 399)

---

18 On the significance of this concept in the special theory of relativity, see Heydenreich 2021a and Heydenreich 2021b. For a cultural studies perspective on this concept, see: Gamper and Hühn 2014; Gamper, Hühn and Richter 2020; Gamper 2020.

Exotic matter should be understood as a functional concept (Cassirer 2000), which is conceptualized epistemically in order to enable further theoretical modeling.

3.5 Focalization as an interformative technique

The narratological concept of focalization finds its interformative correlate in the perception from different spacetime frames provided by the special theory of relativity. Thorne and Morris also argue in this way. At this point, literary and relativity-theoretical modeling approaches are superimposed:

> If wormhole travel is to be at all convenient for human beings, the traveler's journey must satisfy [. . .] constraints: (i) the entire trip should require less than or of order 1 year as measured both by the traveler and by people who live in the stations, [. . .] (ii) the acceleration **a** felt by the traveler must not exceed by much [. . .] Earth gravity; [. . .] As an aid to discussing the accelerations felt by the traveler, we introduce the orthonormal basis of her own reference frame, $e_{\hat{0}'}$, $e_{\hat{1}'}$, $e_{\hat{2}'}$, $e_{3'}$. (WiS, 403, emphasis original text)

On the one hand, there is the character Ellie, for whom a specific reference system is specified as her *frame of reference*. She internally focalizes the progress of the thought experiment. This also means that: a) the time it takes to cross the wormhole, b) the acceleration that the traveler experiences, and c) the tidal forces that her body feels (cf. WiS, 403), are always calculated relative to her frame of reference, which is the inner perspective within the wormhole. On the other hand, there are external observers "who live in the stations" (WiS, 403) and who focalize the experiment externally: "Denote by *v(r)* the radial velocity of the traveler as *she* passes radius *r*, as measured by a static observer there". (WiS, 403) The space and time coordinates of Ellie's internally focalized reference system are indicated by Arabic numerals as indices: $e_{\hat{0}'}$, $e_{\hat{1}'}$, $e_{\hat{2}'}$, $e_{3'}$ (cf. WiS, 403). The coordinates of the external focalizers and of the external reference system are recorded with Greek indices: $e_{\hat{t}}$, $e_{\hat{r}}$, $e_{\hat{\theta}}$, $e_{\hat{\phi}}$.

Ellie as an internal focalizer and the observers from the space-time base on the earth as external focalizers would measure the three criteria mentioned above: a) the time it takes to cross the wormhole, b) the acceleration that the traveler experiences, and c) the tidal forces that her body feels. The results of their measurements will not coincide, but this is a fact postulated by special relativity theory. Each focalization frame has its own spacetime structure and hence its own proper measurement-perspective. But this does not mean, that "everything is relative" because their measurements can be correlated through the Lorentz transformation. The space and time coordinates of both reference systems can be transformed from one reference system to the other using the Lorentz transformation.

> Expressed in terms of the orthonormal basis of the static observers, $e_{\hat{t}}$, $e_{\hat{r}}$, $e_{\hat{\theta}}$, $e_{\hat{\phi}}$, the traveler's basis is given by the standard special relativity Lorentz transformation [. . .]. (WiS, 403–404)

> By transforming the components [. . .] of the Riemann tensor from the static observers' frame $e_{\hat{t}}$, $e_{\hat{r}}$, $e_{\hat{\theta}}$, $e_{\hat{\phi}}$ to the traveler's frame $e_{\hat{0}'}$, $e_{\hat{1}'}$, $e_{\hat{2}'}$, $e_{\hat{3}'}$ [a special relativity type transformation (Lorentz transformation) since both sets of basis vectors are orthonormal], [. . .] (WiS, 404)

While the perspectives of the two observers must be regarded as relative to their own reference system, the Lorentz transformation is covariant.[19] This argument is important when it comes to the conceptualization of exotic matter. How do Thorne and Morris argue to introduce this epistemic object? By presenting it as differently observable and measurable from different focalizing perspectives – relative to the frame of reference of the observer.

The problem here is that exotic matter violates a fundamental principle of physics: that of the conservation of energy. The observer traveling through the spacetime tunnel at a very high speed will observe a negative energy density in her own spacetime reference system. "If such an observer moves sufficiently fast [. . .], the observer will see a negative density of mass-energy!" (WiS, 405). However, the external static observers do not observe any violation of the principle of conservation of energy in their own spacetime reference system: "One might wish to require $\rho \geq 0$ everywhere (static observers see nonnegative mass-energy density)". (WiS, 399). Morris and Thorne use narrative focalization techniques, to argue that the measurements from the external reference system, that is for the externally focalized observers, there are no violations of the principle of energy conservation.

In the third phase of exploratory modeling, the style of scientific argumentation changes significantly. Thorne and Morris justify this: "In the remainder of this section we shall ponder the possible existence of the exotic material necessary for wormhole construction. Our pondering will lead us to the forefront of current research – and in doing so will necessitate a change of style [. . .]" (WiS, 405). Thorne and Morris point out that in the 1960s and 1970s most physicists were convinced that no observer would ever be able to measure negative energy

---

19 On the importance of the Lorentz transformation for the conceptualization of Einstein's special relativity see Heydenreich 2021b, 122–133. On the epistemic functions of the Lorentz transformation for the process of interformation see Heydenreich 2021a, 79–82.

density. However, they point to a few exceptions that are supported by some significant research developments: Stephen Hawking's discovery of thermal radiation from black holes, known as Hawking radiation (Hawking 1975), is interpreted by Thorne and Morris as a possible indication that energy conservation constraints can be briefly violated.

In addition, they point to quantum field theory and Dirac's creation and annihilation operators: "In fact, particle creation always entails a violation of the energy conditions" (WiS, 405). Finally, Thorne and Morris refer to the following line of argument: it could be that the laws of nature do not allow the existence of exotic matter. But this prohibition is not yet definitive – because the non-existence of exotic matter has not been theoretically proven. In this respect, both lines of research would be promising: both the theoretical search for exotic matter and the attempt to prove that it does not exist.

> It may well be that the fundamental laws of physics forbid exotic material on the macroscopic scales required for wormhole building; but the authors know of no way to prove so and, in fact, would not be extremely surprised if a quantum-field-theoretic example of such material were found in the near future. The search for such an example, or an impossibility proof, is an interesting challenge. (WiS, 406)

Thorne and Morris also point out the fallibility of the validity of theoretical assumptions: "Similarly, the beliefs of the 1960s and early 1970s that matter must always possess positive energy density and satisfy $|\tau| \leq \rho c^2$, [. . .], have been supplanted more recently by the realization that this is not so" (WiS, 406).

We can conclude here that the interformation between literary and astrophysics modeling practices does not lead to the solution of the problem, but to the reorganization of the research question for the purpose of enabling more flexible approaches to theoretical modeling in astrophysics. Thorne and Morris ultimately argue for an explorative form of theoretical modeling. In the sense of a justifying eristics, they also argue against the prejudices and epistemic routines of the scientific expert community: "It may well be that today's widely held prejudices [. . .] when averaged over macroscopic scales and over time (exotic material) will also fall when we better understand the laws of physics" (WiS, 406). The problem cannot yet be solved because it is situated precisely at the neuralgic intersection between general relativity and quantum field theory: The two fundamental theories of the twentieth century make contradictory predictions because they have not yet been unified.

And finally, Thorne and Morris try to make their case plausible by recurring to literary textual strategies. At the end of the article, they return to the novel and emphasize one of the most genuine techniques of literary texts, their explorative treatment of epistemic blank spaces. Thorne and Morris also make use of this argument for theoretical modeling. Sagan's *Contact* is cited by Thorne and this quotation exemplifies the epistemic functions of interformation:

> Ellie did not see how this represented much progress, and asked him to clarify. The key problem was holding the wormhole open. Eda had found a class of solutions to his field equations that suggested a new macroscopic field, a kind of tension that could be used to prevent a wormhole from contracting fully. Such a wormhole would pose none of the other problems of black holes; it would have much smaller tidal stresses, two-way access, quick transit times as measured by an exterior observer, and no devastating interior radiation field. (*C*, 406; quoted in WiS, 397)

The scientific paper and the literary text cite each other in order to legitimize both the literary and the scientific modeling processes. The novel thus attempts to make its storyworld more plausible by presenting Thorne's model in the words of Eda's character as a theorist of relativity. The scientific paper, in turn, quotes the novel in order to make plausible a greater openness to new questions and ways of thinking in physics research. Thorne and Morris explicitly emphasize the epistemic function of the imagination (see Badura and Kind 2021) for physics research:

> Note the emphasis on the word *imagine*; we today are far from being able to analyze such a process theoretically. Any such analysis will require a reliable understanding of quantum gravity. (WiS, 407, emphasis in the original text)

> Such better understanding (the discovery by Eda of a field that produces an anisotropic stress with $\tau > \rho c^2 > 0$ along one direction), is the key in Sagan's novel (Box 1) to an understanding of the characters' wormhole travel experiences. (WiS, 406)

This solution was suggested by Thorne and presented in literary form by Sagan. Thorne worked for three years on his theoretical model, and when he presented it to the scientific public, he relied on interformative practices with clear references to literary discourse to encourage his epistemic community to take the chance of new ways of thinking. In their conclusions, Thorne and Morris state:

> The wormhole solutions to Einstein's equations presented in this article are not only a pedagogical tool for teaching general relativity. [. . .] The topology change required for wormhole formation may not be classically allowed, is not quantum mechanically understood, and might be quantum mechanically forbidden [. . .]. Nevertheless, we do not know today enough to either affirm or refute these difficulties, and we correspondigly cannot rule out traversible spacetime wormholes. (WiS, 407)

4 Scientific impact of the new wormhole model: reorganization of research discourses

With their scientific paper, Thorne and Morris have reconceptualized existing scientific approaches and thus initiated a fruitful new research approach. From then on, their interformative explorative model was to reorganize research discourses – this is impressively made evident from the statistical survey that records the number of citations of Thorne and Morris' wormhole article between 1988 and 2022 (see Fig. 4 below, citation index from the scientific database *Web of Science*).

Thorne and Morris have shown that the mathematics of Einstein's field equations allows the model of traversable wormholes within the framework of general relativity in principle. The research questions that arise following their article are: Is the modification of the spacetime topology, as proposed by Thorne and Morris, allowed (see Barros et al. 2018)? What about the violation of the law of conservation of energy (see Visser 2003, Di Grezia et al. 2017)? Is exotic matter really necessary (see Sahoo et al. 2018)? How could it be modeled (see Mateos et al. 2018)? How much exotic matter is necessary to stabilize the spacetime tunnels? Does the latest research in the field of quantum gravity open up new possibilities for modeling traversable wormholes (see Novikov 2018)?

Various branches of science are involved in continuing the exploratory research, some of which also started from the process of interformation described: Astronomy and Astrophysics (6459 citations) (see Yousaf et al. 2017), Elementary Particle Physics (7014 citations), Mathematical Physics (960 citations) (see Jusufi 2017, Hammad 2018, Sharif and Nawazish 2018), Cosmology (see Rami Ahmad El-Nabulsi 2017, Wang and Meng 2017), Computer Science and Artificial Intelligence (533 citations), Engineering (359 citations), Optics (213 citations) and Earth Sciences (see Willenborg at al. 2017) have taken up the proposal and taken it further theoretically. The following chart (see Fig. 5) provides an overview of the scientific journals in which the articles appeared: These are high-ranking titles such as *Physical Review Letters, Classical and Quantum Gravity, General Relativity and Gravitation, Modern Physics Letters* and *The European Physical Journal.*

To conclude: One of the epistemic functions of interformation in the case of the explorative model of the traversable wormhole is to identify an epistemic gap in order to generate a discursive competition for the plausibilization and legitimization of the new model up to the point of proving its empirical adequacy. In this case, fictional and factual texts build up a textual nexus as a point of intersection. The very fact that the explorative model does not yet have an ontological, empiri-

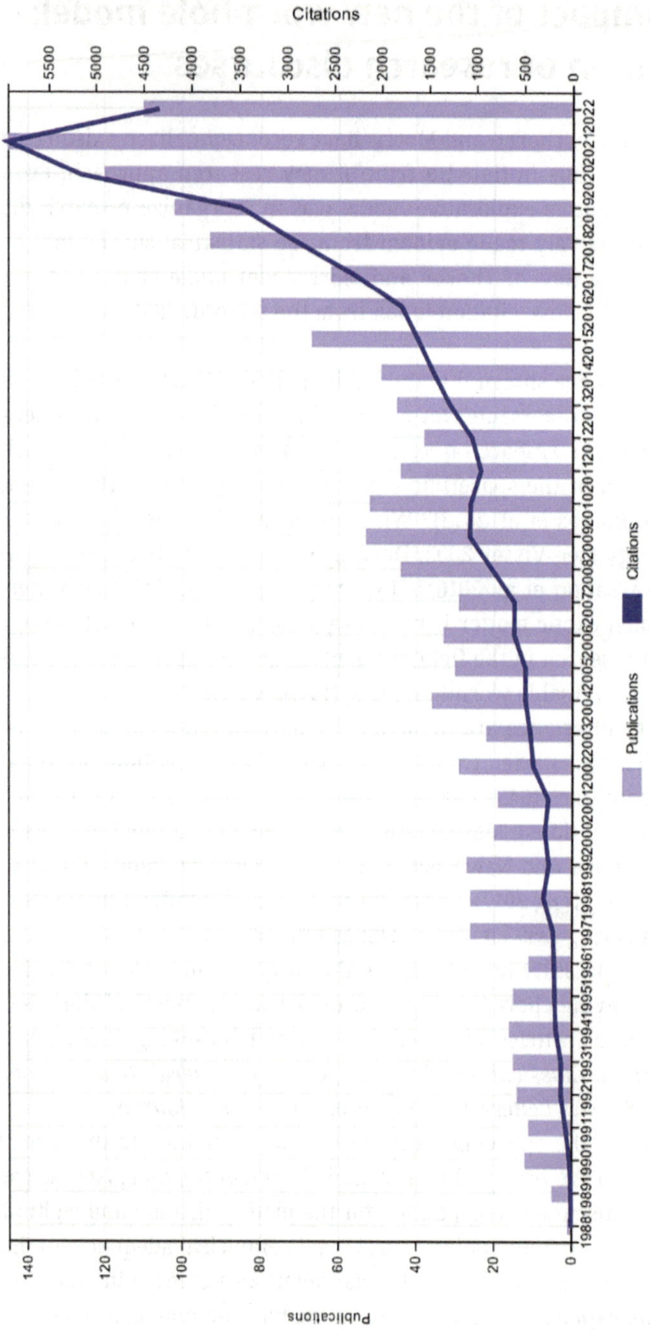

Fig. 4: Citation index for the scientific paper by Thorne and Morris from 1988; source: *Web of Science*, 02.10.2022.

cally verifiable object character, but is still an epistemic object, is a good prerequisite for increasing the complexity of the research discourses.

The system dynamics part of scientific modeling takes into account mathematical operations, physics principles and specific constraints. Without these well-considered *constraints*, the model would not find its way into scientific discourse. The process-dynamic dimension of modeling, on the other hand, introduces explorative leeways. The purposes of these explorative practices are modeling as mediation. Even in this case, not all contradictions are completely resolved. But only explorative modeling enables a reconceptualization of the question and thus an even more differentiated discursive approach to the research problem.

5 The film *Interstellar*: Interformation between science and film knowledge production

The interformative explorative model has also had an impact in other media – for example in the collaboration between Kip Thorne and Christopher Nolan on the film *Interstellar* (2014). There are connections between the film *Interstellar* and the novel *Contact*, because the film represents the continuation of the artistic-scientific interformative collaboration with other means. The film goes back to Kip Thorne's scientific modeling. This time Thorne goes in search of a film director who can work together on the project in an interformative way in order to make the scientific vision aesthetically plausible, visible and perhaps also to legitimize it. The film *Interstellar* made it possible to show the first three-dimensional simulation of a traversable wormhole based on Kip Thorne's calculations and predictions, i.e. on the general theory of relativity, because of the film technology used. Thorne documented this collaboration in his book *The Science of Interstellar (2014)*. One can see the wormhole visualized in the film in Fig. 6).

In 2015, the collaboration on this film resulted in two new scientific publications presenting the scientific findings that emerged from the film work. The first is "Gravitational lensing by spinning black holes in astrophysics, and in the movie 'Interstellar'" by Eugénie von Tunzelmann and Kip Thorne.[20] The second article on the cinematic visualization of the wormhole using three-dimensional film

20 See Oliver James, Kip Thorne et al.: "Gravitational Lensing by Spinning Black Holes in Astrophysics, and in the Movie *Interstellar*." *Classical and Quantum Gravity* 32/6 (2015), Art. 065001, doi:10.1088/0264-9381/32/6/065001.

Fig. 5: An overview of scientific journals that cite the paper by Morris/Thorne on wormholes from 1988. Source: Web of Science 02.10.2022.

Fig. 6: Still from movie *Interstellar*: First simulated three-dimensional image of a wormhole based on Kip Thorne's calculations according to Einstein's field equations.

techniques was published in the same *American Journal of Physics* in which the first model of the traversable wormhole was presented: "Visualizing 'Interstellar's' Wormhole" by Oliver James and Kip Thorne.[21] This is basically the continuation of the interformative project between Sagan and Thorne, now with different means. This would provide a new case for the study of interformation, but this time using other interformative practices between astrophysics, film and media theory. The starting point for future research into the epistemic functions of interformation is indicated by the following quotation:

> In 1988, in connection with Carl Sagan's novel *Contact*, later made into a movie, one of the authors published an article in this journal about wormholes as a tool for teaching general relativity.
>
> This article is a follow-up, a quarter century later, in the context of Christopher Nolan's movie *Interstellar* and Kip Thorne's associated book *The Science of Interstellar*. Like *Contact*, *Interstellar* has real science built into its fabric, thanks to a strong science commitment by the director, screenwriters, producers, and visual effects team, and thanks to Thorne's role as an executive producer.

21 See Oliver James, Kip Thorne et al.: "Visualizing *Interstellar*'s Wormhole." *American Journal of Physics* 83/6 (2015): 486–499, doi:10.1119/1.4916949.

> Although wormholes were central to the theme of *Contact* and to many movies and TV shows since then, including *Star Trek* and *Stargate*, none of these have depicted correctly a wormhole as it would be seen by a nearby human. *Interstellar* is the first to do so. The authors of this paper, together with Christopher Nolan who made key decisions, were responsible for that depiction. (James, Thorne et al. 2015, 486).

6 Appendix

The list of disciplines and branches of research that cite the two Wormhole articles by Morris and Thorne according to the *Web of Science* database.[22]

Web of Science Categories	Record Count	% of 12,892
Physics Particles Fields	7014	54.406
Astronomy Astrophysics	6459	50.101
Physics Multidisciplinary	3078	23.875
Physics Mathematical	960	7.446
Physics Nuclear	912	7.074
Quantum Science Technology	761	5.903
Computer Science and Artificial Intelligence	533	4.134
Engineering Electrical Electronic	359	2.785
Computer Science Interdisciplinary Applications	265	2.056
Optics	213	1.652
Multidisciplinary Sciences	212	1.644
Computer Science Information Systems	202	1.567
Telecommunications	169	1.311
Engineering Multidisciplinary	158	1.226
Physics Atomic Molecular Chemical	139	1.078
Physics Applied	133	1.032
Computer Science Theory Methods	119	0.923
Materials Science Multidisciplinary	109	0.845
Energy Fuels	81	0.628
Operations Research Management Science	79	0.613
Computer Science Software Engineering	74	0.574
Mathematics	73	0.566
Mathematics Applied	71	0.551
Mathematics Interdisciplinary Applications	70	0.543
Engineering Mechanical	69	0.535
Physics Condensed Matter	69	0.535

22 Source: https://www.webofscience.com/wos/woscc/analyze-results/5b185299-e907-451c-a10c-3c9b4ff5cfe3-530291b1 [accessed 2 October 2022].

(continued)

Web of Science Categories	**Record Count**	**% of 12,892**
Mechanics	47	0.365
Education Scientific Disciplines	44	0.341
Automation Control Systems	43	0.334
Mathematical Computational Biology	42	0.326
Environmental Sciences	40	0.310
Civil engineering	38	0.295
Instruments Instrumentation	38	0.295
Neurosciences	35	0.271
Chemistry Multidisciplinary	32	0.248
Geosciences Multidisciplinary	31	0.240
Green Sustainable Science Technology	30	0.233
Nanoscience Nanotechnology	30	0.233
History Philosophy Of Science	29	0.225
Engineering Aerospace	28	0.217
Thermodynamics	26	0.202
Chemistry Physical	25	0.194
Engineering Biomedical	21	0.163
Computer Science Hardware Architecture	20	0.155
Engineering Industrial	19	0.147
Engineering Chemical	18	0.140
Physics Fluids Plasmas	18	0.140
Chemistry Analytical	16	0.124
Engineering Manufacturing	15	0.116
Environmental Studies	14	0.109
Materials Science Characterization Testing	12	0.093
Water Resources	12	0.093
Construction Building Technology	11	0.085
Imaging Science Photographic Technology	10	0.078
Biology	9	0.070
Engineering Environmental	9	0.070
Remote Sensing	9	0.070
Engineering Geological	7	0.054
Statistics Probability	7	0.054
Electrochemistry	6	0.047
Robotics	6	0.047
Education Educational Research	5	0.039
Transportation Science Technology	5	0.039
Acoustics	4	0.031
Computer Science Cybernetics	4	0.031
Engineering Marine	4	0.031
Materials Science Biomaterials	4	0.031
Meteorology Atmospheric Sciences		40.031
Philosophy		40.031

(continued)

Web of Science Categories	Record Count	% of 12,892
Biochemical Research Methods		30.023
Biochemistry Molecular Biology		30.023
Management		30.023
Materials Science Textiles		30.023
Medical Informatics		30.023
Metallurgy Metallurgical Engineering		30.023
Oceanography		30.023
Engineering Ocean		20.016
Geochemistry Geophysics		20.016
Health Care Sciences Services		20.016
Medicine General Internal		20.016
Psychology Experimental		20.016
Radiology Nuclear Medicine Medical Imaging		20.016
Surgery		20.016
Agricultural Engineering		10.008
Agronomy		10.008
Biophysics		10.008
Biotechnology Applied Microbiology		10.008
Chemistry Applied		10.008
Clinical Neurology		10.008
Crystallography		10.008
Economics		10.008
Engineering Petroleum		10.008

Reference list

Aneesh, S., Sukanta Bose, and Sayan Kar. "Gravitational Waves from Quasinormal Modes of a Class of Lorentzian Wormholes." *Physical Review D: Particles and Fields* 97.12 (2018): art. 124004. doi:10.1103/PhysRevD.97.124004.

Balke Friedrich, Bernhard Siegert, and Joseph Vogl, eds. *Modelle und Modellierung*. Paderborn: Fink 2014.

Barros, Bruno J., and Francisco S. N. Lobo. "Wormhole Geometries Supported by Three-Form Fields." *Physical Review D: Particles and Fields* 98.4 (2018): art. 044012. doi:10.1103/PhysRevD.98.044012.

Bokulich, Alisa. "Explanatory Fictions." *Fictions in Science. Philosophical Essays on Modeling and Idealization*. Ed. Mauricio Suárez: New York and London: Routledge, 2009. 91–109.

Cassirer, Ernst. *Substanzbegriff und Funktionsbegriff. Untersuchungen über die Grundfragen der Erkenntniskritik. Gesammelte Werke. Hamburger Ausgabe*. Vol. 6. Ed. Birgit Recki. Hamburg: Meiner, 2000.

Chakraborty, Subenoy. "A New Stability Criterion for General Cylindrical Thin-Shell Wormholes." *General Relativity and Gravitation* 49.3 (2017): art. 47. doi:10.1007/s10714-017-2210-2.

Chew, Xiao Yan, Burkhard Kleihaus, and Jutta Kunz. "Spinning Wormholes in Scalar-Tensor Theory." *Physical Review D: Particles and Fields* 97.6 (2018): art. 064026. doi:10.1103/PhysRevD.97.064026.

Coleman, Douglas W. "So You Want to Communicate with Space Aliens? BEROM." *LACUS Forum* 34 (2009): 55–63.

Contact. Dir. Zemeckis Robert. Warner Home Video, 1997.

Di Grezia, Elisabetta, Emmanuele Battista, Mattia Manfredonia, and Gennaro Miele. "Spin, Torsion and Violation of Null Energy Condition in Traversable Wormholes." *The European Physical Journal Plus* 132.12 (2017): art. 537. doi:10.1140/epjp/i2017-11799-6.

Douglas, Christopher. *If God Meant to Interfere: American Literature and the Rise of the Christian Right*. Ithaca, UK and London: Cornell University Press, 2016.

Echeverria, Fernando, Gunnar Klinkhammer, and Kip S. Thorne. "Billiard Balls in Wormhole Spacetimes with Closed Timelike Curves: Classical Theory." *Physical Review D: Particles and Fields* 44.4 (1991): 1077–1099. doi:10.1103/PhysRevD.44.1077.

Einstein, Albert. "Die formale Grundlage der allgemeinen Relativitätstheorie." *Sitzungsberichte der Preussischen Akademie der Wissenschaften zu Berlin* (1914): 1030–1085.

Einstein, Albert. "Die Feldgleichungen der Gravitation." *Sitzungsberichte der Preussischen Akademie der Wissenschaften zu Berlin* (1915): 844–847.

Einstein, Albert. "Die Grundlage der allgemeinen Relativitätstheorie." *Annalen der Physik* 49 (1916): 769–822.

Einstein, Albert, and Nathan Rosen. "The Particle Problem in the General Theory of Relativity." *Physical Review* 48.1 (1935): 73–77. doi:10.1103/PhysRev.48.73.

El-Nabulsi, Rami Ahmad. "Wormholes in Fractional Action Cosmology." *Canadian Journal of Physics* 95.6 (2017): 605–609. doi:10.1139/cjp-2017-0109.

Elgin, Catherine Z. "Telling Instances". *Beyond Mimesis and Convention: Representation in Art and Science*. Eds. Roman Frigg and Matthew C. Hunter. Dordrecht: Springer, 2010. 1–17. doi:10.1007/978-90-481-3851-7_1.

Erdbeer, Robert Matthias. "Poetik der Modelle. *Textpraxis: Digitales Journal für Philologie* 11.2 (2015). http://www.uni-muenster.de/textpraxis/robert-matthias-erdbeer-poetik-der-modelle (27 August 2024).

Erdbeer, Robert Matthias, Florian Kläger, and Klaus Stierstorfer, eds. *Literarische Form / Literary Form. Theorien – Dynamiken – Kulturen. Beiträge zur literarischen Modellforschung / Theories – Dynamics – Cultures. Perspectives on Literary Modelling*. Heidelberg: Winter, 2018.

Erdbeer, Robert Matthias, Florian Kläger, and Klaus Stierstorfer, eds. "Across Philologies. Modelling and Literary Form". *Literarische Form / Literary Form. Theorien – Dynamiken – Kulturen. Beiträge zur literarischen Modellforschung / Theories – Dynamics – Cultures. Perspectives on Literary Modelling*. Eds. Robert Matthias Erdbeer, Florian Kläger, and Klaus Stierstorfer. Heidelberg: Winter, 2018. 9–31.

Falkenburg, Brigitte, and Wolfgang Muschik, eds. *Models, Theories and Disunity in Physics*. Frankfurt a. M.: Klostermann, 1998.

Fisher, Grant, Axel Gelfert, and Friedrich Steinle. "Exploratory Models and Exploratory Modeling in Science: Introduction". *Perspectives on Science* 29.4 (2021): 355–358.

Flamm, Ludwig. "Beiträge zur Einsteinschen Gravitationstheorie." *Physikalische Zeitschrift* 17 (1916): 448–454.

Flamm, Ludwig. "Contributions to Einstein's Theory of Gravitation." Trans. Jörg Frauendiener. *General Relativity and Gravitation* 47.6 (2015): art. 72. doi:10.1007/s10714-015-1908-2.

Frigg, Roman. "Scientific Representation and the Semantic View of Theories." *Theoria: An International Journal for Theory, History and Foundations of Science* 55 (2006). 49–65. doi:10.1387/theoria.553.

Frigg, Roman. Art. “Models in Physics (2009)”. *Routledge Encyclopedia of Philosophy Online (REP Online)*. London: Routledge. doi:10.4324/9780415249126-Q135-1.
Frigg, Roman. “Models and Fiction”. *Synthese* 172/2 (2010): 251–268.
Frigg, Roman. “Fiction and Scientific Representation.” *Beyond Mimesis and Convention. Representation in Art and Science*. Eds. Roman Frigg and Matthew C. Hunter. Dordrecht: Springer 2010. 97–138, doi:10.1007/978-90-481-3851-7_6.
Frigg, Roman. *Models and Theories. A Philosophical Inquiry*. London: Routledge, 2023.
Frigg, Roman, and James Nguyen. “The Fiction View of Models Reloaded.” *The Monist* 99/3 (2016): 225–242, doi:10.1093/monist/onw002.
Frigg, Roman, and Stephan Hartmann. Art. “Models in Science (2012).” *The Stanford Encyclopedia of Philosophy*. Summer 2018 Edition. Ed. Edward N. Zalta. Stanford, online at: https://plato.stanford.edu/archives/sum2018/entries/models-science/ (8 July 2022).
Gamper, Michael, ed. *Ästhetische Eigenzeiten der Wissenschaften*. Hanover: Wehrhahn, 2020.
Gamper, Michael, and Lukas Wolff, eds. *Physiker lesen, Physiker schreiben. Wissen der Literatur und humanistische Bildung in der modernen Physik*. Göttingen: Wallstein, 2024. 169–226.
Gamper, Michael, and Helmut Hühn, eds. *Zeit der Darstellung. Ästhetische Eigenzeiten in Kunst, Literatur und Wissenschaft*. Hanover: Wehrhahn, 2014.
Gamper, Michael, and Helmut Hühn. *Was sind Ästhetische Eigenzeiten?* Hanover: Wehrhahn, 2014.
Gamper, Michael, Helmut Hühn, and Steffen Richter, eds. *Formen der Zeit. Ein Wörterbuch der ästhetischen Eigenzeiten*. Hanover: Wehrhahn, 2020.
Gao, Ping, Daniel Louis Jafferis, and Aron C. Wall. “Traversable Wormholes via a Double Trace Deformation.” *Journal of High Energy Physics* 12 (2017): art. 151. doi:10.1007/JHEP12(2017)151.
Gelfert, Axel. *How to Do Science with Models. A Philosophical Primer*. Dordrecht and Cham: Springer, 2016.
Gelfert, Axel. “The Ontology of Models.” *Springer Handbook of Model-Based Science*. Eds. Lorenzo Magnani and Tommaso Bertolotti. Cham: Springer, 2017. 5–23.
Gelfert, Axel. “Cultures of Modeling. Rudolf Peierls on Model-Making in Physics”. *Studia Metodologiczne* 39 (2019): 49–71, doi:10.14746/sm.2019.39.2.
Giere Ronald N. “Why Scientific Models Should Not be Regarded as Works of Fiction.” *Fictions in Science. Philosophical Essays on Modeling and Idealization*. Ed. Mauricio Suárez. New York and London 2009. 248–258.
Grant Fisher, Axel Gelfert, and Friedrich Steinle. “Exploratory Models and Exploratory Modeling in Science: Introduction.” *Perspectives on Science* 29 (4) 2021: 355–358.
Godfrey-Smith, Peter. “The Strategy of Model-based Science.” *Biology and Philosophy* 21 (2006): 725–740.
Godfrey-Smith, Peter. “Models and Fictions in Science.” *Philosophical Studies* 143 (2009): 101–116.
Goodman, Nelson. *Languages of Art. An Approach to a Theory of Symbols*. 2nd edition. Indianapolis: Hackett, 2008.
Hammad, Fayçal. “Revisiting Black Holes and Wormholes under Weyl Transformations.” *Physical Review D: Particles and Fields* 97.12 (2018): art. 124015. doi:10.1103/PhysRevD.97.124015.
Hawking, Stephen W. “Particle Creation by Black Holes.” *Communications in Mathematical Physics* 43.3 (1975): 199–220. https://projecteuclid.org/euclid.cmp/1103899181 (8 July 2022).
Hayward, Sean A. “Wormhole Dynamics in Spherical Symmetry.” *Physical Review D: Particles and Fields* 79.12 (2009): art. 124001. doi:10.1103/PhysRevD.79.124001.
Heydari-Fard, Malihe, and Mohaddese Heydari-Fard. “Inhomogeneous Exact Solution in Brane Gravity and its Applications.” *General Relativity and Gravitation* 49.2 (2017): art. 21. doi:10.1007/s10714-017-2190-2.

Heydenreich, Aura. *Literatur und Naturwissenschaft: Interformation und epistemische Transformation zwischen Literatur und Astronomie, Mikrobiologie, Relativitätstheorie und Quantentheorie 1600-2020*. Berlin and Boston: De Gruyter, 2024.

Heydenreich, Aura. "Epistemic Narrativity in Albert Einstein's Treatise on Special Relativity. A Narratological Approach to 'The Electrodynamics of Moving Bodies'. The Process of Interformation (Part I)." *Physics and Literature. Concepts – Transfer – Aestheticization*. Eds. Aura Heydenreich and Klaus Mecke. Berlin and Boston: De Gruyter 2021a. 49–103. https://doi.org/10.1515/9783110481112-003.

Heydenreich, Aura. "Albert Einstein's 'Physics and Reality' and 'The Electrodynamics of Moving Bodies'. The Process of Interformation, Semiologic Foundations and Epistemic Transformations (Part II)." *Physics and Literature. Concepts – Transfer – Aestheticization*. Eds. Aura Heydenreich and Klaus Mecke. Berlin and Boston: De Gruyter 2021b. 105–138. https://doi.org/10.1515/9783110481112-004.

Heydenreich, Aura. „Zur epistemischen Funktion der Interformation zwischen Astrophysik und Literatur. Carl Sagans Roman *Contact* und Kip Thornes theoretisches Modell der durchquerbaren Raumzeittunnel." *Physiker lesen, Physiker schreiben. Wissen der Literatur und humanistische Bildung in der modernen Physik*. Eds. Michael Gamper and Lukas Wolff. Göttingen: Wallstein, 2024. 169–226.

Interstellar. Dir. Christopher Nolan. Paramount Pictures, 2014.

James, Oliver, Eugénie von Tunzelmann, Paul Franklin, and Kip S. Thorne. "Visualizing *Interstellar*'s Wormhole." *American Journal of Physics* 83.6 (2015): 486–499. doi:10.1119/1.4916949.

James, Oliver, Eugénie von Tunzelmann, Paul Franklin, and Kip Thorne. "Gravitational Lensing by Spinning Black Holes in Astrophysics, and in the Movie *Interstellar*." *Classical and Quantum Gravity* 32/6 (2015), Art. 065001, doi:10.1088/0264-9381/32/6/065001.

Jusufi, Kimet. "Deflection Angle of Light by Wormholes Using the Gauss-Bonnet Theorem." *International Journal of Geometric Methods in Modern Physics* 14.12 (2017): art. 1750179. doi:10.1142/S0219887817501791.

Jusufi, Kimet. "Conical Morris-Thorne Wormholes with a Global Monopole Charge." *Physical Review D: Particles and Fields* 98.4 (2018): art. 044016. doi:10.1103/PhysRevD.98.044016.

Kind, Amy, and Christopher Badura, eds. *Epistemic Uses of Imagination*. New York and London: Routledge, 2021.

Kind, Amy, and Peter Kung. *Knowledge through Imagination*. Oxford: Oxford University Press, 2016.

Knuuttila, Tarja. *Models as Epistemic Artefacts. Toward a Non-Representationalist Account of Scientific Representation*. Helsinki: University of Helsinki Press, 2005.

Kocuper, Ewa, Jerzy Matyjasek, and Kasia Zwierzchowska. "Stress-Energy Tensor of Quantized Massive Fields in Static Wormhole Spacetimes." *Physical Review D: Particles and Fields* 96.10 (2017): art. 104057. doi:10.1103/PhysRevD.96.104057.

Kokubu, Takafumi, Hideki Maeda, and Tomohiro Harada. "Does the Gauss-Bonnet Term Stabilize Wormholes?" *Classical and Quantum Gravity* 32.23 (2015): art. 235021. doi:10.1088/0264-9381/32/23/235021.

Koschorke, Albrecht. "Die Grenzen des Systems und die Rhetorik der Systemtheorie." *Widerstände der Systemtheorie: Kulturtheoretische Analyse der Werke von Luhmann*. Eds. Albrecht Koschorke and Cornelia Vismann. Berlin: Akademie Verlag, 1999. 49–60. doi:10.1524/9783050077857.49.

Koschorke, Albrecht. "Codes und Narrative. Überlegungen zur Poetik der funktionalen Differenzierung." *Grenzen der Germanistik. Rephilologisierung oder Erweiterung?* Ed. Walter Erhart. Stuttgart and Weimar: Metzler, 2004. 174–185.

Koschorke, Albrecht. “Zur Funktionsweise kultureller Peripherien.“ *Explosion und Peripherie. Jurij Lotman's Semiotik der kulturellen Dynamik revisited*. Eds. Susi K. Frank, Cornelia Ruhe, and Alexander Schmitz. Bielefeld: transcript, 2012. 27–40.

Koschorke, Albrecht. *Fact and Fiction. Elements of a General Theory of Narrative*. Transl. by Joel Golb. Berlin and Boston: De Gruyter, 2018.

Kuhfittig, Peter K. F. “Conformal Symmetry Wormholes and the Null Energy Condition.” *Journal of the Korean Physical Society* 70.11 (2017): 962–966. doi:10.3938/jkps.70.962.

Link, Jürgen. “Diskursanalyse unter besonderer Berücksichtigung von Interdiskurs und Kollektivsymbolik.“ *Handbuch sozialwissenschaftliche Diskursanalyse*. Eds. Reiner Keller et al. Vol. 1: Theorien und Methoden. Wiesbaden. 2nd. ed. 2006. 407–430.

Link, Jürgen. “Literaturanalyse als Interdiskursanalyse. Am Beispiel des Ursprungs literarischer Symbolik in der Kollektivsymbolik“. *Diskurstheorien und Literaturwissenschaft*. Eds. Jürgen Fohrmann and Harro Müller. Frankfurt a.M.: Suhrkamp, 1988. 284–307.

Lotman, Jurij. *Universe of the Mind. A Semiotic Theory of Culture*. London: Tauris, 2000.

Lotman, Jurij M. “On the semiosphere”. *Sign Systems Studies. 33*.1 (2005). 205–229. https://doi.org/10.12697/SSS.2005.33.1.09

Magnani, Lorenzo, Nancy J. Nersessian, and Paul Thagard, eds. *Model-Based Reasoning in Scientific Discovery*. New York: Kluwer Academic/Plenum Publishers, 1999.

Mateos, Jesús, and Carlos Sabín. “Quantum Simulation of Traversable Wormhole Spacetimes in a Bose-Einstein Condensate.” *Physical Review D: Particles and Fields* 97.4 (2018): art. 044045. doi:10.1103/PhysRevD.97.044045.

Mazharimousavi, S. Habib, and M. Halilsoy. “Wormhole Solutions in *f(R)* Gravity Satisfying Energy Conditions.” *Modern Physics Letters A* 31.34 (2016): art. 1650192. doi:10.1142/S0217732316501923.

Mecke, Klaus. “Zahl und Erzählung. Metaphern in Erkenntnisprozessen der Physik“. *Quarks and Letters. Natural Sciences in Contemporary Literature and Culture*. Eds. Aura Heydenreich and Klaus Mecke. Berlin and Boston: De Gruyter 2015. 31–85. doi:10.1515/9783110406542-002.

Mishra, Anuj, and Subenoy Chakraborty. “On the Trajectories of Null and Timelike Geodesics in Different Wormhole Geometries.” *The European Physical Journal C* 78.5 (2018): art. 374. doi:10.1140/epjc/s10052-018-5854-5.

Misner, Charles W., Kip S. Thorne, and John A. Wheeler. *Gravitation*. San Francisco: Freeman, 1973.

Moraes, P. H. R. S., and P. K. Sahoo. “Nonexotic Matter Wormholes in a Trace of the Energy-Momentum Tensor Squared Gravity.” *Physical Review D: Particles and Fields* 97.2 (2018): art. 024007. doi:10.1103/PhysRevD.97.024007.

Moraes, P. H. R. S., R. A. C. Correa, and R. V. Lobato. “Analytical General Solutions for Static Wormholes in *f(R,T)* Gravity.” *Journal of Cosmology and Astroparticle Physics* 7 (2017): art. 029. doi:10.1088/1475-7516/2017/07/029.

Morgan Mary S., and Margaret C. Morrison. “Models as Mediating Instruments.” *Models as Mediators. Perspectives on Natural and Social Sciences*. Eds. Mary S. Morgan and Margaret C. Morrison. Cambridge and New York: Cambridge University Press, 1999. 10–37.

Morgan Mary S., and Margaret C. Morrison, eds. *Models as Mediators. Perspectives on Natural and Social Sciences*. Cambridgeand New York: Cambridge University Press, 1999.

Morris, Michael S., and Kip S. Thorne. “Wormholes in Spacetime and Their Use for Interstellar Travel: A Tool for Teaching General Relativity.” *American Journal of Physics* 56.5 (1988): 395–412. doi:10.1119/1.15620.

Morris, Michael S., Kip S. Thorne, and Ulvi Yurtserver. “Wormholes, Time Machines, and the Weak Energy Condition.” *Physical Review Letters* 61.13 (1988): 1446–1449.

Morris, Michael S., Kip S. Thorne, and Ulvi Yurtserver. "Traversible Wormholes, Closed Timelike Curves, and the Averaged Weak Energy Condition." *Abstracts of Contributed Papers: Twelfth International Conference on General Relativity and Gravitation (GR12)*. Boulder, CO, 2–7 July 1989. 247.

Morrissey, Thomas J. "Teaching the Alphabet to the Ants: *The Adventures of Pinocchio* and Carl Sagan's *Contact*." *Extrapolation: A Journal of Science Fiction and Fantasy* 43.2 (2002): 163–170. doi:10.3828/extr.2002.43.2.04.

Morrison, Margaret C. "Modelling Nature. Between Physics and the Physical World." *Philosophia Naturalis* 35 (1998): 65–85.

Morrison, Margaret C. "Fictions, Representations and Reality." *Fictions in Science. Philosophical Essays on Modeling and Idealization*. Ed. Mauricio Suárez. New York and London: Routledge, 2009. 110–135.

Novikov, I. D., D. I. Novikov, and N. S. Kardashev. "The Role of Pressure as a Source of Gravity and Wormholes." *Astronomy Reports* 62.7 (2018): 421–425. doi:10.1134/S1063772918070028.

Ono, Toshiaki, Asahi Ishihara, and Hideki Asada. "Deflection Angle of Light for an Observer and Source at Finite Distance from a Rotating Wormhole." *Physical Review D: Particles and Fields* 98.4 (2018): art. 044047. doi:10.1103/PhysRevD.98.044047.

Övgün, Ali, and Kimet Jusufi. "Stability of Effective Thin-Shell Wormholes under Lorentz Symmetry Breaking Supported by Dark Matter and Dark Energy." *The European Physical Journal Plus* 132.12 (2017): art. 543. doi:10.1140/epjp/i2017-11829-5.

Petrilli, Susan. *Sign Studies and Semioethics. Communication, Translation and Values*. Berlin and Boston: De Gruyter Mouton, 2014.

Rheinberger, Hans-Jörg. *Experiment, Differenz,Schrift. Zur Geschichte epistemischer Dinge.* Marburg an der Lahn: Basilisken-Presse, 1992.

Rheinberger, Hans-Jörg. *Experimentalsysteme und epistemische Dinge. Eine Geschichte der Proteinsynthese im Reagenzglas*. Göttingen: Wallstein, 3rd edition. 2019.

Ryan, Marie-Laure. *Narrative as Virtual Reality: Immersion and Interactivity in Literature and Electronic Media*. Baltimore: Johns Hopkins University Press, 2001.

Ryan, Marie-Laure. "Cognitive Maps and the Construction of Narrative Space." *Narrative Theory and the Cognitive Sciences*. Ed. David Herman. Stanford: CSLI, 2003. 214–242.

Sabín, Carlos. "One-Dimensional Sections of Exotic Spacetimes with Superconducting Circuits." *New Journal of Physics* 20.5 (2018): art. 053028. doi:10.1088/1367-2630/aac0db.

Sagan, Carl. *Contact.* New York: Simon and Schuster, 1985.

Sahoo, P. K., P. H. R. S. Moraes, Parbati Sahoo, and G. Ribeiro. "Phantom Fluid Supporting Traversable Wormholes in Alternative Gravity with Extra material terms." *International Journal of Modern Physics D* 27.16 (2018): art. 1950004. doi:10.1142/S0218271819500044.

Schaeffer, Jean-Marie. *Pourquoi la fiction?* Paris: Seuil, 1999.

Schaeffer, Jean-Marie. *Why Fiction?* Trans. Dorrit Cohn. Lincoln: University of Nebraska Press, 2010.

Schwarzschild, Karl. "Über das Gravitationsfeld eines Massenpunktes nach der Einsteinschen Theorie." *Sitzungsberichte der Preussischen Akademie der Wissenschaften zu Berlin* 1 (1916): 189–196.

Sharif, M., and Iqra Nawazish. "Wormhole Geometry and Noether Symmetry in *f*(*R*) Gravity." *Annals of Physics* 389 (2018): 283–305. doi:10.1016/j.aop.2017.12.012.

Shinkai, Hisa-aki, and Sean A. Hayward. "Fate of the First Traversible Wormhole: Black-Hole Collapse or Inflationary Expansion." *Physical Review D: Particles and Fields* 66.4 (2002): art. 044005. doi:10.1103/PhysRevD.66.044005.

Steinle, Friedrich. *Explorative Experimente. Ampère, Faraday und die Ursprünge der Elektrodynamik.* Stuttgart: Steiner, 2005.

Stone, Bryan P. "Religious Faith and Science in Contact." *Journal of Religion and Film* 2.2 (1998). https://digitalcommons.unomaha.edu/jrf/vol2/iss2/6/ (8 July 2022).

Suárez, Mauricio. "Scientific Representation. Against Similarity and Isomorphism." *International Studies in the Philosophy of Science* 17 (2003): 225–244.

Suárez, Mauricio, ed. *Fictions in Science. Philosophical Essays on Modeling and Idealization.* New York and London: Routledge, 2009.

Tamang, Amarjit, Alexander A. Potapov, Regina Lukmanova, Ramil Izmailov, and Kamal K. Nandi. "On the Generalized Wormhole in the Eddington-Inspired Born-Infeld Gravity." *Classical and Quantum Gravity* 32.23 (2015): art. 235028. doi:10.1088/0264-9381/32/23/235028.

The Nobel Foundation. "Press Release: The Nobel Prize in Physics 2017." *The Official Website of the Nobel Prize.* 3 October 2017. https://www.nobelprize.org/uploads/2018/06/press-40.pdf (8 July 2022).

Thorne, Kip S. "The Dynamics of Spacetime Curvature: Nonlinear Aspects." *Nonlinear Phenomena in Physics.* Ed. Francisco Claro. Berlin and Dordrecht: Springer, 1985. 280–291. doi:10.1007/978-3-642-93289-2_11.

Thorne, Kip S. *Black Holes and Time Warps: Einstein's Outrageous Legacy.* New York: Norton & Company, 1994.

Thorne, Kip S. *The Science of "Interstellar".* New York: Norton & Company, 2014.

Van Fraassen, Bas C. *Scientific Representation: Paradoxes of Perspective.* Oxford: Clarendon Press, 2013.

Visser, Matt. *Lorentzian Wormholes: From Einstein to Hawking.* New York and Berlin: Springer, 1996.

Visser, Matt. "The Quantum Physics of Chronology Protection." *The Future of Theoretical Physics and Cosmology.* Celebrating Stephen Hawking's 60th birthday. Eds. Gary W. Gibbons, E. Paul S. Shellard, and Stuart J. Rankin. Cambridge: Cambridge University Press, 2003. 161–176.

Wang, Deng, and Xin-He Meng. "Wormholes from Cosmological Reconstruction Based on Gaussian Processes." *Physics of the Dark Universe* 16 (2017): 81–86. doi:10.1016/j.dark.2017.04.006.

Wang, Deng, and Xin-He Meng. "Traversable Braneworld Wormholes Supported by Astrophysical Observations." *Frontiers of Physics* 13.1 (2018): art. 139801. doi:10.1007/s11467-017-0701-y.

Wendler, Reinhard. *Das Modell zwischen Kunst und Wissenschaft.* Paderborn: Fink, 2013.

Weyl, Hermann. "Feld und Materie." *Annalen der Physik* 65/14 (1921): 541–563. doi:10.1002/andp.19213701405.

Wheeler, John A. "Geons." *Physical Review* 97.2 (1955): 511–536. doi:10.1103/PhysRev.97.511.

Wheeler, John A. "On the Nature of Quantum Geometrodynamics." *Annals of Physics* 2.6 (1957): 604–614. doi:10.1016/0003-4916(57)90050-7.

Wheeler, John A. *Geometrodynamics.* New York: Academic, 1962.

Willenborg, Felix, Saskia Grunau, Burkhard Kleihaus, and Jutta Kunz. "Geodesic Motion around Traversable Wormholes Supported by a Massless Conformally Coupled Scalar Field." *Physical Review D: Particles and Fields* 97.12 (2018): art. 124002. doi:10.1103/PhysRevD.97.124002.

Woods, John, ed. *Fictions and Models. New Essays.* Munich: Philosophia, 2010.

Yousaf, Z., M. Ilyas, and M. Zaeem-ul-Haq Bhatti. "Static Spherical Wormhole Models in $f(R,T)$ Gravity." *The European Physical Journal Plus* 132.6 (2017): art. 268. doi:10.1140/epjp/i2017-11541-6.

Zwaan, Rolf A., Gabriel A. Radvansky, Amy E. Hilliard, and Jacqueline M. Curiel. "Constructing Multidimensional Situation Models during Reading." *Scientific Studies of Reading* 2.3 (1998): 199–220. doi:10.1207/s1532799xssr0203_2.

List of figures

Elisabeth Haefs

"Long ago, deep in space"

Narrating Gravitational Waves

Abstract: This paper illustrates how metaphors and other figurative devices can encode narrativity in contemporary scientific articles.[1] Since narrative – in the common sense of "storytelling" – is one of the oldest forms of structuring knowledge, it does not come as a surprise that even the most factual reports seem imbued with narrative elements. Many scientific papers contain plots or narrative patterns that appear virtually literary. Generally, academic interest in the relation between science and literature has, for a long time, focused on the echoes of scientific ideas in literary texts. The reverse activity, namely the study of literary devices in scientific texts (especially research articles and papers) is less common. To address this shortage, the following analysis focuses on two different texts about the first discovery of gravitational waves in 2015. These are the news article "Gravitational waves, Einstein's ripples in spacetime, spotted for first time" from the online platform of *Science,* and the "original" paper from *Physical Review Letters,* "Observation of Gravitational Waves from a Binary Black Hole Merger." A common form of (subdued) narrativity in scientific articles is the quest narrative. The pervasiveness of narrative and its fundamental cognitive ordering function thus becomes evident. Especially in accounts of gravitational waves, a quest form emerges, wherein the scientists or the gravitational waves themselves function as heroes. In this vein, this paper aims to illustrate the narrative elements that permeate two early accounts of this scientific breakthrough.

1 Stories and scientific journals

The search for a proof of Einstein's gravitational-wave theory (see Einstein 1916) has spanned around a hundred years and recently led to a veritable breakthrough and the 2017 Nobel Prize in physics. It is thus no coincidence that the history of gravitational-wave science is titled a "quest" in *Traveling at the Speed of Thought: Einstein and the Quest for Gravitational Waves* (Kennefick 2007). Since

1 In this paper, which contains research conducted until 2022, "scientific" refers to the natural sciences. The designation "scientific article" encompasses scientific research papers as well (Gross et al. 2002, 3).

https://doi.org/10.1515/9783110782844-007

narrative – in the common sense of "storytelling" – is one of the oldest forms of structuring knowledge, it does not come as a surprise that even the most factual reports seem imbued with narrative elements.[2] Many scientific papers contain plots or literary patterns of narrative, as do the reports on LIGO's[3] gravitational-wave discovery, which is why this paper not only consults narratology, but also Joseph Campbell's model of the heroic journey. In numerous accounts of gravitational waves, a quest form emerges, with the scientists or the gravitational waves themselves as heroes, depending on interpretation. This shows the pervasiveness of narrative and its fundamental cognitive ordering function, which informs even "objective" accounts. This article aims to illustrate the narrative elements that permeate the early accounts of this breakthrough: the slightly popularized but professional news article "Gravitational waves, Einstein's ripples in spacetime, spotted for first time" (Cho 2016) from the online platform of *Science*,[4] with a brief comparison to the "original" paper from *Physical Review Letters*,[5] "Observation of Gravitational Waves from a Binary Black Hole Merger" (Abbott et al. 2016).

In order to convince readers, scientific journal articles[6] need some sort of coherent story, since they are not mere "glorified laboratory notebook[s]" (Souza 2013). The main function of narrative in scientific journals is the communication of results to peers in a coherent and comprehensible way. It is needed as an organizing force that enables observation, establishes interconnections and produces presentations of knowledge that make sense, as in cohesive stories instead of singular bits of information (Koschorke 2018, 268). Consequently, texts from scientific journals also employ rhetorical[7] devices such as metaphors, and as we shall see, it is not always possible to fully distinguish between aspects such as metaphor and storytelling.

According to Gross et al., scientific journals in the twentieth century display a "virtual absence of narrative elements" (2002, 224). This paper is an attempt to show that narrativity does exist in contemporary scientific articles of the twenty-first century, on the story and content level as well as on the discourse level, en-

---

2 "After seminal studies such as Paul Ricœur's *Time and Narrative* it does not seem at all exaggerated to view humans as narrative animals, as *homo fabulans* – the tellers and interpreters of narrative" (Currie and Sterelny 2017, 2).

3 Laser Interferometer Gravitational-Wave Observatory, USA.

4 *Science* is a peer-reviewed academic journal, not a popular science magazine such as *Scientific American* or its German equivalent *Spektrum der Wissenschaft.*

5 Hereafter abbreviated as *PRL*. This article can be called a standard scientific research paper.

6 Naturally, the degree of narrativity can vary within journals, depending on the type of article or report. Basic narrative structure can also be found in calculations (Meuter 2013, 31–25; note that the numbers in citations from the *Living Handbook of Narratology* refer to the numbered online paragraphs instead of pages).

7 For narrative as rhetoric, see Phelan (2005).

coded within metaphors and other figurative devices. The argument is posited in the field of Literature and Science studies, with its background of the well-known "Two Cultures" debate, as well as important academic developments relating to the sociology of science, such as the "Science Wars" and the "Sokal Hoax."[8]

Generally, academic interest in the relation between science and literature has long focused on the echoes of scientific ideas in literary texts. The reverse activity – the tracing of literary devices in scientific texts – has only been attempted in rare cases (Weininger 1989, xiv). Existing research on scientific articles usually concentrates on the early stages of modern science or on linguistic aspects, not on plot or narrative in contemporary journals (Gross et al. 2002, viii), although there are some exceptions to the rule.[9] Still, science and literature need each other, especially in the public discourse on new discoveries or ideas (Vanderbeke 2004, 439). Rhetorical analysis can unmask relationships between scientific and literary discourses and yield fruitful interaction between different disciplines – without declaring a supremacy of rhetoric (Vanderbeke 2004, 130–132). The same could then be applied to narrative analysis.

However, the approach employed here has one limitation: The news-like properties of the *Science* article – written by a scientist-journalist – stand in contrast to the *PRL* research paper, in which scientists present their own work. Although the *Science* author Adrian Cho is a particle physics scientist, his news article is a form of science communication and not a research paper written by a scientist on their own research. This distinction is important because there are obvious differences between the narrativity of popular science, science communi-

8 For an account of the changing disciplines and the cultural significance of the increasing specialisation of knowledge, see Collini (1998, xlii–lxi). For the "Science Wars" and their precedents, see Carrier et al. (2004), Sokal (2000a; 2000b), Robbins and Ross (2000), and Gross and Levitt (1998). For the construction of scientific theories, as well as radical and social constructivism, see Vanderbeke (2004, 29–56), Latour and Woolgar (1986), and Bloor (1991); see also Kuhn (1970) and Popper (1973). For the question of objectivity in science, see Anderson (2008) and Hayles (1991).

9 Regarding the analysis of scientific books, Beer's extensive work on Darwin's *On the Origin of Species*, its language and influence on Victorian literature, and vice versa, is exemplary (see Beer 2000). Evolutionary theory also forms the basis of the frequently analyzed scientific journal article on spandrels by Gould and Lewontin (1979), which stands out because of its rhetoric and its figurative nature. Another article that has been analyzed several times is Watson and Crick's announcement of the precise build-up of DNA structure, the double helix. Watson also wrote a book on his personal experience of the same scientific "quest" fifteen years after the discovery (Kretzenbacher 1994, 96; Locke 1992, 126–129; Watson and Crick 1953; Watson 1980). See Gross et al. (2002) for an extensive rhetorical analysis of scientific articles ranging from the seventeenth to the twentieth century. For largely linguistic approaches, see Gläser (1990), Kretzenbacher (1994), and Niederhauser (1999). For narrative in scientific models and simulations, see Wise (2011).

cation, and scientific journal articles. A main difference is in the more or less overt quality of narrativity. Whereas narrativity is more likely to be concealed in scientific journal articles, popular science communication often uses obvious storytelling. Especially in entertainment media or popular science platforms, communication relies more heavily on narrative devices due to their simplifying and condensing properties, as well as the emotional response these platforms often aim to evoke.[10] This dynamic is evident in an exemplary *Scientific American* article on a coronavirus researcher, which is titled "How China's 'Bat Woman' Hunted Down Viruses from SARS to the New Coronavirus" (Qiu 2020).

Other types of science communication are carried out by agents that are closer to science than to journalism, such as the National Academy of Sciences or similar scientific institutions. Although these types of communication certainly show a reduced amount of narrativity in comparison to entertainment media or popular science, they also depend on emotional engagement and thus on varying degrees of narrativity as well (see Dahlstrom 2014, 13614). For instance, in a detailed research agenda, NASEM proposes "[s]imply to share the findings and excitement of science" as the first goal of science communication (National Academies of Sciences, Engineering, and Medicine 2017, 2).

Science communication is usually aimed at the general public and non-experts. However, in the case of *Science*, differentiation is necessary: Although the news platform of an academic journal like *Science* certainly also addresses the public, it probably addresses scientists to an equal measure. Since the articles in *Science* encompass all scientific research areas, the news section needs to use more "open" and thus narrative forms of communication. For instance, a specialist in volcanology reading this platform might not have the background needed to understand news articles that require expert knowledge of astrophysics. This explains why *Science* is listed as belonging to the "interspecialist" level of science communication (Cloître and Shinn, quoted in Bucchi 2008, 61).[11]

---

10 Although "the dissemination of science via entertainment does indeed have the potential to distort its content, it also has the power to capture attention, increase engagement, and promote the understanding and enjoyment of science" (Kaplan and Dahlstrom 2017, 312). Additionally, it may "instill wonder" (Kaplan and Dahlstrom 2017, 317). According to Kaplan and Dahlstrom, there are three formats for the communication of science: the "expository format" and the "argumentative format" mostly used in academic journals, as well as the "narrative format." For narrative as argument, see Goodrich (2005).

11 Journals such as *Nature*, *Science* and papers "given at meetings of researchers belonging to the same discipline, but working in different areas" (Bucchi 2008, 61) constitute interdisciplinary, but not popular, science platforms. These journals also serve as an orientation medium for regular journalism covering scientific topics (see Niederhauser 1999, 77). In a "continuity model" of scientific communication, these journals are placed at the "interspecialist" level, situated be-

Consequently, the *Science* news article is science communication written by a scientist via the online platform of an academic scientific journal, and thus it is inherently different from forms of communication that prevail in popular science formats, magazines, or entertainment media. Therefore, this text from *Science* will be used as a lens that helps bring into focus the narrative pattern that also exists within the original research article from *PRL*. The underlying narrativity of the research paper will be shown to largely correspond with that of the news article.

2 Narrative, analogy, and metaphor in scientific writing

Scientists who write research papers often uphold their claims to truthfulness and serious research by trying to avoid the use of figurative language. According to Kretzenbacher, the "ideal," objective scientific language is described with optical metaphors of transparency (1994, 100).[12] The same happens nowadays in reference to objective language and acoustic metaphors, for instance in the "noise" of language (Kretzenbacher 1994, 101). This indicates the difficulty of establishing an unambiguous language: even its self-description depends on figurative devices, and the process of writing always implies some form of crafting. Locke's analysis of a guide on how to write a scientific paper shows that leaving out "unimportant" details is crucial in terms of readability and clarity (1992, 7); he calls the scientific paper a "crafted, shaped object" (Locke 1992, 8).[13]

The process of scientific research always manifests itself in the resulting article or report, in the form of an underlying research narrative. Holmes claims that while the "modern scientific paper is perceived as coldly analytical, scientific

tween the "intraspecialist" (represented in this case by *PRL*) and the "pedagogic" level, which is then succeeded by the "popular" level (Cloître and Shinn, quoted in Bucchi 2008, 61).

12 The exemplary windowpane-theory was formulated by Gusfield in 1976, according to which scientific language should act like a "pane of clear glass" (1976, 17). He calls this the "intrinsic irrelevance of language to the enterprise of Science [sic]" (Gusfield 1976, 16–17). The author J. L. Borges observes that even 'plain style' is metaphorical: "For example, if I say, 'Style should be plain,' then I don't think we should remember that 'style' (*stylus*) meant 'pen,' and that 'plain' means 'flat,' because in that case we would never understand it" (2000, 106).

13 Locke observes the causal relations which regulate scientific language: "To claim that the language of science is [. . .] objective because science itself is objective is to confuse cause and effect. The apparent objectivity is the result of the decision made about language usage, not the cause of it" (1992, 91).

writing of the past [is perceived] as free to tell adventurous 'stories' about exploring nature" (1991, 164). However, some of his observations apply to all scientific papers. He maintains that all scientific research papers have a "past history of thought, of gathering either ideas or information" (Holmes 1991, 179). In addition, he claims that the intrinsic chronology of "experimentation [. . .] makes narrative the natural mode in which to describe it. To incorporate what is salient from the investigative 'story' into a paper structured as a critical argument requires transformations that are as central to the creativity of science as are the investigations themselves" (Holmes 1991, 179). This "creative" act is especially traceable in the first parts of a scientific article: Kretzenbacher states that the parts with the highest amount of narrativity are usually the "materials and methods"-sections (1994, 93–94). This corresponds with the narrativity of experimentation – because materials and methods are part of the experiment – and the paper becomes a "palimpsest, the underlying narrative structure ([. . .] the history of the research) being overwritten by a totally reformulated non-narrative structure in order to meet the stylistic requirements of the scientific texts" (Kretzenbacher 1994, 94).[14] The scientist Kevin Padian calls this an "anti-narrative" (2018, 1231). Additionally, such a process of rewriting applies to most kinds of literary writing as well.[15]

Analogies form a crucial part of the cognitive processes that lead to the formation of scientific theories. The importance of analogy in scientific modeling has been prominently shown by scholars like Mary Hesse (1970) and Max Black (1972). Significant examples are wave models regarding the wave's original element, water, and its analogous use in the field of sound and light (Hesse 1970, 11 and 60). Hesse defends the use of analogy in scientific theory as necessary and descriptive, even if analogies are not "true": "The logic of analogy [. . .] may be descriptive without being justificatory" (Hesse 1970, 56). Thus, analogy helps illustrate aspects of theory, instead of falsifying or distorting scientific facts. Since such rhetorical devices are inherent parts of theory, they must also be part of the textual communication of science. Analogy, in general, is necessary for cognitive processing: "We understand the new by reference to the already known" (Beer 2000, 76). Analogy, then, is a rather logical and complex argument, which differs from the way in which a direct and sometimes new meaning is created by metaphor.

14 This led the scientist Medawar to the conclusion that the scientific paper "may be a fraud" (qtd. in Vanderbeke 2004, 108).

15 This applies to both the process of writing performed by one author (as in revising), and the notion that "there is no original text" as most prominently posited by Roland Barthes: "a text is not a line of words releasing a single 'theological' meaning [. . .] but a multi-dimensional space in which a variety of writings, none of them original, blend and clash" (Barthes 1977, 146).

Similar to analogy, however, metaphor is central to the process of rational explanation and the understanding of new concepts or ideas in all areas that are influenced by language – including science. In fact, as Hesse describes, "the deductive model of scientific explanation should be modified and supplemented by a view of the theoretical explanation as metaphoric redescription of the domain of the explanandum" (Hesse 1970, 157). Correspondingly, Hörisch ascribes to metaphor the power to cross or build bridges, because it connects separate entities (Hörisch 2007, 44). Thus, metaphors found within a text not only help cognitive processing; they also form a web of instantly evoked interconnections. In a linguistic and cognitive-science approach, Lakoff proposes that metaphor – in the form of "cross-domain mappings" (Lakoff 1993, 203) – permeates everyday language as much as it does figurative or poetic language.[16] It is possible to argue that the abstraction and inference that characterize the cognitive process of conceptual metaphors – the mapping of one "image structure" onto a "target domain"[17] – are comparable to the stages of abstraction (see Kennefick 2007, 5–6) in theoretical scientific models. If this is true, these similar processes reveal a common ground for metaphorical thinking, science, and narrative: scientific models have been found to contain and/or induce narrative thinking as well, as proposed by Fuchs (2015, 945–946).

The established common ground of science, metaphor and narrative is the principal reason why metaphor constitutes an important part of this paper. Paul Ricœur proposes an affinity between metaphor and narrative, because of their respective "synthesis of the heterogeneous" (Ricœur 1990, ix). According to him, both metaphor and narrative belong to the "same basic phenomenon of semantic innovation" on the discourse level (Ricœur 1990, ix). With metaphor, this innovation happens through "impertinent attribution," with narrative, it happens in the

16 Lakoff's approach to metaphor illustrates what could be called basic metaphors that structure all forms of discourse, such as "love-as-journey" (1993, 206–210) and the "event-structure" metaphor (Lakoff 1993, 219–229). The event-structure metaphor is used to describe how most events are conceptualized in terms of physical space and movement – such as "states as locations," "change as movement," "causes as forces" and "means as paths" (Lakoff 1993, 220). In this case, the boundaries between metaphor and narrative – taking narrative as a time sequence marked by significant, connected events that cause change – seem to blur.

17 "Spatial inferences are characterized by the topological structure of image-schemas. We have seen cases such as CATEGORIES ARE CONTAINERS and LINEAR SCALES ARE PATHS where image-schema structure is preserved by metaphor and where abstract inferences about categories and linear scales are metaphorical versions of spatial inferences about containers and paths" (Lakoff 1993, 216; capital letters in original).

"inventing of another work of synthesis – a plot" (Ricœur 1990, ix).[18] Consequently, both processes can be "carried back to the productive imagination" (Ricœur 1990, ix). In this way, Ricœur illustrates the properties of renewing, mapping and fusing that connect narrative and metaphor.

As Lakoff states, conceptual metaphors permeate all variations of language. Giving an overview of academic discourse forms structured by such metaphors, Lakoff explains the high occurrence of "the guided tour, the heroic battle, and the heroic quest" (1993, 243). Interestingly, the word "scientist" only appears when he mentions quest form: "The heroic quest discourse form is based on the metaphor that knowledge is a valuable but elusive object that can be 'discovered' if one perseveres. The scientist is the hero on a quest for knowledge, and the discourse form is an account of his difficult journey of discovery" (Lakoff 1993, 244).[19] This shows how metaphor structures events and makes them comprehensible, and how "real" entities are conveyed via metaphor (Lakoff 1993, 244). More importantly, Lakoff's observations reveal how fuzzy the boundaries are between conceptual metaphors and narrative templates. Furthermore, his use of the heroic quest not only coincides with the concept of the research narrative underlying all scientific articles, it also explains why Campbell's notion of the heroic monomyth appears highly useful for an analysis of scientific narrative.

Scientific writers diverge in their opinions on the use of narrative. There are several texts that provide insight regarding scientists' views of their own craft. While some scientists deem narrative to be a rather "frivolous"[20] affair, and unfit for the scholarly communication of science, others insist that narrative is a valuable device to convey scientific information, that it makes use of cognitive processes which result in the storing of knowledge. Neither literary critics nor science writers are free from ideological assumptions about certain kinds of texts and writing methods. To address this problem, two neuroscientists proclaim that using communication tactics from the humanities might help science. They also

---

18 Ricœur describes the innovative process of metaphor formation as follows: "The displacement in meaning the words undergo in the metaphorical utterance [. . .] to which ancient rhetoric reduced metaphor, is not the whole of metaphor. It is just one means serving the process that takes place on the level of the entire sentence, whose function it is to save the new pertinence of the 'odd' predication threatened by the literal incongruity of the attribution" (1990, ix). See also Ricœur (1978).

19 Padian gives this story a new twist: "When scientists do research, it can be thought of as a quest, and the 'gift' of funding (NSF, NIH, etc.) to develop our research data 'transforms' our question into a 'triumph' of an answer" (Padian 2018, 1226).

20 "Some in the scientific community view narratives and entertainment media as frivolous or even antithetical to science, and there are many examples of narratives that attempt to persuade audiences to hold views unsupported by scientific consensus" (Kaplan and Dahlstrom 2017, 316).

conclude that "plot-building and research narratives share certain parallels [. . .]. We care not only about when things happen [. . .] but why they happen. If we are so lucky as to find out the why, then we have a tale to tell" (Martinez-Conde and Macknik 2017).[21] Storytelling, in fact, is explicitly recommended by participants in a 2013 debate in the journal *Nature Methods*, which focuses on writing techniques for life scientists and chemists.[22]

Another perspective encouraging narration is introduced by scientist and filmmaker Randy Olson, who provides templates for science writers. He presents a strategy to communicate science effectively, using what he calls "ABT": always connecting information by "and," "but" and "therefore," an approach that he describes as borrowed from the creators of *South Park* and screenwriting instructor Frank Daniel (Olson 2015, 102).[23] To illustrate this, he shows how Watson and Crick used ABT in their well-known and concise paper (Olson 2015, 109–110). Furthermore, he discusses the IMRAD formula (introduction, methods, results and discussion), whose popularity in scientific writing, especially biomedicine, has risen steadily since the 1960s (Olson 2015, 7). The linguist William Labov's model of the progression of narrative is reminiscent of the IMRAD template (Olson 2015, 3): "Labov distinguishes five stages (on the discourse level): orientation, complication, evaluation, conclusion, coda" (Hühn 2008, 142; see Koschorke 2018, 48–50). It is moreover a structure common to many journal articles, not only in science. While Olson certainly exaggerates at this point, he states that the simplicity of

---

21 For regular science journalism and narrative, as well as science communication, see Blum (2008), Campenhausen (2014), Jones and Crow (2017), and Nisbet and Fahy (2017). For narrative in non-scientific journalism, see Früh and Frey (2014). The biologist Rafael Luna provides detailed storytelling instructions for scientists writing research papers. He proclaims that in science, "the moral of the story is the hypothesis, which is substantiated or refuted. In scientific storytelling, a logical series of experiments are performed to support the main hypothesis" (Luna 2013, 2; see Clemens 2018 and Montagnes et al. 2022). Interestingly, Luna also compares the scientific method with the dramatic arc (Luna 2013, 15–16). Holmes observes a similar analogy regarding a research paper's structure: "We can, however, view these standard sections of the paper also as a framework within which to fit a heavily stylized, formulaic version of a story" (Holmes 1991, 179).

22 See Krzywinski and Cairo (2013a; 2013b); see Katz (2013) for the counterpart "against" storytelling. Katz suggests that storytelling might restrict the amount and breadth of information "to that agreeable with the story," leaving out material that might become relevant from a different angle. He maintains that stories have to be complete and without alternative narratives and are therefore not well-equipped to convey experimental data. This, however, might be considered a slightly narrow view of storytelling.

23 Interestingly, this mirrors Mandler and Johnson's concept of tree-like narrative nodes that create episodes or sequences and are always connected by the conjunctions "and," "then," and "because" (Koschorke 2018, 55; see Mandler and Johnson 1977). Olson does not mention this concept, even though it illustrates the relevance of storytelling in cognitive processes.

IMRAD, and its form, make it "essentially identical to the three-act structure that is at the heart of virtually every movie or play written today. It is the structure of a story, which has a beginning (I), middle (M&R), and end (D)" (Olson 2015, 6). This three-act structure can be traced back to the *Epic of Gilgamesh* as well as to the Hegelian Triad and Aristotle's *Poetics* (17, 33–34). Thus, Olson aims to indicate how fundamentally narrative structure influences scientific thought and writing, not necessarily because scientific activity needs it, but because narration is a human activity and human thinking depends upon it. It is probably safe to say that most scientists do not naturally utilize the "logico-scientific" mode,[24] they learn to use it during their training. In a response to Olson, as well as to other scientists and narratologists, Newman posits narrative as "common ground" (Newman 2018, 278) of science and literature. At the same time, criticizing Olson, he warns of seeing narrative as a set structure to fit the science into (Newman 2018, 279).[25] A focus on partial narrativity (see White 1980) or narrative elements instead of "whole" narratives seems to Newman more fruitful than "importing canonical forms such as the quest and the 'three-act-structure' from literature" (Newman 2018, 280). This seems to be a valid point for Newman, because facts might disappear in the vague imposing of narrative on a scientific text; still, as shown by Lakoff, at least the quest motif is a form inherent to much scientific research without having to be "imported." However, the focus on narrative elements instead of complete narratives enables an approach that also includes postclassical narratology[26] (Newman 2018, 282), as well as "big" models or genres such as Campbell's monomyth.

Consequently, scientists appear to need narrative elements in order to render their research comprehensible – to themselves and their colleagues as much as to non-experts,[27] and – presumably – to the institutions that control the allocation of

24 "At a gross level, the logico-scientific mode (I shall call it paradigmatic hereafter) deals in general causes, and in their establishment, and makes use of procedures to assure verifiable reference and to test for empirical truth. Its language is regulated by requirements of consistency and noncontradiction" (Bruner 1986, 13). For a concise comparison between the logical-scientific mode and the narrative mode, see Dahlstrom (2014, 13614–13615).

25 However, Olson more than once warns of misusing narrative and "bend[ing] the science" (Olson 2015, 134) by using simplifications that make scientific intricacies and important points disappear.

26 For postclassical narratology, see Herman (1997).

27 Julika Griem notes that an emphasis on storytelling can cover up areas of science that might seem less "spectacular" or "adventurous" than others. Especially the image of the scientist as heroic adventurer, she says, brings about a problematic focus on one single person instead of the group of scientists that take on the less "exciting" work. She proposes to maintain science's inherent challenge and sophistication when communicating it (Griem, quoted in Krieger 2019). How-

third-party funding: doing (paid) research often depends on the ability to convincingly "sell" its necessity. The imaginative and emotional dimensions of science indicate the importance of storytelling abilities as well.[28] Comprehensibly formulated and "tellable" science is an important part of intellectual discourse, which is why narratology and science can constructively interact with one another.

Let us consider narratological concepts that enable or support the analysis of non-literary texts.[29] Narrative is a "cognitive schema and discourse type manifested in both literary and non-literary forms of expression" (Herman et al. 2005, ix).[30] The sometimes disputed term "narrativity" indicates a certain grade or degree. Measuring is complicated in this context, but as H. P. Abbott notes, narrativity is widely applicable because of its "grammatical status as a reference to a property [. . .] rather than to a thing or class" (2014, 3). Fludernik proposes that narrativity is not a text-inherent property, but an "attribute imposed on the text by the reader who interprets the text *as narrative*, thus *narrativizing* the text" (2003, 244).[31] These three examples show that narrativity can be found in non-literary discourse. In a more radical approach, narrativity can even be said to arise from one of the most basic, everyday processes, which is motion: Plotnitsky maintains that "the very idea of motion, which grounds all physics, cannot be conveyed without a narrative" (2005, 515). Correspondingly, and indicating the alignment – or even overlap – of narrative and metaphor, Lakoff conceptualizes motion as an integral part of various metaphors that inform common, non-poetic language regarding time (1993, 217).

ever, keeping science inaccessible to those who do not actively seek to learn about it will not change the current climate of antiscientific trends and skepticism (Boetius, quoted in Krieger 2019; cf. Dean 2017, 3–11).

28 It is worth emphasizing that science depends on imagination (see Otis 2002, xxi; see Vanderbeke 2004, 110–111; see Black 1972, 242–243).

29 The term "narratology" was coined by Todorov in 1969 to denote the science of narrative based on Saussure's structural linguistics (Herman et al. 2005, ix). Among the important theorists of the field one usually counts Lotman, Barthes, Genette and Greimas, as well as Lyotard, Bal and Ricœur (McQuillan 2000, 311–313).

30 Cognitive operations regarding narrative are conveyed by the following questions: "[I]n what order did the represented events occur; [. . .]; what motivates actions and how does the outcome of these actions compare to the intent of the agent. If a text confronts us with such questions [. . .], we read the text as a story, [. . .] whether or not we are aware of what we are doing" (Ryan 2007, 33).

31 Additionally, Knoespel claims that "forms of discourse in the natural and human sciences are themselves ordered as narratives. [. . .] For theoreticians, the examination of such narrative networks offers a means for detecting how individuals and disciplines account for themselves, as well as how entire periods have become framed through metanarratives" (1991, 101). This observation echoes Lyotard's account of "metanarrative" (Lyotard 1984, 33–34).

A major part of narrative structure consists of "events." Within such a structure, events can be divided into types I and II. The first type designates a "simple" event or change of state[32] (Hühn 2013, 1). A type II event, however, forms if it "is accredited – in an interpretive, context-dependent decision – with certain features such as relevance, unexpectedness, and unusualness" (Hühn 2013, 1), which accounts for the usefulness of the concept in the analysis of scientific texts. Eventfulness then denotes the property of a text with many type II events, constituting a high degree of narrativity.

Events are fashioned into a narrative sequence via "emplotment." The term was coined by Hayden White, to illustrate how historians transform chronological events into narratives: it is a structure that produces plot[33] by embedding events in a causal time sequence with beginning, middle and end: "[T]he historian must first prefigure as a possible object of knowledge the whole set of events reported in the documents. This prefigurative act is poetic inasmuch as it is precognitive and precritical in the economy of the historian's own consciousness" (1987, 30–31).[34] White discerns four modes of emplotment, building on the theory of myths developed by Northrop Frye (2000, 33–67): romance, tragedy, comedy and satire (White 1987, 7). These come with corresponding figures of speech as modes of consciousness (White 1987, 426) – metaphor, metonymy, synecdoche and irony (White 1987, 31) – and likely ideological implications. Emplotment thus not only fashions meaningful stories out of events, it illustrates how a narrative can weave underlying concepts and dichotomies into a text.[35] In general, underlying

---

32 Bal simply calls it "the transition from one state to another state, caused or experienced by actors" (1985, 13).

33 E. M. Forster's well-known summation of plot reads as follows: "'The king died and then the queen died' is a story. 'The king died, and then the queen died of grief' is a plot. The time-sequence is preserved, but the sense of causality overshadows it" (Forster 2000, 45). Fludernik considers this example a paradigm and not useful to explain narrativity (1996, 14). It is, however, a precise definition of plot.

34 Currie and Sterelny emphasize the epistemic value of narrating the historical past: "Speculation is often productive: it furthers our epistemic reach by enabling us to identify the diverse lines of evidence knowledge of the past requires" (2017, 21).

35 For figures of thought that conceptualize dichotomies, see Gurr (2013). Vanderbeke also acknowledges E. C. White's treatise on cosmology and narrative (2004, 139). Drawing on Hayden White's "metahistory" and Lyotard's "metanarrative," E. C. White claims that, in natural history, narrative in the form of farce or picaresque might constitute a "legitimate," non-teleological representation of cosmology (1990, 107). A farcical emplotment of natural history is, however, hardly found within science itself.

oppositions[36] propel a story, as most prominently argued by Lévi-Strauss and, subsequently, Greimas.[37]

Not unlike White, Ricœur defines emplotment ("mise en intrigue") as "the operation that draws a configuration out of a simple succession" (1990, 65). His approach has a strong emphasis on temporality and the human perception of time: "Time becomes human time to the extent that it is organized after the manner of a narrative; narrative, in turn, is meaningful to the extent that it portrays the features of temporal experience" (Ricœur 1990, 3). Thus, the relationship between temporality[38] and narrativity is circular (Ricœur 1990, 54); they are dependent upon each other, with the plot showing their "crossing point" (Ricœur 2000, 257). Narrative, then, is how humans make sense of experience. This concept is embodied in Ricœur's threefold concept of mimesis:[39] "Mimesis I" mirrors the "prefigurative act" noted by White, it is the prefiguration and the "composition of plot [. . .] grounded in a preunderstanding of the world of action, its meaningful structures, its symbolic resources, and its temporal character" (Ricœur 1990, 54). The central "mimesis II" denotes the configuration of plot – emplotment – which is the "poetic act" itself, a "grasping together" of events (Ricœur 1990, 66). This configuration happens "between a stage of practical experience that precedes it and a stage that succeeds it" (Ricœur 1990, 53), which illustrates the circularity[40] of Ricœur's mimesis and the mediating function of configuration (Ricœur 1990, 65). The succeeding "refiguration" ("mimesis III") happens in the process of reading; it constitutes a transformative experience: "What a reader receives is not just the sense of the work, but, through its sense, its reference, that is, the experience it brings to language and [. . .] the world and the temporality it unfolds in the face

36 In his treatise on forms of eventfulness in fiction, Hühn sums up how most structuralist narratological approaches rely on a system of perspectivity and temporality (2008, 141), and notes moreover that these two exist in the "binary opposition [. . .] underlying the majority of narratological models such as *histoire/récit*, story/discourse, *fabula/sjužet*, story/text" (Hühn 2008, 141). Binary oppositions structure the oldest stories of humanity, as this partial reiteration of the Zoroastrian *Bundahish* shows, for example: "The Two Creators through whose dialectic everything in this world has been wrought, so that it shows in every part a nature that is both good and evil, [. . .] were existent from all eternity" (Campbell 1974, 201).

37 In his structural analysis of myth, Lévi-Strauss proposes that myth is constructed of the tensions within "a gathering of contradictions" (Greimas 1983, 243, see Lévi-Strauss 1955). Similarly, the "semiotic square" developed by A.-J. Greimas shows how "basic semantic elements [are] arranged by contrary and contradictory opposition" (Johansen 2005, 524).

38 In addition, narratives usually feature a temporal duality: that of "narrated time" and that of "narrating time," as proposed by Günther Müller (Scheffel et al. 2014, 18; see Müller 2011).

39 Ricœur builds on Aristotle's concepts of "mimesis" (imitation) and "muthos" (plot) (Ricœur 1990, 31–51) and Augustine's concept of time (Ricœur 1990, 5–30; see Aristotle 2013).

40 Ricœur compares threefold mimesis directly to the hermeneutic circle (1990, 86).

of this experience" (Ricœur 1990, 78–79). This phenomenological approach not only applies to the process of reading literature, it also explains the basic processes that condition what White calls the historian's emplotment, and the apparent diffusion of narrative structures into all areas where humans try to make sense of what they see, observe or experience.

Emplotment and eventfulness are closely related to "tellability," which consists of the "features that make a story worth telling" (Baroni 2014, 1).[41] This concept is versatile and easily adaptable to the analysis of a scientific text. A scientific paper or journal article always has a certain "point" to make. Tellability is also closely related to narrativity, even if they are sometimes seen as contrasting (Ryan 2005, 589). In a scientific text, tellability can for instance be posited as an indicator of the content's "newsworthiness."[42]

Fludernik sees "experientiality," which is related to tellability, as a major basis of narrativity. As Fludernik surmises, events "become tellable precisely because they have started to mean something to the narrator on an emotional level. It is this conjunction of experience reviewed, reorganized, and evaluated ('point') that constitutes narrativity" (2003, 245). The ordering of the text, as described by Fludernik, coincidentally mirrors the description of an experiment: "experience reviewed, reorganized, and evaluated." Whereas emotion is obviously less important in a scientific context than in conversational storytelling, personal thoughts, viewpoints and emotions possibly still influence the way a scientific text is ordered or written, at least the way in which a scientist regards his or her work.[43]

In contrast to these rather "positive" aspects of narration, the act of storytelling is also suspected of bending the truth – and not only by Plato. In his essential treatise on narrative theory, Koschorke explains that narration has been allocated a separate cultural space because stories are not necessarily "true" (2018, 9).[44] Still, narration does not adhere only to the spheres of fictional arts and entertainment; nor does it fit into the societal division of labor and its respective spaces, as can be seen in the realm of narrative theory, where narrative is no longer just part of literary criticism but of other disciplines as well (Koschorke 2018,

41 The tellability concept was mainly developed by Labov and Harvey Sacks; it originally stems from the investigation of conversational storytelling (Baroni 2014, 6).

42 "The tellability of a story can [. . .] rely on purely contextual parameters (e.g. the newsworthiness of an event)" (Baroni 2014, 1).

43 For instance, Holub describes the emotions arising in laboratory activity from a personal point of view (2001, 61).

44 In this case, Koschorke refers to an assumed "ontological indifference" of fiction and the arts (2018, 7–9). The arts can be an imaginative playground, enabling "simulation at a reduced risk" (Wellershof 1973, 57; trans by EH).

10). According to Koschorke, the active part of narration – the intervention of narrative into the world and its act of “naming” – is not a mere mirror but also an act of creation.[45] This creative act is closely related to questions of epistemology (Koschorke 2018, 273–277). The acknowledgement of narrative’s important role in the organization of knowledge shows that knowledge and narrative are not antithetical (Koschorke 2018, 268). Echoing Beer, Koschorke maintains that evolutionary theory is based on narrative structure[46] and that narrative often constitutes more than a random ingredient of scientific processes (2018, 268).[47]

The important concept of fiction in science (“Fiktionsbegriff”; Heydenreich and Mecke 2015, 23) illustrates Koschorke’s statements here, namely that the imagination or the idea of a thing or a theory strongly propels research – that imagination drives science.[48] Kennefick, the physicist telling the story of gravitational-wave science, stated the following before the 2015 discovery: “[T]he phenomenon has never been detected and has been without even indirect evidence for decades, yet widespread belief in its existence persists. [. . .] [T]he force behind this extraordinarily tenacious scientific belief is that of analogy [between the gravitational and the electromagnetic field]” (2007, 4). Thus, the “quest,” as Kennefick calls it as well, is strongly influenced by analogous thinking (2007, 7) and correspondingly, by imagination.

According to Koschorke, the blurring of boundaries between fact and fiction begins with the border between “concept” and “narrative,” because concepts can already contain “miniaturized or [. . .] immobilized narratives”[49] (2018, 269). His observations are helpful in disentangling the significance of narrative in science,

---

45 This creative act of narration is reminiscent of Ricœur’s statements regarding the innovative power of metaphor, building on Aristotle: “[M]etaphor is the rhetorical process by which discourse unleashes the power that certain fictions have to redescribe reality. [. . .] From this conjunction of fiction and redescription I conclude that the ‘place’ of metaphor [. . .] is [. . .] the copula of the verb *to be*. The metaphorical ‘is’ at once signifies both ‘is not’ and ‘is like.’ If this is really so, we are allowed to speak of metaphorical truth, but in equally ‘tensive’ sense of the word ‘truth’” (Ricœur 1978, 7).

46 On the other hand, Abbott maintains that evolutionary theory, and natural selection in particular, are not easily narratable – partly because the latter is too dependent on coincidence. This, he suggests, complicates the narration of causal processes and agencies (2003, 146–147).

47 “In such cases narration is not addendum to ‘pure’ science but one of its data-synthesizing procedures; accordingly genuinely scientific and scholarly knowledge-centered processes can be at least partially reformulated in terms of narrative theory” (Koschorke 2018, 268).

48 Science fiction as a genre is another interface where science and literature meet. For the potential of science fiction as mediator between science and literature, see Vanderbeke (2004, 202–206). See also Gräfrath (1996).

49 “As we have seen, many concepts contain miniaturized or as it were immobilized narratives, as conceptual tools indispensable for understanding the world” (Koschorke 2018, 269).

since he directly relates narratology to epistemology (Koschorke 2018, 273–277). This concept of a mute narrative within a term seems to correlate with Ricœur's approximation of metaphor and narrative. Simultaneously, Koschorke defends the idea of an objective form of truth, meaning that the question of reference – to a thing outside of a narrative's own system, as well as an "object" of study as being on the "outside" of scientific enquiry itself – is relevant and cannot be "disputed out of existence through any relativism of cultural theory" (2018, 276).[50] The fundamental ambiguities that inform the zone between science and narrative thus come to light.

As a means of narrative emplotment, quest stories are travel narratives. They are often found in speculative fiction – e.g. as a subgenre of fantasy (see Senior 2012) – and they are described by Campbell in his study of myths.[51] Criticized as "too grandly deductive" (Miller 2000, viii) or having "plundered" the world's mythologies (Miller 2000, 68), Campbell is often accused of simplifying mythological intricacies to fit them into his cyclical monomyth that sums up the heroic journey.[52] Though simplifying in structure, it might just be its abstracting, universal property that makes Campbell's model applicable to a wide range of texts, certainly to many scientific articles. According to Campbell, the abstract monomyth template entails the "Adventure of the Hero" (1973, 49–251):

> Separation – initiation – return: [. . .] might be named the nuclear unit of the monomyth. *A hero ventures forth from the world of common day into a region of supernatural wonder: fabulous forces are there encountered and a decisive victory is won: the hero comes back from this mysterious adventure with the power to bestow boons on his fellow man.* (Campbell 1973, 30)

---

50 This could be a direct contribution not only to the "Science Wars" but also to age-old philosophical questions. Naturally, Koschorke mentions Kant here, as well as Roy Bhaskar's "central paradox of science" (2018, 277–278; see Bhaskar 2008), wherein Bhaskar refers to the contradiction between the fact that scientific knowledge is a product of human beings, and that this "knowledge is *of* things which are not produced by men at all [. . .]. None of these 'objects of knowledge' depend upon human activity" (Bhaskar, quoted in Koschorke 2018, 278).

51 Campbell was influenced by Propp's study of folktales (see Propp 1968). See Greimas (1983, 197–221) for "actantial models" of narratives. Greimas builds on Propp as well. Additionally, Campbell's model is not only connected to early mythological stories, but also to modern literature known for its mythological allusions, as can be seen, for instance, in Campbell's work on James Joyce (see Campbell and Epstein 1993). The hero's journey is thus also tied to more contemporaneous stories that probably were familiar to Campbell's audience.

52 "Campbell added large parts of Jung's theory of archetypes, enabling him to fill out the 'monomyth' in which the eternal heroic pattern is made congruent with a cosmic *Überbild.* [. . .] Moreover, he relentlessly inflated, desecularized, and apotheosized the heroic role" (Miller 2000, 68). To many screenwriters, Campbell's work was a revelation: "Campbell had broken the secret code of story" (Vogler 2007, xxviii).

This description reveals how easily the hero's journey can be used to represent – or glorify – scientific research, primarily in popular science and science communication. Campbell's model narrative has markers such as the call to adventure, helper figures, abduction, battle, flight, trials and eventual return (1973, 245). Campbell also posits Daedalus as a possible model of the hero scientist: "He is the hero of the way of thought – singlehearted, courageous, and full of faith that the truth, as he finds it, shall make us free" (1973, 24). The Hero's Journey is also discussed as a "paragraph template" by Olson (2015, 127–140). He explains how recognizing narrative problems and incongruities can be trained, and how variations of Campbell's heroic journey stages can be helpful for developing narrative structure in a scientific article – at the same time, he also warns of making the science fit the story. Olson identifies markers such as the "problem-solution binary," the "flawed character" and "near failure" as most "valuable" points to develop narrative structure while writing a scientific text (2015, 134–140).

The concepts introduced from narratology will be employed in the following section, which attempts to illustrate how emplotment, a "grasping together of events," appears in the texts that announced the detection of gravitational waves, and how, on various levels, Ricœur's configurative plot as well as Campbell's quest narrative can be observed.

3 A quest through spacetime, or: Narrative and gravitational waves

The two selected texts from *Science* and *Physical Review Letters*, like narrative in general, rest on oppositional systems: the content level houses an opposition between new and previous research, while the discourse level is characterized by narration – especially in the case of *Science* – on the one hand, and language that is as objective as possible, on the other. Thus, the interdisciplinary nature of journals like *Science* is translated into the textual structure: their mediatory function between specialist levels is mirrored in the way they seem to resolve the often-presumed dichotomy between science and narrative. Although a great part of the analysis focuses on the *Science* news article, the narrative correspondence of the original report in *PRL* (Abbott et al. 2016) to the news article will be investigated through frequent comparisons. Abbott et al. will also be consulted to convey certain points of analysis and narrativity more clearly.

Gravitational waves are undulations or "ripples" in the fabric of spacetime, caused by events such as the collision of two black holes or the merging of neutron stars. They were predicted by Einstein – although he was not always convinced of their existence (Kennefick 2007, 81), and they were not proven to exist until 2015,[53] earning members of LIGO (US) and Virgo (Italy) the 2017 Nobel Prize for physics. The corresponding news article on *Science*'s online platform was written by scientist and staff writer Adrian Cho.[54] Proving the existence of gravitational waves has been a long-standing goal in science, not least because "from the beginning, [they] were seen as an element of a possible unified theory of electromagnetism and gravity" (Kennefick 2007, 8).

To begin with, the word "gravitational wave" is a metaphor and relates to the wave analogy that is a part of theoretical models of light and sound.[55] However, unlike such waves, "gravitational waves propagate freely through the Earth" (Abbott et al. 2016, Fig. 3), without encountering the planet as an obstacle. This description, though accurate scientifically, creates an image of almost infinite (large) and liberated power in motion; it also coincides with Lakoff's conception of how everyday language casts "external events" as "large, moving objects" (Lakoff 1993, 220). The apparent motion of gravitational waves through space – or rather, their propagation through the medium of spacetime – thus already in itself constitutes an event.

The article's headline incorporates figures of speech: "Gravitational waves, Einstein's ripples in spacetime, spotted for first time." The waves are obviously not Einstein's own ripples; metonymy makes the discovery immediately relatable and classifiable. Furthermore, the verb "to spot" in connection with a gravitational wave seems a rhetorical understatement, it renders the discovery small. The beginning of Cho's article reads as follows:

> Long ago, deep in space, two massive black holes – the ultrastrong gravitational fields left behind by gigantic stars that collapsed to infinitesimal points – slowly drew together. The stellar ghosts spiraled ever closer, until, about 1.3 billion years ago, they whirled about each other at half the speed of light and finally merged. The collision sent a shudder through the universe: ripples in the fabric of space and time called gravitational waves. Five months

53 The detection itself was not announced until 2016, which makes it exactly a hundred years since Einstein's prediction (see Einstein 1916). The connection between these two points in time is constantly reinforced. For instance, Stephen Colbert described the discovery as part of Einstein's "homework from a hundred years ago" on television ("Gravitational Waves Hit the Late Show" 2016, 01:00–01:04).

54 See Cho (2016); instead of page numbers, the numbers for quotations from Cho (2016) and Abbott et al. (2016) indicate paragraphs.

55 To emphasize figuration in this context, the wave model – because of its complementarity to the particle model in quantum mechanics – could be considered one half of a "both/and"-figure of thought in physics, in terms of common conceptions in dealing with oppositions (Gurr 2013, 63). For the gravitational wave analogy, see Kennefick (2007, 1–17).

> ago, they washed past Earth. And, for the first time, physicists detected the waves, fulfilling a 4-decade quest and opening new eyes on the heavens. (Cho 2016, 1)

The first five words evoke the classical fairy tale opening. "The Frog King," for example, starts in a similar way: "Long long ago, when wishing could still lead to something" (Grimm, quoted in Campbell 1973, 49). This exposition also calls to mind the well-known *Star Wars* opening crawl.[56] In terms of narrative, this paragraph tells the story of a signal from far away that was "meant" to reach Earth, and it corresponds with the "4-decade quest"[57] in the last sentence. Evidently, cosmology, astrophysics and astronomy are especially well-equipped for storytelling.

Quest, as a means of emplotment, is traceable throughout Cho's article and its structure. This quest has two possible plots: the scientists are on a quest that leads to discovery, or the wave is on a quest to reach earth and "fulfil" Einstein's prophecy. Exemplary type I events are the initial collapse of "gigantic stars" and the black holes figuratively "whirling about each other," as if they had been choreographed. The central event is the merging of the black holes, which eventually causes a significant discovery on earth, making the collision a type II event: it is the focal point which connects all ends of the story. The relevance of this event is accentuated by the adverb in "finally merged," which signals a goal-oriented process and posits the merging as desirable goal.[58] Therefore, the result of the collision – the wave – can be considered the interface which connects the opposites cosmos and humanity. Read this way, the significance of the wave emerges from the tension between opposites, as narrative does. The binary black hole pair is then a central dichotomy,[59] and in this story, they merge to initiate narrative – and they merge to create the secondary dichotomy between cosmos and humanity via the "traveling" wave.

Similarly, in the *PRL* abstract, the two black holes come across as goal-oriented entities. They are described as "pair" going through the "inspiral and merger [. . .] and the ringdown of the resulting single black hole," endowing the process with teleological implications, the new black hole being something like a final product of the process ("subsequent *final* black hole ringdown," Abbott et al. 2016, 8; emphasis by EH) – but there is never a "final" product in nature. The merger and ringdown of the black holes signify an attraction of two powerful bi-

56 "A long time ago in a galaxy far, far away . . ." (see Lucas 2006).

57 The "four decades" indicates that the first experiments to detect gravitational waves started in the 1960s.

58 This does not mean that the natural event has teleological qualities; it is merely its narration that is of interest here.

59 However, they do not signify opposites in meaning, but two opposite points in space.

nary entities (see Abbott et al. 2016, Fig. 2). It is certainly coincidental, but still unexpectedly symmetrical, to have the signal produced by the merging of two entities result in such a triumph via merging the signals registered in two other entities, namely the interferometers in Washington and Louisiana: because of the distance between the two, the respective signals are shifted to lose the time that it takes the wave to travel between the interferometers – "intersite propagation time" (Abbott et al. 2016, 21; see Fig. 1). The meaning that emerges from the tension between binary opposites in narrative is thus mirrored in the new meaning (discovery) produced by the interplay between two merging black holes and two merging interferometer signals, wherein the black holes and the interferometers are "connected" by the gravitational wave.

Furthermore, the research narrative presented in *PRL* is a narrative, not a simple enumeration of events, precisely because it is provided with an origin (Einstein's prediction), obstacles (doubted existence, debate, technical challenges) and a definite meaning or "victorious" moment of final discovery (Abbott et al. 2016, 5). The (removed) obstacles can also be found in the extensive validation of the findings, listing all alternative possible phenomena that could produce a similar signal and undermine the findings (Abbott et al. 2016, 16–18). Anticipating the article's first paragraph, its abstract directly connects the discovery to prediction, like the pieces of a puzzle falling into place: "It matches the waveform predicted by general relativity [. . .]." Thus, the research narrative always seems to mirror what Ricœur calls the "grasping-together" of events as the core of emplotment. Accordingly, traces of the quest motif can be found in the abstract as well, in the "source [that] lies at a luminosity distance of [. . .]," like a beckoning signal calling forth the hero or the voyager.

The parenthesis in Cho's first sentence is an explanation of what black holes are (Cho 2016, 1). However, even this "dry" description has poetic dimensions, beginning with the seemingly hyperbolic quality of the technical term "ultrastrong" and the personification that lies dormant in "left behind by," as if the astrophysical object had turned into a black hole by choice or consciousness. The contrast between parenthesis and main sentence immediately implies two ways of narrating this story, both of which are used in the article. One of them is embodied in the way the two quest-like stories are told, and the other is the more factual report of discovery. The research narrative emerges in the last sentence, introducing the physicists and their quest.

Without the parenthesis, the first exposition sentence lays the groundwork for plot: "Long ago, deep in space, two massive black holes [. . .] slowly drew together." This placement in an unknown past and place increases suspense, the image of a slow process adds a layer of mystery and intrigue, leaving the reader free to imagine what draws these two objects together. The collision can also be read as a type I event which turns into a type II event via emplotment: the event is made important, even mythical, from a human perspective. Therefore, this collision establishes tellability, the point of the narrative being the arrival of the signal on Earth.

The paragraph's second sentence displays motion and metaphor: the inserted temporal specification "1.3 billion years" contrasts with the main clause, just as the explanatory parenthesis does in the first sentence. To describe the remains of a star as a ghost is metaphorical and adds to the mystery established in the first sentence. The two objects "whirling about each other" might suggest a (slightly frantic) dance and form an analogy between natural gravitational pull and the pull of attraction between individuals. "Ever closer" implies a rise of tension which culminates in increased speed ("whirl at half the speed of light") and – lastly – in the collision. The following sentence presents the outcome of this encounter: a "shudder" is "sent" through the universe; the collision "radiates ripples" (caption in Fig. 2; Cho 2016). This not only increases the teleological undertone ("sent"), it also evokes varying connotations. "To shudder" is usually connected with feelings of fear or disgust. However, it also alludes to seismic signals or induces erotic associations. The shudder becomes more intense, as it turns into "ripples" and then into "waves." To name them "ripples in the fabric of space and time"[60] is an accurate physical description, but it also signifies a process that is fundamental and "earth-shattering" in the literal sense, as explained later in the article (Cho 2016, 3).

After this background scene-setting, the actual announcement follows: "Five months ago, [the waves] washed past Earth." The distant and barely tangible cosmological event is endowed with actuality and imminence. While other waves might have gone past Earth undetected, this one was registered. The narrative and its protagonists emerge: while being the common scientific term and not an explicit literary device, "to detect" can also imply that the waves are hunted or do not want to be found – apparently coincidentally, Cho calls the history of gravitational wave science a "history of hunting" (2016, 3) – thus, the waves are portrayed as elusive and mysterious and thereby personified. Stories of detection can be found quite often in scientific activity: "The scientific community wants to know 'how it works' or *in the literary sense 'who-done-it,'* hence scientific research shares similarities with literary narratives" (Luna 2013, 3; emphasis by EH). Put bluntly, the research narrative itself equals a story of detection, of unravelling a mystery. Since detective stories – traditionally dependent on one detecting, investigating character – make up no small part of fiction, this might not be a coinci-

60 See Kennefick (2007, 7 and 11–12) for a discussion of the "ripples in spacetime" metaphor.

dence after all.[61] If read this way, the other protagonists of Cho's text are the "detective"-scientists[62] on their quest, finally reaching their goal. Thus, the classic detective trope and the hero's journey seem to merge quite seamlessly.[63] Personifications of natural entities can, apart from religious and mythological associations, also be explained by the "perceptual gestalt" which Fuchs calls "Force of Nature," meaning that certain conceptual models endow objects or entities in the natural world with agency: "The perception of forces of nature in large-scale events [. . .] lets us construct the figure of *natural agents* central to stories (this is a process of *narrative perception*).[64] This, in turn, allows us to tell stories about forces of nature (as an act of *narrative production*)" (Fuchs 2015, 935). Fuchs argues that these stories are powerful enough to "propose concepts and models and, therefore, elements of theory" (2015, 947). This "force of nature" gestalt also coincides with what Lakoff calls the "cause as force" metaphor (1993, 220).

The detectors come across as less personified in *PRL*, but still as acting entities: "Only the LIGO detectors were observing at the time of GW150914" (Abbott et al. 2016, 7). Once again, though at a subconscious level, this creates the impression of spies or detectives "lying in wait" at just the right moment. However, the overall impression of detectors is a more passive one: "Gravitational-wave astronomy exploits multiple, widely separated detectors to distinguish gravitational waves from local instrumental and environmental noise" (Abbott et al. 2016, 9). If the detectors are posited as more passive agents, this reinforces the reading of the wave as an active cosmic messenger – or message – that influences the detectors. Consequently, the overall impression of the detectors remains slightly ambiguous, oscillating between actively detecting and passively receiving. By contrast, black holes are continuously personified as "candidates [that] have now been identified" (Abbott et al. 2016, 2) or by having their orbiting "evolution [. . .] characterized by the chirp mass" (Abbott et al. 2016, 8). They are also found "relaxing to a final stationary Kerr configuration" (Abbott et al. 2016, 8).

61 See, for instance, an analysis of emplotted detective narratives in modern clinical case reporting and stories of diagnosis: "Detective and clinical case work share imaginative, interrogative and inferential moves inflected in appeals to puzzle out and fit together diverse pieces of information in 'a single connected narrative'" (Hurwitz 2017, 67).

62 The detectors themselves could also be interpreted as protagonists in this case.

63 Opposing Griem's notion that the narrative of an adventurer-scientist suppresses the scientist collective, this article posits the whole group as heroes, not one single scientist. The same is true for Abbott et al., where the rhetorical first-person plural is used to represent the scientists. For the use of "we" in scientific texts, see Harré (2000, 166).

64 Thus, reading gravitational waves as a personified, hunted force does not seem an overstatement.

The physicists' role is elevated in the last sentence of Cho's introductory paragraph, making them actors in a mythical tale, "fulfilling a 4-decade quest and opening new eyes on the heavens." If the scientific discovery was to be communicated in an objective way, "sky" or "universe" instead of the allusive and transcendental "heavens" would have sufficed. This sentence concludes the emplotment and the dramatic arc, its significance increased by allusions that range from mythical quest, fantasy or fairy tales to biblical references. The metaphorical "new eyes" opened by this discovery intensify these allusions, and could also mirror the first-time use of glasses, or, in this case, of a telescope. This optical metaphor illustrates how the wave, being the provider of these "new eyes," resembles a text that can be read, like a translator between the cosmos and human science. Thus, the wave is text and tool at the same time, it functions like a lens or a magnifying glass and sharpens the perspective. At this point it becomes evident how, according to Ricœur, metaphor resembles narrative in terms of "semantic innovation" (see section 2). With the wave posited as message-like text and tool, the question of medium arises. Since spacetime is the medium of gravitational waves (Kennefick 2007, 9), and spacetime "only" propagates the gravitational wave, the "medium is the message" (see McLuhan 2003): Instead of seeing space as the wave's container, the wave is part of the medium, it induces movement – an undulation[65] – in spacetime but it does not contain physically moving elements. In this reading, the medium of spacetime is an instrument of communication, establishing a wave-formed connection between the scientists and the black holes.

All in all, this first paragraph displays many ingredients of speculative fiction: the mysterious beginning, powerful and destructive protagonists (the black holes) and a message that is sent – in this case, across time and space – to another protagonist party (the physicists) that must analyze the message. This is only one way of interpreting the paragraph. From a different angle, the physicists could be the only protagonists, and the wave and black holes mere objects. Notwithstanding these different perspectives, the exposition establishes narrativity and meaning. In this way, this text emplots a process that is "strictly" astrophysical science.

In *PRL*, the discovery is also embedded into a century-long research narrative of prediction and obstacles. Comparable to Cho's, this text immediately historicizes the discovery: "In 1916, the year after the final formulation of the field equations of general relativity, Albert Einstein predicted the existence of gravitational waves" (Abbott et al. 2016, 1). Prediction carries a faint ring of prophecy with it,

65 At this point, the grounding of the metaphorical concept of "wave" becomes clear; its structure, like most metaphors, relies on the embodied experience (Gibbs et al. 2004, 1207; see Lakoff 1993, 220) that human beings have when they encounter waves in an ocean or another body of water. This also explains the explanatory force of the term "gravitational wave."

once again elevating a mathematical calculation to a more sublime level. Not only does this formulation provide the background for establishing the discovery's tellability, it also positions the findings as something like a prize, a final goal reached – in other words, the end of a quest. Conflict, the central event of narrative structure, is also evident in this paragraph: "there was significant debate about the physical reality of gravitational waves" (Abbott et al. 2016, 1). This beginning resembles a passage from a history book more than a scientific article. However, this article shows the significant difference between the scientist-journalist's rendering and the direct report by the scientists who did the science: it adheres closely to traditional scientific writing standards, and the narrative elements it contains are less obvious. Still, they can be observed.

Following this introduction in *PRL*, the next paragraphs provide a list of significant steps in gravitational wave science, culminating in the third paragraph, which displays a type II event, the discovery of a binary pulsar system: "This discovery, along with *emerging* astrophysical understanding, *led to the recognition* that direct observations of the amplitude and phase of gravitational waves *would enable studies* of additional relativistic systems and *provide new tests* of general relativity" (Abbott et al. 2016, 3; emphasis by EH). Moreover, the historical significance of the discovery at times seems to overrule objectivity: "A century after the *fundamental* predictions [. . .], *we* report the *first direct* detection of gravitational waves [. . .]. Our observations provide *unique* access to the properties of space-time" (Abbott et al. 2016, 5; emphasis by EH). This linkage between the old predictions and the authors has the rhetorical effect of establishing a direct connection between Einstein and the LIGO/Virgo teams, intensified by words such as "unique" and "first direct," creating a subtext of cooperative exploration and frontier. All the observed elements add to the impression of a quest or an endeavor that spans centuries. Thus, the quest narrative is detectable in *PRL* as well as in *Science*.

Returning to the *Science* text, the third paragraph of Cho's article contains a significant aspect of the detection: "LIGO researchers sensed a wave that stretched space by one part in 10^{21}, making the entire Earth expand and contract by 1/100,000 of a nanometer, about the width of an atomic nucleus" (Cho 2016, 3). The metonymy in "LIGO researchers sensed" increases the accessibility of the event; it would have been less tangible if the Laser Interferometer Gravitational-Wave Observatory had registered the wave. Furthermore, "sense" is more likely to evoke emotion than "detect" or "register," because it implies a physical sensation felt by humans. The literal impact that this wave had on Earth and its inhabitants is accentuated by the expansion and contraction of the whole planet. "Expand" and "contract" have various connotations, reaching from physiological and chemical processes to breathing, childbirth or orgasm. Put more objectively, spacetime was minimally altered or stretched and then returned to its previous state, but the article's formulation renders Earth a living, breathing organism. A strong opposition between the hardly imaginable smallness of "the width of an atomic nucleus" and the expansion of "the entire Earth" intensifies this impression. This temporary expansion is a direct consequence of the wave, and it again shows that the black hole collision is a type II event. The expansion of Earth, then, is the indirect consequence of the collision and spawns further events, such as the discovery and the impact it has on all of Earth. The subsequent sentence provides the wave with another layer of meaning, or indicates a deciphered part of the wave if it is read as text: "The observation tests Einstein's theory of gravity, the general theory of relativity, with unprecedented rigor and provides proof positive that black holes exist" (Cho 2016, 3).

Narrativity is also found in the precise but fairly colloquial description of the detector's workings:

> LIGO watches for a minuscule stretching of space with what amounts to ultraprecise rulers: two L-shaped contraptions called interferometers with arms 4 kilometers long. Mirrors at the ends of each arm form a long "resonant cavity," in which laser light of a precise wavelength bounces back and forth, resonating just as sound of a specific pitch rings in an organ pipe. Where the arms meet, the two beams can overlap. If they have traveled different distances along the arms, their waves will wind up out of step and interfere with each other. That will cause some of the light to warble out through an exit called a dark port in synchrony with undulations of the wave. (Cho 2016, 4)

"LIGO watches" is another metonymy, the "ultraprecise rulers" a metaphor, while "arms" is an exemplary dead metaphor. The organ pipe serves as useful analogous explanation of how the system works, while "contraptions" seems slightly mocking and colloquial. The "warble" of light through the ominous "dark port" – a name probably invented by the scientists who developed LIGO – concludes the paragraph in an epitome of figurative language. "Warble" is onomatopoeic, mimicking the movement of the light, shudderingly escaping through the exit. Usually, "warble" relates to sound, e.g. the sound of birds. However, in physics, a "warble tone" can refer to an oscillating frequency, which is probably where this light-related use originates.[66] The lines of the luminous movement contain a story as well: light traveling inside the interferometer (exposition), being disturbed by the wave (event type I), causing an interference and the proliferation of an important signal (event type II), making it readable and resulting in detection (tellability). The meeting of the two arms then conditions the formation of narrative. The altering of the light beams thus constitutes the decisive point in the story of travel-

66 According to the OED, a "warble tone" is a "constant amplitude tone whose frequency is cyclically varied between certain limits, used in acoustic measurement to avoid irregularities associated with the use of single frequencies" ("Warble, n.1.").

ing light. Fuchs's account of narrative in science corresponds to this situation: he shows that scientific models indirectly "relate to storyworlds" (Fuchs 2015, 936) in his description of how the model of boiling water in a kettle (Fuchs 2015, 943, Fig. 1) contains narrative elements that clearly emerge in the simulation of the model: "When we (mentally) simulate the model, or if we recount a simulation in natural language, we are setting up a concrete scenario in the world of the figures of mind (in the storyworld) of the model and then follow the agent(s) through a sequence of events" (Fuchs 2015, 945). This process can be observed in Cho's figurative explanation of the luminous movement within the interferometer, taking the simplified diagram of the detector in Abbott et al. (2016, Fig. 3) as the model that contains the story for Cho's narrative simulation.

In contrast, the detailed description of the detector's inner workings in *PRL* is an exemplary "methods" section. It contains a list of the detectors' parts and their functions. Apart from situating the techniques used in the detectors in the field of gravitational-wave astronomy, and emphasizing the extraordinary nature of the new findings, it displays next to no narrativity (Abbott et al. 2016, 9–15). However, part IV., "Detector Validation" (Abbott et al. 2016, 16–18), does display narrativity, since it gives a detailed description of the various requirements to detect a gravitational wave amid the background noise that feeds into the highly sensitive detectors. It correlates with the conjured image of a spy "lying in wait" (see Cho 2016), while at the same time it shows a seemingly more fragile side, namely the immense sensitivity of these instruments. This passage also demonstrates that the "dry" punctuation often required in scientific writing can also create dramatic suspense. After a build-up illustrating the search for disturbances that might produce false signals, two significant sentences end the paragraph: "Special care was taken to search for long-range correlated disturbances that might produce nearly simultaneous signals at the two sites. No significant disturbances were found" (Abbott et al. 2016, 17). In general, though, most of the narrativity in the *PRL* article seems to be distributed among the abstract, the opening paragraphs and the conclusion paragraphs.

After the detectors' description, a figure appears in Cho's article, illustrating how LIGO works: "Catching a wave" refers to surfing, emphasizing the role of the physicists as active protagonists. The figure also pictures the gravitational wave "[z]ipping along at light speed," like an arrow with a precise target: "[A] wave stretches space in one direction and squeezes in the perpendicular direction, then reverses the distortions" (Cho 2016, Fig. 2). This linear movement of "zipping along" with different reactions of space during the journey, as depicted, could represent the way the wave is read continuously, like a text sequence with different significant parts. As Ricœur notes, "[t]o follow a story is to actualize it by reading it" (1990, 76).

The discovery is the first black-hole evidence that "does not depend on watching hot gas or stars swirl around them at far greater distances" (Cho 2016, 8). This endows black holes with a new kind of reality that is less hypothetical and more "touchable." In addition, "swirl" echoes "whirl" in the first paragraph, connecting different ends of the text through onomatopoeic and rhyming verbs. The circular, vortex-like movement of mass and stars is thus suggestively mirrored. A dramatizing simile occurs after the "astounding, invisible" explosion that is described as follows: "Modeling shows that the final black hole totals 62 solar masses – 3 solar masses less than the sum of the initial black holes. The missing mass vanished in gravitational radiation – a conversion of mass to energy that makes an atomic bomb look like a spark" (Cho 2016, 9). Not only is the sense of mystery heightened because of the "invisible explosion" and the "vanished" mass; there is also the comparison between the collision and an atomic bomb. The atomic bomb, epitome of mass destruction, is rendered trivial by this comparison: the scale of astrophysical processes, the "sublime incomprehensibility of the universe" (E. C. White 1990, 107) and the insignificance of humankind could hardly be communicated in a more exemplary fashion.

The flow of the narrative is delayed through an insertion of additional information regarding the detection system and its way of ruling out false signals (Cho 2016, 11–12). However, since the content of this insertion proves that the detection cannot be a misinterpretation, the narrative weight of this sequence is reinstated because it once more validates the discovery and confronts the scientific "heroes" with obstacles to overcome. Afterwards, the aspect that makes up a major part of the article's tellability is introduced, namely future science and all its possibilities. This is described with powerful vocabulary: "It is the prospect of the science that might be done with gravitational waves that really *excites* physicists. [. . .] [T]he first LIGO result shows the *power* of such radiation to *reveal unseen* astrophysical objects like the two *ill-fated* black holes" (Cho 2016, 14; emphasis by EH). Mystery is now partly lifted or "revealed"; it is lifted in parallel with the continuously revealed importance of the collision. The "ill-fated black holes" bring forward a new referent – that of romantic love. The expression could be read as follows: it was the black holes' fate to collide and die under an unfortunate star; they are the "star-cross'd lovers."[67] This analogy has been noted by other scientists as well.[68]

---

67 In the prologue of Shakespeare's *Romeo and Juliet*, the characters are presented as a "pair of star-cross'd lovers" (Shakespeare 1969, 3).

68 For instance, an astronomer calls the collision of two neutron stars, discovered by Virgo and LIGO in 2017, a "pair of neutron-star-cross'd lovers" in her blog (Delhaize 2017). Fittingly, Abbott et al. mention "[a] pair of neutron stars" (2016, 8).

Related powerful terminology continues in the next paragraph, which echoes the extremity of the previous atomic bomb comparison: "By studying gravitational waves, [physicists] can now *explore extreme* conditions in which the energy in an object's gravitational field accounts for most or all of its mass – the *realm* of *strong* gravity *so far explored* by theorists *alone*" (Cho 2016, 15; emphasis by EH). This pioneer exploration of a heretofore unknown "realm" is an overt frontier reference.[69] This reference concurs with the idea of scientific exploration and discovery and thus fits scientific research.[70] Besides, it is a variation of the quest narrative that dominates the first paragraph – the word "quest" is repeated as well (Cho 2016, 17). This reintroduction of quest and exploration builds a bridge between the middle section and the final paragraphs of the article. Exploration, detection and quest are embodied in the recurrent surfing metaphor of the last paragraph (Cho 2016, 19). It begins with "meanwhile," which brings the narrative back to a current state of events, adding nuance to the text's tellability. Therefore, the narrative focus now rests on the present, embedded in time between past and future: this makes the last part of this article another example of Ricœur's emplotment, producing causality and narrative meaning out of temporal succession.

At the end of their *PRL* paper, Abbott et al. introduce binary black hole systems at larger distances: "If the signal from such a population were detected, it would provide information about the evolution of such binary systems over the history of the universe" (Abbott et al. 2016, 42). Although black hole "populations" seems like a personification, it is a common scientific term. However, this sentence shows a high degree of narrativity due to its reference to the origin and development of the universe, and the place of binary black holes in this (hi)story. It also adds to the overall tellability and the research narrative of the paper, casting a prognosis of what future research might bring. Thus, the research narrative weaves past, present and future into a comprehensible order.

Concerning the future, a prediction is made in Cho's final paragraph as well: "Meanwhile, other detections may come quickly. LIGO researchers are still analysing data [. . .] and they plan to start taking data again in July. [. . .] Physicists eagerly await the next wave" (Cho 2016, 19). This phrase, which likens physicists to surfers (rather than to the detectives from the first paragraph), underlines the aspect of competition – in sports as well as in scientific research – and it accentuates the "intimate" relationship between observer and the observed: if the physicist is a surfer, catching a gravitational wave, he or she can ride it and is directly

69 The use of the frontier as metaphor usually brings about problematic connotations. It implies, for instance, an "empty space" to be "conquered."

70 For discovery narratives in accounts written by quantum physicists, see Gencarelli (2017).

connected to it, carried by it, only separated by a surf board, or a detection device. Additionally, the last sentence does not end the narrative, but leaves it open for more events – more waves – to come. In this sense, the discovery utterly transforms the scientists, and transformation is another basis of narrative.[71] This also corresponds with the power of the wave: if the wave is a text to be read, this transformation could be a representation of Ricœur's "mimesis III" – in refiguration (after reading the wave), the scientists are altered by the experience (Ricœur 1990, 78–79).

In conclusion, two basic narrative directions can be discovered in Cho's article. The first one incorporates the natural or cosmological part, namely the black hole collision, the traveling waves, their arrival in the solar system and on Earth. The second conveys the cultural aspect and the research narrative, from a human point of view, beginning with Einstein's prognosis, the development of wave-sensitive instruments and the first discovery, in addition to the possibilities opened by it. Both stories work with waves, the first one literally and the second one in a figurative way, seen as waves of scientific development. Both are threaded into this article, and the narrative thus emplots these two stories in a temporal order which symbolizes a long journey (past), a fulfilled quest (present) and new possibilities (future). As noted, the core of this structure is mirrored in the *PRL* paper, albeit less obviously.

The two juxtaposed narrative directions are reiterated on the textual level, where the figurative first and final paragraphs frame the detailed description of LIGO, the research history[72] and relevant scientists. These two "styles" are held together by the wave: one could imagine the journey of the wave and the text as one graph, with multiple aspects to be deciphered and interpreted. The "zipping along"-caption (Cho 2016, Fig. 2) and the respective illustration exemplify this idea. This emplotment of the wave as central metaphor shows how the human perception of a cosmic event influences the way it is presented. It also demonstrates how the cosmic insignificance of humans, as opposed to an egocentric perspective of humanity, results in a narrative that imposes a goal-oriented process onto something that does not originate from nor is in any way related to Earth – in *Science* as well as in *PRL*. The article's structure – the exposition paragraph, the quest history, the detailed discovery report and the concluding surfer para-

---

71 Transformation in narrative equals a "syntagmatic unfolding of the difference between an initial and a subsequent state" (Panier 2005, 377), it happens on the level of the "semantic universe that underlies all discourse" (Panier 2005, 377). This opposition between state and transformation induces narrativity (Panier 2005, 377). For models of transformation, see Greimas (1983, 222–256).

72 The research history is also recapitulated in a detailed timeline below the article.

graph – mirrors this teleological process. The metaphor of scientists as heroes bringing home the Campbellian "boon" bridges the gap between "cold" stellar matter and human experience.[73]

4 Conclusion

While science and literature often appear as opposites, this chapter – as a contribution to interdisciplinary research – has shown that texts from scientific journals display narrativity. Narrative organizes content in a way that is beneficial to cognitive processing; it transforms chronological research into a comprehensible format. However, narratological analysis is seldom applied to texts from scientific journals. As previously established, a great number of scientific articles exist in the twofold tension between new and old research aspects, and between narrativity and the rather non-narrative objectivity expected from the scientific community. The investigated journal texts do not correspond to the "windowpane"-ideal of absolute clarity in scientific language (see section 2). Additionally, the results of this analysis demonstrate that the supposed "virtual absence of narrative elements in the twentieth century" (Gross et al. 2002, 224) within scientific journals, if accurate at all, has not continued into the new millennium.

The most frequent form of narrative in science is the research history, which must appear in almost every scientific article. While overt narratives can be detected in Cho's article, the text from *PRL* contains more understated narrative elements. This is hardly surprising, because the news-like properties of the *Science* article – written by a scientist-journalist – stand in contrast to the *PRL* research report, in which scientists present their own work. Consequently, research reports are more strongly shaped by the conventions of scientific writing, unlike articles that, for instance, include interviews. However, as demonstrated, narrativity and traces of the quest do also exist in the *PRL* research report.

The way that metaphor is employed in these texts illustrates why it is an important device with descriptive and rational properties (see Hesse 1970, 176–177), a tool of "semantic innovation" (Ricœur 1990, ix) and the embodiment of a bridge (see Hörisch 2007, 44), as seen in the mediating function of the gravitational wave. Other significant and often-employed devices are metonymies and personifications or similar forms of agency attribution. The idea of the gravitational wave as a mediating entity that connects two separate elements – in this case,

73 The emphasis on scientists waiting for waves also distantly echoes Kuhn's conceptualization of "scientific revolutions."

cosmos and self-conscious humanity – is captured in Cho's article, which interweaves two writing styles: the research history about LIGO is framed by a highly figurative beginning and ending – the near-mythical opening that mirrors Campbell's monomyth, and the corresponding final paragraph which pictures the scientists as surfers waiting to catch the next wave. Both narrative threads depend on the wave, are carried by it – in this sense, the wave resembles a continuing stream of letters.

The analysis illustrates that narratological, rhetorical and linguistic analyses can be utilized to discover narrative and ambiguities in texts that profess to be objective. The overall results illuminate the techniques that make articles from *Science* comprehensible to non-experts, although they primarily address general scientists, unlike popular science. It is also important to acknowledge that the narrative and rhetorical elements analyzed here do not seem to distort the factual authenticity of the presented scientific findings, as feared by Katz (see section 3), or to turn the scientific results into a mere construction. As Vanderbeke explains, rhetorical analysis potentially enables the investigation of borders, interfaces and continuities between sciences and humanities (2004, 130–131). This positive impact can just as well be ascribed to an analysis of narrative as a building block of both disciplines, as this essay has attempted. Thus, narratives in science writing might play a crucial role in interdisciplinary work to come.

Reference list

Primary texts

Abbott, B. P., R. Abbott, T. D. Abbott, M. R. Abernathy, F. Acernese, K. Ackley, et al. (LIGO Scientific Collaboration and Virgo Collaboration). "Observation of Gravitational Waves from a Binary Black Hole Merger." *Physical Review Letters* 116.6 (2016): art. 061102. doi:10.1103/PhysRevLett.116.061102.

Cho, Adrian. "Gravitational Waves, Einstein's Ripples in Spacetime, Spotted for First Time." *Science.org* (11 February 2016). doi:10.1126/science.aaf4041.

Secondary material

Abbott, H. Porter. "Unnarratable Knowledge: The Difficulty of Understanding Evolution by Natural Selection." *Narrative Theory and the Cognitive Sciences*. Ed. David Herman. Stanford: CSLI, 2003. 143–162.

Abbott, H. Porter. “Narrativity” (2014 [2011]). *The Living Handbook of Narratology*. Eds. Peter Hühn et al. Hamburg: Hamburg University. https://www-archiv.fdm.uni-hamburg.de/lhn/node/27/revisions/280/view.html (3 August 2022).
Anderson, James A. “Objectivity in Science.” *The International Encyclopedia of Communication*. Vol. 8. Ed. Wolfgang Donsbach. Malden: Blackwell Publishing, 2008. 3351–3354.
Aristotle. *The Poetics*. Trans. Anthony Kenny. Oxford: Oxford University Press, 2013.
Bal, Mieke. *Narratology: Introduction to the Theory of Narrative*. Trans. Christine van Boheemen. Toronto: University of Toronto Press, 1985 [1978].
Baroni, Raphaël. “Tellability” (2014 [2011]). *The Living Handbook of Narratology*. Eds. Peter Hühn et al. Hamburg: Hamburg University. https://www-archiv.fdm.uni-hamburg.de/lhn/node/30/revisions/338/view.html (3 August 2022).
Barthes, Roland. “The Death of the Author” (1968). Trans. Stephen Heath. *Image – Music – Text*. By Roland Barthes. London: Fontana Press, 1977. 142–148.
Beer, Gillian. *Darwin's Plots: Evolutionary Narrative in Darwin, George Eliot and Nineteenth-Century Fiction*. 2nd edition. Cambridge: Cambridge University Press, 2000 [1983].
Bhaskar, Roy. *A Realist Theory of Science*. London and New York: Routledge, 2008 [1975].
Black, Max. *Models and Metaphors: Studies in Language and Philosophy*. Ithaca and London: Cornell University Press, 1972 [1962].
Blum, Deborah. “Narrative Style in Science Journalism.” *WissensWelten: Wissenschaftsjournalismus in Theorie und Praxis*. Eds. Holger Hettwer et al. Gütersloh: Verlag Bertelsmann Stiftung, 2008. 550–554.
Bloor, David. *Knowledge and Social Imagery*. Chicago and London: University of Chicago Press, 1991 [1976].
Borges, Jorge Luis. *This Craft of Verse*. Ed. Călin-Andrei Mihăilescu. Cambridge, MA: Harvard University Press, 2000.
Bruner, Jerome. *Actual Minds, Possible Worlds*. Cambridge, MA: Harvard University Press, 1986.
Bucchi, Massimiano. “Of Deficits, Deviations and Dialogues: Theories of Public Communication of Science.” *Handbook of Public Communication of Science and Technology*. Eds. Massimiano Bucchi and Brian Trench. London and New York: Routledge, 2008. 57–76.
Campbell, Joseph. *The Hero with a Thousand Faces*. 2nd edition. Princeton: Princeton University Press, 1973 [1949].
Campbell, Joseph. *The Masks of God: Occidental Mythology*. London: Souvenir Press, 1974 [1964].
Campbell, Joseph, and Edmund L. Epstein. *Mythic Worlds, Modern Words: On the Art of James Joyce*. New York: Harper Collins, 1993.
Campenhausen, Jutta von. *Wissenschaft vermitteln*. Wiesbaden: Springer VS, 2014.
Carrier, Martin, Johannes Roggenhofer, Günter Küppers, and Philippe Blanchard, eds. *Knowledge and the World: Challenges Beyond the Science Wars*. Berlin and Heidelberg: Springer, 2004.
Clemens, Anna. “Writing a Page-Turner: How to Tell a Story in Your Scientific Paper.” *LSE Impact Blog*. 21 May 2018. https://blogs.lse.ac.uk/impactofsocialsciences/2018/05/21/writing-a-page-turner-how-to-tell-a-story-in-your-scientific-paper/ (25 March 2020).
Cloître, Michel, and Terry Shinn. “Expository Practice: Social, Cognitive and Epistemological Linkages.” *Expository Science: Forms and Functions of Popularization*. Eds. Terry Shinn and Richard Whitley. Dordrecht: Reidel, 1985. 31–60.
Collini, Stefan. “Introduction” (1993). *The Two Cultures* (1959). By C. P. Snow. Cambridge: Cambridge University Press, 1998 [1964]. vii–lxxi.
Currie, Adrian, and Kim Sterelny. “In Defence of Story-Telling.” *Studies in History and Philosophy of Science Part A* 62 (2017): 15–21. doi:10.1016/j.shpsa.2017.03.003.

Dahlstrom, Michael F. "Using Narratives and Storytelling to Communicate Science with Nonexpert Audiences." *Proceedings of the National Academy of Sciences* 111.4 (2014): 13614–13620. doi:10.1073/pnas.1320645111.

Dean, Cornelia. *Making Sense of Science: Separating Substance from Spin*. Cambridge, MA and London: The Belknap Press of Harvard University Press, 2017.

Delhaize, Jacinta. "A Pair of Neutron-Star-Cross'd Lovers." *Travelling Light: The Home Page and Blog of Dr Jacinta Delhaize*. 17 October 2017. https://jacintadelhaize.com/2017/10/17/a-pair-of-neutron-star-crossd-lovers/ (12 February 2018).

Einstein, Albert. "Näherungsweise Integration der Feldgleichungen der Gravitation." *Sitzungsberichte der Königlich Preußischen Akademie der Wissenschaften* (1916): 688–696. https://ui.adsabs.harvard.edu/abs/1916SPAW ... ... .688E (2 January 2019).

Fludernik, Monika. *Towards a "Natural" Narratology*. London and New York: Routledge, 1996.

Fludernik, Monika. "Natural Narratology and Cognitive Parameters." *Narrative Theory and the Cognitive Sciences*. Ed. David Herman. Stanford: CSLI, 2003. 243–267.

Forster, E. M. "'The Story' and 'The Plot'" (1927). *The Narrative Reader*. Ed. Martin McQuillan. London and New York: Routledge, 2000. 44–46.

Früh, Werner, and Felix Frey. "Narration und Storytelling." *Narration und Storytelling*. Eds. Werner Früh and Felix Frey. Cologne: Herbert von Halem Verlag, 2014. 63–119.

Frye, Northrop. *Anatomy of Criticism: Four Essays*. Princeton: Princeton University Press, 2000 [1957].

Fuchs, Hans U. "From Stories to Scientific Models and Back: Narrative Framing in Modern Macroscopic Physics." *International Journal of Science Education* 37.5–6 (2015): 934–957. doi:10.1080/09500693.2015.1025311.

Gencarelli, Angela. "'Entdeckungsgeschichten' der Quantenphysik: Zur narratologischen Erforschung einer naturwissenschaftlichen Erzählform." *Diegesis: Interdisciplinary E-Journal for Narrative Research* 6.1 (2017): 1–19. https://www.diegesis.uni-wuppertal.de/index.php/diegesis/article/view/260 (25 March 2018).

Gibbs, Raymond, Paula Lima, and Edson Francozo. "Metaphor is Grounded in Embodied Experience." *Journal of Pragmatics* 36 (2004): 1189–1210. doi:10.1016/j.pragma.2003.10.009.

Gläser, Rosemarie. *Fachtextsorten im Englischen*. Tuebingen: Gunter Narr Verlag, 1990.

Goodrich, Peter. "Narrative as Argument." *Routledge Encyclopedia of Narrative Theory*. Eds. David Herman et al. London and New York: Routledge, 2005. 348–349.

Gould, Stephen Jay, and Richard C. Lewontin. "The Spandrels of San Marco and the Panglossian Paradigm: A Critique of the Adaptationist Programme." *Proceedings of the Royal Society of London*. 205.1161 (1979): 581–598. doi:10.1098/rspb.1979.0086.

"Gravitational Waves Hit the Late Show." *YouTube*. Uploaded by The Late Show with Stephen Colbert. 25 February 2016. https://www.youtube.com/watch?v=ajZojAwfEbs (22 December 2018).

Gräfrath, Bernd. "Lem und Golem: Erkenntnis durch Science Fiction." *Philosophie in Literatur*. Eds. Christiane Schildknecht and Dieter Teichert. Frankfurt a. M.: Suhrkamp, 1996. 342–364.

Greimas, Algirdas Julien. *Structural Semantics: An Attempt at a Method*. Trans. Daniele McDowell et al. Lincoln and London: University of Nebraska Press, 1983 [1966].

Gross, Alan, Joseph E. Harmon, and Michael S. Reidy. *Communicating Science: The Scientific Article from the 17th Century to the Present*. Oxford and New York: Oxford University Press, 2002.

Gross, Paul, and Norman Levitt. *Higher Superstition: The Academic Left and Its Quarrels with Science*. Baltimore and London: The Johns Hopkins University Press, 1998 [1994].

Gurr, Jens Martin. "'Without contraries is no progression' – Emplotted Figures of Thought in Negotiating Oppositions, *Funktionsgeschichte* and Literature as 'Cultural Diagnosis.'" *Text or*

Context: Reflections on Literary and Cultural Criticism. Eds. Rüdiger Kunow and Stephan Mussil. Würzburg: Königshausen & Neumann, 2013. 59–77.

Gusfield, Joseph. "The Literary Rhetoric of Science: Comedy and Pathos in Drinking Driver Research." *American Sociological Review* 41.1 (1976): 16–34. doi:10.2307/2094370.

Harré, Rom. "Some Narrative Conventions of Scientific Discourse." *The Narrative Reader*. Ed. Martin McQuillan. London and New York: Routledge, 2000. 165–166.

Hayles, N. Katherine. "Introduction: Complex Dynamics in Literature and Science." *Chaos and Order: Complex Dynamics in Literature and Science*. Ed. N. Katherine Hayles. Chicago and London: The University of Chicago Press, 1991. 1–33.

Herman, David. "Scripts, Sequences, and Stories: Elements of a Postclassical Narratology." *PMLA* 112.5 (1997): 1046–1059. doi:10.2307/463482.

Herman, David, Manfred Jahn, and Marie-Laure Ryan. "Introduction." *Routledge Encyclopedia of Narrative Theory*. Eds. David Herman et al. London and New York: Routledge, 2005. ix–xii.

Hesse, Mary B. *Models and Analogies in Science*. Notre Dame: University of Notre Dame Press, 1970 [1966].

Heydenreich, Aura, and Klaus Mecke. "Literatur- und Naturwissenschaften – Wissenskulturen in Wechselwirkung: Plädoyer für eine interaktionale Diskurszone zwischen den epistemischen Gemeinschaften." *Quarks and Letters: Naturwissenschaften in der Literatur und Kultur der Gegenwart*. Eds. Aura Heydenreich and Klaus Mecke. Berlin and Boston: De Gruyter, 2015. 1–28. doi:10.1515/9783110406542-001.

Hörisch, Jochen. *Das Wissen der Literatur*. Munich: Wilhelm Fink Verlag, 2007.

Holmes, Frederic L. "Argument and Narrative in Scientific Writing." *The Literary Structure of Scientific Argument: Historical Studies*. Ed. Peter Dear. Philadelphia: University of Pennsylvania Press, 1991. 164–181. doi:10.9783/9781512801590-008.

Holub, Miroslav. "Poetry and Science: The Science of Poetry / The Poetry of Science." *The Measured Word*. Ed. Kurt Brown. Athens: University of Georgia Press, 2001. 47–68.

Hühn, Peter. "Functions and Forms of Eventfulness in Narrative Fiction." *Theorizing Narrativity*. Eds. John Pier and José Ángel García Landa. Berlin and New York: De Gruyter, 2008. 141–164.

Hühn, Peter. "Event and Eventfulness" (2013 [2011]). *The Living Handbook of Narratology*. Eds. Peter Hühn et al. Hamburg: Hamburg University. http://www.lhn.uni-hamburg.de/node/39.html (29 November 2017).

Hurwitz, Brian. "Narrative Constructs in Modern Clinical Case Reporting." *Studies in History and Philosophy of Science Part A* 62 (2017): 65–73. doi:10.1016/j.shpsa.2017.03.004.

Johansen, Jørgen Dines. "Semiotics." *Routledge Encyclopedia of Narrative Theory*. Eds. David Herman et al. London and New York: Routledge, 2005. 522–527.

Jones, Michael D., and Deserai Anderson Crow. "How Can We Use the 'Science of Stories' to Produce Persuasive Scientific Stories?" *Palgrave Communications* 3 (2017): art. 53. doi:10.1057/s41599-017-0047-7.

Kaplan, Martin, and Michael F. Dahlstrom. "How Narrative Functions in Entertainment to Communicate Science." *The Oxford Handbook of the Science of Science Communication*. Eds. Kathleen Hall Jamieson et al. New York: Oxford University Press, 2017. 311–319.

Katz, Yarden. "Against Storytelling of Scientific Results." *Nature Methods* 10.11 (2013): 1045. doi:10.1038/nmeth.2699.

Kennefick, Daniel. *Traveling at the Speed of Thought: Einstein and the Quest for Gravitational Waves*. Princeton and Oxford: Princeton University Press, 2007.

Knoespel, Kenneth J. "Instability and Narrative Order." *Chaos and Order: Complex Dynamics in Literature and Science*. Ed. N. Katherine Hayles. Chicago and London: The University of Chicago Press, 1991. 100–124.

Koschorke, Albrecht. *Fact and Fiction: Elements of a General Theory of Narrative*. Trans. Joel Golb. Berlin and Boston: De Gruyter, 2018 [Ger: 2012].

Kretzenbacher, Heinz L. "'Just Give Us the Facts!': The Connection between the Narrative Taboo, the Ego Taboo and the Metaphor Taboo in Scientific Style." *Lingua e Stile* 29.1 (1994). 91–106.

Krieger, Anja. "Storytelling in der Wissenschaft: Der Forscher als Held und Abenteurer." *Deutschlandfunk Kultur* (3 January 2019). https://www.deutschlandfunkkultur.de/storytelling-in-der-wissenschaft-der-forscher-als-held-und.976.de.html?dram:article_id=437369 (7 January 2019).

Krzywinski, Martin, and Alberto Cairo. "Points of View: Storytelling." *Nature Methods* 10.8 (2013a): 687. doi:10.1038/nmeth.2571.

Krzywinski, Martin, and Alberto Cairo. "Reply to: Against Storytelling of Scientific Results." *Nature Methods* 10.11 (2013b): 1046. doi:10.1038/nmeth.2700.

Kuhn, Thomas S. *International Encyclopedia of Unified Science*. Vol. 2, No. 2: *The Structure of Scientific Revolutions*. 2nd edition. Chicago: The University of Chicago Press, 1970 [1962].

Lakoff, George. "The Contemporary Theory of Metaphor." *Metaphor and Thought*. Ed. Andrew Ortony. Cambridge: Cambridge University Press, 1993. 202–251.

Latour, Bruno, and Steve Woolgar. *Laboratory Life: The Construction of Scientific Facts*. 2nd edition. Princeton: Princeton University Press, 1986 [1979].

Lévi-Strauss, Claude. "The Structural Study of Myth." *The Journal of American Folklore* 68.270 (1955): 428–444. doi:10.2307/536768.

Locke, David. *Science as Writing*. New Haven and London: Yale University Press, 1992.

Lucas, George, dir. *Star Wars Episode IV: A New Hope*. 20th Century Fox, 2006 [1977]. DVD.

Luna, Rafael E. *The Art of Scientific Storytelling: Transform Your Research Manuscript: Using a Step-by-Step Formula*. Lexington: Amado International, 2013.

Lyotard, Jean-François. *The Postmodern Condition: A Report on Knowledge*. Trans. Geoff Bennington and Brian Massumi. Manchester: Manchester University Press, 1984 [1979].

Mandler, Jean M., and Nancy S. Johnson. "Remembrance of Things Parsed: Story Structure and Recall." *Cognitive Psychology* 9.1 (1977): 111–151. doi:10.1016/0010-0285(77)90006-8.

Martinez-Conde, Susana, and Stephen L. Macknik. "Opinion: Finding the Plot in Science Storytelling in Hopes of Enhancing Science Communication." *Proceedings of the National Academy of Sciences of the United States of America* 114.31 (2017): 8127–8129. doi:10.1073/pnas.1711790114.

McLuhan, Marshall. *Understanding Media: The Extensions of Man*. Critical ed. Ed. W. Terrence Gordon. Corte Madera: Gingko Press, 2003 [1964].

McQuillan, Martin. "A Chronology of Narrative Theory in the Twentieth Century." *The Narrative Reader*. Ed. Martin McQuillan. London and New York: Routledge, 2000. 311–313.

Meuter, Norbert. "Narration in Various Disciplines" (2013 [2011]). *The Living Handbook of Narratology*. Eds. Peter Hühn et al. Hamburg: Hamburg University. http://www.lhn.uni-hamburg.de/node/49.html (29 November 2017).

Miller, Dean A. *The Epic Hero*. Baltimore and London: Johns Hopkins University Press, 2000.

Montagnes, David J. S., E. Ian Montagnes, and Zhou Yang. "Finding Your Scientific Story by Writing Backwards." *Marine Life Science & Technology* 4 (2022): 1–9. doi:10.1007/s42995-021-00120-z.

Müller, Günther. "The Significance of Time in Narrative Art" (1947). *Time: From Concept to Narrative Construct. A Reader*. Eds. Jan Christoph Meister and Wilhelm Schernus. Berlin: De Gruyter, 2011. 67–84.

National Academies of Sciences, Engineering, and Medicine. *Communicating Science Effectively: A Research Agenda*. Washington, DC: The National Academies Press, 2017. doi:10.17226/23674.

Newman, Daniel Aureliano. “Narrative: Common Ground for Literature and Science?” *Configurations* 26.3 (2018): 277–282.

Niederhauser, Jürg. *Wissenschaftssprache und populärwissenschaftliche Vermittlung*. Tuebingen: Gunter Narr Verlag, 1999.

Nisbet, Matthew C., and Declan Fahy. “New Models of Knowledge-Based Journalism.” *The Oxford Handbook of the Science of Science Communication*. Eds. Kathleen Hall Jamieson et al. New York: Oxford University Press 2017. 273–281.

Olson, Randy. *Houston, We Have a Narrative: Why Science Needs Story*. Chicago and London: The University of Chicago Press, 2015.

Otis, Laura. “Introduction.” *Literature and Science in the Nineteenth Century: An Anthology*. Ed. Laura Otis. Oxford and New York: Oxford University Press, 2002. xvii–xxviii.

Padian, Kevin. “Narrative and ‘Anti-narrative’ in Science: How Scientists Tell Stories, and Don’t.” *Integrative and Comparative Biology* 58.6 (2018): 1224–1234. doi:10.1093/icb/icy038.

Panier, Louis. “Narrative Transformation.” *Routledge Encyclopedia of Narrative Theory*. Eds. David Herman et al. London and New York: Routledge, 2005. 377.

Phelan, James. “Rhetorical Approaches to Narrative.” *Routledge Encyclopedia of Narrative Theory*. Eds. David Herman et al. London and New York: Routledge, 2005. 500–504.

Plotnitsky, Arkady. “Science and Narrative.” *Routledge Encyclopedia of Narrative Theory*. Eds. David Herman et al. London and New York: Routledge, 2005. 514–518.

Popper, Karl. *Logik der Forschung* (1934). 5th edition. Tuebingen: J. C. B. Mohr (Paul Siebeck), 1973 [1971].

Propp, Vladimir. *Morphology of the Folktale*. 2nd edition. Trans. Laurence Scott, rev. Louis A. Wagner. Austin: University of Texas Press, 1968 [1928].

Qiu, Jane. “How China’s ‘Bat Woman’ Hunted Down Viruses from SARS to the New Coronavirus.” *Scientific American* (11 March 2020). https://www.scientificamerican.com/article/how-chinas-bat-woman-hunted-down-viruses-from-sars-to-the-new-coronavirus1/ (30 March 2020).

Ricœur, Paul. *The Rule of Metaphor: Multi-Disciplinary Studies of the Creation of Meaning in Language*. Trans. Robert Czerny et al. London and Henley: Routledge & Kegan Paul, 1978 [1975].

Ricœur, Paul. *Time and Narrative*. Vol. 1. Trans. Kathleen McLaughlin and David Pellhauer. Chicago and London: University of Chicago Press, 1990 [1984, ger. 1983].

Ricœur, Paul. “Narrative Time” (1980). *The Narrative Reader*. Ed. Martin McQuillan. London and New York: Routledge, 2000. 255–261.

Robbins, Bruce, and Andrew Ross. “Response: Mystery Science Theater” (1996). *The Sokal Hoax: The Sham that Shook the Academy*. Eds. The Editors of *Lingua Franca*. Lincoln and London: University of Nebraska Press, 2000. 54–58.

Ryan, Marie-Laure. “Tellability.” *Routledge Encyclopedia of Narrative Theory*. Eds. David Herman et al. London and New York: Routledge, 2005. 589–591.

Ryan, Marie-Laure. “Toward a Definition of Narrative.” *The Cambridge Companion to Narrative*. Ed. David Herman. Cambridge: Cambridge University Press, 2007. 22–35.

Scheffel, Michael, Antonius Weixler, and Lukas Werner. “Time” (2014 [2013]). *The Living Handbook of Narratology*. Eds. Peter Hühn et al. Hamburg: Hamburg University. http://www.lhn.uni-hamburg.de/node/106.html (17 March 2018).

Senior, W. A. “Quest Fantasies.” *The Cambridge Companion to Fantasy Literature*. Eds. Edward James and Farah Mendlesohn. Cambridge: Cambridge University Press, 2012. 190–199.

Shakespeare, William. *Romeo and Juliet*. Ed. John Dover Wilson. Cambridge: Cambridge University Press, 1969 [1955, orig. 1597].

Sokal, Alan. "Revelation: A Physicist Experiments with Cultural Studies" (1996). *The Sokal Hoax: The Sham that Shook the Academy*. Eds. The Editors of *Lingua Franca*. Lincoln and London: University of Nebraska Press, 2000a. 49–53.

Sokal, Alan. "Transgressing the Boundaries: Towards a Transformative Hermeneutics of Quantum Gravity" (1996). *The Sokal Hoax: The Sham that Shook the Academy*. Eds. The Editors of *Lingua Franca*. Lincoln and London: University of Nebraska Press, 2000b. 11–45.

Souza, Natalie de. "Should Scientists Tell Stories?" *Nature Methods* 10.11 (2013): 1037. doi:10.1038/nmeth.2726.

Vanderbeke, Dirk. *Theoretische Welten und literarische Transformationen: Die Naturwissenschaften im Spiegel der 'science studies' und der englischen Literatur des ausgehenden 20. Jahrhunderts*. Tuebingen: Niemeyer, 2004.

Vogler, Christopher. *The Writer's Journey: Mythic Structure for Writers*. 3rd edition. Studio City: Michael Wiese Productions, 2007 [1998].

"Warble, n.1." *OED Online*. Oxford University Press, January 2018. www.oed.com/view/Entry/225600 (13 March 2018).

Watson, James D. *The Double Helix: A Personal Account of the Discovery of the Structure of DNA*. Ed. Gunther S. Stent. New York and London: W. W. Norton, 1980 [1968].

Watson, James D., and Francis H. C. Crick. "Molecular Structure of Nucleic Acids: A Structure for Deoxyribose Nucleic Acid." *Nature* 171.4356 (1953): 737–738. doi:10.1038/171737a0.

Weininger, Stephen J. "Introduction." *Literature and Science as Modes of Expression*. Ed. Frederick Amrine. Dordrecht et al.: Kluwer Academic Publishers, 1989. xiii–xxiii.

Wellershof, Dieter. *Literatur und Lustprinzip*. Cologne: Kiepenheuer and Witsch, 1973.

White, Eric Charles. "Contemporary Cosmology and Narrative Theory." *Literature and Science: Theory and Practice*. Ed. Stuart Peterfreund. Boston: Northeastern University Press, 1990. 91–112.

White, Hayden. "The Value of Narrativity in the Representation of Reality." *Critical Inquiry* 7.1 (1980): 5–27. https://www.jstor.org/stable/1343174 (7 January 2019).

White, Hayden. *Metahistory: The Historical Imagination in Nineteenth-Century Europe*. Baltimore and London: The Johns Hopkins University Press, 1987 [1973].

Wise, M. Norton. "Science as (Historical) Narrative." *Erkenntnis* 75.3 (2011): 349–376. doi:10.1007/s10670-011-9339-2.

Hans U. Fuchs, Elisabeth Dumont, Federico Corni

Narrative Minds in the Construction and Use of Theories of Forces of Nature – A Model of Experience at Different Scales

Abstract: If the body of formal theories known as physics were the result of imaginative rationality, we might be inclined to think of physics rather differently than is traditionally the case. We could then easily identify conceptual metaphoric structures and narrative forms and processes in our formal theories, and we would consider them constitutive of the scientific enterprise and not just superficially imposed linguistic phenomena.

We shall argue here that continuum physics is precisely this: the result of human imagination and imaginative rationality. We shall present linguistic and methodological evidence that modern classical macroscopic physics is metaphoric, analogical, and narrative all the way to its conceptual core. We will then outline a model of experience (action/perception/conception) at different temporal, spatial, and systemic scales – but all within the range of human mesoscale worlds – that explains how metaphoric and narrative processes create the concepts of modern continuum physics.

1 Introduction

The title of this chapter combines both a claim and a suggestion of how we want to go about building our argument. We claim that macroscopic[1] physical science is an outgrowth of human imaginative powers expressed through the stories we tell as a people about *forces of nature*. The second point suggests that a model of

Acknowledgements: We would like to thank the editors of this volume for their extensive, in-depth and most valuable feedback to an earlier version of our chapter. It has helped us clarify many centrally important elements of our presentation.

1 Macroscopic physics is here understood in contrast to physical theories created on the basis of microscopic models (naively put, the assumption that the physical world can be explained in terms of the motion of little particles in fields or otherwise empty space). Macroscopic physics in the form of *continuum physics* (CP) uses a rather different set of explanatory structures. We restrict our treatment of imaginative structures in physics to continuum physics (for a brief description of continuum physics, see footnotes 8 and 6).

https://doi.org/10.1515/9783110782844-008

experience (of nature and fellow humans) working at different temporal, spatial, and systemic scales ranging from small to large will provide arguments in support of the main claims made here.

1.1 Background and motivation

Physics, as it is traditionally presented,[2] is either a formal construct quite inaccessible to most of us if we are not working in the field, or, where it is made more accessible, a verbal and pictorial display of its results, i.e., of what we perceive as our knowledge of the physical world. We rarely assume that a science such as physics is an imaginative affair that makes use of cognitive tools such as metaphor and narrative that are important in human life and culture and, in turn, contributes to the development of these same tools.

Our analysis of conceptual structures in physical science and our work on models of embodied cognition grew out of applications in science and education. More than 20 years ago, a dynamical theory of heat was developed using a methodology of analogical mappings between models of fluids, electricity and heat (Fuchs 2010 [1996]). This work proved to be the stepping-stone for our investigations into metaphoric and narrative structures of macroscopic physical science (Fuchs 2006; 2015; Fuchs et al. 2018).

The design of courses for engineering students at Zurich University of Applied Science in Winterthur (Dumont et al. 2014) and student teachers at the Universities of Modena and Reggio Emilia and of Bolzano (Corni and Fuchs 2020), and of a laboratory for mechanics at the Ducati Foundation in Bologna (Corni et al. 2018), taught us how to apply narrative approaches to the teaching and learning of physical sciences. One of the outcomes of practical design and theoretical model building is a laboratory where future kindergarten and primary school teachers learn how to write stories of *forces of nature* (Corni et al. 2014). Moreover, stories of forces of nature have already been written for young students and used in primary school in Modena (Fuchs 2013).

In this chapter, we would like to demonstrate how macroscopic physical science is a fundamentally metaphoric and narrative affair. We want to show how metaphoric and narrative processes contribute to both conceptual structures and scientific methodologies in physics.

2 Examples of what we might call "traditional" expositions in physics can be found in books ranging from school or university physics texts to technical memoirs (see, for example, Halliday et al. 2013; Feynman et al. 2011, and Jou et al. 1996).

1.2 Usage of some terms

Before we begin, we want to briefly mention how we are going to use a small number of important terms and concepts since our usage may differ from that of other investigators and authors.

We use the term *experience* in Dewey's (1925) sense to mean the (emotional and cognitive) *result of feedback* in *action-perception loops*, occurring between an organism and its physical, social-psychological and cultural-linguistic environments. Alternatively, we might call experience the *unified action of perception and conception*. It is assumed that this unified action can lead to the construction of *experiential* or *perceptual units* (gestalts) that, upon analysis, show some internal structure. We are mainly concerned with two forms of experience and their interaction: *physical* and *linguistic*.

We take the term *scale* to apply to *temporal*, *spatial* and *systemic* "size" or extent. An experience that is recognized as such and forms an experiential unit can be brief or long lasting, it can occupy small to large spaces, and it may involve a smaller or larger (or rather, simpler or more complex) system.

All scales considered here – from small to large – are assumed to be in the range of "*human scale*," i.e., they apply to physical and social worlds that are more or less directly accessible to experience without the need for complex intervening technology. This is sometimes called the *mesoscale world*. The different scales we talk about all fall within this mesoscale world.

Force[3] *of Nature*[4] is our term for denoting a class of phenomena that present themselves as *causative* or *agentive*[5] to the human mind. It is assumed that the

3 Note that *force of nature* is *not* the narrow concept of *force* as it is used in mechanics. Force of nature is a subspecies of the category we call *Force* that comprises psychological, social and cultural forces such as pain, justice or music. See also footnote 4.

4 Forces of nature are natural phenomena. Here are reasons for using the term *force of nature* rather than a word such as *natural phenomenon*. The term *phenomenon* denotes a category that encompasses just about any natural scene and not just phenomena capable of causing change – the category phenomenon is too large and diffuse. When analyzed, we see that our mind categorizes phenomena such as wind, water, ice, lightning or fire as belonging to a domain of experience we might call *Force* (see the analysis presented in section 2). This, and the use of the term *Forces of Nature* in earth science for natural phenomena such as wind, water, ice, fire, volcanoes or earthquakes, let us believe that the term *Forces of Nature* is apt for what we have in mind.

5 Even though this might lead to misunderstandings if we take a narrow view of how much imagination is "allowed" in scientific conceptualization of causative phenomena, we use the term *agentive* alongside the more neutral sounding *causative* for a simple reason. As we show in this chapter, language use, both natural and formal, betrays imaginative forms in experience that can only be called *agentive*: we see *agents* at work behind activities in nature (see the argument based upon figure-ground reversal between groups of metaphors discussed in section 2). We do

class and its members form experiential units in one's mind. At the human scale, and in everyday prescientific experience, examples of *forces of nature* are water, light, wind, soil, food and medicines, heat and cold (or fire and ice), lightning and thunder, gravity, moving bodies, etc. In formal renderings of macroscopic physics, this long and heterogeneous list is reduced to fluids, electricity and magnetism, heat, substances, gravity and linear and rotational motion; these constitute the fields of inquiry of continuum physics.[6]

In both physics and everyday language use, this experiential unit is characterized by three basic aspects: quality, quantity and power.[7] Our experience of such phenomena suggests notions of quality, intensity, tension (wind can be stronger or weaker, it creates felt tensions); of quantity or size (there can be more or less wind); and of power (wind, a storm, can be more or less powerful).

---

not need to attribute intentionality to these agents, but we clearly need to be able to "see more" in nature than can be expressed by "ball A strikes ball B." The sense of agency entertained by the human mind is one of the expressions of narrative understanding (see sections 3 and 5).

6 The notion of *forces of nature* may seem to arise from pre-scientific approaches to the experience of nature and may therefore be alien to or at least unnecessary in formal (physical and chemical) science. Indeed, when we first explore our direct physical experience, we arrive at a heterogenous list of candidates that create a category held together by the still vague expression "phenomena presenting themselves as causative or agentive to the human mind." Aborigine cultures and children, but also the non-scientific lay-public, make use of this form of categorization of dynamical phenomena. It may therefore come as a surprise that there exists a scientifically "sanitized" list of what we have called forces of nature. The fields ("phenomena") of continuum physics (CP, see footnote 8) – fluid dynamics, electrodynamics, thermodynamics, chemical dynamics, gravity, and translational and rotational mechanics – are treated analogously to a fairly large degree, which demonstrates the existence of a "medium-scale" theoretical structure – this is what we call "scientific" forces of nature (the scientific conceptualization of the causative phenomena presented in the previous list). Its elements (laws of balance of fluid-like quantities, potential landscapes, gradients and flows, capacitive, resistive and inductive constitutive relations, etc.; see Fuchs 2010) are found as well in qualitative form in pre-scientific conceptualizations of forces of nature.

7 In everyday usage, the notion of *power* is not clearly defined and is often conflated with *force*. In this chapter, we use *force* in conjunction with force of nature (or other experiential forces) and *power* as a term denoting the third aspect of such forces. Note that the problem of not clearly distinguishing between force and what later became known as power existed for centuries in physics, where force was used as referring to either force proper in mechanics or to energy. In standard representations of formal physics, the term force is related to transfers of momentum whereas power denotes the rate at which energy is exchanged in the interaction between forces of nature.

1.3 Outline

In section 2, we shall analyze metaphoric and analogical structures in continuum physics and demonstrate the existence of a *medium-scale cognitive structure called Force of Nature*; this will be done from the vantage point of embodied cognitive science. We limit our analysis and description to *modern continuum physics* and *dynamical systems*.[8] These fields constitute the largest and most important body of formal theories of macroscopic physical nature and phenomena and their applications in areas of engineering, environmental science and medicine. The term *modern* refers to the extended theories, models and applications that have been created over the last 60 years. This includes, in particular, a modern *dynamical theory of heat* (Truesdell and Toupin 1960; Truesdell and Noll 1965; Jou et al. 1996; Fuchs 2010 [1996]).

Subsequently, we show how narrative processes create large-scale synthetic structures that let forces of nature appear as agents in stories told about dynamical physical systems and the processes they undergo (section 3).

To help us understand how precisely metaphor and narrative create the conceptual structures of macroscopic physical science, we propose a model of experience at different temporal, spatial and systemic scales (section 4), and use it to demonstrate how conceptualizations found in modern thermodynamics emerge from the interaction of metaphoric and narrative processes that result from mappings across scales of experience (section 5). We conclude by briefly summarizing results and suggesting avenues for further investigations (section 6).

2 Metaphor and analogy in macroscopic physics

We begin by describing observations concerning concepts and relations used in models and theories of macroscopic physical systems and processes. We first collect examples that are related to metaphoric processes before we clarify the important distinction between metaphor and analogy. In section 3, we shall examine the use of narrative forms in science.

8 *Continuum physics* (CP, see also footnote 6) is the most comprehensive form of the physics and chemistry of macroscopic systems and processes. In CP, systems and processes are described as spatially continuous (variables change along the three spatial directions; there are no point masses or particles) and are generally dynamic (there is temporal change as well). If we integrate variables over regions of space containing matter or fields and describe what is the case in a finite volume in terms of single values of variables, we arrive at what are commonly called (spatially uniform) *dynamical systems*.

The material that allows us to describe the phenomenology of scientific conceptualization and explanatory practice is at least threefold: There are natural linguistic products, visualizations and equations. In the following, we limit ourselves to the use of natural language; but see Fuchs (2014) and Corni et al. (2019) for descriptions of figurative forms expressed in diagrams (visual metaphors) and equations (see also Sherin 2001).

2.1 Examples of metaphoric expressions concerning temperature

We take examples from the field of thermal phenomena, but we shall make an important excursion into the world of music[9] to demonstrate the extent of our analysis. As we shall see later in this section, different types of phenomena related to different forces of nature are all treated analogously.

There are two basic concepts needed to talk about thermal phenomena: *temperature* and *heat*. We begin with expressions (found online and in books) that demonstrate how we deal with temperature and related terms:

(1) . . . pretend you are in a warmer place!
(2) [Weather . . .] going from hot to cold and back again so quickly . . .
(3) Global temperature rise is fastest in at least 11,000 years, study says . . .
(4) This huge temperature difference makes OTEC possible.
(5) . . . your body temperature is at the lowest level of the day.
(6) . . . it still takes my body forever to warm up in the mornings.
(7) Severe heat places unique demands on LED lighting systems and can cause poor performance [. . .] if [. . .] the system is not designed for high temperature use.
(8) The outdoor – indoor temperature difference drives heat through an opaque panel.
(9) [. . .] bodies of water [. . .] differ in their temperatures, providing a thermal landscape which might serve as an orienting cue in fish migration.
(10) The truth is, all of these thermal ups and downs are perfectly normal . . .

9 By drawing upon, or actually creating, a parallel between thermal phenomena and music, we claim that there exists an experiential unit larger than *Forces of Nature* which we like to simply call *Force*. We claim that there are natural, psychological and social and cultural "forces" that are treated, at least to some extent, analogously by the human mind. The claim is supported by analysis of metaphoric webs used to understand these diverse instances of *Force*.

What arise here are images that suggest an interpretation of temperature as an *intensity* (5, 7), as a *locality in a thermal landscape* (1, 2), as a *level* (3, 5, 8, and 10), or, by combining the schemas of locality and level, as a *thermal landscape* through which we or other bodies move at various speeds (3, 6). Differences of temperature are felt as *tensions* that serve important orienting and causal functions (4, 7–9).

In these examples, we see features familiar from studies of cognitive linguistics (Geeraerts and Cuyckens 2010): The expressions are all *figurative* (Gibbs 1994), and they make use of simple schemas (essentially *image schemas*: see Johnson 1987; Hampe 2005) that are projected *metaphorically* (Lakoff and Johnson 1980, 1999) upon our perception of the HOT ↔ COLD *polarity*; in other words, schematizing action and metaphoric processes lead to figurative structuring of what we might want to call the *domain of hotness*.[10] Put differently, experience produces some of the imaginative structures we use to understand and talk about temperature and related linguistic forms (see section 5, Fig. 4, for a detailed study of linguistic forms having to do with the concept of temperature).

2.2 Schematic constructs in descriptions of heat

The experience of heat is not limited to our perception of the HOT ↔ COLD polarity. Through more general exposure to thermal phenomena, which may be accompanied by narrative experience, and through the action of figure-ground reversal (see below), our mind becomes ready to infer the existence of an invisible something usually called *heat* or, more formally, *quantity of heat*.[11] This is the famous *caloric* of early investigations into phenomena associated with hot and cold.[12] In

10 *Hotness* is the term used in continuum thermodynamics for the one-dimensional manifold from which the concept (and measure) of temperature is derived. The term is perfectly suited to describe the experiential unit created by perception of hot and cold.

11 The first part of the sentence seems to insinuate that the concept of quantity of heat arises purely "naturally." However, didactic experiments in kindergarten and early primary school education suggest that concepts such as quantity of heat or electricity arise through the interplay of direct physical and narrative experience. Narrative experience can be created through stories of forces of nature used by teachers in parallel to exposure to nature (Corni and Fuchs 2020).

12 A theory of thermodynamics based upon caloric as the *extensive* thermal quantity, and temperature as its *thermal potential*, effortlessly leads to a dynamical theory of heat (Fuchs 2010). In extended form, caloric is analogous to entropy in macroscopic thermodynamics.

Tab. 1, we present examples of figurative expressions involving the notion of heat, plus a list of schematic constructs apparent in these expressions.[13]

Tab. 1: Examples of expressions involving heat and schematic constructs used.

Metaphoric expressions	Schematic constructs used
(11) All bodies contain heat.	Container, store, hold, accumulate; lack of, abundance of; collect
(12) How do you collect heat in a passive solar house?	Flow, transport, extract emit/absorb, exchange; heat moves
(13) Electricity is used to produce heat.	Balance (law of balance of . . .)
(14) This means heat flows "downhill" from hot to cold.	Use, produce, generate heat
(15) . . . heat is an agent of vast importance in chemical reactions and engineering processes.	Heat as location, landscape; level, intensity, degree, sensation of heat along a scale
(16) Law of the dependence of the active force of heat upon the temperature . . . (Clausius)	Balance of heat and cold, hot and cold; thermal tension
(17) This exterior heat lets the crust become crispy.	Power, force of heat
(18) Heat makes me dizzy . . .	Heat is an agent: Heat causes, drives, makes, counteracts, lets, balances
(19) Clouds and storms follow the warm water, pumping heat and moisture high into the atmosphere . . .	Heat is a patient: Pump, force, make, counteract, block, hold (back), enable, prevent, oppose, let/allow heat
(20) Heat must balance cold . . .	

In expressions (11–20), we recognize metaphorically projected schemas such as CONTAINER, FLOW, BALANCE, FORCE, plus the entire list of *force dynamic schemas* that include LET, MAKE, or BLOCK (Talmy 2000); force-dynamics involves the interaction of participants and therefore the schema of CAUSE. We make out a fairly large web of metaphors that supports our understanding of the notion of heat as the *fluid-like* thermal quantity. This is how our experience of heat as a force of nature obtains an object-like or fluid-substance-like aspect in addition to the qualitative aspect brought about by our direct experience of hot or cold. In formal science we say that we have introduced the concept of the *extensive* thermal quantity (called

13 The two lists do not correspond one-to-one. Rather, the list on the right is condensed from linguistic experience and research in cognitive linguistics. Short expressions involving schemas from the list have been used to conduct Internet searches that yielded all the expressions listed in the left column. Example of a collocation: "heat makes" yields over 750,000 entries in a Google search (7 January 2020) from which Expression (18) has been taken.

caloric or *entropy*)[14] that complements the *intensive* quantity called *temperature* (Fuchs 2010 [1996], ch. 4).

2.3 Johnson's analysis of our experience of music

Before we continue with our investigation of figurative structures in physical science, we shall take an important detour and briefly describe Mark Johnson's analysis of our experience of music (Johnson 2007, ch. 11). Johnson demonstrates how understanding of music is constituted in terms of three aspects: *object, landscape* and *moving force* (or power). These aspects arise in three groups of metaphors he calls the MOVING MUSIC metaphor, the MUSICAL LANDSCAPE metaphor and the MUSIC AS MOVING FORCE metaphor, respectively.

Our experience of music is embodied, and its structure is made apparent through figurative language and thought. We experience music as an object (example expression: *The music goes faster here*) that moves through a musical landscape (example expression: *The melody rises up ahead*) and moves us like any other force moves us (example expression: *This song always makes me smile*).

2.4 A metaphoric web underlying forces of nature

Johnson does not call music a (cultural) *force*, but that's exactly what it is if we compare its experiential content with that of our understanding of phenomena such as heat, electricity, water, wind and so on. There is certainly more to the results of our analysis of linguistic expressions (1) through (20) than a long list of otherwise unrelated figurative forms. The schematic constructs appearing in our examples (see the list on the right in Tab. 1) profile a new and larger unit, that of a *force of nature* (Fuchs 2006; 2013).

14 In macroscopic thermodynamics, entropy is simply the extensive thermal quantity. As such, it satisfies a law of balance (it can be stored, can flow and can be produced). It relates to temperature by making bodies warm or letting them expand, and to temperature differences as driving forces of its flow. It relates to energy in physical processes just as any other extensive quantity does (see Fuchs 2010, ch. 2, 4 and 5). The macroscopic concept needs to be distinguished from entropy as the number of possible "states" or "arrangements" (in statistical physics) and from entropy in information theory. The comparison of entropy in macroscopic models to the role of caloric in earlier theories has led some thermodynamicists to propose a modern extended version of the concept of caloric to make understanding of thermodynamics more natural (Mareš et al. 2008; Fuchs 2010 [1996]; Fuchs et al. 2022).

A summary of schematic forms and metaphoric projections paints an image of a perceptual unit called HEAT as having the following three main aspects. There is a *quantity of heat* (caloric or entropy) that is contained in objects (11, 12), and can be produced (13) and transported (14). The quantity of heat is related to the intensive aspect of the phenomenon as implied in expressions (11–12), (14) and (16): it creates and is subject to the *intensity of heat* (temperature) and its differences (thermal *tension*). And heat has the *power* to influence and cause other phenomena or is subject to other forces (as in (15–19)). The first time an explanation of thermal processes was made formal along these lines was in Sadi Carnot's *La puissance du feu* (1824; see Fuchs 2010 [1996], in the Introduction).

Clearly, Johnson's first and second groups of metaphors correspond to our understanding of quantity of heat (caloric) and intensity of heat (temperature; remember how temperature traces a thermal landscape). In other words, we could have identified metaphoric groups for our experience of heat and called them the MOVING HEAT metaphor and the THERMAL LANDSCAPE metaphor, respectively.

The third metaphoric group exists as well – witness the examples of expressions given in (15–19) that suggest the metaphor HEAT IS A MOVING FORCE. Moreover, our identification of schematic structures in Tab. 1 shows that heat takes on roles of *agent* and *patient* in our imaginative understanding of thermal processes. In section 3, we shall see how narrative completes what is only hinted at in metaphoric expressions: *forces become agents (*expressed in form of a metaphor, HEAT IS A POWERFUL AGENT).

2.5 Figure-ground reversal in metaphoric rendering of forces of nature

Lakoff and Johnson (1999, 195–198) discuss the notion of dual metaphoric groups and show how they arise in figure-ground reversal (1999, 198–200). Take, for example, two groups of metaphors, CAUSATION IS THE FORCED MOVEMENT OF AN ENTITY and CAUSATION IS THE TRANSFER OF A POSSESSABLE OBJECT: "These two conceptualizations have opposite figure-ground orientations." (Lakoff and Johnson 1999, 199). Note that this applies to the MOVING MUSIC and the MUSICAL LANDSCAPE metaphors as well.

It should therefore not come as a surprise that the MOVING HEAT metaphor and the THERMAL LANDSCAPE metaphor, which we have seen in our examples, are dual in the same sense (Fuchs et al. 2018; 2021). If we assume experience of hot-cold to be prior to that of (quantity of) heat, we can explain how the notion of heat (as a quantity) arises: it is created by imagination.

A common modern and supposedly "fundamental" form of describing physical dynamical processes focuses upon bodies (as *figures*) and attributes change (their dynamics) to energy passed on to or taken away from these bodies; this then changes their states (temperature, speed, electric or chemical potential, pressure, etc.). These states are described by intensive quantities which, in our imagination, can be seen to form landscapes – they are the *ground*(s) upon which bodies perform their metaphoric movements.

When physicists create formal dynamical models of processes, however, they make use of a group of quantities that seem to appear "out of the blue" on the stage of scientific conceptualization. These are the *extensive* (or additive) quantities having names such as volume, electric charge, entropy (or caloric), amount of substance, momentum, etc. (they are the duals of pressure, electric potential, temperature, chemical potential, velocity, etc.; see Callen 1985; Fuchs 2010 [1996]). So, what is the "blue" they arise from?

Consider again the case of thermal phenomena. In the first view, bodies are figures that move against the ground of thermal landscapes: they get hotter or colder. Now imagine bodies "fading" into the background (to become the *ground*), and a thermal landscape with its hills and valleys "morphing" into *figures* or shapes that preserve the qualities of the thermal landscape they come from (i.e., temperature). At least two distinct types of shapes or figures come to mind: "fluids" and shape-preserving individualized "bodies;" both, fluids and bodies, will be tensed or relaxed, preserving the intensive aspect of experience originally embodied as "potential landscapes." The former are the physicists' *extensive quantities* – space-filling, flowing, and, in some cases, appearing and disappearing, fluid-like quantities. The latter can be imagined as ghosts, spirits, or, generally speaking, *agents* that become characters of stories we tell of events happening in nature (see footnote 5).[15]

2.6 Blending as compression over phenomena: Origins of analogy

We can now sketch how *analogies* between our descriptions of different forces – fluid, thermal, electrical, chemical, etc. – arise. We use a methodology developed in the model of conceptual integration (Fauconnier 1994; Fauconnier and Turner

15 This imaginative approach has been used to create a vivid product – an animated story of an inventor trying to build a perpetuum mobile – where spirits (young children call them "little monsters") appear after figure-ground reversal that represent the agentive aspect of forces of nature (Deichmann 2014).

2002; Turner 2006) and show how a blend of figurative renderings of fluids and heat can emerge (Fig. 1).

As we see in *input spaces*[16] of heat and fluids, experience of fluids and heat makes use of already existing *abstractions* we know as (image) schemas; these populate a *generic space* that is created by abstraction over the input spaces. These abstractions are projected onto the input spaces leading to metaphoric expressions of the kind we have analyzed in this section. By blending our experience (including linguistic experience) with fluids and heat, a *blended space* arises in which heat behaves like a fluid that can be generated.[17] We literally learn to "see" heat as a fluid substance concentrated more or less densely in materials, letting them appear warm or cold, flowing spontaneously from hotter to colder places, and being pumped or generated in non-spontaneous processes.[18]

The process creating the blend is called *compression*. In Turner's words: "The blend is a compression of the input spaces, while the generic space is an abstraction over them" (2006, 98) and "[a]dditionally, the blend develops emergent properties that are not possessed by any of the input [spaces]" (2006, 96). In summary, through the act of compression of different types of experience, the blend makes a figure emerge that did not exist before.

When we compare the input spaces, we see basically the same type of mechanism at work behind conceptualizations of different forces of nature. This allows us to deliberately produce analogical mappings that support both informal and formal reasoning about a phenomenon. We can use formal structures from theories of fluids in order to create formal models and theories of thermal processes. We can do the same with fluids and electricity, and we can finally use electrical formalisms in order to create thermal or fluid models (such as for heat transfer networks or the blood flow system of mammals, respectively; see Fuchs 2010 [1996]).

---

16 A mental space is a cognitive model of a section of reality often arising spontaneously such as in conversation or thinking about a subject (Fauconnier 1994).

17 Note that the direction of projection can be turned around: water as a force of nature behaves, at least to a large degree, like heat. We can learn about a new field from one we have mastered.

18 This is the imaginative structure created by and used in modern forms of the caloric theory. The analogy between heat and fluids (and electricity) has been the guiding heuristic in the creation of a dynamical theory of heat (Fuchs 2010 [1996]). It clashes with the traditional pronouncement that "heat is (a form of) energy" known from classical thermodynamics (see Fuchs 2010 [1996], Introduction, for an in-depth discussion of this point). We might say that the traditional form of conceptualizing "heat" as "energy" is stuck in the bodies-and-energy form; figure-ground-reversal has been denied (even though it clearly existed in Carnot's thermodynamics before 1850).

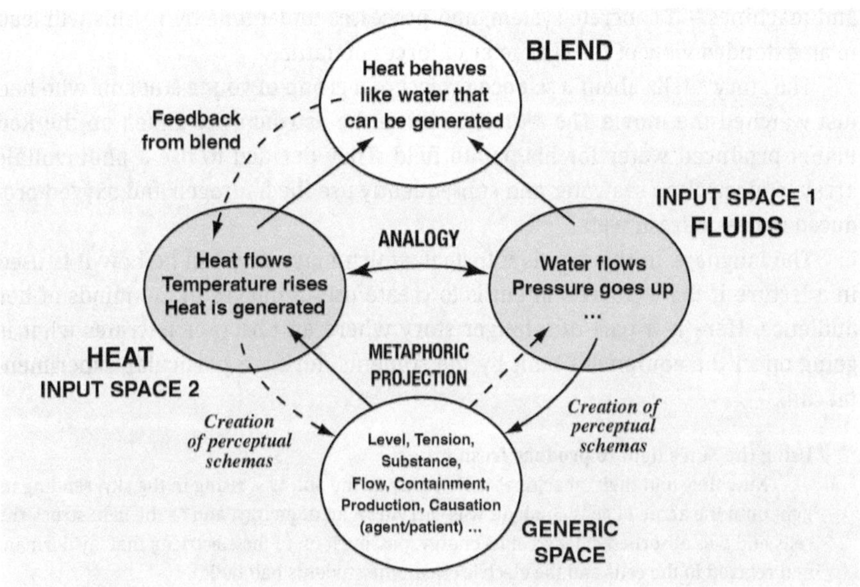

Fig. 1: Diagram showing conceptual integration at work in fluids and heat. Fluids and, to a lesser extent, heat, create elements of a generic space that is used in metaphoric renderings of theories of fluids and heat. Since both types of experience use mostly the same webs of metaphors, input spaces 1 and 2 appear similar to the human mind.

3 Narrative framing in macroscopic physics

We have seen some of the linguistic evidence of metaphoric forms used in formal theories of continuum physics. Metaphoric expressions are typically short, but they combine and create a larger web of figurative forms that helps us understand the logic inherent in conceptualizations of forces of nature (section 5). We now turn to an example of a linguistic product that exists as a fairly large unit, a story-like text[19] describing a scene, i.e., a situation involving humans and nature

19 It is important to note that narrative is a fairly large category encompassing different types of text forms ranging from encyclopedia entries, through narrative explanations, to accounts of events, and finally to *stories*. Herman (2009) calls stories the *prototypical members* of the *radial category* called *narrative*. Stories are narratives that include all of the following elements: (1) events; (2) (conscious) experiencing of events by agents; (3) tension for creating events; and (4) reason or occasion for telling by a narrator (Fuchs 2015).

and machines – a concrete system, and processes undergone by it. This will lead to an extended view of the character of forces of nature.

The story[20] tells about a science project of a group of young students who had just watched the movie *The Martian*, where the astronaut stranded on the Red Planet produced water for his potato field. They decided to use a photovoltaic array to electrolyze seawater and subsequently use the hydrogen and oxygen produced to create fresh water.

The language in this text is informal, which may very well be how it is used in a lecture if the lecturer's intent is to create useful images in the minds of her audience. Here is a part of a longer story where an observer narrates what is going on in the equipment built by the students during a particular experimental run.

Using the Sun's light to produce fresh water

Now they and their "machine" were ready as the Sun was rising in the sky, sending its light onto the array of cells. Sunlight was carrying a lot of energy, and as the light struck the cells and was absorbed, it let go of its energy, passing it on to the electricity that lay dormant and relaxed in the cells and the electric circuit the students had built.

As the electricity received the energy of the disappearing sunlight, its tension went up. It was as if it woke up and started flowing through the cells and the rest of the circuit. Climbing through cells that were connected one after another, and continuously receiving the energy of the Sun's light, its tension went up and up; the charge went to higher and higher electric levels. And the currents of electricity coming from cells connected next to each other combined to make it a strong flow through the wire toward the tank with seawater and from there through the water and back through the second wire.

It was actually fortunate that the water in the tank was salty – salty water lets electric charge pass quite easily. As the electricity flowed through the water, it did its job of making hydrogen and oxygen out of the water in the seawater tank. It let go of the energy that had been handed to it by sunlight, and its tension dropped. The electricity then returned to the cells, fully relaxed and ready to tense up again with the help of the energy of sunlight.

The water in the seawater tank was "relaxed," i.e., at a low chemical level. As energy was made available by the electricity flowing through it, it disappeared, producing two new substances whose combined chemical tension was way higher than that of the original substance. Together, hydrogen and oxygen were ready to do their job in turn, using the energy they had collected to let the chain of events continue.

Witness what this text does. First of all, if we take a small-scale view and concentrate upon single expressions or small sections of the text, we see that it makes use of the same figurative elements we have encountered above in section 2. We might say that the text is assembled bottom-up from small-scale linguistic units

20 The story is not of a real case. It has been created to teach students the use of stories for generating story-worlds which can be transformed into formal dynamical models (Fuchs 2015).

such as metaphoric expressions. And again, we see experiential units emerge that we have called forces of nature.

Secondly, and this is our main point, if we let the text impress us as a narrative unit, a new and different feeling emerges – there is a chain of events initiated by tensions involving characters or agents. Furthermore, the story causes what Caracciolo (2014) calls *narrative experience* and creates a story-world in our mind (Herman 2002; 2009). The part of the text shown here creates a scene, a natural or technical world, in which *agents* act and suffer. In short, the agents appearing in this story-world are sunlight, electric charge, water, hydrogen and oxygen (heat has been left out, but it should be included in a longer and more complete text). Their environment is constituted by Sun and Earth, solar cells, wires, tanks with seawater, cables, graphite electrodes, etc. "Landscapes" exist in this environment in which the agents move up, down, or horizontally, interact and hand off energy, i.e., make it available and use it for their purposes (Fuchs 2013; 2015). Moreover, details of the event narrated suggest properties of agents and their environments and how these impact agent-agent and agent-environment interactions.

Becoming aware of and learning to move in this scene has at least two important consequences: one relates to products of our imagination, the other relates to scientific methodology.

3.1 Time, process, agency: Forces of nature as agents in narratives

Narrative has an important cognitive role to play (Herman 2013); stories support processes of imaginative rationality in a manner that goes beyond what metaphor and related cognitive tools can do by themselves (Fuchs et al. 2018).

If we read a story, it creates images larger and richer than any single figurative (metaphoric or other) expression. This is not simply a consequence of its length; a story makes use of abstractions that are not usually or directly part of single expressions. Chief among these are TIME and PROCESS (dynamics) and, emerging from their combination and the schema CAUSE, *agency* as a fundamental feature of forces (of nature).

Time, process, or causation may appear separately in single expressions (as in examples (6), (14), or (18)), but the power of their combination is uniquely narrative. This combination creates in us images of *powerful agents* causing other agents to suffer consequences or themselves suffer consequences of actions initiated by other agents. Furthermore, narratives are the proper repositories for multiple interactions by several agents whereas single expressions typically connect no more than two agents in a single act (see example expression (13): *Electricity produces heat*). In short, water, wind, heat, and all the other forces emerge as agents in story-worlds.

3.2 Narrative and the process of science

Apart from their part in human experience and their schematizing power that helps new images arise in our mind, texts of this form, both orally narrated and written, play an important practical and methodological role in natural and social science and medicine. Our example suggests what it is that connects narrative and (the practice of) science.

Medical practitioners use narratives for diagnostic purposes. Using narratives lets images arise in the mind of a medical doctor, helping direct attention to symptoms and suggest mechanisms that might be responsible for what is observed (Hurwitz 2006). Quite the same is true for economists who tell stories of phenomena that carry meaning suggestive of economic processes, helping greatly in the formation of formal models (Morgan 2001). Physicists and chemists using computational tools see unexpected mechanisms hinted at by bundles of simulated trajectories showing the behavior of physical systems (for example, computer models of growth of snowflakes are of this type: their simulations "grow" rather than "derive" explanation; Wise 2011). This is what our little story does as well – it hints at mechanisms behind the phenomena being observed and suggests relations that are needed to create a dynamical model of the system built by the human protagonists.

This leads us to the following analogy. Like a simulation of a (mathematical) model, a story told about a system and its processes is expected to give rise to a mental model in speakers and listeners that evokes knowledge of scenes (spatial constructs, temporal settings and agents and their properties and abilities; in short, *story-worlds*) that is used in the construction of a formal model. In summary, simulations are like stories told, and formal models are like story-worlds arising in the mind of a reader (Fuchs 2015).

3.3 Summary: What is unique about narrative?

Notions of time, agency and dynamics properly belong to the narrative realm. When applied to a force, these schemas let the image of a powerful agent and what it is going through arise in the imagination: We arrive at a metaphor HEAT IS A POWERFUL AGENT (Fuchs et al. 2018). Experience of larger scale phenomena (those which we relate to narrative) lets us see agents through the process of narrative framing (Fuchs 2015).

Having used the notion of small or large *scales* informally a couple of times so far, it is time to formalize what we mean by this term, make it a part of a model of experience, and use the result in an explanation of the roles of metaphor and narrative in a science such as physics.

4 A model of embodied experience at different scales

We now endeavor to construct a model of experience of the physical world that helps explain what we have observed. The notion of *experience* describes a number of feedback processes between an individual organism and its environments (physical, social, linguistic, artistic, literary, scientific, etc.) that lead to the creation of perceptual units and their conceptual analysis. Importantly, the experience we allude to involves joint physical experience by two or more humans including their (linguistic) interactions.

We want to understand exactly how metaphor and narrative help create imaginative structures that constitute informal understanding of physical systems and processes and formal physical science. The construction of the model in this section will proceed in two steps. An important consequence of the model – illuminating how formal conceptualization of forces of nature comes about in both bottom-up and top-down processes – will be presented in section 5.

4.1 A summary model of experience

In a first step, we simply render concrete and useful for our purpose the type of interpretation of cognition discussed in research in *extended* or *enactive embodied cognition* (Varela et al. 1991; Chemero 2009; Noë 2004; Stewart et al. 2010). Interactions with our physical (natural and technical), social, cultural and personal inner environments can be represented as action-perception loops (Fig. 2). The totality of these interactions is assumed to allow an embodied mind of an individual organism (IOEM) to emerge with all its physical abilities, conceptions, models and knowledge.

Here, we restrict attention to physical and linguistic interactions with the outside physical world and other human individuals, respectively.[21] At least two indi-

21 Linguistic interactions go beyond standard language, but in this chapter the evidence we consider is only of this natural (spoken and written) type.

vidual organisms interact physically with the same natural environment and interact with each other linguistically (Fig. 2). The growth of a mind is assumed to proceed through imaginative processes (such as metaphor and narrative) and to lead to (informal or formal) conceptions.

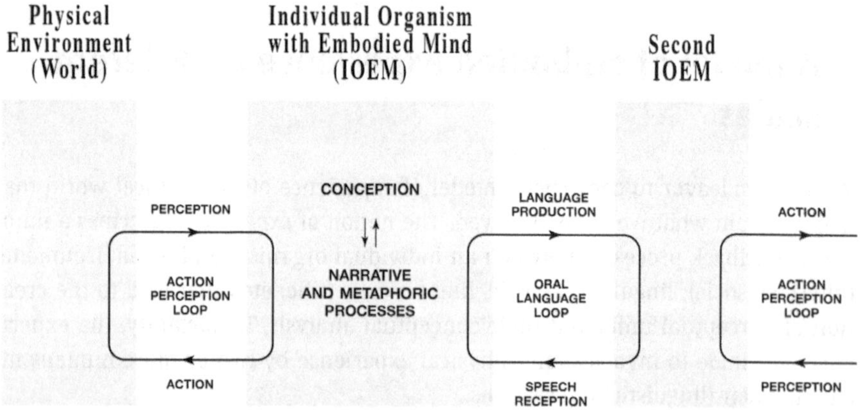

Fig. 2: Model of the interaction of an Individual Organism having an Embodied Mind (IOEM) with its physical environment and a second IOEM. Interactions take the form of physical and linguistic action-perception feedback loops and lead to conceptions in the mind of an individual.

4.2 A model of scale of experience and linguistic products

Experience of our natural environment and of the linguistic acts between individuals that join together in physical experience can take many different forms. We wish to point to a particular aspect of this diversity we might call "size" or *scale*. To repeat what we said in the introduction, an experience that is recognized as such and forms an experiential unit can be brief or extended, it can occupy small to large spaces, and it may involve a smaller or larger (or rather, simpler or more complex) system. In other words, it makes sense to speak of *temporal, spatial and systemic scales of experience.*

Unaided by instruments, perception and conception happen in a limited range of temporal, spatial and systemic scales. Time scales range from seconds to maybe years, and spatial scales from fractions of a millimeter to some hundreds of kilometers.

It is more difficult to describe what is meant by systemic scale. We could speak of structural scale – how simple or complex a physical or social structure appears to be – but that would leave out the aspect of dynamic complexity. That is why we have chosen the term systemic scale, to denote the degree of both structural and dynamic complexity.

A simple division of scales into *small*, *medium*, and *large* is useful for our purpose. In section 2, we identified a type of recurrent experience that leads to the construction of small and basic (image) *schematic structures* such as CONTAINER, SCALE, (fluid) SUBSTANCE, BALANCE and VERTICALITY. We can see very well how experience that is of fairly short duration, happens in small spaces, and is relatively simple from a systemic viewpoint can be the source of such abstractions. We shall call such experience a form of *incident* (see Fig. 3, bottom left). (An incident might be stumbling over a stone or seeing a short flash of light.)

Our physical environment affords us a different class of experience as well, the product of which we termed *forces of nature*. When we experience a force such as wind or heat, we are typically exposed to more complex phenomena that can last longer and cover a larger space. This is why we want to describe such experience as being of medium scale, and we use the term *episode* to speak of phenomena leading to this type of experience (see Fig. 3, middle left). Examples of thermal episodes include opening the door of a furnace and getting a blast of heat in one's face; running warm water over cold hands and feeling them slowly warm up; watching a marathon in winter from the roadside and feeling cold creeping up one's legs; seeing magma running down the side of a volcano glow brightly in the dark; or getting burned by the flame that consumes a match.

What happens when several forces interact in phenomena that last longer and take up a much larger space in order to play out in a complex system? Examples of such *events* (Fig. 3, top left) abound: think of a hurricane hitting a city, a garden bringing forth flowers and vegetables, the birth of a child, a glacier sculpting a landscape, or a forest fire engulfing a town. Events clearly are a type of large-scale experience in the sense described in this chapter.

The following step is important for the continuation of our argument: we postulate that large-scale experience leads to the formation of experiential units we might call *worlds*. Just as we know when we have a problem with our balance, or when we are exposed to a force such as wind, heat or water, we immediately recognize a large-scale *event* when we experience it.

Our observation regarding experience at different scales also applies to linguistic interactions. Note that we commonly use different linguistic products to communicate small, medium or large-scale (physical) experience. An exclamation or a single sentence, i.e., an *expression* (Fig. 3 bottom right), may suffice to relate some small-scale experience such as stumbling over a stone and regaining one's balance. We may need a few sentences (a *passage*, for lack of a better term) to tell what happens when the wind picks up and drives a sailboat forward. And we

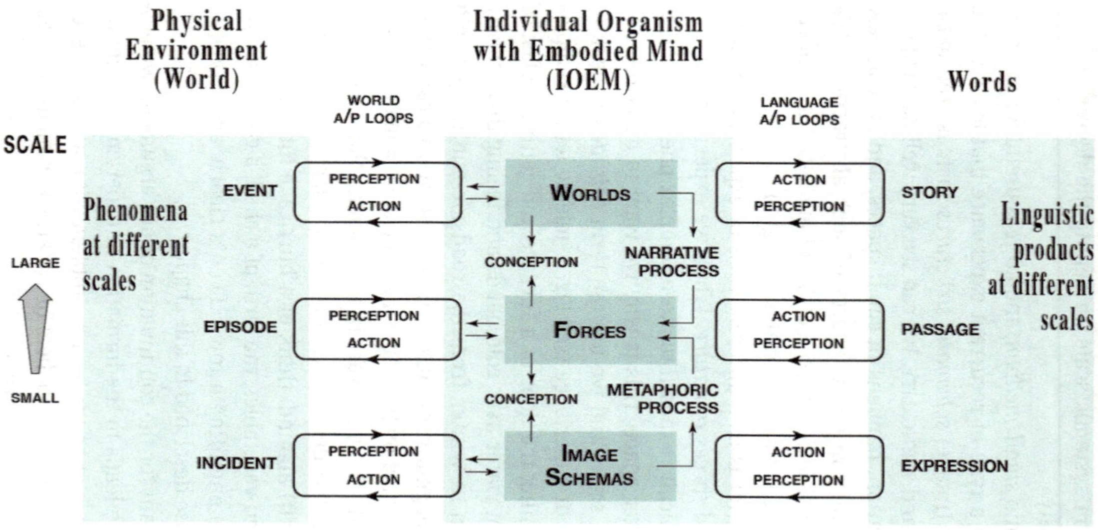

Fig. 3: A diagram sketching a model of physical and linguistic experience at different scales. Note the forms of physical experience (incident, episode, event) and of linguistic products (expression, passage, story). These observations are used to infer the existence of figurative mental structures and metaphoric and narrative processes. The occurrence of metaphoric and narrative processes is represented in a manner that does not capture its full complexity (see section 5).

properly use full *stories* to communicate all the aspects of how we lived through a forest fire.[22]

An important problem still needs to be addressed: how are the products of experience working at different scales related to each other? This question applies particularly to the mental products that we believe form in experience: image schemas, the gestalt of force(s), and worlds (see the central part of the diagram in Fig. 3). Whereas we have some detailed knowledge of how our understanding of forces such as heat or music is shaped when metaphoric webs form as a result of projections of small-scale schemas upon medium-scale perceptual units (remember the outline in section 2), it is still less clear how narrative figures in this picture.

5 Scientific concepts as imaginative structures

We have arrived at a point where we can ask how experience at different scales and the imaginative processes we have talked about – schematizing action of an embodied mind, metaphor, analogy, figure-ground reversal and narrative (in the form of prototypical stories) – let scientific conceptualizations of forces of nature arise. We shall explore this issue by creating concept maps of linguistic forms related to temperature and heat that show how these notions arise from metaphoric and narrative processes (Fig. 4 and 5). The maps are used to suggest that some concepts belong to the realm spanned by small-scale and medium-scale experience, whereas others properly belong to the world between medium-scale and large-scale experience.

We shall see that much of the load of creating concepts rests on the shoulders of schematizing processes, figure-ground reversal, metaphor and analogy (section 2). The role of narrative in creating conceptual meaning is directly important for some concepts, especially those related to the *power of a force*, and indirectly for influencing the meaning of metaphors (Fuchs et al. 2018). However, its power lies with other than purely conceptual aspects of science – it is found mostly in the framing of natural and technical scenes (Fuchs 2015) of large-scale phenomena. Stories of such scenes produce narrative experience (Caracciolo 2014) from which

22 It is easy to think of exceptions to the rule that small-scale experience is expressed in short expressions and large-scale experience in (long) stories. We can tell a story in a few words, and we can go on talking forever about a small incident. However, that would not generally be the norm.

suggestions for *relations* between concepts, and the construction of *models* from such relations, emerge (section 3).

5.1 Conceptual processes mediating between small and medium-scale structures

Before we begin, let us briefly describe some terminology we want to use (Figs. 4 and 5). We postulate that experience at three different scales produces *perceptual units* (PU) we shall call *polarity*[23] (such as the HOT ↔ COLD polarity), *force of nature* (such as heat) and *world* (for example, scenes where heat and other forces interact), respectively. As outlined before, our analysis of physical science and conceptualizations of it reveals that a force of nature is understood in terms of the aspects of *intensity*, *amount* (size) and *power*. In Figs. 4 and 5, we shall identify these three *aspects* as *domains* (D). Furthermore, we make use of *perceptual schemas* (such as CONTAINER, BALANCE, TIME, POLARITY) whose projections create webs of figurative structures. The linguistic terms created by the interaction of physical and linguistic experience and used in informal and formal science, shall be called *linguistic forms* or symbolic units (Evans 2009).

Figure 4 presents a map of concepts related to the notion of *temperature*. Note that more is involved than this single concept, i.e., temperature: the domain of hotness is populated, at the level of linguistic forms, by a good number of formal and informal terms such as *hot, thermal level, temperature difference, rate of change of temperature*, etc. Clearly, temperature concepts go beyond merely making it possible to tell how warm or cold something is.

Much of the meaning of the temperature concepts listed in Fig. 4 is related to basic schemas such as SCALE, PATH or VERTICALITY. Expressions used to speak about concepts related to temperature suggest the metaphor TEMPERATURE IS THE THERMAL

23 Is the notion of polarity (or, for that matter, binary opposition) as basic as some (image) schemas are assumed to be? We could argue that polarity is "created" in joint perception of what leads to the schemas of SCALE and SOURCE-PATH-GOAL. It could also be the result of the (more basic?) experience of TENSIONS (which otherwise would be differences of two points along a path connecting the poles of a polarity). However, there is linguistic evidence that polarity (not just scale) is one of the fundamental and simple experiential units: little children first use a single word for polarities (cold for hot-cold) or binary opposites (open for open-closed); and, early in the development of human language, single short words have been used to denote polarities (on the last point, see Győri and Hegedűs 1999). Actually, asking which of the schemas is "more" basic may be the wrong path taken; as with metaphors, the structures created by experience may be complex webs rather than hierarchies (where, in a strict bottom-up process, simple elements create bigger ones and so on).

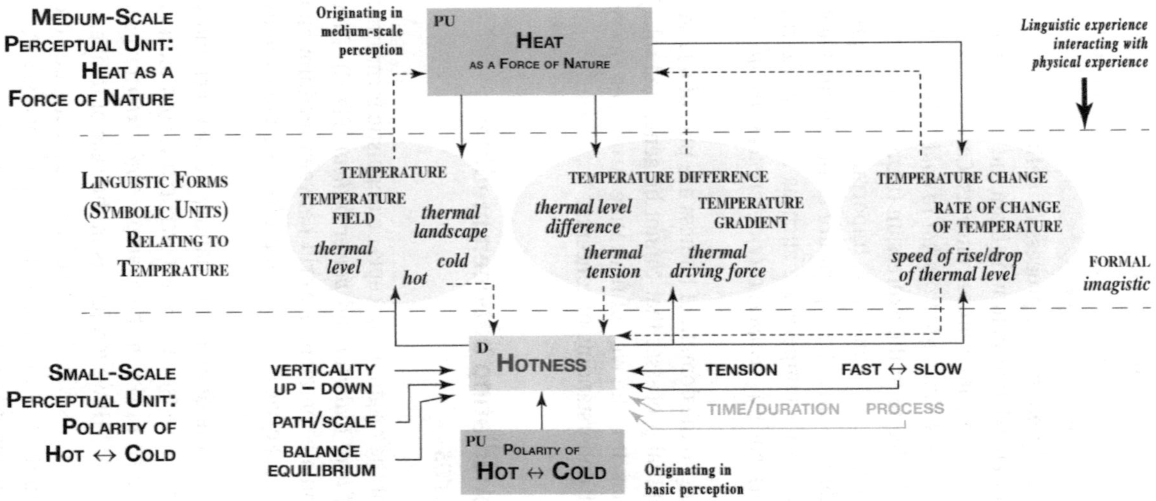

Fig. 4: A map of concepts (linguistic forms or symbolic units) related to temperature partly resulting from metaphoric processes between small-scale and medium-scale perceptual units. PU: Perceptual units; D: Domains; single capitalized terms: Schemas (grayed out schemas stem from experience at a different scale). Dashed arrows: Feedback from linguistic experience to domains of physical experience.

LEVEL. Note, however, that some of these linguistic forms, such as rate of change of temperature, use schemas and abstractions such as TIME or PROCESS that are related to our medium and large-scale forms of experience.

There is no such thing as a purely bottom-up process of defining concepts. There are instances of projection of knowledge stemming from the larger perceptual unit of *heat as a force of nature* onto temperature concepts. Consider expression (14) in Tab. 1: "*This means heat flows 'downhill' from hot to cold.*" The particular property of (quantity of) heat alluded to in the expression (namely, that it can flow) is part of the metaphor HEAT IS A FLUID SUBSTANCE. Importantly, the expression sheds additional light upon the meaning of temperature as the thermal level (and the rest of imagery associated with temperature such as tension, driving force, thermal landscape, gradient), such as *temperature differences are driving forces of flows of caloric.*

In sum, linguistic forms relating to the domain of hotness are part of a larger structure, a web created by conceptual processes that cannot be generated solely by projection of small-scale schemas. It is the web of figurative structures emerging in this manner that creates our understanding of concepts.

5.2 Conceptual processes mediating between medium and large-scale structures

We should now turn to the issue of the origin of concepts or linguistic forms that belong to the domains *quantity of heat (caloric)* and *power of heat* (Fig. 5). They are made up of concepts that populate the mental world between medium and large-scale experience. For brevity's sake, we shall limit our discussion to *quantity of heat.*

Schemas that are constitutive of the domain of quantity of heat are those known also from smaller-scale experience (such as SUBSTANCE, FLOW, CONTAINER or SIZE), plus *force-dynamic schemas*, such as LET in the example "*this exterior heat lets the crust become crispy*" (Table. 1).[24] However, just as in the case of temperature, meaning from experience at larger scales is infused in structuring our understanding of caloric. According to the metaphor HEAT IS A FLUID SUBSTANCE ("*. . . heat flows 'downhill' . . .*" or "*. . . pumping heat and moisture high into the atmosphere . . .*"; Tab. 1), we understand caloric (quantity of heat) as an entity that can

24 Force dynamic schemas do not play a role in constituting the domain of hotness, except indirectly when we hold the level of heat responsible for what is normally the work of the force of heat.

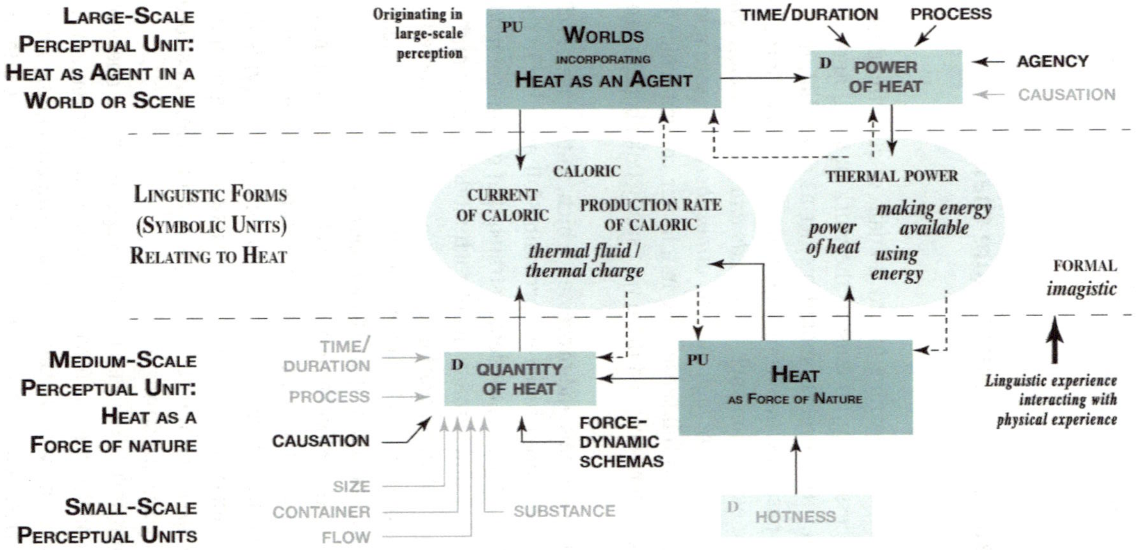

Fig. 5: A map of concepts (linguistic forms or symbolic units) related to heat partly resulting from narrative processes between large-scale and medium scale perceptual units. PU: Perceptual units; D: Domains; single capitalized terms: Schemas (grayed out schemas stem from experience at a different scale). Dashed arrows: Feedback from linguistic experience to domains of physical experience.

flow. This involves notions of PROCESS and TIME which we have associated with narrative experience before.

5.3 Narrative understanding structures more than conceptual maps

Much of the heavy lifting in creating meaning for the domain of *quantity of heat* can be done by cognitive processes that do most of the work in small and medium-scale experience as well.[25] However, it should be just as clear that, in expressions (see Tab. 1) or short passages, important meanings of heat as a force of nature do not emerge. A single expression such as "*Clouds and storms follow the warm water, pumping heat and moisture high into the atmosphere . . .*" says a lot about "heat," but the sentence makes full sense only in a larger context which is related to events in large-scale experience; and these we usually talk about in prototypical narrative form.

Concrete real-life dynamical systems and processes, which, in the case of science, are treated in formal models, are needed for providing full meaning to the experience of physical nature. In section 3, we demonstrated that dynamical models are the analog of story worlds, and simulations of such models correspond to stories told against the backdrop of these worlds. In stories, a sense of forces of nature as agents that act and suffer through events emerges; some of this meaning can be expressed through the metaphor HEAT IS A POWERFUL AGENT. At first, this may seem to be equivalent to HEAT IS A (MOVING) FORCE; on closer inspection, however, it becomes clear that it is not – the sense embedded in the former is wider and deeper than that furnished by the latter. Power (and with it, energy) is derived from a force of nature understood as *agentive*. *Agents* can be big or small and relaxed or tense. *Moving forces*, on the other hand, are too anemic to afford such important properties; they are just strong or weak.

Furthermore, the metaphor HEAT IS A POWERFUL AGENT suggests that it does not suffice to know the properties of a particular force, represented as an agent, independently of others; we need to know how agents interact. Understanding interactions rests upon the character(s) of all of the agents involved, and their interaction with the physical scene they are embedded in. All this understanding derives from story-worlds that give us agents, their tensions, and their power, and that

25 The case of the domain of power (see Fig. 5) is a little different; narrative understanding will play a more prominent role. This will be hinted at in the next paragraph, but a full discussion of the origin of the sense of power in physical and narrative experience and its transfer to formal science would require a longer text.

lay out their physical environment. Remember how the sense of physical and chemical agency emerges when we hear or read the small story presented in section 3.

6 Conclusion

Our formal theories of macroscopic physical systems and processes are shaped in a most profound way by human experience. We have argued that experience works at different scales ranging from small to large, and integrates interactions between individuals and their natural, linguistic, social and cultural environments. This makes the concepts, models and theories of a science such as physics a result of imaginative rationality operating through metaphoric and narrative processes.

On these assumptions, we can expect a number of important consequences for science and the learning of science. First, we can understand why and how narrative is a constitutive element of science and not just a pretty and motivating form of packaging for scientific content. Second, there should be a certain element of continuity of conceptualization of the physical world between children and grownups, and laypeople and scientists. Scientists' conceptions cannot be so totally different from those of a child encountering nature; put differently, the difference between a child's and a scientist's understanding of nature cannot lie in the mental processes and abstractions we all have access to; it must be found elsewhere.

Third, and this may be most important, our understanding of nature cannot be so totally different from our understanding of other domains accessible to experience. We can hope that learning about nature does not proceed along paths that are so totally divergent from those we travel when we learn how to understand our fellow human beings. If children (and adults) learn about other people through the stories told in our cultures (Hutto 2007), they may very well learn to understand and predict the behavior of natural agents through stories told about forces of nature (Fuchs et al. 2018).

Reference list

Callen, Herbert B. *Thermodynamics and an Introduction to Thermostatistics*. 2nd edition. New York: Wiley and Sons, 1985.

Caracciolo, Marco. *The Experientiality of Narrative: An Enactivist Approach*. Berlin: De Gruyter, 2014.

Chemero, Anthony. *Radical Embodied Cognitive Science*. Cambridge, MA: The MIT Press, 2009.

Corni, Federico, and Hans U. Fuchs. "Primary Physical Science for Student Teachers at Kindergarten and Primary School Levels: Part I – Foundations of an Imaginative Approach to Physical Science." *Interchange* 51 (2020): 315–343. doi:10.1007/s10780-019-09382-0.

Corni, Federico, Hans U. Fuchs, and Elisabeth Dumont. "Conceptual Metaphor in Physics Education: Roots of Analogy, Visual Metaphors, and a Primary Physics Course for Student Teachers." *Proceedings of the GIREP-ICPE-EPEC 2017 Conference*. 3–7 July 2017, Dublin, Ireland. Eds. Eilish McLoughlin and Paul van Kampen. *Journal of Physics: Conference Series* 1286 (2019): art. 012059. doi:10.1088/1742-6596/1286/1/012059.

Corni, Federico, Hans U. Fuchs, and Giovanni Savino. "An Industrial Educational Laboratory at Ducati Foundation: Narrative Approaches to Mechanics Based Upon Continuum Physics." *International Journal of Science Education* 40.3 (2018): 243–267.

Corni, Federico, Enrico Giliberti, Hans U. Fuchs. "Student Teachers Writing Science Stories: A Case Study." *Ebook Proceedings of the ESERA 2013 Conference: Science Education Research for Evidence-Based Teaching and Coherence in Learning*. Eds. Constantinos P. Constantinou, Nicos Papadouris, and Angela Hadjigeorgiou. Part 13. Co-eds. Lucy Avraamidou and Marisa Michelini. Nicosia, Cyprus: European Science Education Research Association, 2014. 2494–2505. http://urn.kb.se/resolve?urn=urn:nbn:se:hkr:diva-13272.

Deichmann, Marion. *Im übertragenen Sinne: Metaphern und Bildvergleiche in der Wissernschaft*. Bachelor thesis. Zürcher Hochschule der Künste, Zurich, 2014. Animation at http://vimeo.com/98311515 (20 August 2021).

Dewey, John. *Experience and Nature* (1925). *The Later Works, 1925–1953*. Vol. 1: *1925* . Ed. Jo Ann Boydston. Carbondale, IL: Southern Illinois University Press 1981.

Dumont, Elisabeth, Hans U. Fuchs, Werner Maurer, Francesca Venturini. "From Forces of Nature to the Physics of Dynamical Systems." *The 9th International Conference on Conceptual Change*. Bologna, Italy, 26–29 August 2014.

Evans, Vyvyan. *How Words Mean: Lexical Concepts, Cognitive Models, and Meaning Construction*. Oxford, UK: Oxford University Press, 2009.

Fauconnier, Gilles. *Mental Spaces: Aspects of Meaning Construction in Natural Language*. 2nd edition. Cambridge, UK: Cambridge University Press, 1994.

Fauconnier, Gilles, and Mark Turner. *The Way We Think: Conceptual Blending and the Mind's Hidden Complexities*. New York, NY: Basic Books, 2002.

Feynman, Richard, Robert Leighton, and Matthew Sands. "The Feynman Lecture on Physics." New York, NY: Basic Books, 2011.

Fuchs, Hans U. "From Image Schemas to Dynamical Models in Fluids, Electricity, Heat, and Motion: Examples, Practical Experience, and Philosophy." *Proceedings of the 2006 GIREP Conference*. University of Amsterdam, 2006.

Fuchs, Hans U. *The Dynamics of Heat: A Unified Approach to Thermodynamics and Heat Transfer*. 2nd edition. New York: Springer, 2010 [1996].

Fuchs, Hans U. "Il significato in natura [Meaning in Nature – From Schematic to Narrative Structures of Science]." *Le scienze nella prima educazione: Un approccio narrativo a un curricolo interdisciplinare*. Ed. Federico Corni. Trento, Italy: Erickson, 2013.

Fuchs, Hans U. "The Narrative Structure of Continuum Thermodynamics." *Ebook Proceedings of the ESERA 2013 Conference: Science Education Research for Evidence-Based Teaching and Coherence in Learning*. Eds. Constantinos P. Constantinou, Nicos Papadouris, and Angela Hadjigeorgiou. Part 1. Co-eds. Marios Papaevripidou and Roser Pinto Casulleras. Nicosia, Cyprus: European

Science Education Research Association, 2014. 50–61. http://urn.kb.se/resolve?urn=urn:nbn:se:hkr:diva-13272.

Fuchs, Hans U. "From Stories to Scientific Models and Back: Narrative Framing in Modern Macroscopic Physics." *International Journal of Science Education* 37.5/6 (2015): 934–957.

Fuchs, Hans U., Annamaria Contini, Elisabeth Dumont, Alessandra Landini, and Federico Corni. "How Metaphor and Narrative Interact in Stories of Forces of Nature." *Narrative and Metaphor in Education: Look Both Ways.* Eds. Michael Hanne and Anna A. Kaal. London, UK: Routledge, 2018. Ch. 7.

Fuchs, Hans U., Elisabeth Dumont, and Federico Corni. "Narrativity in Complex Systems." *Science, Environment, Health: Towards a Science Pedagogy of Complex Living Systems.* Eds. Albert Zeyer and Regula Kyburz-Graber. 2nd edition. Cham, Switzerland: Springer, 2021. 31–50.

Fuchs, Hans U., Michele D'Anna, and Federico Corni. "Entropy and the Experience of Heat." *Entropy* 24.5 (2022): art. 646. doi:10.3390/e24050646.

Geeraerts, Dirk, and Hubert Cuyckens, eds. *The Oxford Handbook of Cognitive Linguistics.* Oxford, UK: Oxford University Press, 2010.

Gibbs, Raymond W. *The Poetics of Mind: Figurative Thought, Language, and Understanding.* Cambridge, UK: Cambridge University Press, 1994.

Győri, Gabor, and Irén Hegedűs. "Is Everything Black and White in Conceptual Oppositions?" *Issues in Cognitive Linguistics: 1993 Proceedings of the International Cognitive Linguistics Conference.* Eds. Leon de Stadler and Christoph Eyrich. Berlin and New York: Mouton de Gruyter, 1999. 57–74.

Halliday, David, Robert Resnick, and Jearl Walker. *Fundamentals of Physics.* 10th edition. New York: Wiley, 2013.

Hampe, Beate. *From Perception to Meaning: Image Schemas in Cognitive Linguistics.* Berlin and New York: Mouton de Gruyter, 2005.

Herman, David. *Story Logic: Problems and Possibilities of Narrative.* Lincoln, NE: University of Nebraska Press, 2002.

Herman, David. *Basic Elements of Narrative.* Chichester: Wiley-Blackwell, 2009.

Herman, David. *Storytelling and the Sciences of Mind.* Cambridge, MA: MIT Press, 2013.

Hurwitz, Brian. "Form and Representation in Clinical Case Reports." *Literature and Medicine* 25.2 (2006): 216–240.

Hutto, Daniel D. "The Narrative Practice Hypothesis: Origins and Applications of Folk Psychology." *Royal Institute of Philosophy Supplement* 82.60 (2007): 43–68.

Johnson, Mark. *The Body in the Mind: The Bodily Basis of Meaning, Imagination, and Reason.* Chicago, IL: University of Chicago Press, 1987.

Johnson, Mark. *The Meaning of the Body: Aesthetics of Human Understanding.* Chicago, IL: University of Chicago Press, 2007.

Jou, David, José Casas-Vázquez, and Georgy Lebon. *Extended Irreversible Thermodynamics.* 2nd edition. Berlin: Springer, 1996.

Lakoff, George, and Mark Johnson. *Metaphors We Live By.* Chicago, IL: University of Chicago Press, 1980.

Lakoff, George, and Mark Johnson. *Philosophy in the Flesh: The Embodied Mind and Its Challenge to Western Thought.* New York: Basic Books, 1999.

Mareš, J. J., P. Hubík, J. Šesták, V. Špička, J. Krištofik, and J. Stávek. "Phenomenological Approach to the Caloric Theory of Heat." *Thermochimica Acta* 474.1–2 (2008): 16–24. doi:10.1016/j.tca.2008.05.001.

Morgan, Mary S. "Models, Stories, and the Economic World." *Journal of Economic Methodology* 8.3 (2001): 361–384.

Noë, Alva. *Action in Perception*. Cambridge, MA: MIT Press, 2004.

Sherin, Bruce L. “How Students Understand Physics Equations.” *Cognition and Instruction* 19.4 (2001): 479–541.

Stewart, John, Olivier Gapenne, Ezequiel A. Di Paolo, eds. *Enaction: Toward a New Paradigm for Cognitive Science*. Cambridge, MA: MIT Press, 2010.

Talmy, Leonard. *Toward a Cognitive Semantics*. Vol. I and II. Cambridge, MA: MIT Press, 2000.

Truesdell, Clifford A., and Walter Noll. *The Non-Linear Field Theories of Mechanics*. Ed. Siegfried Flügge. *Encyclopedia of Physics*. Vol. III/3. Berlin and Heidelberg: Springer, 1965.

Truesdell, Clifford A., and Richard A. Toupin. *The Classical Field Theories*. Ed. Siegfried Flügge. *Encyclopedia of Physics*. Vol. III/1. Berlin and Heidelberg: Springer, 1960. 226–793.

Turner, Mark. “The Art of Compression.” *The Artful Mind: Cognitive Science and the Riddle of Human Creativity*. Ed. Mark Turner. Oxford, UK: Oxford University Press, 2006.

Varela, Francisco J., Eleanor Rosch, and Evan Thompson. *The Embodied Mind: Cognitive Science and Human Experience*. Cambridge, MA: MIT Press, 1991.

Wise, M. Norton. “Science as (Historical) Narrative.” *Erkenntnis* 75.3 (2011): 349–376.

Part III: **Scientific Approaches to Narrative Cognition in Science**

Sébastien Lemerle

A Fortress Filled with Highways

A Study on the Uses of Metaphors in Popular Neurosciences

Abstract: One of the most common features of science popularization is its use of metaphors and analogies. This fact is linked to the need to translate abstract concepts and styles of thought into a language comprehensible to anyone, and to the fact that metaphors may be at the heart of our conceptual system, as suggested by George Lakoff. This study aims to explore this hypothesis, from the example of the French edition of "Brain Awareness Week," an annual event of neuroscience popularization organized in North America and Europe since the 1990s. It will show that metaphors, along with other means, are indeed central to the discourses of the neuroscience popularizers. Consequently, they have a great influence on our basic extra-scientific narratives concerning the brain. But they are not so numerous and diverse. They can be gathered in a brief list: brain as a black box or a fortress that has to be opened; brain organization as neural highways; brain as a supercomputer, etc. This narrow range of metaphors can be explained by several factors constraining their use in popularization discourses: the sociocultural and scientific contexts, the popularizers' skills, cultural capital and positions, and the characteristics of the public or the representation that popularizers have of it. All of these elements extensively shape popularization narratives, and consequently the lay understanding of neuroscience in today's society. Hence, a narratology of science popularization could draw great benefit from the study of the social conditions of metaphor formation and use.

1 Introduction

Popular science is often considered to be a practice aimed at adapting scientific thought for extra-scientific audiences (Jeanneret 1994).[1] Whether it is described in terms of translation, alteration (Whitley 1985), rewording (Jacobi 1999) or paraphrase of scientific discourses (Mortureux 1982), it is also usually seen as an attempt to bridge a gap between specific cognitive worlds, in so far as

1 Many thanks to Michael Sinding and Aura Heydenreich for their reviews, corrections and very helpful comments on a first draft of this paper.

https://doi.org/10.1515/9783110782844-009

> the expansion of scientific research and diversity of specialised interests of researchers have meant that scientists seeking reputation from a broader audience than their immediate specialist colleagues have to transform their ideas and results into a form that other groups can understand. (Whitley 1985, 6)

This is particularly true of neuroscience popularization. For more than half a century, modern neuroscience, the "Sistine Chapel of science" according to James G. Ballard (Ballard 2000, 326), has continuously drawn public attention, both in Europe and the US. As brain science has become more and more sophisticated, it has generated a vast amount of print, radio, TV and, more recently, online programs designed to fulfill a growing interest, which has also been carefully nurtured by scientific writers, publishers, journalists, TV producers, Youtube channel managers, etc.

In France, neuroscience has been receiving much public attention since *L'Homme neuronal* (*Neuronal Man)* by Jean-Pierre Changeux (Changeux 1983). On the cultural side, numerous popular neuroscience essays have become best-sellers for more than three decades (Lemerle 2013) and neuroscience has been displayed more and more not only on air but also in museums, exhibitions and public events. In the intellectual field, several public figures from all parts of the ideological spectrum rely on neuroscience in their stances on school reforms, terrorism, and gender or racial issues (Lemerle 2009; 2013). More generally, social scientists such as Edgar Morin, Dan Sperber and Gérald Bronner have been advocating for taking neuroscience into account in order to refound their disciplines, as well as in their public discussions on human nature, culture or socialization (Morin and Piatelli-Palmarini 1974; Sperber 1996; Bronner 2010). Authorities have also tried to draw inspiration from neuroscience to shape policies in the realms of education, health, the fight against crime, and so on for at least the last 15 years (Lemerle 2016). In 2018, for instance, Education Minister Jean-Michel Blanquer created the *Conseil scientifique de l'Éducation nationale* (*Scientific Counsel of National Education*), whose president is Stanislas Dehaene, a renowned neuroscientist and a professor at the Collège de France. The mission of the *Conseil scientifique de l'Éducation nationale* is to provide advice drawn mainly from neuroscientific advances in order to fight education inequalities. In 2019, almost half of the 24 members of its board were neuroscientists or cognitive scientists.

Given this specific context, studying neuroscience popularization in contemporary France appears to be an important issue, especially "live" popularization, which has attracted notice since the early 2000s. This kind of popularization, which includes, for instance, public conferences and interactive sessions with commentaries from lecturers, is a powerful way to promote neuroscientific ideas and their authors (scientists or journalists), who come to introduce their books to a vast and devoted audience. The most important of these events, the “Semaine du Cerveau,” is the subject of this study. It is a very telling example of the attempts made by neuroscientists to deliver what could be a “relevant” popularization of neuroscience – that is, one that would not be affected by the multiple biases carried, according to them, by mass-media popularization, which often result in erroneous reports, damaging misconstruals and wild extrapolations. However, these neuroscientists-turned-popularizers seemingly cannot help having recourse to some of the frames and methods used in the kinds of popularization they want to avoid. This chapter will show how they try to maintain a balance between their original purpose and the means that they are led to resort to, by focusing on a specific technique, the use of metaphors and analogies.

Thus, after having introduced the fieldwork, I will deal with the hypothesis that one of the most common features of neuroscience popularization at the “Semaine du Cerveau” lies in its use of metaphors and analogies. I will also describe the metaphors and analogies that recur most frequently, and discuss some consequences induced by their utilization in terms of public understanding of neuroscience. However, the metaphors of the “Semaine du Cerveau” are less numerous than one would expect. In the final section, I will explain this surprisingly narrow range of metaphors by identifying several constraints on their use in popularization discourses, which will highlight the ambivalence of the popularization process, even when performed by the scientists themselves.

2 The “*Semaine du Cerveau*”

The “Semaine du Cerveau” is the French edition of “Brain Awareness Week,” an annual event of neuroscience popularization, organized in North America and Europe since the 1990s. “Brain Awareness Week” was launched in 1996 by the *Dana Foundation* in the United States, as a campaign dedicated to advancing public awareness about the progress and benefits of brain research. The French edition was created in 2000, as “La Semaine du Cerveau” (“Brain Week”), run by the *Société des Neurosciences*, one of the largest neuroscience societies in Europe.[2]

The event really took off in 2010 and it has gained more and more public attention every year since then. Hosted now by the *French National Institute of Health and Medical Research*, high profile research centers such as the *Brain and Spine Institute* or the *Institut de la Vision*, the *Collège de France, the Pasteur Insti-*

2 https://www.neurosciences.asso.fr/qui-sommes-nous/presentation/ (21 September 2018).

tute and numerous universities, the “Semaine du Cerveau” is held every March. In 2016, it involved 34 French cities and more than 700 volunteers and attracted 35,000 people (8,500 people in Paris alone) (Société des neurosciences 2016).

The “Semaine du Cerveau” consists of open houses at neuroscience labs such as the *Brain and Spine Institute,* lectures at the *Collège de France* or the *École Normale Supérieure,* exhibitions and workshops at the *Palais de la Découverte* and the *Cité des Sciences et de l’Industrie,* displays at libraries, movies and debates in theaters and community centers and classroom presentations. It provides a rich source of material to study the ways in which neuroscience is popularized. The “Semaine du Cerveau” has another point of interest, which is that it is an event entirely managed by scientists and scientific institutes’ staff, with almost no involvement from the media or the publishing world. The core of the “Semaine du Cerveau” initiative lies in the fact that it is thought of as an encounter between researchers and the public. In brief, no cultural go-betweens, except some moderators in a few debates.

I attended the “Semaine du Cerveau” in Paris from 2015 to 2017, and I conducted thirty interviews with people involved in its organization, mostly researchers, scientific mediators and communication managers. I have also attended the work meetings of the “Semaine du Cerveau” steering comittee for more than three years. In that context, I have had access to some material used by the “Semaine du Cerveau” speakers (Powerpoint presentations, videos) and in 2016 and 2017, I carried out two audience surveys in order to gain a better knowledge of the audience of the “Semaine du Cerveau.”

3 Popularizing implies using metaphors

3.1 What “*Semaine du Cerveau*” popularizers do

The scientists of the “Semaine du Cerveau” aim at reconciling scientificness and attractiveness to a large audience. They are very concerned about the quality of their lectures, displays and workshops. Sometimes they even quarrel when they discuss future events. One senior member of the steering comittee keeps on saying: “It has to be scientific!” as a motive for dismissing several projects proposed by outside partners, for instance theatrical companies that would like to get the “Semaine du Cerveau” label. The “Semaine du Cerveau” intends to show the researchers’ work, the way they work (one of them even talked about their “sacrifices”), their results, how these results can be useful to society, and so on. One can see the “Semaine du

Cerveau" as an initiative of self-justification and self-promotion of research in neuroscience.

Between 2010 and 2017, more than a hundred events were organized by the "Semaine du Cerveau" in Paris, dealing with several dozen topics. Health is the most frequent subject (mental health, neurodegenerative diseases, etc.), followed by presentations on perception (the five senses, the inner ear, etc.) and general presentations on the history of the neurosciences. We can also mention several presentations on memory and aging, sleep, language, cognition, and more general topics such as human nature, tolerance, justice and transhumanism.

The "Semaine du Cerveau" organizers think that they must avoid two things. On the one hand, they do not want to be associated with fundraising for medical research. On the other hand, most of the time they dismiss spectacular popularization, which they associate with looseness and with some despised form of journalism. Most of them consider that their work at the "Semaine du Cerveau" differs in numerous ways from journalism, but as a matter of fact, the situation is a bit more complex.

When one attends the "Semaine du Cerveau" or talks with the researchers involved in it, one may find four main popularizing methods.[3]

One can mention first the use of narrative telling, for instance based on a historical perspective, as with the long arc of research on memory from Descartes to Pavlov. This historical frame setting can be applied to the history of a discovery, a theory or an institution. This is a proven technique, playing with widely shared cultural habits, to introduce a topic: the current situation of neuroscience is represented as the result of a string of *events*, which is the minimal definition of narration, according to Gérard Genette (1969). These events are depicted as having origins, meaning and an evolution that can be explained.

This very simple kind of narrative is not the only kind that one can encounter at the "Semaine du Cerveau." There are of course various sorts of narrative, from the aforementioned minimal historical introduction to more complex discourses. The decision to resort to one sort of narrative or another can depend on the popularizing skills and the experience of the speakers. It can also depend on the nature of the subject of the session and on the knowledge of the public concerning the matter to be discussed. Supposedly lesser-known topics such as the functions of the cerebellum or the biological bases of addiction may need a more attractive introduction than a mere historical presentation of these fields of research.

3 These results are based on a larger study of the popularization of neuroscience made for my habilitation thesis. For a detailed presentation, see Lemerle (2019, 2021b).

Hence, the storytelling at the "Semaine du Cerveau" actually encompasses numerous methods appealing to very powerful cognitive schemes that are articulated to the notion of an event, such as the inquiry ("What is the problem with the cerebellum?"), the revelation of a hidden truth ("The second brain located in our stomach") or the surprising, marvelous anecdote ("The sheep who fell in love with a goat"). The purpose is to put the audience in an emotional state which will put them at ease, in a familiar environment which will facilitate the introduction of more complex notions. However, this kind of strategy is not specific to the "Semaine du Cerveau" popularizers. Playing on emotions, on the sense of mystery and wonder, are well-known professional gimmicks among scientific journalists, as Françoise Tristani-Potteaux has shown in the French case (Tristani-Potteaux 1997, 45–46).

Telling stories of anorexia, insomnia or addiction proves to work better than describing the action of a specific protein in some remote and tiny area of the brain. As one of the scientists puts it, "depression talks to people."

Second, another method often followed at the "Semaine du Cerveau" is the use of pictures and sketches to introduce a concept or a process: they can make understanding easier or at least create the impression of understanding. Such a method also plays on cultural habits rooted in the fascination created by brain scans:

> [. . .] the images produced by brain scans are certainly central to the new ways in which this issue of mind and brain is being posed, not just in the laboratory, or in psychiatry, but in the popular media, in the arts, and in many other areas of human life. (Rose and Abi-Rached 2013, 54).

Third, popularizers can also use objects or games, insofar as it appears easier to learn something when one is active, rather than when one listens passively to a lecture. Hence, the "Semaine du Cerveau" public may be asked to manipulate microscopes, brain models, and electrode headsets, or to play video games.

Fourth, another very common strategy consists in using specific rhetorical techniques, which can be combined with the three previous methods. Speaking of their work as popularizers, the "Semaine du Cerveau" scientists often employ the words "parallels" or "images," which clearly refer to the use of analogies and metaphors.

One of the main results of the study that I have carried out on the "Semaine du Cerveau" is that analogies and metaphors are indeed crucial in order to produce narratives on neuroscience topics in extra-scientific contexts. This idea had already been formulated in various studies of popular science (McRae 1993). In his book about the writing of popular science, Yves Jeanneret stressed that metaphors are useful to scientific popularizers because they allow the public to grasp scientific concepts thanks to daily, ordinary ways of talking (Jeanneret 1994). But he also stated that metaphors are in any case necessary because of the way our minds operate. In this claim, he draws on joint work by American linguist George Lakoff and philosopher Mark Johnson.

3.2 The centrality of metaphors

The Lakoff-Johnson framework, known as conceptual metaphor theory or the cognitive theory of metaphor (CMT), is currently the most influential approach to the study of metaphors. According to it, our conceptual system works on a metaphorical basis (Lakoff and Johnson 1980). Far from being an ornament or an embellishment of language, conceptual metaphors are "mappings across conceptual domains that structure our reasoning, our existence, and our everyday language" (Lakoff and Johnson 1999, 47). Such an assertion is based on the fact that the interactions of our bodies with the outside world structure our mental experience: "The same neural and cognitive mechanisms that allow us to perceive and move around also create our conceptual systems and modes of reason" (Lakoff and Johnson 1999, 4). Thus, our vocabulary for expressing causality is, in the words of Lakoff and Johnson, "embodied," that is to say, based on models generated by our sensori-motor systems: "Our bodies define a set of fundamental spatial orientations that we use not only in orienting ourselves, but in perceiving the relationship of one object to another" (Lakoff and Johnson 1999, 34). Here lies the reason why metaphorical thinking appears at the root of our conceptual systems. Humans have a natural tendency to conceive of things in the same terms that they use to describe their physical apprehension of reality. This is the reason why something "important" is typically referred to as "big."

Moreover, Lakoff and Johnson distinguish between "primary" and "complex" metaphors. A "primary" metaphor is a "neurally instantiated correlation between a sensorimotor operation [. . .] and a subjective experience or judgement" (Lakoff and Johnson 1999, 54–55). Thence, a metaphor is to be thought of as a "mapping" from one domain of experience (the "source domain") to another (the "target domain"). For instance, the idea of time (target domain) is typically based on the primary experience of motion (source domain) – for example, as we say in English, "time flies." A "complex" metaphor results from the combination of two or more primary metaphors. Of course, these complex metaphors differ according to language and culture, as shown by Zoltán Kövecses (2005).

Analyzing science popularization with such an approach can be very fruitful since it helps us to understand how scientific discourses actually access the public sphere. It explains the well-known fact, underlined by the German linguist Andreas Musolff, that

> Media discourse continuously makes creative and eclectic use of science terminology – often to the point of contradicting the views of scientific experts on the subject matter in question. (Musolff 2007, 67)[4]

Indeed, the reason for such a distortion lies in the fact that the media, and almost any cultural intermediary, work with the metaphors of popular science, rather than with the genuine scientific concepts, which often cannot easily be grasped by the layman. Facing the necessity of rewording esoteric scientific discourses, resorting to metaphors seems unavoidable insofar as, according to CMT, they are the most natural way to perceive relations between entities or conceptual domains. On the other hand, this can lead to more uncontrolled entailments than popularizers may have imagined at first. We are now going to see this with examples taken from the "Semaine du Cerveau."

3.3 The "*Semaine du Cerveau*" metaphors

As mentioned earlier, metaphors are heavily used by "Semaine du Cerveau" popularizers, and most of those that I have listed, originally formulated in French, have their counterparts in English. At the "Semaine du Cerveau," five recurring metaphors can be noted:

- The brain described as a *black box* ("boîte noire") or a *fortress* ("forteresse") that has to be opened; that is to say: full of secrets that must be discovered. This metaphor has been especially apparent during each inaugural lecture of the "Semaine du Cerveau" that I attended, as well as during the interviews with the speakers. At the beginnings of lectures, workshops or exhibitions, the brain is often introduced as the most mysterious thing in the world. This is rather traditional rhetoric in popular science: the wonder or the mysteries of nature.
- Nerve impulses as flows of *electricity*, related to other metaphors from the same realm: axons and dendrites as electric wires, myelin as insulating material which makes electric circulation easier.
- Nerve impulses are also described through the metaphors of *messages* and *information*, which can be used in relation to chemical processes as well.
- Brain organization as neural *roads*, with nerve impulses circulating on these roads, following signs, etc. The roads often turn into two-lanes, then four-lanes and finally highways, to characterize paths more and more commonly used.

4 See also, in the French case, Lemerle (2013; 2014).

- Brain as a *machine*, most often a *computer*, analyzing data taken from the environment in order to carry out thoughts and actions. Brains can also be depicted as global positioning systems.

In light of a recent content study of *Scientific American*, one can remark that some frequent metaphors, at least in English (for instance "a neuron is a tree" or "the nervous system is circuitry") did not emerge during this research (Shuttleworth 2017, 86).

These metaphors can be analyzed in relation to several conceptual metaphors identified by Lakoff and Johnson (Lakoff and Johnson 1999, 45ff.) First, the ORGANIZATION IS PHYSICAL STRUCTURE metaphor. The target domain is "relationships unifying the brain." It is expressed with the source domain of a building (a fortress) or an object (a black box, a computer) or a network of roads.

In the cases of the "fortress" and the "black box," another conceptual metaphor can be underlined: UNDERSTANDING IS GRASPING, but in a specific way: they stress the difficulty of reaching what is to be understood. The fortress and the black box are things that hinder access to some kind of treasure, or at least to a hidden thing. Moreover, whereas a box can actually be grasped, speaking of the brain as a fortress implies that it is something to besiege and seize, suggesting a more agonistic view of scientific work, in terms of conquest and surrender. It also conveys the idea that researchers are facing something bigger than themselves and offers potential links to the aforementioned idea of complexity.

This leads to another primary metaphor: KNOWING IS SEEING. It can be inferred from the fortress metaphor: once you break in, you will see the truth. This is also the case for the black box metaphor: it is necessary to open the box and light up what is plunged into darkness, in order to expose the functioning of thought, the emergence of consciousness, and so on.

Another dimension may be implied by this metaphor: "black box," both in French and English, also means a flight recorder in an aircraft. At stake here is a complex mapping with two conceptual metaphors. First, THE MIND IS A MACHINE, which can also be found with the metaphor of the brain as a computer. Second, THINKING IS MOVING (Lakoff and Johnson 1999, 236–237). Here, the brain is figured as a device attuned to the trajectory of human thinking and behavior; this can also be expressed with the GPS metaphor, which accentuates the idea of behavioral regulation. Finally, all the metaphors related to the idea of moving are especially consistent with the realm of roads and highways constantly referred to in the presentations.

3.4 The social and cultural impact of metaphors used in popular science

If one wants to consider the cultural and social reach of neuroscience, one must pay attention to these metaphors because they will have a crucial impact on mundane, extra-scientific narratives concerning the brain. Playing on metaphorical primary schemes or, in other words, metaphorical schemes shared on a cultural scale, they will be widely understood and appropriated. Their reuse will be either conventional or creative, but they will constitute the basis for the spreading of neuroscientific ideas in the social world, in ways that scientists obviously will not control. Indeed, latent meanings of a metaphor can be exploited in order to convey interpretations going beyond the original purposes of its first users, as we can see in the case of the "reptilian brain" coined by US neuroscientist Paul D. MacLean and nowadays used as a synonym for any archaic trait of human behavior (Lemerle 2021a). Think also of the computer metaphor, which has been developed according to the idea of self-empowerment: the exploitation of all the potentialities of the machine that is in our skulls, the conviction that we only use 10% of our brain, that we can program or re-program our brain, and the like. The highway metaphor is another good example of that kind of extrapolation. It is frequently used during presentations as a source domain for the idea of habits that are hard to escape from. Talking of "neuronal highways," one speaker explains that "the more a road is taken, the more we are likely to take it," in order to suggest that changing behavior is very hard: once you have taken the highway, it is almost impossible to change direction.

My hypothesis is that these metaphors set a large part of the layman's understanding of neuroscience and of the basic principles of brain operations. But on the other hand, the meaning of these metaphors is not fixed and may vary according to specific contexts, or, to use Hans Robert Jauss's words, according to the public's horizon of expectations (*Erwartungshorizont*).

4 Social determinations of the uses of metaphors

4.1 The hypothesis of constraints in metaphorizing

The "Semaine du Cerveau" popularizers are aware of the risks inherent in the use of metaphors. They are in fact rather cautious regarding the matter. Although exceptions can be found, they usually do not like going too far when leaving the well-known boundaries of scientific discourses, especially in terms of specific science terminology. The fact that metaphors used in the "Semaine du Cerveau" are indeed neither numerous nor diverse, suggests the existence of constraints on the use of metaphors among the speakers involved in the event.

Metaphors are always more than a simple substitution of one word for another one. According to the philosophers Max Black and Paul Ricœur, metaphors must be understood in relation to their context of utterance, especially the system of commonplaces associated with this context, which leads us to consider metaphors from a sociocultural perspective (Black 1968, 40; Ricœur 1997, 87ff.). Zoltán Kövecses (2005) also reminds us that each social, cultural, historical context has its own metaphorical conventions in terms of mapping between source domain and target domain, which overdetermines the use of certain words and expressions. In the case of the "Semaine du Cerveau," at least four factors influence the popularizers' recourse to metaphors.

4.2 The sociocultural context

Cultural habits certainly affect the popularizers' choice and use of metaphors. As many historians of ideas and philosophers put it (Eamon 1994; Pesic 2000; Hadot 2004), the idea of nature as a hidden reality which has to be unveiled and deciphered can be traced from antiquity, and especially the Hellenistic period, through the Middle Ages and its "literature of secrets," to the eighteenth century. This idea can be studied along with another deeply rooted cultural scheme, the metaphor of the legibility (*Lesbarkeit*) of the world, as the German philosopher Hans Blumenberg showed, especially in the case of the "decryption" of the DNA code (Blumenberg 2008, 377–412).

Nowadays, such a pattern can survive through the cultural clichés used and reused by the media, to which neuroscientists and popularizers are exposed as much as anybody else. For instance, using the "black box" metaphor is something very common in French newsmagazines (see tab. 1).

Another social determinant of metaphor use is the influence of the technical context, as already evoked with the reference implied by the black box metaphor to the aircraft flight recorder.[5] The influence of the technological environment is obvious when one takes a look at the various metaphors used to describe the brain during the twentieth century: telegraph, bell, telephone, radio, computer, GPS, etc. (Borck 2012). Therefore, one can recognize that terms like "black box," "brain ma-

5 For other examples of metaphors related to technical context, see Schlanger (1971).

Tab. 1: "Black Box" metaphor in the French press.[6]

Le mystérieux contenu de la boîte crânienne. (*Le Point*, 6 October 2005)	The mysterious contents of the cranial box
Les neurosciences analysent la "boîte noire" des 12–18 ans. (*Le Nouvel Observateur*, 15 September 2005)	Neurosciences analyze the teenagers' "black box"
On est loin d'avoir sondé tous les mystères de la boîte noire. (*L'Express*, 30 August 2004)	One is far from having probed all the mysteries of the black box.

chine" ("cerveau machine"), and "brain computer" ("cerveau ordinateur") refer to a certain technical reality, which belongs to a given historical moment.

4.3 The scientific context

Kövecses argues that if there are universal metaphors as "underlying" layers of cultural experience, some metaphors may also be culture-specific. Among others, "the subculture of science and its various subgroups provide an interesting example of how metaphorical source domains change over time. This is one of the most obvious cases of metaphor variation" (Kövecses 2005, 104). As a matter of fact, the historian Cornelius Borck, in his study of the metaphors used in neuroscience, notes that "the brain appears to be an unstable organ that has been compared to a wide and variable range of objects" (Borck 2012, 118). This instability could explain the gap between the media and neuroscience research when it comes to brain metaphors. The media can keep on using metaphors that have been abandoned by neuroscience. This is the reason why, at the "Semaine du Cerveau," the popularizers most often choose their metaphors not from media discourse, but according to the current state of the art in neuroscience, as it is reflected in journals like *Science, Nature, Scientific American* and *La Recherche* or in researchers' blogs such as "Aux frontières du cerveau" ("At the borders of the brain"), which are the speakers' main sources for preparing their presentations.

Their terminology depends on what is considered legitimate in this field: at the moment, that means electricity and highways rather than computers. Out of the 25 events that I attended at the "Semaine du Cerveau" between 2015 and 2017, "electricity" metaphors are found in at least five presentations, as well as the idea of transportation (including the metaphors of path, road, highway, network, GPS and compass), whereas the computer metaphor surfaced only three times. During

6 These data are extracted from Lemerle (2014).

the 30 interviews that I conducted with "Semaine du Cerveau" organizers and speakers, the computer metaphor was used only twice, whereas the road and highway metaphors have benefited from rather elaborate developments, such as the following one:

> Regarding axonal regeneration, either there are molecules that trigger or there are molecules that will inhibit. Therefore, I made little red lights and green lights. I try to translate the thing! [. . .] It's rather about paths and roads, etc., because neurons must find their way during brain development. (*CNRS Senior researcher*, interview, May 2015)

There can of course be common ground between the media and popularizaring experts. For instance, the metaphors of the black box and the fortress are still used, not only because of a general cultural context inherited from an ancient past. Nikolas Rose and Joel Abi-Rached remind us that since the second half of the nineteenth century, brain research mutated fundamentally "when the gaze of the doctor focused beneath the skin, into the interior of the patient's body" (Rose and Abi-Rached 2013, 59). From the works of Gall and other phrenologists to those of Broca or Flechsig on cerebral localization, from the studies of neuroanatomists such as Fritsch and Hitzig to the attempts of Golgi and Ramón y Cajal to visualize the brain's architecture using the method of staining, brain research has continuously been trying to understand unknown brain mechanisms by rendering them visible. This goal has been pursued during the twentieth century thanks to new methods allowing the study of living brains rather than dead ones – methods using scanning technologies such as X-rays and functional magnetic resonance imaging – to such an extent that nowadays one can speak of an "industry of visualization" (Rose and Abi-Rached 2013, 74) which fosters this neuroscientific "visual imaginary." As Rose and Abi-Rached put it, brain visualization is at the root of the idea of neuroscience progress and effectiveness (even though this idea raises a lot of difficulties). Resorting to the metaphor of the black box participates in this narrative of the unknown brain reality which must and will be rendered visible.

In most cases, however, "Semaine du Cerveau" popularizers tend to select metaphors in relation to what they think can be said or not from a scientific perspective. Thus, their metaphorical creativity is self-restrained, inasmuch as they are usually concerned about the scientific plausibility and the legitimacy of the metaphors, images, parallels and analogies to which they resort. One can mention two main types of figures, according to the categories of the philosopher Richard Boyd: pedagogical metaphors and theory-constitutive metaphors. Pedagogical metaphors play a role in the teaching and the popularization of theories that already possess non-metaphorical formulations. Theory-constitutive metaphors are used by scientists to express "theoretical claims for which no adequate literal paraphrase is known" (Boyd 1993, 485–486). They are to be found in younger disciplines, for instance in molecular biology in the first half of the twentieth century (Keller 2004) or in computer science and cognitive psychology in the 1960s and after, from which Boyd quotes, among others, the example of the brain being termed "a sort of 'computer'" (Boyd 1993, 486).

The computer metaphor constitutes an interesting case of a theory-constitutive metaphor turned into a pedagogical one. At first, its development is closely related to the development of cognitive neuroscience. As Cornelius Borck shows, the computer has been "the most powerful brain model of the twentieth century," since Alan Turing's works and the interdisciplinary evolution of cybernetics, so fashionable shortly after the end of World War II (Borck 2012). According to Howard Gardner, "the invention of computers in the 1930s and 1940s, and demonstrations of 'thinking' in the computer in the 1950s, were powerfully liberating to scholars concerned with explaining the human mind" (Gardner 1987, 40). It has provided a model of human thought:

> It was because the activities of the computer itself seemed in some ways akin to cognitive processes. Computers accept information, manipulate symbols, store items in 'memory' and retrieve them again, classify inputs, recognize patterns and so on. (Neisser 1976, 5; quoted in Gardner 1987, 119)

Since then, the brain has commonly been compared to a computer in neuroscience, especially in popularized versions. As an example, Paul D. MacLean's theory of the triune brain in the 1960s relied heavily on the metaphor of the brain as an assemblage of three "biological computers" (Lemerle 2017). But this computer model had vanished by the end of the 1990s, to the benefit of other models such as neural networks and plasticity. At the beginning of the 2010s, in a historian's eye,

> it is already hard to reconstruct how it was that the computer so easily assumed the role of central metaphor in brain research over such a long period of time. Maybe one day we will look back on the computer as the most convenient and common form of misunderstanding the brain in modern history. (Borck 2012, 127)

Its lasting theoretical plausibility explains why the computer metaphor is still used by "Semaine du Cerveau" popularizers, in a minor mode, despite its obsolescence. Universally known, it still works to set the audience at ease, though without being exploited beyond the stage of introductory formulas. It has turned pedagogical.

The metaphor of electricity can also be mentioned, since its influence is enduring throughout the history of neuroscience. As Borck notes, the "metaphorical appropriation of the body by electrical tools began with the simple analogy between cable and nerve fibre and provided the material basis for comparing the nervous system with telegraphy" (Borck 2012, 119). The metaphor has evolved along with technology, turning into comparisons with radio etc., but it has never been refuted and is still pertinent to neuroscience popularization, especially at the "Semaine du Cerveau."

One last striking example is the highway metaphor, which must be linked to studies on the mapping of the circuitry of the brain, described as "a grid-like pattern of fibers crossing at right angles," and even as "superhighways" which "form a three dimensional analog of the streets and avenues in New York City, running at right angles to each other" (Mitra 2012; see also Wedeen et al. 2012). Although the use of the highway metaphor at the "Semaine du Cerveau" has not been formally related by the speakers to its theory-constitutive version, their temporal coincidence is a clue to the legitimacy of this metaphor in the current neuroscience vocabulary.

4.4 The popularizers' skills, cultural capital and positions

Sociocultural and scientific contexts can be considered as the expression of social structures that frame the activity of popularizing. Far from denying the creativity of popularizers, the analysis of these contexts allows us to take into account the several constraints impinging on their work. The effects of these social factors can be seen in the popularizers' performances, although these performances also depend on more personal parameters such as skills, training and biographical trajectories.

Popularizers are more or less comfortable with the use of metaphors depending on whether they are researchers, trained scientific mediators working in science museums, or communications managers working for research institutes. Researchers usually want to be scientifically accurate and are rather sober on that subject. All of them use metaphors, without necessarily theorizing about it. They cling to the few controlled images that have already been mentioned. Of course, the longer they have popularized neuroscience, the better they can talk about the drawbacks of having recourse to metaphors. Mediators from the Palais de la Découverte and the Cité des Sciences et de l'Industrie have a professional point of view on science popularization. They know all of the different techniques and often participate in or organize training sessions about them. Among the available resources, employing metaphors is one of the most cited and used. Sometimes the mediators' lack of involvement in actual research (even though they all have an advanced scientific background) may drive them to be less cautious in their choice of terms, because their main purpose is to interest their audience, to make their presentations, as one of them put it, "catchy."[7] As an example, Marie G., a mediator at the Cité des Sciences et de l'Industrie, told me that she once spoke of the human microbiota as a zoo where each bacteria needs to be fed according to its specific dietary regime. Hence, she could justify how important it is to have a varied diet, in order to keep all of the "species" in the "zoo" in good health.[8] As one can see, the message does not need to be more precise. What is almost as important to them is preventing large parts of the public from getting up shortly after the conference has begun in order to go to another attraction in the museum. As a consequence, these mediators are often more innovative than scientists and sometimes argue with some of them who find their ways of popularizing somehow weird: relying on flashy metaphors, but also using toys, models, plasticine, plush items, electric circuits with strange bulbs, etc.

Returning to the popularization of neuroscience at the "Semaine du Cerveau," one of the best examples in that respect is a game called "Cervo-mix" created by Scube, an association dedicated to scientific popularization in the Paris area,[9] which can be seen as a variant of the highway metaphor. In this game designed for children, the various stages of brain development are represented by tasks that people have to perform to have the right to access the next level. For instance: striking a little ball into small holes, as on a mini golf course, with each hole representing a certain stage in brain development. The two persons who created this game, Florian D., a scientific mediator, and Amandine G., a Ph.D. student, told me that they encountered resistance from the members of the directorial board of their institution, who were rather skeptical about the project. In fact, these two mediators were also significantly younger than the members of the board. Along with different conceptions of popularization, there may also be a generation gap issue here. Because they were young, these popularizers seemed all the more willing to make neuroscience accessible in a playful manner that older scientists found frivolous and of low interest.

For both of these reasons (professional skills, age), it appears that the social position of the popularizers also affects the way they engage in their activity. In Pierre Bourdieu's words, one could say that the positions and dispositions of the popularizers clearly affect their statements (Bourdieu 1992, 321–322). Thanks to these various parameters, the outlines of a model can be proposed, which must be resituated in the context of a public event of neuroscience popularization, dealing above all with the public image of neuroscience in a specific country, France. Depending on whether or not they are researchers, whether or not they

7 Interview with Sarah G., a Ph.D. student at the École des Neurosciences de Paris and a popularizer at the Semaine du Cerveau, Paris, 28 March 2017.

8 Interview, Paris, 27 January 2017.

9 http://location.partageonslessciences.com/realisations/cervo-mix (14 October 2019).

are at the beginning of their career, and whether or not they are dominant in their field, popularizers may be more or less audacious, and their metaphors more or less abundant and more or less commonplace. The closer they are to the research, the more they will tend to be cautious concerning the use of metaphors and to mainly resort to tried and tested ones. The more respected they are as members of the scientific establishment, the less they will tend toward extravaganza. Such an outline would need to be integrated in a broader reflection on the specificities of French popularization, as compared with the situations in Germany, the United Kingdom or North America.

4.5 An implicit representation of the audience

The fourth and last factor concerning the use of metaphors in science popularization is related to the previous one. It concerns the representation of the public among science popularizers. A satisfactory understanding of the metaphors and analogies used at the "Semaine du Cerveau" often requires some prior knowledge. If a speaker compares the brain to a computer, he or she expects people to know roughly how a computer works. The neural highways metaphor also plays with a semantic network depending on the notions of message, source, sender, channel, receiver, destination, and so on. That is to say that it echoes basic aspects of theories of communication, even though this is not exactly what is targeted, and its use implies that the audience is familiar with them. This is what Hugh Petrie and Rebecca Oshlag call *interactive* metaphors, in contrast to mere comparative ones (Petrie and Oshlag 1993). Talking of electricity in terms of flow is a comparative metaphor. An interactive metaphor creates similarities between prior conceptual schemes and the new scheme to be taught, as in the case of the computer metaphor. In popular neuroscience, even the metaphor of electricity is interactive, because it is based on a comparison between nerve impulses and electricity, itself intertwined with another metaphor, that of electric flow, which is the main framework of our concept of electricity.

In other words, the effectiveness of metaphors depends in large part on the "cultural capital" (Bourdieu 1979) of the audience. Yet this characteristic is, most of the time, taken for granted by the popularizers, especially the scientists. Their speeches are implicitly aimed at a certain type of public who can have an interest in or at least a bit of curiosity about the brain's "secrets." In fact, sociological studies have long established that, in France, this interest and curiosity are diffused mainly in portions of the upper and middle classes who have important aspects of academic and cultural capital in common (see Donnat 2009; Ministère de la Culture et de la Communication 2018).

These results have been confirmed by two surveys that I conducted of the "Semaine du Cerveau" audience in 2016 and 2017: two-thirds of the more than 400 persons who completed the survey each time had a university degree, and around 50% had at least a master's degree (Lemerle 2019, 2022). Most of these people worked in the health sector or were teachers, managers, or students in these realms. Of course, this does not mean that popular neuroscience is for educated people only, but rather, in a more Bourdieusian approach, that this kind of science popularization is more likely to attract people with certain important cultural backgrounds, preferably scientific, and that this reality is implicitly taken into account by the science popularizers, who, by choosing their style of popularization, also choose or anticipate, somehow, a certain type of audience.

5 Conclusion

As shown by the "Semaine du Cerveau" case, one of the major ways in which neuroscience popularization can achieve its goal lies in its use of metaphors. On the one hand, this technique shapes many popularization narratives, and consequently the layman's understanding of neuroscience as well as discourses on the brain in today's society. Metaphors appear to be a major tool used to cover the "cognitive distance" between neuroscientific experts and the general public. Such a pattern can be found in other areas of knowledge. On the other hand, the use of metaphors is a way to endorse the fact that scientific discourse cannot simply be delivered in the public space in the same terms used in specialists' circles. Resorting to metaphoric reformulation means acknowledging the existence of a gap between the scientific field and the rest of the society.

We have seen that the centrality of metaphors may have cognitive origins, while their content and significance are basically sociocultural. We have also stressed the multiple social constraints on the use of metaphors, related to the general cultural context, to the scientific field, notably in its relation to the media, to the personal dispositions of the popularizers, and to the characteristics of the public or the representation that popularizers have of it.

From all of these elements, one can conclude that a narratology of science popularization could benefit greatly from the study of the social conditions of the formation and use of its most prominent metaphors, especially since a simplified neuroscientific discourse, based only on a few efficient metaphors, is always at risk of conveying a message that is counterproductive vis-à-vis the original intent of popularization: spreading trustworthy knowledge to citizens.

Reference list

Ballard, James Graham. *Millénaire mode d'emploi: Essais et critiques*. Auch: Tristram, 2006.

Black, Max. *Models and Metaphors, Studies in Language and Philosophy*. Ithaca, NY: Cornell University Press, 1968.

Blumenberg, Hans. *La lisibilité du monde*. Paris: Cerf, 2008.

Borck, Cornelius. "Toys Are Us: Models and Metaphors in in Brain Research." *Critical Neuroscience: A Handbook of the Social and Cultural Contexts of Neuroscience*. Eds. Suparna Choudhury and Jan Slaby. Chichester, West Sussex: Wiley-Blackwell, 2012. 113–133.

Bourdieu, Pierre. *La Distinction: Critique sociale du jugement*. Paris: Minuit, 1979.

Bourdieu, Pierre. *Les règles de l'art: Genèse et structures du champ littéraire*. Paris: Seuil, 1992.

Boyd, Richard. "Metaphor and Theory Change: What Is 'metaphor' a Metaphor for?" *Metaphor and Thought*. 2nd edition. Ed. Andrew Ortony. Cambridge: Cambridge University Press, 1993. 481–532.

Bronner, Gérald. "Cerveau et socialisation: Quelques éléments de discussion." *Revue française de sociologie* 51.4 (2010): 645–666.

Changeux, Jean-Pierre. *L'Homme neuronal*. Paris: Fayard, 1983.

Donnat, Olivier. *Les pratiques culturelles des Français à l'ère numérique: Enquête 2008*. Paris: La Découverte/Ministère de la Culture et de la Communication, 2009.

Eamon, William. *Science and the Secrets of Nature: Books of Secrets in Medieval and Early Modern Culture*. Princeton, NJ: Princeton University Press, 1994.

Gardner, Howard. *The Mind's New Science: A History of the Cognitive Revolution*. New York: Basic Books, 1987.

Genette, Gérard. *Figures II*. Paris: Seuil, 1969.

Hadot, Pierre. *Le voile d'Isis: Essai sur l'histoire de l'idée de nature*. Paris: Gallimard, 2004.

Jacobi, Daniel. *La communication scientifique: Discours, figures, modèles*. Grenoble: Presses universitaires de Grenoble, 1999.

Jauss, Hans Robert. *Pour une esthétique de la réception*. Paris: Gallimard, 1978.

Jeanneret, Yves. *Écrire la science: Formes et enjeux de la vulgarisation*. Paris: Presses universitaires de France, 1994.

Keller, Evelyn Fox. *Expliquer la vie: Modèles, métaphores et machines en biologie du développement*. Paris: Gallimard, 2004.

Kövecses, Zoltán. *Metaphor in Culture: Universality and Variation*. Cambridge: Cambridge University Press, 2005.

Lakoff, George, and Mark Johnson. *Metaphors We Live By*. Chicago: University of Chicago Press, 1980.

Lakoff, George, and Mark Johnson. *Philosophy in the Flesh: The Embodied Mind and Its Challenge to Western Thought*. New York: Basic Books, 1999.

Lemerle, Sébastien. "Les habits neufs du biologisme en France." *Actes de la recherche en sciences sociales* 176/177.1/2 (2009): 68–81.

Lemerle, Sébastien. *Le singe, le gène et le neurone: Du retour du biologisme en France*. Paris: Presses universitaires de France, 2013.

Lemerle, Sébastien. "Le neurone des *newsmagazines*: Le traitement des neurosciences dans la presse française des années 2000." *Neurosciences et société: Enjeux des savoirs et pratiques sur le cerveau*. Eds. Brigitte Chamak and Baptiste Moutaud. Paris: Armand Colin, 2014. 81–108.

Lemerle, Sébastien. "Tout est dans la tête: Les sciences du cerveau, nouveau savoir légitime." *La vie intellectuelle en France*. Vol. 2. Eds. Christophe Charle and Laurent Jeanpierre. Paris: Seuil, 2016. 551–556.

Lemerle, Sébastien. “Les mues du cerveau reptilien: Aspects cognitif et social d’une métaphorisation.” *Penser le vivant*. Eds. Laurence Dahan-Gaida, Christine Maillard, Gisèle Seginger, and Laurence Talairach. Paris: Éditions de la MSH, 2017. 222–233

Lemerle, Sébastien. *Neurotransmission: Deux études sur la vulgarisation des sciences du cerveau*. Habilitation thesis. Université de Paris 1 Panthéon-Sorbonne, June 2019.

Lemerle, Sébastien. *Le Cerveau reptilien: Sur la popularité d’une erreur scientifique*. Paris: CNRS éditions, 2021a.

Lemerle, Sébastien. “La Semaine du cerveau à Paris (2014–2017): Étude sur les contraintes de la vulgarisation des neurosciences.” *Sociologie* 12.2 (2021b): 129–147.

Lemerle, Sébastien. “Le goût des neurosciences: Les publics de la Semaine du cerveau (2016–2017).” *Revue européenne des sciences sociales* 60.2 (2022): 95–127. doi: https://doi.org/10.4000/ress.9144.

McRae, Murdo William, ed. *The Literature of Science: Perspectives on Popular Scientific Writing*. Athens, GA.: University of Georgia Press, 1993.

Ministère de la Culture et de la Communication. Département des Études, de la Prospective et des Statistiques. *Les pratiques culturelles des Français: Enquêtes 1973, 1981, 1988, 1997, 2008*. http://www.pratiquesculturelles.culture.gouv.fr (1 October 2018).

Mitra, Partha. “The Brain’s Highways: Mapping the Last Frontier.” *Scientific American*. 22 May 2012. https://www.scientificamerican.com/article/brains-highways-mapping-last-frontier (7 October 2016).

Morin, Edgar, and Massimo Piattelli-Palmarini, eds. *L’Unité de l’homme*. Paris: Seuil, 1974.

Mortureux, Marie-Françoise. “Paraphrase et métalangage dans le dialogue de vulgarisation.” *Linguistique française* 53 (février 1982): 48–61.

Musolff, Andreas. “Popular Science Concepts and Their Use in Creative Metaphors in Media Discourse.” *Metaphorik.de* 13 (2007): 67–86. https://www.metaphorik.de/de/journal/13/popular-science-concepts-and-their-use-creative-metaphors-media-discourse.html (22 August 2021).

Neisser, Ulric. *Cognition and Reality*. San Francisco: W. H. Freeman, 1976.

Pesic, Peter. *Labyrinth: A Search for the Hidden Meaning of Science*. Cambridge, MA: MIT Press, 2000.

Petrie, Hugh S., and Rebecca S. Oshlag. “Metaphor and Learning.” *Metaphor and Thought*. 2nd edition. Ed. Andrew Ortony. Cambridge: Cambridge University Press, 1993. 579–609.

Ricœur, Paul. *La métaphore vive*. Paris: Seuil, 1997.

Rose, Nikolas S., and Joelle M. Abi-Rached. *Neuro: The New Brain Sciences and the Management of the Mind*. Princeton: Princeton University Press, 2013.

Schlanger, Judith. *Les métaphores de l’organisme*. Paris: Vrin, 1971.

Shuttleworth, Mark. *Studying Metaphor in Translation: An Inquiry into Cross-Lingual Translation Practices*. New York and London: Routledge, 2017.

Société des Neurosciences. *Semaine du Cerveau 14–20 mars 2016: Rapport d’activités*. Paris, 2016.

Sperber, Dan. *La contagion des idées: Théorie naturaliste de la culture*. Paris: Odile Jacob, 1996.

Tristani-Potteaux, Françoise. *Les journalistes scientifiques, médiateurs des savoirs*. Paris: Economica, 1997.

Wedeen, Van J., Douglas L. Rosene, Ruopeng Wang, Guangping Dai, Farzad Mortazavi, Patric Hagmann, et al. “The Geometric Structure of the Brain Fiber Pathways.” *Science* 335.6076 (2012): 1628–1634.

Whitley, Richard. “Knowledge Producers and Knowledge Acquirers: Popularization as a Relation between Scientific Fields and their Publics.” *Expository Science: Forms and Functions of Popularisation*. Eds. Terry Shinn and Richard Whitley. Dordrecht et al.: Reidel, 1985. 3–28.

Monica Gonzalez-Marquez, Anouschka Foltz, Jeffrey K. Bye, Ali J. Fulsher, Matthias Wilde

What Little Red Riding Hood Can Teach Us about Reading Science

Abstract: Why is reading scientific articles so difficult? Many learners experience a deep-rooted anxiety when told they must read scientific texts of any type. When asked, they describe the process as overwhelming, frightening, and intimidating (Negrete and Lartigue 2004; 2010). Yet, when asked to read a novel, such feelings are unlikely to emerge. This is very plausibly because readers know what to expect: a progression of events leading to a conclusion of some type. In stark contrast, most learners never receive much training in how to read science, other than being told to interpret results as facts, not as part of a narrative. We argue that this assessment is incorrect. There is increasing scientific evidence (e.g., Ochs and Capps 1996; Turner 1998; Chow et al. 2014) that story is deeply entwined in our psyches, and that in fact, it may be a primary information-structuring mechanism. If narrative underlies general information structuring, then it should also underlie scientific literature: after all, science is a goal-directed process carried out by humans. As such, we propose that once students learn to read science as narrative, their comprehension will increase significantly. We present preliminary findings of an experiment testing the effectiveness of teaching to read science as narrative. We also discuss the implications of this framing for our understanding of scientific processes in general, in the context of the open science movement.

1 Introduction

Science is a human endeavor and a human endowment. It is knowledge created by humans, for humans. In order to participate fully in civic life, it is imperative that all citizens have a basic understanding of how science creates knowledge. That understanding comes from consuming scientific literature and media more broadly (e.g., documentaries, podcasts). But our understanding of science is only as good as how well we document and communicate it. Even a vague awareness of the replication/reproducibility crises (e.g., Open Science Collaboration 2015; Freedman et al. 2015; Baker 2016; Shrout and Rodgers 2018; but see Fanelli 2018) currently sweeping the sciences is enough to suggest that scientific literature is in big trouble. To best understand the issues, we should begin with first principles: What is the purpose of the scientific journal article?

https://doi.org/10.1515/9783110782844-010

The purpose of a scientific journal article (henceforth "paper") is to document a scientific process and its findings in such a way that they can be reproduced. (We are aware that there are legion reasons why an actual reproduction may be impossible. What is meant here is that readers should be able to accurately retrace the steps of the original researchers, if only as a mental exercise.) Attempts to reproduce studies, and difficulties conducting meta-analyses, have shown repeatedly that the paper in its current form, and as the principal source of information, fails to serve its purpose, despite being subjected to the peer review process (Lakens et al. 2017; Polanin et al. 2020; Errington et al. 2021; López-Nicolás et al. 2021). We argue that there are two reasons for this failure:

1. Learners are not taught systematically what science papers are vis-a-vis the science that was done, or how to use the information they contain to reconstruct scientific processes (even hypothetically).
2. Papers do not contain enough information to make this reconstruction possible.

Given these realities, it's not surprising that people struggle to read and understand science, nor that peer review, as the accepted assessment procedure, is so unreliable, as has been long and broadly acknowledged (e.g., Ernst et al. 1993; Rubin et al. 1993; Fox et al. 2017; Kelly et al. 2014; Vallée-Tourangeau et al. 2022). There is a substantial literature on differences between novice and expert readers, ranging from academic expertise differences (e.g., Afflerbach 1986; Shanahan et al. 2011) to reading development in children (e.g., Castles et al. 2018). We argue that this literature is not directly relevant here, given that regardless of the level of expertise, the current situation with scientific literature is that scientific processes are generally too incompletely documented to be well understood, as shown by the problems with peer review and meta-analyses (but see Shanahan et al. 2011 for their description of biases used by scientists when selecting and reading the literature). So how do we begin to address these problems? We argue here that the key is teaching *Science Reading*.

We introduce the term Science Reading[1] and define it as the act of conceptually reconstructing a scientific process from available documentation. It is a perspective as much as a process, beginning with a view of the scientific process as a sequence of humanly-orchestrated events, starting with a question and outlining the attempt to find an answer. The purported goal is for the reader to understand

1 The earliest known use of the term "Science Reading" is found in this book: A Guide to Science Reading. Compiled and edited by Hilary J. Deason. New American Library, New York, 1963. It is an annotated bibliography of general audience books on a broad spectrum of scientific topics. Its goal is scientific literacy, as described in our main text, and not the reconstruction of scientific processes, as we propose.

what the researchers did and why, well enough to accurately assess the reliability of the science that was done, and to be able to use all the information it generated.

Science Reading uses narrative (Fludernik 2009; Onega and García Landa 2014 [1996]), and specifically cognitive narratology (Jahn 1997; Turner 1998; Tucan 2013; Herman 2014), as the foundation of its approach. Its goal is to uncover and interpret the inherent narrative of the scientific process, as a means to understanding the scientific process itself, and specifically to understanding science as human experience (Herman 2014). We define narrative as "a discourse unit that includes events and one or more characters," where "our understanding of a given event often depends on other events in the narrative" (Willems 2020, 271). The main characters in science are researchers, who orchestrate and/or observe events that occur in a particular context and in a particular relative order. (Other characters may include participants.) Science is a quest for meaning and sense, and the scientific method centers around the outcomes of human actions: humans conducting an experiment, making observations, and interpreting these observations. As such, the scientific method involves what Bruner (2020 [1986]) calls "narrative thinking," where we construct "plots with motivated characters and inter-woven events," and are "guided by a quest for meaning and sense, attempting to view events as outcomes, intended or unintended, of human actions, emotions and intentions" (Gabriel 2013, 105). Scientists *do* science (Chalmers 2013; Dunbar 1999), and as such, science has motivated characters and inter-woven events, and is best conceptualized as a narrative.

It should be clear that Science Reading is a specialized skill. It may be related to but is distinct from science literacy in that Science Reading focuses on the conceptual reconstruction of a specific scientific process, while science literacy encompasses understanding science in general terms with a focus on scientific outputs, e.g. the atomic structure of carbon. This also differs from science communication (e.g., Burns et al. 2003; Martinez-Conde and Macknik 2017), which uses storytelling superimposed as a secondary narrative for a non-scientific audience, rather than the *original* narrative of the science itself.

We argue that training students in Science Reading can provide them with a means to better understand the science that was done as well as possible given the available documentation, and a means to identify what relevant information is missing. For researchers, it is a tool to salvage what we can from the extant scientific literature, as well as a framework for the documentation we need to create a reliable scientific knowledge base. In other words, it can help us understand the knowledge we have, with all its limitations, and how to make any knowledge we create in the future, better.

2 Why do we need to read science well?

Reading is the cornerstone of education. It is an old, apt adage that first we learn to read, and then we read to learn. Science Reading, the term we have coined for this process, is specifically concerned with accurately constructing the narrative of how the scientific research was carried out (its characters, events, and motivations) from how it is presented and organized in the article and any other available documentation, i.e., data, code, etc.

Unfortunately, most students are never explicitly taught how to read the scientific material they use to build their own scientific understanding or produce their own scientific writing. We surveyed 125 university students in Germany and the Netherlands on whether they had been trained before in how to read science papers (Gonzalez-Marquez et al. 2018); as shown in Fig. 1, most never received any training. The students were surveyed as part of invited training sessions. (To clarify, when university instructors heard through word of mouth that we were conducting these studies, we received several invitations to come teach students how to read science papers. Much of our data, including the data set presented here, were gathered in this way.)

Such difficulties in reading science hinder science learning, and in the long term, may dissuade students from pursuing science as a career or having science self-efficacy (Chen and Usher 2013) – to say nothing of the need for knowledge producers themselves to be able to navigate scientific knowledge. In fact, the lack of training becomes particularly problematic among researchers in that, given the intense competition (Carson et al. 2013) and the high prevalence of impostor syndrome (Jaremka et al. 2020), researchers are likely disinclined to acknowledge their lack of ability, instead assuming it to be a personal failing that must be kept hidden. Fig. 2 supports this claim by presenting data from an anonymous Twitter poll suggesting that the paucity in training persists into professional science. The poll was run in 2018 while MGM served as a visiting Twitter curator for "Science for Progress" (https://www.scienceforprogress.eu), a science advocacy organization. Although the poll was anonymous, almost all retweets were made by professional scientists, as shown by their Twitter profiles, thus suggesting that a substantial proportion of the discussion participants were also professional scientists.

Scientists may object that, since disciplines vary widely in methods and scope, each field has a customized way to read science and to teach how to read it. Of course, different fields will have their own idiosyncrasies. However, we argue that 1) *all* scientists and science students would benefit from a *general* understanding of how to read science, 2) there is little evidence of substantive attempts to teach Science Reading, even in specialized fields, and 3) even when attempts are made to address the need for training, the existing approaches to such

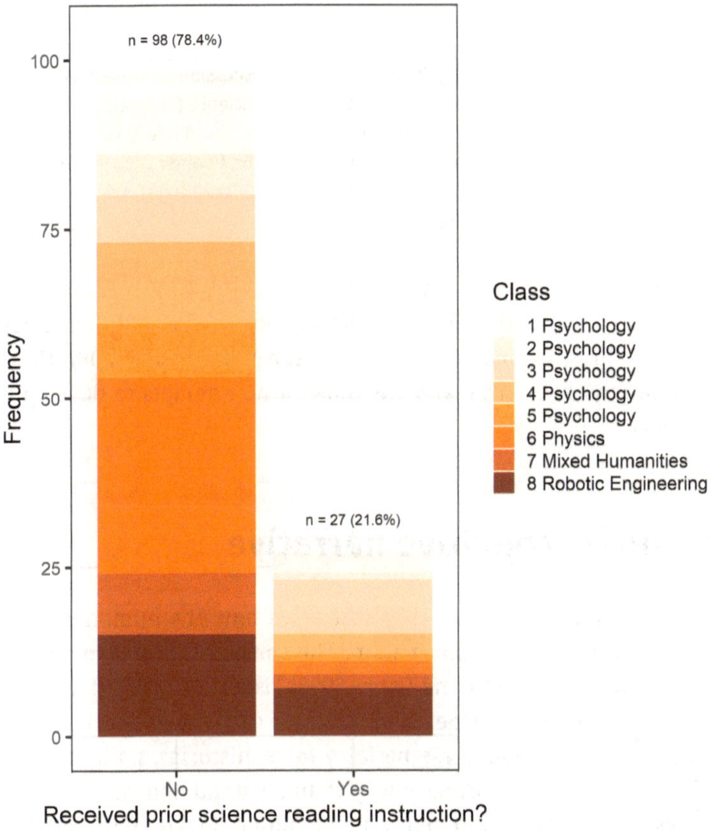

Fig. 1: Science Reading instruction self-report, data from eight training sessions in different disciplines conducted at Dutch and German universities, where the narrative method was taught.

teaching fail to acknowledge the centrality of science as a *process*, instantiated by discrete events, conducted by humans with individual and collective goals, and hence as having an inherent narrative quality. Scientists make decisions about how to address the questions they ask. The things they do are *events* and *plot points* that are interconnected; they motivate each other as part of the narrative constructed to arrive at an answer to the original question, becoming the narrative's conclusion.

In the present chapter, we are specifically concerned with assessing how to teach university students to read science. Although there have been various attempts to teach Science Reading in the past (see Glynn and Muth 1994 for an overview of long-standing general approaches), few have taken advantage of narrative as a natural, intuitive information structuring mechanism, and none have applied this framework to the dry writing typically found in scientific research

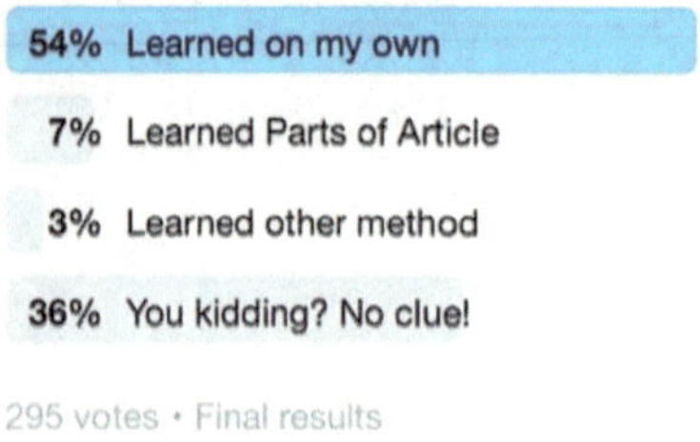

Fig. 2: Informal poll of discussion participants during a visiting curation of the Science for Progress Twitter account. Participants were asked how they learned to read a journal article. https://twitter.com/SfPRocur/status/1039841593140948993?s=20&t=jAT-msvRLeThUYVpZs0B7A.

articles. Note, however, that there is an extensive literature on the narratives found in science, with concentrations in semiotics and in science and technology studies (e.g. Croissant 2003; Herman 2003; Lowrey and Venkatesan 2008). However, this work is largely theoretical, and we know of no attempts to develop it into pedagogical tools.

3 Background on cognitive narrative

There is a growing scientific literature that addresses how the human species comes to structure and understand information. The consensus seems to be that we are a species of storytellers (Ochs and Capps 1996; Fisher 2021 [1987]). Human "existence is inherently storied" (Kearney 2002, 130). We exchange information by telling each other stories, whether these be fairy tales, histories, political platforms or scientific reports. As our primary way to understand and structure the lived world, stories are essential to human understanding (Lewis 2007; van den Broek 1997). Further, evidence from cognitive psychology and neuroscience indicates that we are wired for narrative: Narratives "tap into the brain's native machinery for representing the world;" they are "the human brain's way of consolidating and conveying the temporally-evolving world we live in" (Willems 2020, 271; see also Bruner 2020 [1986]). Thus, stories are essential for our conceptualizations of events. Science is no exception here. In fact, narrative structure is essential for understanding the highly complex data and situations that we typically find in science because it "provides a scaffold for representing complex relations among agents" (Willems 2020, 271). Tapping into science's inherent narrative lowers (unnecessary) cognitive load (Willems 2020) and allows us to understand complex relationships without dumbing down the science. In short, if you want someone to understand something, tell them a story.

Using narrative in science is not a new idea. Teachers may take scientific literature and transform it into a story, which they then use to teach their students content knowledge. This approach is immensely valuable for introducing students to science. A story about the spread of polio during the early twentieth century, and how Jonas Salk worked tirelessly to develop a vaccine, is undoubtedly fascinating reading.

However, there are several problems with this approach when it is applied to teaching Science Reading. First, by receiving this information from a teacher, who is an authority figure, students are less prone to question their teacher's interpretation, thus taking away a chance for students to sharpen their critical thinking skills. After all, the goal of Science Reading is for the reader to be able to critically assess scientific documentation on their own, not to be told what to think. Second, and most importantly, this method fosters a model where students are passive, instead of teaching them how to critically engage in the knowledge discovery process. What scientifically literate people need are the skills to construct the story themselves. And third, introducing stories only as a means to add interest can even interfere with learning of the scientific content (Harp and Mayer 1998).

Narrative should undoubtedly be used in the science classroom, but not merely as packaging provided by teachers for a difficult subject. Instead, it should be used as the natural scaffolding provided by human cognition to bring structure to challenging information. The study we describe below is a partial and hence, preliminary analysis of our attempt to develop evidence-based pedagogy using cognitive narrative to teach Science Reading.

4 The current study

We tested the efficacy of a narrative teaching method for Science Reading in a three-condition, between-subjects pilot experiment with bilingual students at a Dutch university (N=77). In the current study, we focus on the results from the post-test and on changes from pre- to post-test to answer the following research questions:

RQ1: How does teaching narrative vs. teaching article structure vs. a baseline of no instruction affect students' ability to determine what research question a scientific article is asking?

RQ2: How does teaching narrative vs. teaching article structure vs. a baseline of no instruction affect students' ability to determine how a research question was asked and what the answer to the question was?

RQ3: When answering these questions about a research article, to what extent do students copy words from the article itself vs. use their own wording and interpretation (as a function of the teaching they received)?

General Procedure for all conditions, and Treatments (below): Source of original research articles: Psychological Science - Short Reports ◦ Approximately 90 minutes. ◦ Pep talk ◦ Read pre-test article ◦ Answer comprehension questions ◦ Treatment (see below) ◦ Read post-test article ◦ Answer comprehension questions		
Narrative ◦ Identify narrative in: ◦ Images ◦ Social contexts ◦ Objects ◦ Deconstruct Little Red Riding Hood ◦ Deconstruct article as a narrative using structure insights from the fairy tale (focus on events, universal interrogatives)	**Parts of the Article (POA)** Review per APA Style Manual (most common method taught) ◦ Title ◦ Authors and affiliations ◦ Place of publication ◦ Year of publication ◦ Abstract ◦ Introduction ◦ Methods (participants, apparatus, procedure) ◦ Results ◦ Discussion References	**Doing Nothing** (Participants received no instruction, common for the vast majority of students) • Read treatment article • Answer comprehension questions

Fig. 3: Schematic of the experimental procedure, with general outline for all conditions (top) and the unique components for each of the three conditions (bottom).

4.1 Methods

4.1.1 Participants

A total of 77 students were recruited from Vrije Universiteit Amsterdam. All were native Dutch-speaking, although fluent in English, per university requirements for participation in coursework for a B.A. The students participated in the study as part of a regular university class. All classes took place within a 2-week period at the end of the normal semester, were taught by different instructors and covered coursework in either computer science, marketing or the humanities. Each class was randomly assigned to a condition, such that each participant was in one of three treatment groups: 25 participants (22 women, 3 men; *Mean* age = 23.2 years; *SD* = 3.0) formed the narrative group, 29 participants (9 women, 20 men; *Mean* age = 20.4 years; *SD* = 2.2) were in the article structure group, and 23 participants (1 woman, 22 men; *Mean* age = 20.3 years; *SD* = 2.0) were in the no instruction group, used as a baseline comparison.

4.1.2 Materials

Three short scientific articles were taken from *Psychological Science*'s short reports for this study. This paper format, now defunct, included papers that were shorter than 4 printed pages in length. For this study, only papers of no more than 2 printed pages were considered so that they could accommodate the limited time frame. The paper topics were chosen to be as interesting and as accessible as possible to readers with limited experience with the literature in experimental psychology. This was done to encourage optimal learning conditions. One article reported on a study about how pain can reduce guilt (Bastian et al. 2011), another focused on how prior beliefs affect people's evaluation of scientific evidence (Bastardi et al. 2011), and the last article covered a study on mental time travel (Miles et al. 2010). The Results sections of each article were removed to avoid inducing anxiety in participants when expected to interpret statistics; the Discussion sections were retained, allowing readers to still receive the primary results themselves along with the authors' interpretations. The remaining text from each article was copied into a word processing document so that it fit onto one A4-sized page.

The same comprehension questions, given in (1), were used for all three articles. The results we present here focus on questions (1a) and (1b), as questions (1c) and (1d) are beyond the scope of this chapter.

(1) Comprehension questions used in the current study:
   a) What is the question the article is asking?

b) How was the question asked and what was the answer?
c) Why is the finding important?
d) What are some possible weaknesses of the study?

4.1.3 Procedure

Each experimental session took approximately 90 minutes, a typical class period. The sequence of events for each session occurred as follows (see Fig. 3). The class instructor introduced the instructional researcher as a guest who would be leading the class for that day. The guest instructor then distributed consent forms and informed the students that participation in the study was voluntary. However, since the material of the class session was designated as course material by the normal instructor, this meant that although the students could choose not to participate in the study, their non-participation was limited to withholding their data. They were otherwise expected to remain and participate in the training session as part of their typical class participation. The study was exempt from institutional ethics approval because the session took place as part of a normal class and involved pedagogical material relevant to the normal coursework. No student chose to withhold their data. Upon completion of the consent form, the students were given a packet containing basic demographics questions as well as the readings and reading comprehension question sets. They received no financial compensation for participation.

The instructional researcher was a native English speaker and one of the authors, who led the class sessions using an active teaching methodology developed specifically for this study. This method is based on the Socratic method. Briefly, the Socratic method involves asking students probing questions to allow for self-discovery. This method is very effective in this setting given that training progresses by asking the students pointed questions instead of simply providing them with information. The guest instructor began the class with brief information about what students would be doing during the class session. Students then read a pre-test article and answered the comprehension questions about the pre-test article.

For the narrative treatment group, the general goal was to first teach the participants to recognize the ubiquity of narrative structuring in both their physical and social environments. The premise was that the greater their ease in "seeing" narrative everywhere, the more easily they would be able to recognize the narrative inherent in papers, and thus to use it to understand the science the paper described. The procedure was as follows: Participants were first shown a black and white image and asked to say what they saw (see Fig. 4 below).

Fig. 4: Image used to introduce participants in the narrative treatment to the ubiquity of narrative in human cognition. By Susan Barstow. https://www.summitpost.org/disappearing-rabbit-trick/185785.

Participants were guided, using the Socratic method, to describe the events implicit in the image, i.e., a bird of prey capturing its meal. Then they were informed that in fact all they saw was a collection of black and white pixels, and that the events they had described were the direct product of their mind's propensity to understand the world in narratives. Participants were then asked to identify other places in their lives where narrative also provided the guiding structure and to describe how they were narratives. Participants were given an example if they were unable to produce one, in this case job interviews, after which they were able to offer their own, i.e., catching up with friends, gossip, newspaper articles, etc.

The next step was to deconstruct a fairy tale of their choice. Little Red Riding Hood was the most typical choice. Deconstruction involved first having a participant recount the fairy tale. This was followed by using the universal interrogatives, who, what, where, why, when, how, so named because they appear in some form in all human languages, to make sure the story was complete. Once those details were laid out, an arc was drawn on the board to signify a narrative arc, and the participants were asked to list the events produced by the information extracted using the universal interrogatives. These events were positioned along the arc in a progression from initial event to final event. For the highest point in the arc, participants were asked to identify the event or events that made the conclusion of the story possible. For Little Red Riding Hood, that would have been the little girl realizing the wolf was disguised as her grandmother and the appearance of the huntsman. Once participants were satisfied all key events were included in the arc, and understood how the events had led from the beginning to the end of the fairy tale, they moved on to reading an actual paper. Participants were given about 10 minutes to read the paper, after which they repeated the process undergone with the fairy tale, albeit this time using the information in the paper.

In the "article structure" group, the instructor reviewed the structure of a scientific research article with students, using the parts of a research article given in the APA Style Manual (APA 2020): title, authors and affiliations, place of publication, year of publication, abstract, introduction, methods (participants, apparatus/materials, procedure), results, discussion, references. The lesson involved a brief lecture describing the contents of each section. For example, the introduction was explained as the first step of the paper that is meant to set the stage for the study and that should describe the relevant literature as well as clearly state the research question, hypotheses and motivations, etc.

In the "no instruction" group, the instructor provided no Science Reading instruction and students used the time to read an article and practice comprehension questions for it. Following the teaching treatment, participants in all groups read a post-test article and answered the comprehension questions about the post-test article. Articles were counterbalanced across treatment groups in a Latin squares design, such that each group of participants received the articles in a different order (1–2–3, 2–3–1 or 3–1–2).

4.1.4 Data coding and analysis

The same researcher who taught the treatment groups also coded students' responses to questions (1a) and (1b). Responses contained neither any identifying information about students nor information about whether answers were from the pre- or post-test, so that the researcher was not aware as to which treatment group individual students' responses belonged to. The researcher used the three following independent coding criteria (Norris and Phillips 2003). They were evaluated using two different measures. The first indicated whether correctness/copying/interpretation had occurred. The second, measured on a Likert scale, indicated the extent to which copying or interpretation has occurred. Correctness was not assessed in this manner. An answer was coded as *correct* if the student's response accurately answered the question, regardless of whether the response was copied directly from the text or not. A correct answer was further coded as *copied* if a student's response contained at least three content words in a row that were copied directly from the text. Content words exclude articles, prepositions and the like. A correct answer was further coded as *interpreted* if a student's response contained at least three words in a row that were unique to the student, i.e., not found in the text. Samples of correct, copied and interpreted participant responses can be found in the tables below:

Tab. 1: Question 1 – "What was the question?" Correct responses.

Copied	How is temporal information processed when one revisits the past or anticipates the future.
Interpreted	The variation between desires and beliefs, whether they trump each other under various circumstances.

Tab. 2: Question 2 – "How was the question asked?" Correct responses.

Copied	Participants were given an outline of two fictional studies, one in favor of each care type. They were asked to evaluate the validity and outcome of the studies.
Interpreted	Participants were divided into two groups, conflicted and unconflicted. They had to evaluate two studies, to see if their beliefs changed.

The criteria "copied" and "interpreted" were used to discern different levels of understanding of the text. In the reading comprehension literature, the product of successful reading comprehension is a coherent mental representation of the text's meaning (van Dijk and Kintsch 1983). This mental representation is distinct from the exact wording of the text, which is the "most superficial and short-lived" representation of knowledge (van Dijk and Kintsch 1983; Fletcher 1994). Recalling an idea in the reader's own words indicates that they have constructed, and are using, this higher level mental representation of the knowledge to answer the question (Rapp et al. 2007). Copying indicates a shallower processing of the text, and therefore lower quality of learning (King 2007), or merely locating the information in the text that appears to answer the target question, rather than answering the question based on their understanding of the text.

Finally, a correct answer was coded for the extent to which it involved students' own interpretation (*interpreted score*) by using a 6-point Likert scale ranging from *0 = no interpretation (entire response is copied directly from the text)* to *5 = entire response consists of unique words*. Only the end points of the scale were marked, with the coder using approximate percentages, at 20 point increments, e.g., 20%, 40%, etc. We did not further code incorrect answers. We also clarify that participants were given no instruction as to whether copying was allowed or putting things in their own words was necessary. They were simply instructed to answer the questions to the best of their ability. Finally, the copied/interpreted criteria were not mutually exclusive. It was possible for a response to contain both copied and interpreted text.

We analyzed the data in R using binomial generalized linear models, which are appropriate for binary data, and cumulative link models, which are appropriate for ordered data. The independent variable in all models was treatment group. We used binomial generalized models to analyze correct, copied and interpreted responses from the post-test. We used cumulative link models to analyze correct, copied and interpreted responses from the difference in scores from the pre-test and the post-test. We also used cumulative link models to analyze the interpreted scored responses. We chose treatment coding for all analyses, with the no instruction group as the reference level. Since incorrect responses were not coded for being copied, interpreted or interpreted scored, these were treated as missing data points (NA) in these analyses. Data and analysis scripts are available on the Open Science Framework at https://osf.io/9cv7f/.

5 Results

5.1 Correct responses

We begin with RQ1 and RQ2 and explore how having been in one of three treatment groups affected students' abilities to determine what research question a scientific article is asking, how a research question was asked, and what the answer to the question was. We first look at the results from the post-test to see how well participants in the three groups performed after their instruction (or the extra practice in the case of the no instruction group). Figure 5 shows the proportion of correct responses for how students answered the questions *What is the question the article is asking?* (1a) and *How was the question asked and what was the answer?* (1b) during the post-test. Over 80% of students in all groups answered question (1a) correctly, with numerically the highest number of correct responses in the narrative group. The statistical results suggest that both the narrative group (β = 17.67; *SE* = 2150.80; z = 0.01; p = 0.993)[2] and the article structure group

2 For this particular comparison only, the analysis result is not reliable and needs to be treated with caution because the narrative group has only correct responses for (1a) and no incorrect responses. In such cases, the data lack the necessary variation for the model to estimate the probability of success (or failure). For all reported statistical results, β is the coefficient estimate, which specifies the average change of the response variable (in log odds) when the predictor variable increases by one unit. *SE* is the standard error, which provides information about the variability inherent in the coefficient estimate. The z value is the coefficient estimate divided by the standard error. The p value provides a probability for the z value. Here, we follow the convention that a p value that is smaller than 0.05 indicates a statistically significant effect.

($\beta = -0.33$; $SE = 0.79$; $z = -0.42$; $p = 0.678$) had similar amounts of correct answers compared with the no instruction group. A total of 76% of students in the narrative group answered question (1b) correctly, whereas only 30% in the no instruction group and 11% in the article structure group provided correct responses. The statistical analyses show that the proportion of correct responses is similar in the article structure group compared to the no instruction group ($\beta = -1.29$; $SE = 0.76$; $z = -1.70$; $p = 0.089$), but that the narrative group provided significantly more correct responses than the no instruction group ($\beta = 1.98$; $SE = 0.65$; $z = 3.04$; $p = 0.002$).

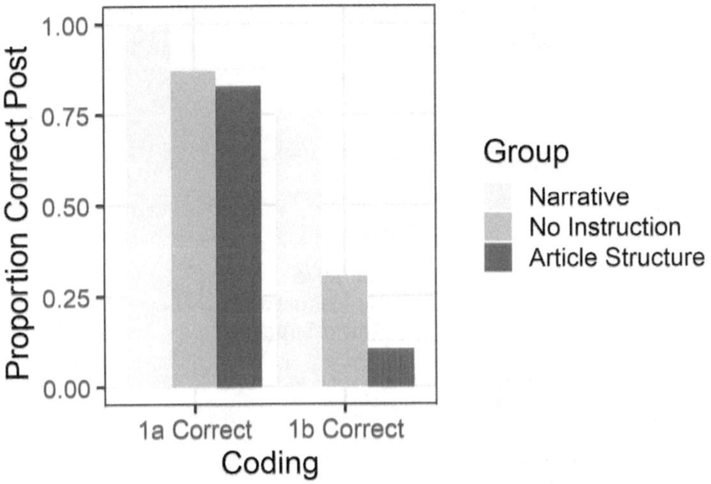

Fig. 5: Proportion of correct post-test responses to the questions *What is the question the article is asking?* (1a) and *How was the question asked and what was the answer?* (1b). Correct = answer was accurate.

We now turn to potential improvement from the pre-test to the post-test to assess whether performance in the post-test was actually improved or whether students in one of the groups simply performed better than students in other groups from the beginning. For that, we look at the difference in performance between the post-test and the pre-test for each treatment group. Figure 6 shows the difference in results for how students answered the questions *What is the question the article is asking?* (1a) and *How was the question asked and what was the answer?* (1b) during the post-test as compared to the pre-test. For question (1a), the figure shows an improvement in response accuracy of 20% in the narrative and 22% in the no instruction groups. In contrast, the percentage of accurate responses decreased by 7% in the article structure group. The cumulative link model analyses treating improvement as an ordinal variable reveal that the narrative group showed improvement similar to the no instruction group ($\beta = -0.14$; $SE = 0.67$; $z = -0.21$; $p = 0.833$), whereas the article structure group's improvement was significantly worse than that of the no instruction group ($\beta = -2.02$; $SE = 0.88$; $z = -2.29$; $p = 0.022$). For question (1b), the figure shows an improvement in correct responses of 58% in the narrative group and an improvement of 7% in the article structure and 4% in the no instruction groups. In line with this observation, the statistical analyses show that the narrative group showed significantly stronger improvement compared to the no instruction group ($\beta = 2.40$; $SE = 0.71$; $z = 3.40$; $p < 0.001$), whereas the article structure group's improvement was similar to that of the no instruction group ($\beta = 0.15$; $SE = 0.69$; $z = 0.22$; $p = 0.828$).

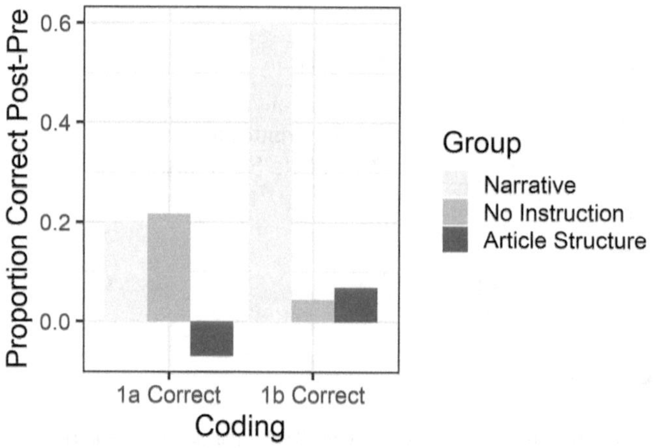

Fig. 6: Difference in proportion of correct responses to the questions *What is the question the article is asking?* (1a) and *How was the question asked and what was the answer?* (1b) between the post-test and pre-test. Correct = answer was accurate.

5.2 Copied, interpreted and interpreted scored responses

We now move on to RQ3 and explore how having been in one of three treatment groups affected to what extent students copied words from the research articles in question or used their own wording and interpretation when answering questions about the research article. This analysis only considers correct responses. Since there were not enough correct responses in the article structure and no instruction groups for question (1b), we focus our analysis on question (1a) only. Again, we first look at the results from the post-test. Figure 7 shows whether and to what extent students copied from the text (copied) or used their own words (interpreted and interpreted scored) when answering the question *What is the question the article is asking?* during the post-test. The figure shows that in each group between 60% and 80% of students copied at least three words (1a copied) in a row directly from the research article, with numerically the highest number of students copying words directly from the text in the article structure group. The statistical results suggest that the amount of copying compared to the no instruction group is similar for both the narrative group (β = 0.05; *SE* = 0.70; *z* = 0.08; *p* = 0.938) and the article structure group (β = 2.04; *SE* = 1.14; *z* = 1.78; *p* = 0.075). The figure also shows that the proportion of students who had at least three words that did not occur in the text in a row in their response (1a interpreted) varied widely across treatment conditions. Specifically, while only about 10% of students taught article structure used their own words in their answer, about 30% of students without intervention and 60% of students in the narrative group did so. The statistical results suggest that the no instruction group did not differ statistically significantly from the article structure group (β = −1.33; *SE* = 0.77; *z* = −1.71; *p* = 0.087) or the narrative group (β = 1.02; *SE* = 0.62; *z* = 1.65; *p* = 0.099). Finally, the figure shows to what extent, as measured on a Likert scale from 0 to 5, students copy words from the research articles in question or use their own wording and interpretation when answering question (1a) during the post-test (1a interpreted scored). The average score across all groups was below 2 in the post-test, suggesting that all groups used their own words to a limited extent even after treatment. In line with this, the statistical analyses suggest that both the narrative (β = 0.62; *SE* = 0.59; *z* = 1.06; *p* = 0.290) and the article structure (β = −1.46; *SE* = 0.77; *z* = −1.89; *p* = 0.059) groups had similar amounts of own wordings compared to the no instruction group. This suggests that while the narrative group performed best numerically in terms of interpreting the article and using their own words to talk about it in the post-test, this difference was not statistically significant.

We now turn to potential improvement from the pre-test to the post-test by looking at the difference in performance between the post-test and the pre-test for each treatment group. Figure 8 shows the difference in results for how students answered the question *What is the question the article is asking?* The figure shows that copying from the text (copied) slightly increased in the no instruction and article structure groups. The statistical analyses reveal that the change in copying from pre- to post-test was similar for both the narrative group (β = −1.01; *SE* = 0.85; *z* = −1.19; *p* = 0.234) and the article structure group (β = −0.14; *SE* = 0.79; *z* = −0.18; *p* = 0.861) as compared to the no instruction group. The figure also shows that students using their own words (interpreted) increased in the narrative group, but decreased in the article structure group, and especially in the no

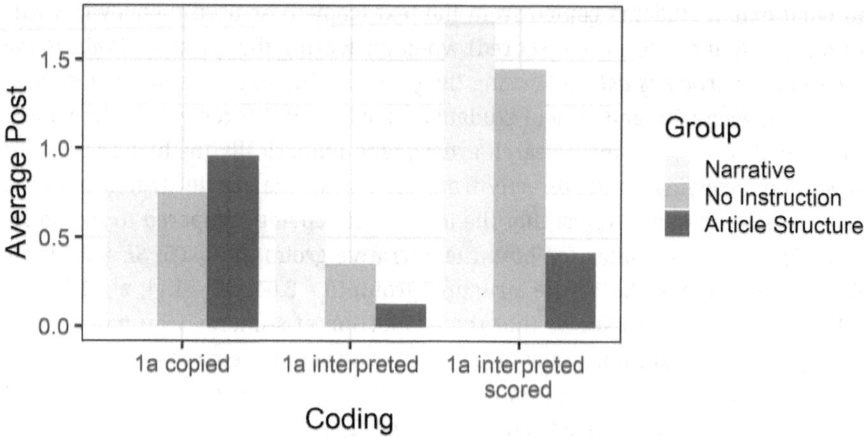

Fig. 7: Coding of correct post-test answers to the question *What is the question the article is asking?* (1a). Copied = at least three words in a row copied (binary); interpreted = at least three words in a row novel (binary); interpreted scored = Likert scale with 0 = *no interpretation (entire response is copied directly from the text)* to 5 = *entire response consists of unique words.*

instruction group. The statistical analyses confirm this pattern: Compared to the no instruction group, both the narrative group ($\beta = 3.76$; $SE = 0.93$; $z = 4.07$; $p < 0.001$) and the article structure group ($\beta = 1.79$; $SE = 0.74$; $z = 2.44$; $p = 0.015$) showed significantly more improvement from pre- to post-test. Overall, this suggests that the narrative intervention (as the only intervention that showed an increase in interpreted answers from pre- to post-test) seems to have a positive effect on students' ability to interpret the text and use their own words to talk about it. In contrast, providing no instruction and teaching article structure seem to have negative effects on students' propensity to interpret the text and use their own words to talk about it during the post-test as compared to the pre-test. Finally, the figure shows to what extent the propensity of students' copying words vs. using their own wording and interpretation when answering question (1a) changes from pre- to post-test (1a interpreted scored). The narrative group showed an improvement in scores from pre-test to post-test, but scores declined for both the article structure and the no instruction groups. The statistical analyses confirm this pattern and suggest that improvement in scores was significantly larger for the narrative group compared to the no instruction group ($\beta = 1.95$; $SE = 0.72$; $z = 2.72$; $p = 0.007$), but that both the article structure and no instruction groups showed a similar decline in scores from pre- to post-test ($\beta = 0.61$; $SE = 0.64$; $z = 0.95$; $p = 0.342$).

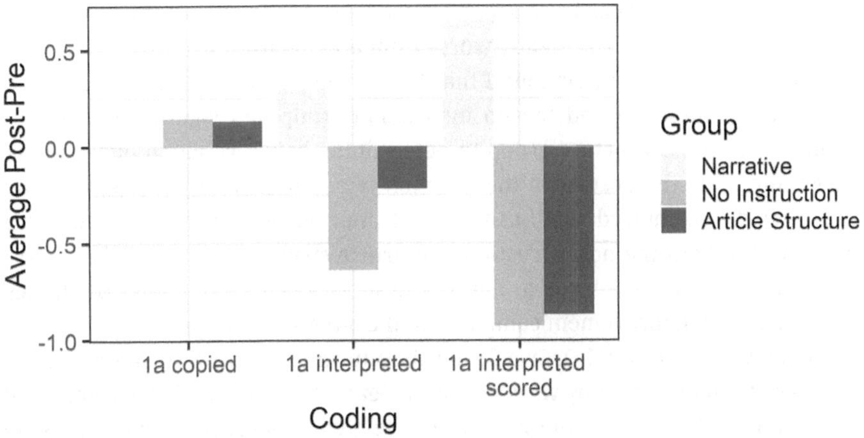

Fig. 8: Coding of correct answers to the question *What is the question the article is asking?* (1a) Difference between answer scores in the post-test and pre-test. Copied = at least three words in a row copied (binary); interpreted = at least three words in a row novel (binary); interpreted scored = Likert scale with 0 = *no interpretation (entire response is copied directly from the text)* to 5 = *entire response consists of unique words.*

6 Discussion

We assessed students' ability to understand scientific articles across three conditions: the narrative group where students were taught to identify narratives across various situations, the article structure group where students learned the structure of a scientific research article, and a group who received no additional instruction but did receive extra practice. We hypothesized that due to the importance of cognitive narrative in structuring human understanding, the narrative group would perform better than the article structure group in interpreting and describing the researchers' questions and answers. The no instruction group was included as a baseline comparison to the two instructional groups.

With respect to RQ1, at post-test we found that all conditions were largely successful in correctly determining what research question a scientific article was asking, with the narrative group numerically outperforming the other two groups. The data also suggest that the narrative and no instruction groups showed similar marked improvement from pre- to post-test. This suggests that teaching narrative does affect students' ability to determine what research question a scientific article is asking, but that this improvement is similar to the baseline. In contrast, teaching article structure did not affect students' ability to determine what research question a scientific article is asking. In fact, students in the article structure group got slightly worse from pre- to post-test.

With respect to RQ2, we found that during the post-test, the narrative group numerically outperformed the no instruction group and significantly outperformed the article structure group in determining how a research question was asked and what the answer to the question was. The data also suggest that the narrative group showed significantly more improvement from pre- to post-test compared to both the no instruction and article structure groups. This suggests that teaching narrative leads to the best post-test results and shows the largest pre- to post-test improvement compared to the other two groups.

Finally, considering RQ3, we found that during the post-test, students in all three groups tended to copy words from the reading more than they paraphrased. The amount of copying from pre- to post-test was relatively stable across the three groups, with slightly more copying from pre- to post-test in the no instruction and article structure groups and a very slight decline in copying in the narrative group. These slight differences were not statistically significant. However, differences emerged in terms of students using their own wording. Here the narrative group outperformed the other two groups during the post-test and in the pre- to post-test comparison. Overall, the narrative group performed numerically best (i.e., copied less and used their own wording more) out of the three conditions, but the post-test comparisons were not statistically significant.

Thus, we found that overall, participants in the narrative condition performed best on the post-test and had the largest pre-post gains, both in terms of accuracy and paraphrasing of their responses. This suggests that narrative structuring measurably improves comprehension of papers. We hypothesize that this improvement stems from the reframing of scientific research as inherently narrative, which lends a more intuitive genre/schema to structure readers' understanding of a paper.

We also found that the article structure condition performed somewhat worse than the no instruction group, which we did not predict. It is possible that the reason the "parts of the article" condition appears counterproductive to comprehension is because it focuses on individual sections of an article, instead of on how those sections work together to provide a cohesive message. On the other hand, the slight decrease could have been because they performed numerically best during pre-test. Future research should examine reading instruction that emphasizes how the parts of an article are interrelated.

Regardless, these results offer a cautionary tale on implementing pedagogical methods simply because they seem like common sense. Though most learners never receive formal science reading instruction, if they do receive training, it is often in the article structure form. As suggested here, however, this form of instruction does not necessarily improve learning, and may in fact interfere with it. This finding thus reinforces the need for all pedagogical methods to be tested for efficacy – and not just assumed to be better than nothing.

These results also support the need for universities to allocate resources to teaching how to read science. At present, most universities have some version of a writing center. This is good, but their presence does not seem to have alleviated any of the problems in scientific documentation. Teaching to write well-formed sentences is not the same as documenting comprehensible and usable science. These results also underline how important the humanities are to helping create a reliable scientific knowledge base by showing how acknowledging narrative structure leads to better comprehension, and can potentially lead to better documentation standards for science.

Beyond the immediate concerns of the study are implications for the reconceptualization of science as narrative – a reconceptualization that should actually be considered a repeat conceptualization, given that papers were originally designed specifically to describe scientific processes as narratives so that they could be better understood by readers. This structure, called IMRAD for "Introduction, Methods, Results and Discussion" became the accepted format by the 1970s (Sollaci and Pereira 2004). However, with the subsequent shift in academia to publication and grants as the currency for career advancement, the motivations for the use of IMRAD were lost. Current practice has instead become to write sections separately as quickly as possible to satisfy journal requirements for brevity and novelty. A return to the conceptualization of science as narrative allows us to focus on the process of doing science instead of on its outputs. To quote Carl Sagan, "The method of science, as stodgy and grumpy as it may seem, is far more important than the findings of science" (1996, 22). Narrative redirects attention to methods, because those are the events comprising the scientific process, and away from science as bean-counting, i.e. "Here's a shiny new finding, oh look! There's another one over there!" etc.

Science as narrative also has deep implications for Open Science and the Science Reform movements. As we noted above, a persistent problem with papers in their current form is that they do not contain enough information to allow evaluation of the processes they purport to document. Much effort and funding has gone to remedying the problem largely by focusing on making data and analysis code available along with papers. Although this step is clearly needed to ensure more complete documentation, neither movement has stopped to ask exactly what information readers need to truly understand a scientific process, i.e., what information is needed to make Science Reading a success. Narrative, by its very nature as an information structuring mechanism for human-like cognition, provides the needed framework and toolset to understand what information is needed. It is key to understanding the unknown unknowns in the documentation process. By examining a scientific process as a sequence of events orchestrated to ask a question, and then asking what is needed to reproduce those events, we arrive at concrete information about what needs to be documented – specifically, what information is needed to reproduce each of the events comprising a scientific process. Analyzing those events leads us to see the labor involved in doing science, to a deep understanding of science as a verb, as something people do, and to understanding scientific documentation as records of things people did to answer questions.

Seeing science in these terms raises issues in many areas, though here we will briefly describe only three: Infrastructure, attribution, and scientific credibility.

The current infrastructure revolves around papers, and more recently, also the archiving of data sets and analysis code (e.g., Leeper 2014; Roche et al. 2015). As we noted above, this information is insufficient for successful Science Reading. Conceptualizing science as narrative, and using the tools narrative provides to understand the event structure of scientific processes, can provide the framework we need to understand what infrastructure is required to document science for successful peer review, replication and use into posterity. Since the goal is understanding each scientific process, and the paper is too constrained to provide the needed information, it then follows that additional forms should be developed. For example, the paper can be conceived of as a type of extended summary with pointers to extended documentation. Methods processes can generate their own output and contain all the required detail, unhindered by space constraints. The same can be done for other parts of the scientific process, i.e. the literature review can be expanded, and now also contain information about all the literature examined, as well as justifications for what was used and what was discarded. This last point would particularly address the systematic exclusion of literature from the Global South by the Global North, for example.

This last point raises the issue of attribution. The CRediT Taxonomy (Allen et al. 2014) was introduced as a means to better acknowledge the significant invisible labor involved in doing science. Although authors are the only people formally acknowledged, the unspoken truth is that typically many more people than are named are involved in making any given scientific process possible. The CRediT Taxonomy is an admirable attempt, although, as with archiving infrastructure, it suffers from a lack of a framework to understand what needs to be attributed. And the same as with infrastructure, narrative can serve that purpose. By helping identify events, a narrative lens can also identify the labor that was done to make that scientific process possible, and specifically, who did that labor.

Finally, one of the challenges of our day is the growing mistrust of science (e.g., Rosman et al. 2022; Erduran 2022). In public discourse, science has been construed as collections of objective facts about the world. The reality of science is that it is an activity engaged in by mere mortals, with all of the weaknesses, fallacies and biases of mere mortals. Understanding science as narrative opens the door to acknowledging our humanity as scientists. This is key to conceptualizing that neither the events, nor their documentation, nor their interpretation can ever be objective because they are all constrained by the inherent biases involved with being human, the same as with writing histories and the logic behind the adage that "History is written by the victors." This humanity, warts and all, is not a bug in the system, but a feature of being humans who do science. Does this mean that the products of this process cannot be trusted? No. It means they must be understood as products of rational minds in the context of specific arguments in specific contexts. This view stands in opposition to the current framing of scientific literacy. Despite not providing the security of science as an immovable rock, the strength of this view lies in showing science as it is, without delusions, while also providing the means for anyone to understand how it creates knowledge. Some readers will note we make no claims about scientific objectivity. We argue that before we can claim objectivity of any type, first we must ensure that science is documented well enough to allow for rigorous evaluation, and consequently for use. Second, we believe the objectivity/relativity dichotomy is a distraction. The real question is whether the products of science are useful. Do they help us navigate the world effectively? We went to the moon with Newtonian physics, not modern physics, because it was most useful to the situation. How objective it was or not was irrelevant. What mattered was that it provided a way to understand physical forces such that we could get a rocket up off the ground and onto the moon. For science, that is what matters, and that is the guiding principle behind this work: is the information provided complete enough to be useful in furthering our ability to navigate the world? That alone renders discussions about objectivity moot. We can have trust in knowledge for its utility, without needing to conflate trust with objectivity (Bruner 1991; Hawking and Mlodinow 2010; Hoffman 2019).

6.1 Limitations and future research

The current study is exploratory and therefore presents quite preliminary and tentative results. In this section, we will discuss limitations of the current work with a view to needed future research. First, only one instructor taught all the students. It is thus not clear whether the effects found are due to this particular instructor's teaching and potential prior beliefs about teaching narrative vs. teaching article structure or whether the effects generalize to different instructors. Furthermore, future research should try to replicate the current results with native speakers of English, the language of the articles, to rule out language-related confounds. That said, English is currently the language of science, whereas globally most researchers and students reading research articles are non-native English speakers. As such, identifying the methods that help non-native English-speaking students and researchers to understand English-language scientific literature is important in itself and our results from non-native English speakers have value independently of whether the results would replicate in native English speakers.

The same researcher who taught the classes also coded the data from the current study. While the researcher was unaware as to which treatment group individual participants belonged and it is therefore unlikely that the coding was biased by treatment group, the coder was nevertheless heavily involved in the data collection. Therefore, data from a second independent coder without any knowledge of the study design is needed to check inter-coder reliability. Furthermore, the coding itself was qualitative, and bias is therefore possible. For example, responses from participants with overall better English writing skills may have inadvertently been coded more positively than responses from participants with overall weaker English writing skills. The use of software to limit access to participants' demographic information, as well as to prior responses, to code this kind of data is therefore desirable. To this end, we have developed software, MsquaredC (Bergman et al. 2022) to create a substantially more unbiased qualitative data evaluation environment. The data set presented here, as well as all additional data sets and future data sets, will be assessed using this software.

Furthermore, we decided to remove the results section from the research articles that our study participants read. In our experience, the results section is typically the least accessible to students, given the predominance of statistical notation and jargon, and the comprehension questions used in the current study could be answered without access to the results, as they were more clearly summarized in the discussion section. That said, we propose there is a narrative inherent to statistical information presented in the results section. However, it will require its own pedagogy, and as such was beyond the scope of this study. However, an important part of reading and understanding scientific articles, especially as students advance, is to understand the connection between the actual results and how the authors of an article interpret those results. In other words, students should be able to understand the results in the context in which they were created and assess for themselves to what extent the authors' interpretations of their data and the conclusions they draw are warranted. We have ignored this important aspect of understanding scientific articles in the current study, and future studies should explore how narrative teaching vs. teaching article parts may affect students' comprehension of results sections.

Finally, the three groups differed substantially in which articles they received as pre- and post-test articles, as well as in their gender distributions. We selected the articles to be similarly difficult, and there is some evidence we did this successfully: the percentage of combined correct responses to questions (1a) and (1b) during the pre-test ranged from 45.6% for the article on how prior beliefs affect people's evaluation of scientific evidence to 47.4% for the article on mental time travel. These pre-test results suggest that the articles were similarly difficult. However, it is still possible that one of the articles was easier than others or that one of the articles spoke more to the demographics or characteristics of a specific class. For example, participants in the article structure group were mostly lifestyle informatics students, while students in the no instruction group were mostly marketing students and students in the narrative group were mostly language and/or literature students. It is thus important to replicate our results with other participant groups as well as articles. Furthermore, it is important to replicate our results more systematically, for example, by counterbalancing the particular articles across and within intervention groups in a Latin square design. Despite being preliminary, the findings support the narrative method as a promising approach to teaching Science Reading skills.

7 Conclusions and future directions

Although highly encouraging, these results must be treated with appropriate caution. Our future goals involve conducting a full replication with the greatest possible rigor, which means:

1. completing analysis of sets of additional already collected data;
2. testing native speakers of English, the language of the articles, to circumvent language-related confounds;
3. employing multiple instructors naive to aims and hypotheses to reduce possible confounds resulting from effects due to the teaching of the single instructor used in the pilot;
4. employing naive response coders;
5. using the novel software "MsquaredC" (https://msquaredc.netlify.app), developed for this project to reduce bias when evaluating qualitative responses.

Regardless of the possible results of our studies, this research has already benefited science by drawing attention to a key weakness in scientific training.

Even if ultimately all our hypotheses cannot be confirmed, our study has moved us one step closer to providing students and teachers of science with a much-needed evidence-based method for learning how to understand scientific documentation.

Reference list

Afflerbach, Peter. "The Influence of Prior Knowledge on Expert Readers' Importance Assignment Processes." *National Reading Conference Yearbook* 35 (1986): 30–40.

Allen, Liz, Jo Scott, Amy Brand, Marjorie Hlava, and Micah Altman. "Publishing: Credit Where Credit Is Due." *Nature* 508 (2014): 312–313. doi:10.1038/508312a.

American Psychological Association. *Publication Manual of the American Psychological Association: The Official Guide to APA Style*. 7th edition. Washington, DC: American Psychological Association, 2020.

Baker, Monya. "Is There a Reproducibility Crisis? A *Nature* Survey Lifts the Lid on How Researchers View the 'Crisis' Rocking Science and What They Think Will Help." *Nature* 533 (2016): 452–454. doi:10.1038/533452a.

Bastardi, Anthony, Eric Luis Uhlmann, and Lee Ross. "Wishful Thinking: Belief, Desire, and the Motivated Evaluation of Scientific Evidence." *Psychological Science* 22.6 (2011): 731–732. doi:10.1177/0956797611406447.

Bastian, Brock, Jolanda Jetten, and Fabio Fasoli. "Cleansing the Soul by Hurting the Flesh: The Guilt-Reducing Effect of Pain." *Psychological Science* 22.3 (2011): 334–335. doi:10.1177/0956797610397058.

Bergman, Jerome, Monica Gonzalez-Marquez, and Simon Volker. *MsquaredC: Multi-Method Qualitative Data Coding Software to Reduce Researcher Bias*. 2022 https://docs.google.com/document/d/1efVGrk-lCDzV4rqFZt1eSE8L_kT28XbNGQPYkIzLhsI/edit (28 October 2022) and https://msquaredc.netlify.app (28 October 2022).

Bruner, Jerome. "The Narrative Construction of Reality." *Critical Inquiry* 18.1 (1991): 1–21. doi:10.1086/448619.

Bruner, Jerome. *Actual Minds, Possible Worlds*. Cambridge, MA and London, UK: Harvard University Press, 1986. Reprint: De Gruyter, 2020. doi:10.4159/9780674029019.

Burns, T. W., D. J. O'Connor, and S. M. Stocklmayer. "Science Communication: A Contemporary Definition." *Public Understanding of Science* 12.2 (2003): 183–202. doi:10.1177/09636625030122004.

Carson, Lydia, Christoph Bartneck, and Kevin Voges. "Over-Competitiveness in Academia: A Literature Review." *Disruptive Science and Technology* 1.4 (2013): 183–190. doi:10.1089/dst.2013.0013.

Castles, Anne, Kathleen Rastle, and Kate Nation. "Ending the Reading Wars: Reading Acquisition From Novice to Expert." *Psychological Science in the Public Interest* 19.1 (2018): 5–51. doi:10.1177/1529100618772271.

Chalmers, Alan F. *What Is This Thing Called Science?* 4th edition. Indianapolis: Hackett Publishing, 2013.

Chen, Jason A., and Ellen L. Usher. "Profiles of the Sources of Science Self-Efficacy." *Learning and Individual Differences* 24 (2013): 11–21. doi:10.1016/j.lindif.2012.11.002.

Chow, Ho Ming, Raymond A. Mar, Yisheng Xu, Siyuan Liu, Suraji Wagage, and Allen R. Braun. "Embodied Comprehension of Stories: Interactions between Language Regions and Modality-

Specific Neural Systems." *Journal of Cognitive Neuroscience* 26.2 (2014): 279–295. doi:10.1162/jocn_a_00487.
Croissant, Jennifer L. "Theory, Narrative, and Discipline at the Intersections of Science and Technology Studies and History." *Bulletin of Science, Technology & Society* 23.6 (2003): 465–472. doi:10.1177/0270467603259870.
Dunbar, Kevin. "How Scientists Build Models in Vivo Science as a Window on the Scientific Mind." *Model-Based Reasoning in Scientific Discovery*. Eds. Lorenzo Magnani, Nancy J. Nersessian, and Paul Thagard. Boston, MA: Springer, 1999. 85–99. doi:10.1007/978-1-4615-4813-3_6.
Erduran, Sibel. "Trust in Science and Science Education – Part 1." *Science & Education* 31 (2022): 1101–1104. doi:10.1007/s11191-022-00386-4.
Ernst, E., T. Saradeth, and K. L. Resch. "Drawbacks of Peer Review." *Nature* 363 (1993): 296. doi:10.1038/363296a0.
Errington, Timothy M., Alexandria Denis, Nicole Perfito, Elizabeth Iorns, and Brian A. Nosek. "Reproducibility in Cancer Biology: Challenges for Assessing Replicability in Preclinical Cancer Biology." *eLife* 10 (2021): art. e67995. doi:10.7554/eLife.67995.
Fanelli, Daniele. "Is Science Really Facing a Reproducibility Crisis, and Do We Need It To?" *Proceedings of the National Academy of Sciences* 115.11 (2018): 2628–2631. doi:10.1073/pnas.1708272114.
Fisher, Walter R. *Human Communication as Narration: Toward a Philosophy of Reason, Value, and Action*. Columbia, SC: University of South Carolina Press, 1987. Reprint: University of South Carolina Press, 2021.
Fletcher, Charles R. "Levels of Representation in Memory for Discourse." *Handbook of Psycholinguistics*. Ed. Morton Ann Gernsbacher. San Diego, CA: Academic Press, 1994. 589–607.
Fludernik, Monika. *An Introduction to Narratology*. London: Routledge, 2009.
Fox, Charles W., C. Sean Burns, Anna D. Muncy, and Jennifer A. Meyer. "Author-Suggested Reviewers: Gender Differences and Influences on the Peer Review Process at an Ecology Journal." *Functional Ecology* 31.1 (2017): 270–280. doi:10.1111/1365-2435.12665.
Freedman, Leonard P., Iain M. Cockburn, Timothy S. Simcoe. "The Economics of Reproducibility in Preclinical Research." *PLoS Biology* 13.6 (2015): art. e1002165. doi:10.1371/journal.pbio.1002165.
Gabriel, Yiannis. "Researchers as Storytellers: Storytelling in Organizational Research." *Narratives in Academic and Professional Genres*. Eds. Maurizio Gotti and Carmen Sancho Guinda. Bern: Peter Lang, 2013. 105–122.
Glynn, Shawn M., and K. Denise Muth. "Reading and Writing to Learn Science: Achieving Scientific Literacy." *Journal of Research in Science Teaching* 31.9 (1994): 1057–1073. doi:10.1002/tea.3660310915.
Gonzalez-Marquez, Monica, Andrea Philipp, and Matthias Wilde. "Developing Evidence-based Methods to Read Scientific Journal Articles." Poster presented at the *2nd Interdisciplinary REASON Spring School 2018*. 2018. https://www.researchgate.net/publication/324110301_Developing_Evidence-based_Methods_to_Read_Scientific_Journal_Articles (28 October 2022).
Harp, Shannon F., and Richard E. Mayer. "How Seductive Details Do Their Damage: A Theory of Cognitive Interest in Science Learning." *Journal of Educational Psychology* 90.3 (1998): 414–434. doi:10.1037/0022-0663.90.3.414.
Hawking, Stephen, and Leonard Mlodinow. *The Grand Design*. London: Bantam Press, 2010. eBook edition: Random House, 2010.
Herman, David. "How Stories Make Us Smarter: Narrative Theory & Cognitive Semiotics." *Recherches en Communication* 19 (2003): 133–154.

Herman, David. “Cognitive Narratology.” *Handbook of Narratology*. Vol. 1. Eds. Peter Hühn, Jan Christoph Meister, John Pier, and Wolf Schmid. 2nd edition. Berlin and Boston: De Gruyter, 2014. 46–64.

Hoffman, Donald D. *The Case against Reality: Why Evolution Hid the Truth from Our Eyes*. London, UK: WW Norton & Company, 2019.

Jahn, Manfred. “Frames, Preferences, and the Reading of Third-Person Narratives: Towards a Cognitive Narratology.” *Poetics Today* 18.4 (1997): 441–468. doi:10.2307/1773182.

Jaremka, Lisa M., Joshua M. Ackerman, Bertram Gawronski, Nicholas O. Rule, Kate Sweeny, Linda R. Tropp, et al. “Common Academic Experiences No One Talks About: Repeated Rejection, Impostor Syndrome, and Burnout.” *Perspectives on Psychological Science* 15.3 (2020): 519–543. doi:10.1177/1745691619898848.

Kearney, Richard. *On Stories: Thinking in Action*. New York, NY: Routledge, 2002.

Kelly, Jacalyn, Tara Sadeghieh, and Khosrow Adeli. “Peer Review in Scientific Publications: Benefits, Critiques, & a Survival Guide.” *eJIFCC* 25.3 (2014): 227–243.

King, Alison. “Beyond Literal Comprehension: A Strategy to Promote Deep Understanding of a Text.” *Reading Comprehension Strategies: Theories, Interventions, and Technologies*. Ed. Danielle S. McNamara. New York and London: Psychology Press, 2007. 267–290.

Lakens, D., E. Page-Gould, M. A. van Assen, B. Spellman, F. Schönbrodt, F. Hasselman, et al. *Examining the Reproducibility of Meta-Analyses in Psychology: A Preliminary Report*. 2017. https://osf.io/preprints/metaarxiv/xfbjf/ (28 October 2022).

Leeper, Thomas J. “Archiving Reproducible Research with R and Dataverse.” *The R Journal* 6.1 (2014): 151–158.

Lewis, Patrick J. *How We Think, But Not in School: A Storied Approach to Teaching*. Rotterdam, NL: Sense, 2007.

López-Nicolás, Rubén, José Antonio López-López, María Rubio-Aparicio, and Julio Sánchez-Meca. “A Meta-Review of Transparency and Reproducibility-Related Reporting Practices in Published Meta-Analyses on Clinical Psychological Interventions (2000–2020).” *Behavior Research Methods* 54 (2021): 334–349. doi:10.3758/s13428-021-01644-z.

Lowrey, Christopher H., and Priya Venkatesan. “Making Science Accessible: A Semiotics of Scientific Communication.” *Biosemiotics* 1.2 (2008): 253–269. doi:10.1007/s12304-008-9017-1.

Martinez-Conde, Susana, and Stephen L. Macknik. “Finding the Plot in Science Storytelling in Hopes of Enhancing Science Communication.” *Proceedings of the National Academy of Sciences* 114.31 (2017): 8127–8129. doi:10.1073/pnas.1711790114.

Miles, Lynden K., Sarah B. Stuart, and C. Neil Macrae. “Moving through Time.” *Psychological Science* 21.2 (2010): 222–223. doi:10.1177/0956797609359333.

Negrete, Aquiles, and Cecilia Lartigue. “Learning from Education to Communicate Science as a Good Story.” *Endeavour* 28.3 (2004): 120–124. doi:10.1016/j.endeavour.2004.07.003.

Negrete, Aquiles, and Cecilia Lartigue. “The Science of Telling Stories: Evaluating Science Communication via Narratives (RIRC Method).” *Journal of Media and Communication Studies* 2.4 (2010): 98–110. doi:10.5897/JMCS.9000080.

Norris, Stephen P., and Linda M. Phillips. “How Literacy in Its Fundamental Sense Is Central to Scientific Literacy.” *Science Education* 87.2 (2003): 224–240. doi:10.1002/sce.10066.

Ochs, Elinor, and Lisa Capps. “Narrating the Self.” *Annual Review of Anthropology* 25 (1996): 19–43.

Onega, Susana, and Jose Angel Garcia Landa. *Narratology: An Introduction*. London and New York: Longman, 1996. Reprint: Routledge, 2014. doi:10.4324/9781315843018.

Open Science Collaboration. “Estimating the Reproducibility of Psychological Science.” *Science* 349.6251 (2015): art. aac4716. doi:10.1126/science.aac4716.

Polanin, Joshua R., Emily A. Hennessy, and Sho Tsuji. "Transparency and Reproducibility of Meta-Analyses in Psychology: A Meta-Review." *Perspectives on Psychological Science* 15.4 (2020): 1026–1041. doi:10.1177/1745691620906416.

Rapp, David N., Paul van den Broek, Kristen L. McMaster, Panayiota Kendeou, and Christine A. Espin. "Higher-Order Comprehension Processes in Struggling Readers: A Perspective for Research and Intervention." *Scientific Studies of Reading* 11.4 (2007): 289–312. doi:10.1080/10888430701530417.

Roche, Dominique G., Loeske E. B. Kruuk, Robert Lanfear, and Sandra A. Binning. "Public Data Archiving in Ecology and Evolution: How Well Are We Doing?" *PLoS Biology* 13.11 (2015): art. e1002295. doi:10.1371/journal.pbio.1002295.

Rosman, Tom, Michael Bosnjak, Henning Silber, Joanna Koßmann, and Tobias Heycke. "Open Science and Public Trust in Science: Results from Two Studies." *Public Understanding of Science* (2022): art. 09636625221100686. doi:10.1177/09636625221100686.

Rubin, Haya R., Donald A. Redelmeier, Albert W. Wu, and Earl P. Steinberg. "How Reliable Is Peer Review of Scientific Abstracts?" *Journal of General Internal Medicine* 8.5 (1993): 255–258. doi:10.1007/BF02600092.

Sagan, Carl. *The Demon-Haunted World: Science as a Candle in the Dark*. New York: Ballantine Books, 1996. Reprint: Ballantine Books, 2011.

Shanahan, Cynthia, Timothy Shanahan, and Cynthia Misischia. "Analysis of Expert Readers in Three Disciplines: History, Mathematics, and Chemistry." *Journal of Literacy Research* 43.4 (2011): 393–429. doi:10.1177/1086296X11424071.

Shrout, Patrick E., and Joseph L. Rodgers. "Psychology, Science, and Knowledge Construction: Broadening Perspectives from the Replication Crisis." *Annual Review of Psychology* 69.1 (2018): 487–510. doi:10.1146/annurev-psych-122216-011845.

Sollaci, Luciana B, and Mauricio G Pereira. "The Introduction, Methods, Results, and Discussion (IMRAD) Structure: A Fifty-Year Survey." *Journal of the Medical Library Association* 92.3 (2004): 364–367.

Tucan, Gabriela. "The Reader's Mind Beyond the Text – the Science of Cognitive Narratology." *Romanian Journal of English Studies* 10.1 (2013): 302–311. doi:10.2478/rjes-2013-0029.

Turner, Mark. *The Literary Mind: The Origins of Thought and Language*. New York and Oxford: Oxford University Press, 1998.

Vallée-Tourangeau, Gaëlle, Ana Wheelock, Tushna Vandrevala, and Priscilla Harries. "Peer Reviewers' Dilemmas: A Qualitative Exploration of Decisional Conflict in the Evaluation of Grant Applications in the Medical Humanities and Social Sciences." *Humanities and Social Sciences Communications* 9.1 (2022): art. 70. doi:10.1057/s41599-022-01050-6.

van den Broek, Paul. "Discovering the Cement of the Universe: The Development of Event Comprehension from Childhood to Adulthood." *Developmental Spans in Event Comprehension and Representation: Bridging Fictional and Actual Events*. Eds. Paul W. van den Broek, Patricia J. Bauer, and Tammy Bourg. Mahwah, NJ: Erlbaum, 1997. 321–342.

van Dijk, Teun A., and Walter Kintsch. *Strategies of Discourse Comprehension*. New York: Academic Press, 1983.

Willems, Roel M., Samuel A. Nastase, Branka Milivojevic. "Narratives for Neuroscience." *Trends in Neurosciences* 43.5 (2020): 271–273. doi:10.1016/j.tins.2020.03.003.

Paola Pennisi

The Contextual, Enabling, and Constitutive Role of Physical Experience in Narratives

Abstract: Most narratives produced in human history are focused on social contents; moreover, it has recently been posited that the main role of humans' narrative tendency is to sharpen social cognition (Gallagher 2004; Gallagher and Hutto 2008; Hutto 2009). Despite this, a great number of narratives – in particular the scientific ones – are free from any social content.

In this paper I will try to show that, while narratives can refrain from representing social contents, they could never lack physical ones; thus, physical experience has a constitutive role in narratives. I will develop this idea step by step by using De Jaegher et al.'s (2010) distinction among contextual, enabling and constitutive roles. Firstly, I will show the contextual role of humans' physical interaction with the world for the development of narrative skills during childhood, by comparing the features of narratives produced by children with Autism Spectrum Disorders and with Cerebral Palsy. Then I will show the enabling role of human physical interaction with the world in the creation of narratives by analyzing examples of both *natural* and *unnatural narratives*. Finally, I will propose that the physical experience of the world is a constitutive factor for narrative production by showing that the concept of time (which is, by definition, a necessary dimension of narratives) depends on the concept of space.

1 Introduction

In this chapter, I will try to show that humans' interaction with the physical world is a constitutive element of narrative competence.

To make this thesis clearer, I will briefly define the sense in which I use the term *constitutive*. A factor can be *contextual, enabling* or *constitutive* with respect to a specific phenomenon (De Jeagher et al. 2010). A factor is contextual when variations in it will cause variations in the phenomenon; it is enabling if, in its absence, the phenomenon cannot occur; and it is constitutive if it is entirely part of the phenomenon. Consider the example of the phenomenon of a book falling off a table because of a strong push. In this case, the direction of the push has a contextual role, because variations in it will cause different directions in the book's descent. The push itself has an enabling role because in its absence the book won't fall down. Finally, the book is a constitutive factor of the phenomenon because it

https://doi.org/10.1515/9783110782844-011

is an essential part of it. As highlighted by the same authors, the three kinds of roles follow a scale of increased specificity (De Jaegher et al. 2010, 443). A factor that is constitutive of a phenomenon also has a contextual and enabling role toward it; a factor that is enabling for a phenomenon also has a contextual role. Thus, when I posit that human bodily interaction with the physical world is a constitutive element of human narrative competence, I also mean that humanity's attitude to interaction with the physical world also has a contextual and enabling role in the development of narrative competence. In other words, this means that:

- variations in an individual's attitude to interaction with the physical world will cause variations in its narrative skills (contextual role);
- one who cannot interact with the physical world cannot narrate (enabling role);
- interaction with the physical world is a part of our narrative skills, as the book in our example is a part of its descent (constitutive role).

The relationship between the notions of contextual, enabling and constituent roles, in fact, is of increasing inclusion and could be summarized as follows:

- contextual role
- contextual role – enabling role
- contextual role – enabling role – constitutive role (for more details on the model cf. Pennisi and Torricelli 2018).

Hence if, in the genesis of a phenomenon, a factor has a constituent role, this implies that it also has a contextual and an enabling role. Given this, to prove my thesis, it would have been sufficient to focus my attention on the constituent role of the narrator's physical interaction with the world in the generation of narratives. In fact, it would have been possible to settle the matter in a dry and simple way by directly writing Section 4 and the Conclusion. However, this article does not have as its sole objective to demonstrate that the narrator's physical interaction with the world is part of the narrative itself. In fact, this work also has a descriptive aim. Clearly, I do not pretend to fully describe the relationship between bodily interaction with the world and narratives; however, I have opted for a demonstration scheme that slowly explores and describes the contextual role first; then the enabling one; and only in the end the constituent one, because it allows me to illustrate various aspects of the question.

Following Shaun Gallagher and Daniel Hutto (2008), I will use Peter Lamarque's (2004) definition of *narrative*: a narrative is a told or written (or thought, I would add, considering the sense in which the term is used by Gallagher and Hutto; cf. Gallagher 2006; 2008; Gallagher and Hutto 2008; Hutto 2008; 2009) story (not a found story[1]) in which at least two events are linked by a temporal dimension. When I use the expression *narrative competence*, I will mean the ability to put at least two events into relation (whether in a correct or incorrect way is unimportant) through a temporal link, toward a personal interpretative act.

With Mark Turner, I will consider narrative a mental instrument, "a basic principle of mind" (Turner 1996, v.) through which we organize most of our experience and knowledge. According to Turner the difference between a narrative and something that is not a narrative is its *reportability*: "Story depends on constructing something rather than nothing. A reportable story is distinguished from its assumed and unreportable background" (Turner 1996, 145). I agree with this idea, but with a small difference: the difference between a narrative and something that is not a narrative is not its *reportability*, but the existence or the absence of a narrator's performative act. So, coming back to our book example, let's consider the case of a spectator who sees a child who pushes a table and a book that falls down from it. Let's suppose now that this spectator thinks, or says, or writes, "a child pushed the table and the book on it fell down." According to my working definition, this is a narrative. By contrast, all perceptions, sensations, unconscious beliefs or desires that he didn't organize in a temporal form are not a narrative.

Turner's idea was criticized by Ryan (2007) with two arguments: this criterion to distinguish between narrative and non-narrative is "superfluous and impossible: superfluous, because it is no longer necessary to differentiate narrative from any other manifestation of human thought, and impossible, because it is inseparable from a complete theory of mind" (Ryan 2007, 28). I would like to respond to both argumentations. First of all, there are many manifestations of human thought that are not narratives, i.e. sensations, perceptions, emotions on which the thinking subject has not exercised a temporal ordering principle. Also what Turner calls *reportable background*, and what I will call *reported background* exercises a strong influence on our behavior; e.g. patients affected by visual neglect (a neurological condition that prevents the vision of an entire hemispace) are influenced by stimuli that they cannot consciously perceive (Mashall and Halligan 1988; Driver et al. 1992; Halligan et al. 2003).[2] I also disagree with Ryan's second point, according to which this definition of *narrative* is impossible because is inseparable from a complete theory of mind. With what we know today of the mind and with this notion of narratives, we can already make a dis-

1 "Found story" in this context has to be considered as opposed to "told story" (cf. Lamarque 2004, 394).

2 For other examples of unconscious perceptions both in typically developing (TD) subjects and in neuropsychological populations, see Bornstein and Pittman (1992).

tinction between what is and what is not a narrative. It may be that in some extreme cases we will disagree, but this is a problem inherent in all human categorizations. Furthermore, it is not entirely to be excluded that one day (probably far in the future) we will conceive a theory of mind, if not complete, at least quite satisfactory for these purposes. Moreover, if physicists waited to have a complete and satisfactory theory of how the cosmos came to light before elaborating theories on smaller phenomena (even if somehow correlated with the birth of the cosmos), we would know much less today than we do.

2 Ontogenetic development of narrative skills: The contextual role of physical interactions with the world

In this section I aim to show that, from an ontogenetic point of view, the interactions each subject has with the physical world will have a contextual role in their narratives. In other words, variations in the development of sensory-motor interactions with and perceptions of the physical world in children will cause differences in their development of narrative skills.

To develop my thesis, I will consider and analyze data from the study of subjects with Autism Spectrum Disorder (ASD). This clinical population is characterized by "persistent deficits in social communication and social interaction" and by "restricted and repetitive patterns of behaviour, interests or activities" during the developmental age (APA 2013, respectively diagnostic criteria A and B). ASD is irreversible. I will consider the ontogenetic development of children with ASD because cognitive anomalies linked to this neurodevelopmental disorder shed new light on the ontogenetic development of certain cognitive abilities, including the capacity for narrative.

The majority of the literature on narrative in autism highlights lower narrative performances in the clinical population compared to typically-developed (TD) subjects: subjects with ASD, in fact, tend to use fewer words and utterances, to include lower semantic variety, and to produce narratives that are less causally connected and less coherent than TD subjects (for a meta-analysis see Baixauli et al. 2016). There is general agreement on this point, but we need to consider a limitation of the tasks design usually used to study these narratives. For people without ASD, a story is something that always involves characters with intentional and emotional states, and with desires, who perform goal-directed actions. It would be very difficult to find someone happy to read a ten-page narrative on the "adventures" of a non-intentional agent, such as for example a ball that moves over various kinds of terrains on a slope.

Narrative skills can be seen as the last step of a human ontogenetic process of growth of interactive skills, mainly social skills (Gallagher 2004). Let's consider a unique feature of the narrative skills of subjects with autism, one that is usually under-estimated in the literature. Limitations on narrative competence in subjects with autism, although evident in all kinds of texts, seem to be more marked in relation to classical narratives, based on social stories, but are less notable in narratives that exclude emotional and social interpretations of events.

Baron-Cohen et al. (1986) found that children with autism used a more appropriate language to narrate mechanical sequences compared to TD children; specifically, they were more capable than TD children in using *because*-clauses such as "the egg broke because it fell off the table"; *by*-clauses such as "the man was hit by the rock"; and *causal verb*-phrases such as "the rock made the man fall down" (Baron-Cohen et al. 1986, 120). Unfortunately, in this study, children with autism had a higher mental age (in both verbal and non-verbal terms) in relation to the control group of TD children. However, data on adult populations of subjects with ASD showed analogous narrative skills between them and the TD subjects when the focus of narratives is not emotional or social. For example, Castelli et al. (2002), showed three types of silent animations, with triangles as characters, to a group of subjects with Asperger Syndrome (a form of ASD) and the control TD group. In the first type of animation, the triangles' movements suggest the expression of mental states, as described for the Heider and Simmel Illusion (HSI, Heider and Simmel 1944). In the second type of animation, the triangles' movements suggest goal-directed actions, i.e. two triangles dancing together. And finally, in the third kind of animation, triangles moved randomly. After watching each animation, experimenters asked the participants to describe the content of the film. Narratives of both groups were analogous in the last two conditions: same lengths, same level of appropriateness, and same quantity of expressions of intentional states. However, there were differences in the first condition: when describing emotional contents, the narratives of the subjects with autism showed the same biases classically described in the literature.

Klin and Jones (2006) asked adult ASD and TD groups to narrate what they watched, after having shown them two animations. The first animation was the HSI, the second comprised films in which a group of geometrical shapes behaved as physical objects, i.e. rotating like planets. Thus in both animations the characters were geometrical shapes, but in the first the shapes seemed to have social behavior, whereas in the second they had only physical behavior. Narratives produced by the participants were evaluated based on their causal interpretations of various fragments. Narratives of the ASD group received lower scores in the social task than those of the TD group. By contrast, for narratives focused on physical interpretations, both groups performed equally.

I think that anomalies shown in ASD subjects' narratives can be explained by considering their social difficulties. That is, the shortness of the ASD narratives relative to TD subjects' narratives can be linked to their reticence in using emotional explanations for phenomena. In fact, as shown in Castelli et al. (2002), TD subjects produce longer narratives when describing shapes with emotional behaviors than when describing shapes with goal-directed behaviors or with random movement patterns.

Let us now consider the narrative skills of another clinical population: children with cerebral palsy (CP). This clinical population is very limited in their interaction with the physical world.

> Cerebral palsy describes a group of permanent disorders of the development of movement and posture causing activity limitation, that are attributed to non-progressive disturbances that occurred in the developing fetal or infant brain. The motor disorders of cerebral palsy are often accompanied by disturbances of sensation, perception, cognition, communication, and behaviour; by epilepsy, and by secondary musculoskeletal problems (Rosenbaum et al. 2007, 9).

Holck et al. (2011) intensively analyzed the narrative skills of children with CP, comparing them with the narrative skills of a group of TD children matched by mental and verbal age. They showed their participants a storybook with pictures but without words (Bus Story Test, Renfrew 1997; 2002), asking them to re-narrate the story. In general, children with CP performed worse than TD children: both groups were able to keep the narrative on topic and to provide a temporal structure to the narrative, and they used the same numbers of self-repairs, repetitions, fillers (such as *ehm, uhm*), pauses and words, and manifested the same level of fluency. The main difference between the two groups was in (a) explicitness and the (b) use of causal conjunctions. That is, first, despite the same level of comprehension of the story, children with CP were less able to make explicit salient passages of the story than the TD children, making it difficult for the listener to grasp the coherence of events. Explicitness, as assessed by Holck et al. (2011), is a composite measure of three variables: (1) informativeness; that is, the presence of all constitutive facts of the object of narration; (2) explicitness; that is, representation of optional details; and finally (3) the presence of the three following narrative components: description, action and evaluation (Bliss et al. 1998). The second difference was that TD children used significantly more causal conjunctions such as "because," "since" and "therefore" than CP children, who usually preferred the simple conjunction "and."

According to the authors, children with CP are able to extract from the story the same level of information as TD children, but they do not spontaneously grasp the causality of the events in the story. By contrast, subjects with ASD, despite having more trouble with narratives than children with CP, have no difficulty with grasping the causality of the story when the latter is focused on physical laws or goal-directed actions that do not need social explanations.

To sum up, I considered two clinical populations, both affected in neurodevelopmental age. Both clinical populations have anomalies in social and physical interactions, but the core deficit of children with ASD is of a social nature whereas the core deficit of children with CP is sensory-motoric. Both clinical populations show anomalies in narrative skills compared to TD children, but these anomalies are of quite different kinds. Now I will try to show that anomalies in narratives of children with CP could be linked to their sensory-motor anomalies; whereas, in the case of autism, anomalies in narrative skills are linked to social deficits which are actually different from those shown by children with CP. To develop this idea I will briefly make reference to Rizzolatti and Sinigaglia's theory on the role of mirror neurons in the human brain (2019; 2008).

According to Rizzolatti and Sinigaglia, mirror neurons detect the goal of a motor action thanks to their ability to integrate contextual information with the beginning of the motor act of another person. The perception of another person who is performing an action is translated by mirror neurons in our brains into an analogous motoric act that is then inhibited by other nerve impulses. So, when I see another person who is performing an action, in my brain the mirror neuron system (MNS) simulates the same motoric action, e.g. if I see a friend who picks up a cup, in my brain, at a pre-motoric level, I will simulate his action, and this is why I will understand (before the end of my friend's action) that he will bring the cup to his mouth. This is not the only way we have to understand the goals of others but in this context we will focus on it (for more details see Rizzolatti and Craighero 2004). According to this view, a way that we have to understand the causality of others' actions is to simulate the same action in our own body, at a pre-motoric stage. Children with CP cannot perform many of the actions that are represented in narratives; often the sequencing of their movements is highly affected. So what do their mirror neurons simulate while they try to organize a narrative starting from a series of images that represent motoric action? In this view it's plausible that their tendency not to causally correlate the events of a narrative may be linked to their physical inability to perform many movements.

By contrast, subjects with ASD usually are able to understand what an agent is physically doing (e.g. he is opening a door); their major difficulties are related to mental explanations of phenomena. Let's consider the classic case of a child with autism who wants to open a door but cannot reach the handle. Usually in similar cases, children with autism won't ask the adult to open the door. They will grab his hand and rest it on the handle. Among researchers the most common explanation of this behavior is that they know that an adult who has his own hand on the handle will open the door, but they won't consider the adult's mental life: they don't ask the adult to open the door because they don't think or consider that the adult could have the desire to help them; they don't think of causality in mental terms as TD subjects usually do (for more details see Pennisi 2016; 2017).

This observation allows us to conclude that anomalies in the ontogenetic development of sensory-motor skills will cause anomalies in narrative skills.

If this hypothesis is correct, this indicates that, from an ontogenetic point of view, our ability to physically interact with the world has a contextual role in the development of our narrative skills: in fact, if my hypothesis is correct, the characteristic way in which children with CP physically interact with the world alters the way in which they narrate causality. Moreover, the way in which subjects with ASD physically interact with the world (e.g. grabbing the hand of someone to mechanically induce him to open the door instead of asking him to open it) alters the way in which they narrate the world ("an adult's hand on the handle usually opens the door" instead of "the adult will help me to open the door").

We perceive or infer causality through movement. Frequently, we grasp the causality of others' acts by mapping their movement in our body, so if our body alters the way in which we physically interact with the world, this will change our way of narrating the world.

3 The interaction between the narrator and the physical world has an enabling role in the existence of narratives

3.1 Enabling role of bodily experiences to narratives

There can be no act of creation without a physical interaction between the creator and the physical world, and the creation of a narrative cannot be considered an exception to this rule, *ça va sans dire!*

All narratives are reported physically – on a piece of paper, registered in a file in the author's computer, impressed in the brains of someone who is listening to a narrator. Even the narration that we do to ourselves, about our lives in our own minds, could not arise if a part of our body did not undergo changes during this process. This last idea will be argued in section 3.2.

3.2 Enabling role of bodily experiences in autobiographic first-person narratives: the mirror neuron system

Let's suppose that one day I accidentally hit another car while leaving a parking lot. In the confusion of the impact I do not immediately have the ability to understand what happened. If I really would like to understand how things went, what would I instinctively do? The first step would be to relive the episode in my mind, through my imagination. Did I look carefully to see if someone was passing before I left the parking lot? Was I distracted by some thoughts as I was leaving the lot, or was I focused on the maneuver I was making? The only way I can answer these questions is to mentally relive the episode. Imagining something means experiencing it in some way with one's own body, even if only at a premotor level. The MNS in our brain consists of all those neurons that become active both when we observe someone perform an act or feel an emotion, and when we perform the same act or feel the same emotion. The first studies on mirror neurons showed that these cells are engaged in the performance and in the perception of a motor act, and that some of them respond to the goal of the motor act, others to the motoric pattern of the motor act (Fogassi et al. 2005; Umiltà et al. 2008; Rochat et al. 2010). Recently mirror neurons have been shown to be present in cortical regions typically associated with emotive reactions. In 2000, Calder et al. found that the recognition and the experience of disgust were impaired in a patient with insula and putamen damage. Krolak-Salmon et al. (2003) found (using implanted depth electrodes, a methodology that leaves no room for doubt) that the ventral anterior insula responds to the perception of disgust in others; in the same year, the common neural basis of seeing and feeling disgust was discovered (Wicker et al 2003). Other more recent studies support this view (Chen et al. 2009; Papagno et al. 2016). Moreover mirror neurons have also been found in the amygdala, which seems to be involved in the perception and experience of fear (Morris et al 1996; Philips et al. 1998; Dolan et al. 2001; Hadsjikhani and De Gelder 2003; Murray et al. 2014; Meletti et al. 2006; Lanteaume et al. 2007; Inman et al. 2018), and in the cingulate gyrus, which seems to be involved in the perception and experience of laughing and the sense of hilarity (Szameitat et al. 2010; Caruana et al. 2015; Gibson et al. 2017; for a meta-analysis, cf. Fusar-Poli et al. 2009). Recent studies show that a great part of the MNS and of cerebral areas linked to motor activities also fire when we imagine performing an act (Pineda et al. 2000; Glenberg et al. 2008; Nedelko et al. 2010; Battaglia et al. 2011; Oosterhof et al. 2012; Ogoshi et al. 2013; Filimon et al. 2015; Harris and De Jong 2014; Di Cesare et al. 2015; Hardwick et al. 2018; Savaki and Raos 2019). For example, Fortunato Battaglia, Sarah Lisanby and David Freedberg (2011) showed that both the observation and the imagining of Michelangelo's *Expulsion from Paradise* significantly increase corticomotor excitability. Moreover, they found that the phenomenon was absent or not significant (at a motor level) during the observation or the imagining of more static scenes, such as Bellini's *Dead Christ with Angels*. So imagination of something observed or lived activates our MNS, at least at a premotor level. However, to demonstrate that this activation has an enabling role in narration, we should also consider those cases in which what is narrated has never been lived through by the narrator.

In fact, to argue that physical interaction with the world has an enabling role in narrative, it's necessary but not sufficient to show, as I have just done, that all personal and real narratives imply the activation, in such a way, of the motor system in our brain, and so a physical interaction with the world. To reach my goal I must also show that everything that we narrate comes from our physical experience with the world. But how can we support the idea that all narratives are enabled by a physical interaction with the world if there are many narratives that describe unreal worlds?

3.3 Enabling role of the body in fictional narratives

In this section I will try to show that in all kinds of narratives, and not exclusively in those discussed in section 3.2, bodily interaction with the world has an enabling role. Firstly I will argue that a distinction between *natural* and *unnatural* narratives is controversial (section 3.3.1), and so to support the idea that bodily interactions with the world are an enabling factor for the existence of narratives, I will argue that the unreal but realistic, the unrealistic, the possible, and even the impossible, are always part of the personal experience of the narrator (section 3.3.2).

3.3.1 The distinction between natural and unnatural narrative is controversial

In this paragraph my purpose is to show that even those narratives that apparently describe something very different from our physical experience, actually are also enabled by the narrator's physical experience with the world. To reach my goal I will briefly synthesize a recent and interesting debate in narratology.

In 2010, Alber et al. discussed the notion of *unnatural narratives* and proposed them as an exemplum of narrative that cannot be explained through the mimetic principle.[3] In contrast to Alber's idea, I will try instead to show that: (1) all narratives aim to depict and illuminate a part of the physical world experienced by the narrator and that without this physical interaction the narratives do not come to light; (2) even fictional narratives, that apparently are not part of the narrator's experience, actually grasp and represent a truth of their physical experience of the world.

Alber defines a narrative as unnatural if it "violates physical laws, logical principles, or standard anthropomorphic limitations of knowledge by representing storytelling scenarios, narrators, characters, temporalities, or spaces that could not exist in the actual world" (Alber 2013). Alber et al. (2010) propose three kinds of frequent unnatural narratives: *unnatural storyworlds* (i.e. the existence of multiple conclusions for a story with one beginning, as in "The Baby Sitter" by Robert Coover); *unnatural minds*, such as when a story has an omniscient narrator; and finally *unnatural acts of narration* (i.e. the unnaturalness of the alternation between a present and a past during the narration of "The Tell-Tale Heart" by Edgar Allan Poe). Alber et al.'s (2010) aim is to demonstrate that there are narratives that cannot be explained by making reference to the mimetic principle, or – to use the authors' words – that not all narrations are "situated" or "told by someone": "Instead of trying to make all narratives fit into the same scheme, the alternative option is to accept that some narratives are situated, others are not, some narratives are told by someone, others are not – and to analyze the potential consequences" (Alber et al. 2010, 130).

In 2012, Monika Fludernik responded critically to this paper. She used four main arguments:

(1) the first argument is that Alber et al.'s (2010) approach reveals limitations when practically applied in the analysis of narratives. For example, the analysis they propose of *The Tell-Tale Heart* and focuses on the alleged unnaturalness of the times of narration. Fludernik (2012) rightly notes that an alternation between present and past in the flow of consciousness of the narrator is absolutely realistic.

(2) The second argument is that the distinction between natural and unnatural as an attempt to escape the category of mimesis is not a good strategy. In fact

3 The mimetic principle is very complex and has a very long history in aesthetics. For reason of space it is impossible to explore the topic here. The interested reader may consult *The Aesthetics of Mimesis* by Stephen Halliwell (2002).

the notion of *natural* cannot be dichotomically opposed to that of *unnatural* without difficulties.

(3) The third argument is that the link between nature and mimesis implied in Alber et al.'s (2010) analysis does not consider the importance of the aesthetic illusionism of the narrator and his ability to make the impossible plausible, if considered in the context of the story. For example, in Kafka's *Metamorphosis*, we cannot say that the metamorphosis is non-natural and *therefore* non-mimetic.

(4) Finally, Fludernik's (2012) fourth argument concerns the link between conventionalization and narration. In fact, Alber et al. (2010) highlighted that when an act of unnatural narration becomes common and thereby conventional, readers will tend not to perceive it as unnatural. On this point, Fludernik argues that if we confound the realistic and the mimetic in a single category, we will risk forgetting that mimesis can create both natural and fictional-unnatural scenarios. As Aristotle clarified in his *Poetics*, sometimes mimesis can create more natural scenarios than realistic ones. For example, second-person fiction is non-natural, but its increasing use has tended to start a process of conventionalization. Despite its non-naturalness persisting intact, its conventionalization will allow it to become mimetic for readers.

Through some examples, I will try to develop Fludernik's ideas more widely.

Let's consider the following extract from *Born on a Blue Day*, by Daniel Tammet: "Eleven is friendly and five is loud, whereas four is both shy and quiet – it's my favourite number, perhaps because it reminds me of myself. Some are big – 23, 667, 1179 – while others are small: 6, 13, 581. Some are beautiful, like 333, and some are ugly, like 289" (Tammet 2007). According to Alber's definition this piece belongs to an unnatural narrative; i.e. to say that 23 is big and 581 is small clearly violates some logical principles. Is *Born on a Blue Day* an unnatural narrative? No, this is the memoir of Daniel Tammet. In this passage the author, suffering from Asperger's Syndrome and Savant's Syndrome, explains the features with which the numbers emerge to the consciousness of his synesthetic mind. Thanks to Oliver Sacks's (1985) and Vilayanur Ramachandran's (2011) studies of synesthesia, today we know that Daniel Tammet does not lie: he cannot see numbers as we do, he perceives numbers and their personalities together and cannot do otherwise. But if Alber et al. (2010) had written in the nineteenth century, perhaps they would have perceived the first chapter of *Born on a Blue Day* as a classic case of unnatural narrative. Still, all that is described in this chapter is Daniel Tammet's strange but real way of looking at numbers. As explained by Ramachandran (2011), this oddity is called synesthesia and it is a natural and logical consequence of an anomaly in brain connectivity.

A reader who does not know the synesthetic phenomenon will probably interpret Tammet's words in a metaphoric sense and will not consider it as realistic in the strict sense. At the same time, he will not consider the text as natural. Yet this phenomenon is natural and Tammet's description is realistic. By attempting to interpret the narrative in the light of the natural/unnatural categories, we deprive it of its greatest power: that of broadening the human horizons of knowledge and creativity.

Actually, a major awareness of the object represented in the narrative is the effect (and maybe the unconscious purpose) of each narration. An indistinct mass of sensations and other mental events (memories, beliefs, hopes, which in turn have a basis in sequences of sensations) turns into an ordered *kosmos* in a narrative; and this narrative *kosmos* will make the narrator's mental and physical experience intelligible to the future readers.

The first time that Daniel Tammet perceived the number 11 as *friendly* he was not composing a narrative, he was simply experiencing. To narrate this experience, he probably began by re-experiencing it, just as I would have done, if after the car accident I had wanted to reconstruct what happened in more detail.

3.3.2 The experience of living, the experience of narrating

One of the most important European narratives is the *Divina Commedia* by Dante Alighieri. In spite of those who believe that unnatural narratives are the result of recent experiments, it is an incontrovertible example of a narrative certainly not experienced by its narrator in any strict sense. In other words, to compose it, Dante did not really go down to Hell, then up to Purgatory and finally reach Heaven. He remained at his desk for several years, exercising his incredible talent, or – as with false modesty he declared in the twenty-fourth canto of Purgatory – faithfully recording what poetic inspiration whispered to him. I believe that for these reasons the *Divine Comedy* would fully fall into the category of unnatural narrative fiction.

In this section I will try to show that Dante's bodily experience of the object of representation had an enabling role in the elaboration of the sixth canto of Hell. The choice of this passage is not accidental, as it is in fact a canto in which the poet represents a feeling of very strong detachment from the represented sinners and the sin they stain themselves with, suggesting, at a superficial reading, his total estrangement from this guilt.

We are in the sixth canto, in the third circle, where the shadows of the gluttons are tormented by Cerberus, who barks with his three throats on their heads; the beast scratches them, peels them, tears them apart and all this takes place under a pouring rain of fetid water, hail and snow.

Cerbero, fiera crudele e diversa,	*Cerberus, cruel, monstrous beast, with three*
con tre gole caninamente latra	*Throats barks doglike over the people submerged*
sovra la gente che quivi è sommersa.	*There.*
Li occhi ha vermigli, la barba unta e atra,	*His eyes are red, his beard greasy and black, his*
e 'l ventre largo, e unghiate le mani;	*Belly large, and his hands have talons; he claws the*
graffia li spirti, ed iscoia ed isquatra.	*Spirits, flays and quarters them.*
Urlar li fa la pioggia come cani:	*The rain makes them howl like dogs;* [. . .]
(Dante, D.C., Inf. VI, vv.13–19).	(Dante, D.C. Inf. VI, vv. 13–19; trans. by Robert Durling in Alighieri 1996)

Both Cerberus and gluttons are described with the semantic field of animality and repugnance. For Natalino Sapegno "The representation of punishment and the punishing demon takes place in an objective and cruel tone, which underlines the poet's detachment and repugnance towards a sin, which disheartens those who commit it to a bestial condition" (Sapegno 1985, 67; trans. by PP). In the previous circle, the circle of the lustful, the poet had heard Paolo and Francesca's story. In contrast to the heartfelt compassion manifested for them, here in the third circle, Dante now opposes his icy detachment for gluttons.

Through a brief analysis of this canto, I will try to show that, in the *Divine Comedy*, Dante's perceptions are the measure and instrument of representation for all sins.

To express this detachment, Dante chooses to deprive these souls of any bodily appearance, unlike other circles of Hell where souls instead have the appearance and consistency of real bodies. The perception of repugnance and distance towards those souls who have lost their bodies becomes, on the contrary, a strong emotion for those who in Purgatory, stained with the same sin of gluttony in life, now have the opportunity to redeem themselves by re-appropriating their own bodily perceptions and, through them, learning detachment from the earthly world. In the sixth circle of Purgatory, in fact, the gluttons do penance, enduring an insatiable hunger and an insatiable thirst; thin and with scaly skin, they see and smell delicious fruits that they cannot eat. The eternal punishment of the damned in Hell and the temporary expiation of the penitents of Purgatory differ above all in the body-world relationship: the former have lost their bodies due to their inability to regulate the most animalistic instincts with virtue; the latter, on the other hand, are redeemed thanks to the possibility of experiencing the painful bodily perceptions of hunger and thirst that are never satiated.

Dante evokes descriptions of tremendous bodily suffering where he wants to arouse pity, and descriptions of repulsive animalistic details in disembodied souls where he wants to manifest and induce disgust and detachment in the reader, without moving the reader to compassion.

Consider the wisdom of this literary strategy. In Purgatory, the physical signs of punishment are evident and are described with great emotion by the poet. Recent studies of mirror neurons indicate that the physical pain of others activates partially similar brain areas between the observer and the one who experiences (one of most famous is Saarela et al. 2007; for a review see Keysers et al. 2010); therefore it seems likely that in this case too it is the MNS that responds to and mediates perceptions and experiences (Saarela et al. 2007; Giummarra and Bradshaw 2008). Dante empathizes with them and is moved to see the effects of his friend's pain, and the reader is also led to do the same. The emotion is ignited by means of the perception of the physical suffering of punishment.

Even in the case of disgust, the perception experienced with one's own body and the one observed in the body of another partially activate the same brain areas (the area of greatest overlap is the insula; cf. Krolak-Salmon et al. 2003; Wicker et al. 2003) and often induce the partial disinhibition of the same bodily acts typically corresponding to emotion, such as the enlargement of the nostrils (Caruana et al. 2011; Jezzini et al. 2012; Rozin et al. 1999). The poet could have made the bodies of souls disgusting in order to induce disgust in the reader, but instead he makes the choice to leave these souls without bodies, and to induce disgust through some repugnant details of that *gran vermo* (great worm) that tortures them. If the sinner does not have a bodily appearance, in fact, the reader will be much less facilitated in their empathizing process. He loses substance for our cognitive-perceptual system; and if the substance fails, the resonance fails.

Of course, the purpose of the section is not to pay homage to Dante's magnificence. More than anything else, I would like to invite the reader to reflect on one point. Dante certainly did not know the research on mirror neurons. So how did he achieve with this finesse the purpose of empathizing with the souls of Purgatory and making us observe the souls of Hell with detachment? The answer is metaphorically inserted into the general construction of the work itself: Dante uses his own body to understand how to obtain the desired poetic effect.

Of course Dante is not the only narrator who uses his experience of bodily interaction with the world to narrate. Let's consider another example.

I believe that in Robert Louis Stevenson's novel *The Strange Case of Dr. Jekyll and Mr. Hyde*, the physical transformation of Dr. Jekyll into Mr. Hyde can be considered another good example of unnatural narrative, according to the parameters suggested by Alber et al. (2010). Here is the passage in which Stevenson describes the first transformation, via Dr Jekyll's letter describing the moments immediately following the first ingestion of the fateful potion:

> The most racking pangs succeeded: a grinding in the bones, deadly nausea, and a horror of the spirit that cannot be exceeded at the hour of birth or death. Then these agonies began swiftly to subside, and I came to myself as if out of a great sickness. There was something strange in my sensations, something indescribably new and, from its very novelty, incredibly sweet. I felt younger, lighter, happier in body; within I was conscious of a heady recklessness, a current of disordered sensual images running like a mill race in my fancy, a solution of the bonds of obligation, an unknown but not an innocent freedom of the soul. I knew myself, at the first breath of this new life, to be more wicked, tenfold more wicked, sold a slave to my original evil; and the thought, in that moment, braced and delighted me like wine. I stretched out my hands, exulting in the freshness of these sensations; and in the act, I was suddenly aware that I had lost in stature (R. L. Stevenson 2002, 57).

This apparently unrealistic passage is actually the result of a profound reflection on one's own and others' physical experience of the world.

It seems almost pleonastic here to remember that Stevenson wrote well before the birth of psychiatry and the publication of the Diagnostic and Statistical Manual of Mental Disorders (whose first edition dates back to 1952). Despite this, his novel has often been cited in the psychopathological field as a paradigmatic example of some disorders that tend to fragment the self, such as schizophrenia or dissociative personality disorder (McNally 2007; Garcia 1990; Singh and Chakrabarti 2008). And some scholars have hypothesized that Stevenson was partially inspired to write it by the medical literature (Stiles 2006; Mighall 2002). Undoubtedly, despite the use of unnatural artifices, Stevenson's novel illuminates like a stage spotlight a corner of mental life hitherto little explored and little narrated, but which – given the immediate and posthumous success of his novel – evidently attracts because of the universality of the principles it embraces.

For the purposes of my argument – that is, to show that Stevenson's bodily experience with the world has had an enabling role in the composition of *The Strange Case of Dr. Jekyll and Mr. Hyde* – I will try to demonstrate that the transformation described is an attempt to impose an order on a set of personal bodily experiences of the world. In fact, although Stevenson makes use of an apparently unnatural device like a concrete body transformation (which today thanks to plastic surgery is not quite as surreal as then), the sense of wonder for the profound mimetic attitude of this piece in my opinion prevails in the reader's mind over the perception of unnaturalness of the transformation. I think many would share this point of view, but of course it is only a point of view. What interests us most here is how (and if) the narrator came to a narration of the unnatural, or if in reality what he is narrating is a natural phenomenon in a broad sense. Stevenson (1888) partially described how his incredible tale of the human soul came to light in "A Chapter on Dreams," a memoir in which the author confesses to his readers how his process of writing this story started. The author tells us that for a long time he had this story in mind in the form of a sensation about human nature, but that it did not find a body. Dr. Jekyll and his story respectively gave a body and an order to make this set of sensations come to light: "I had long been trying to write a story on this subject, to find a body, a vehicle, for that strong sense of man's double being which must at times come in upon and overwhelm the mind of every thinking creature" (Stevenson 2002, 141). Even an unnatural event such as a bodily transformation derives in part from a physical experience perceived by the narrator: that of the duplicity of the mental. The narration he elaborates of it makes this perception of mental duplicity intelligible to the reader, probably recalling in the reader some of the same sensations that prompted Stevenson to rationalize it through the narrative act. According to Stevenson, the experience of the duplicity of the mental was shared by all human beings, or rather by all thinking creatures. And probably the immediate and posthumous success of his novel shows that he was right.

Of course, to demonstrate that bodily experience always has an enabling role in the creation of narratives without incurring the error of the inductivist chicken (Russell 1912[4]), I should analyze not only the sixth canto of Dante's Inferno or a very brief passage from Stevenson's novel, but all existing narratives and those that may exist in the future, and of course this is impossible. But since what has a constituent role in a phenomenon is also enabling with respect to it, I hope to convince the reader completely with the last section of my argument: I will try to demonstrate that the narrator's bodily experience with the world is a constituent factor of narratives.

4 Interaction with the physical world is a constituent element of human narrative skills

In the preceding section, I tried to demonstrate that the physical interaction of a narrator with the world is a necessary condition for the existence of a narrative. I thus defined the role of interaction between the narrator and the physical world as *enabling* for the existence of narratives.

4 A chicken which always receives its food at 9:00 a.m., through inductive reasoning, will think that every day at 9:00 a.m. its owner will enter the cage to give it food. Unfortunately, one day the owner will enter the cage to eat his chicken and the chicken will discover the fallacies of its reasoning.

To conclude my argument, I will show something more subtle and profound about the relationship between humans' narrative attitude and their physical interaction with the world. In fact, I will demonstrate that the physical interaction of human beings with the world is a *constituent* factor of every narrative.

Research on humans' narrative attitude very often focuses mainly on the human tendency always to favor social interpretations of events (as in the case of HSI). It is rarely remarked that every narrative first requires an experience of the narrator with the object of his narrative. A narrative is an attempt to interpret our observations of the physical world. Probably, we narrate what we want to know, understand and thereby control; and because most of us spend our lives attempting to know, understand, and interact with other people, most of our narratives focus on social contents. But these characteristics of narratives frequently obscure the fact that the main tool for building any kind of interpretation is the perception of physicality.

I mean that humans' physical interaction with the world is part of what we narrate, as the book in the example in section 1 was part of the phenomenon "a book is falling off a table because of a strong push." There can be no narrative in which humans' physical interaction with the world is absent or could theoretically be absent. I'll try to demonstrate this claim by developing the following idea: by definition (section 1), narratives always require that represented events are organized in a temporal dimension, and it is impossible to dispose events in a temporal sequence without implying a spatial dimension.

Although probably the main role of humans' narrative attitude is to sharpen social cognition (Gallagher 2004; Gallagher and Hutto 2008; Hutto 2009), it's nonetheless possible to imagine a narrative without any social content. HSI and all the experiments that followed clearly showed the human tendency always to favor a social interpretation of events in narratives. However, I would like to highlight that it's always possible to narrate a scene while renouncing any kind of social interpretation.

If someone were to ask us to describe the sequence of events represented in HSI without any kind of social content but referring exclusively to spatial movements of triangles, we would produce an unusual narrative, but one still understandable to the reader (or the listener). By contrast, we will never be able to describe this scene or any other sequence of events without making explicit reference to at least one physical component of the scene.

4.1 It is impossible to dispose events in a temporal sequence without recreating, at least implicitly, a spatial dimension

In narratives, the temporal dimension is the main tool for creating an order in the events represented. In fact, there are no narratives without a temporal dimension (section 1).

In this section, I will try to show that the way in which a narrator spatializes time also depends on physical interaction with the world he is living in during the production of the narration, and the one in which his neurodevelopment took place. I will also try to show that each of us has some cultural biases, and these push us to spatialize time in different ways.

Studies of how human beings link events temporally are innumerable (for a survey see Núñez and Cooperrider 2013). In recent decades, the most influential works in cognitive sciences on the notion of time have probably been those of Lakoff and Johnson (1980; 1999; Johnson 1987). They suggest that cognitively, the concept of time can be grasped exclusively in two ways: by correlating it with events through a metonymic process and by mapping it onto space through a metaphoric process. According to their theory, we can never conceive of time as an autonomous entity. In the first case, to think of *time*, we always need to compare different events. We can have an experience of time only by putting events in relation to other events that are culturally canonical, such as the ticking hands of a clock, the movement of the sun or the swing of a pendulum. This metonymical process is cognitively efficient because like time, events are directional, irreversible and cannot "unhappen." In the second case, in the metaphoric process, we use a notion that Lakoff and Johnson (1999) call *Spatial Time*: in English, that is, the present is the location of the observer, the future is in front of the observer, and the past is behind the observer. So, if we want to help a friend who lived through a negative experience, we can tell him to "leave this bad story behind you and move on." Many studies today report the spatialization of time in different cultures, probably even in all cultures (Núñez and Cooperrider 2013). Today, most philosophers, anthropologists, and psychologists agree with the idea that time is cognitively represented in a spatial format (for a review supporting this thesis see Bonato et al. 2012).

Our visuo-spatial habits strongly influence our conception of time and our way of narrating events. Different cultures can have different directions of timelines that can peacefully coexist. In fact, in societies that usually write from left to right, people also organize their timelines from left to right, and vice versa (Boroditsky 2011; Bottini et al. 2015; Fuhrman and Boroditsky 2010; Weger and Pratt 2008). In American English, when speakers are thinking about the sequencing of events, this can evoke both a sagittal (front-back) and a lateral (right-left) timeline

(Walker et al. 2014). But if no particular manipulation is implemented, people have a mental timeline coherent with their cultural direction of writing (Bonato et al. 2012; Núñez and Cooperrider 2013). Bonato et al. (2012) suggest that the physical direction of movement of the writing hand can be considered a link between physical interaction with the world as primed in the narrator and his way of conceptualizing time. Moreover, Chatterjee et al. (1999) showed that in Western countries the action agent is usually mentally represented in sentences and drawings on the left and the action object on the right, whereas countries where people read from right to left show an opposite behavior (Maas and Russo 2003). This means that our visuo-motor habits exert a strong influence on our representation of time and that, by influencing temporal concepts, these habits also influence our narratives. Moreover, left-to-right readers – as they are sequencing a story – seem to develop mental representations that position events from left to right on an analogical mental line (Santiago et al. 2010).

In 2008, Casasanto and Boroditsky showed in six psychophysical experiments that people who are estimating the duration of an event cannot ignore spatial information. For example, in the first experiment they asked participants to estimate and reproduce using a computer mouse the displacement or the duration of some growing lines in a computer screen. They found that a spatial displacement of lines affected participants' estimation of duration, but duration of stimuli did not affect participants' estimation of space. Based on their results, experimenters posited that the metaphorical relationship between time and space highlighted by Lakoff and Johnson exists in humans' more basic representations of time. Specifically, they describe the relationship between space and time in human cognition as *asymmetrically dependent*: the perception of time depends on the (more direct) perception of space. Casasanto and Borodistky (2008) cited many classic studies to support their view (Benussi 2002; Bill and Teft 1969; Cohen 1967; Cohen et al. 1954; Collyer 1977; Helson 1931; Jones and Huang 1982; Price-Williams 1954; Sarrazin et al. 2004), but all of these studies were focused on *space* and *time* as linguistic concepts. On the contrary, all the experiments reported by Casasanto and Borodistky (2008) used non-linguistic stimuli and responses and overall showed that the pattern of cross-dimensional interference between space and time is also asymmetrical when there is no linguistic mediation. Their experiments show that people cannot ignore spatial information during time estimation, and so they conclude: "These findings provide evidence that the metaphorical relationship between space and time observed in language also exists in our more basic representations of distance and duration, and suggest that our mental representations of things we can never see or touch may be built, in part, out of representations of physical experiences in perception and action" (Casasanto and Boroditsky 2008, 591).

Extensive research shows how spatial inputs are used in all cultures to temporally interpret events at a pre-linguistic level. For example, different studies show that the spatial position of a stimulus influences the estimation of its duration at a pre-linguistic level (Conson et al. 2008; Ishihara et al. 2008; Vicario et al. 2008; Vicario et al. 2009; Torralbo et al. 2006; Santiago et al. 2007; Ouellet et al. 2010; Di Bono et al. 2012; Cicchini and Morrone 2009). For example, Vicario et al. (2008) showed that right-handed subjects underestimate the duration of stimuli in a screen if these are presented in the left hemispace and overestimate the duration of the same stimuli if these are presented in the right hemispace. The orientation of spatial attention influences the perception of the duration of events (Frassinetti et al. 2009). A stimulus' dimension influences the perception of its duration at a pre-linguistic level (Xuan et al. 2007; Ono and Kawahara 2007; Matthews et al. 2011). Also, the decision to categorize words as *referring to the past* (i.e. "before") or *referring to the future* (i.e. "after") is influenced by stimulus' collocation in the screen and by buttons' positions in the keypress (Torralbo et al. 2006; Santiago et al. 2007). This trend has also been shown in the sequencing of histories (Santiago et al. 2010).

Many studies show that in many cultures the use of deictic time is associated with the sagittal axis of the body of the percipient (Eikmeier et al. 2013; Hartmann and Mast 2012; Koch et al. 2011; Kranjec and McDonough 2011; Sullivan and Barth 2012; Ulrich et al. 2012). For example, Hartmann and Mast (2012) asked participants to categorize verbal stimuli as in the past or in the future while they were displaced forward or backward. In this way they found that the future was categorized faster in a forward than in a backward position. However, none of these studies asks the participants to actively narrate an event after the stimulus. In fact, the link is mainly shown by priming effects.

However, some studies show that perception of movement influences the way in which a narrator temporally links events during daydream.[5] Miles et al. (2010a) found that the observation of movement through space in one direction or another will influence the temporal locus of mental activity. Specifically, they showed participants an animated star-field display. The stars were animated to move centripetally or centrifugally, corresponding to the perception of backward and forward movement, respectively. Participants were instructed to observe the screen and complete a dull task in which daydreaming was expected to occur. Finally, the experimenters asked the participants if they experienced daydreams and to estimate what proportion of the daydream was related to the past com-

5 Daydreams fully meet our definition of narratives because they are the product of a narrator who puts a series of events into relation (in this case in his mind) by linking them temporally.

pared to the future. Participants who observed the centrifugal movement reported a significantly higher proportion of future-related daydreams than participants who observed the centripetal movement of the stars.

Moreover, Miles et al. (2010b) showed that the link between a temporal trend in mental imagery and a certain bodily attitude of the percipient is bidirectional. They asked participants to (a) recall some past events of their lives, and (b) imagine some future events of their lives, while wearing a blindfold. Meanwhile, the experimenters measured their postural sway by a magnetic motion tracking system. Results showed that participants who recalled past events moved backwards whereas participants who imagined future events moved forwards.

Unfortunately, it is not possible to discuss here all the scientific literature on the connections between the notions of space and time. All the same, I hope the data presented have been sufficient to convince the reader that as human beings we always need space to conceptualize time and that this link is deeply embodied in us. Because narratives are, by definition, a temporal linking of events, it will always be impossible for anyone to arrange events in a narrative sequence without recreating, even if only implicitly, a spatial dimension.

5 Conclusions

In 1997, the French journalist Jean-Dominique Bauby, despite being affected by Locked-in Syndrome,[6] published his autobiographical book *Le scaphandre et le papillon* (*The Diving Bell and the Butterfly*). He wrote a narrative despite his almost total paralysis by blinking when a therapist, who would slowly recite the alphabet, reached the correct letters for each word. Obviously, without his last physical capacity of blinking, he could never narrate (another example of the enabling role of physical world interaction for narrative). But the more important point is that physical interaction with the world was also at the centre of his book:

> An ordinary day. At seven the chapel bells begin again to punctuate the passage of time, quarter hour by quarter hour. After their night's respite, my congested bronchial tubes once more begin their noisy rattle. My hands, lying curled on the yellow sheets, are hurting, although I can't tell if they are burning hot or ice cold. To fight off stiffness, I instinctively stretch, my arms and legs moving only a fraction of an inch. It is often enough to bring relief to a painful limb.

---

6 A condition, usually due to brainstem stroke, in which a perfectly aware patient cannot move any part of his body except for vertical eye movements and blinking.

> My diving bell becomes less oppressive, and my mind takes flight like a butterfly. There is so much to do. You can wander off in space or in time, set out for Tierra del Fuego or for King Midas's court.
>
> You can visit the woman you love, slide down beside her and stroke her still-sleeping face. You can build castles in Spain, steal the Golden Fleece, discover Atlantis, realize your childhood dreams and adult ambitions.
>
> Enough rambling. My main task now is to compose the first of these bedridden travel notes so that I shall be ready when my publisher's emissary arrives to take my dictation, letter by letter. In my head [. . .] (Bauby 2019 [1997], Prologue).

What I have tried to show in this essay is that humans cannot narrate (or imagine, perceive, or think) without involving their experience of the physical world in this cognitive activity. All cognitive activities are physically supported by cellular activities and neuronal discharges, but this enabling role is not sufficient to describe the relationship between thought and the physical world. The interaction of a human being with the physical world is a constituent part of their thought, and all physical experiences that they live through will be part of their mind, of their reasoning, of their narratives.

Reference list

Alber, Jan. "Unnatural Narrative" (2013). *The Living Handbook of Narratology*. Eds. Peter Hühn et al. Hamburg: Hamburg University. https://www.lhn.uni-hamburg.de/node/104.html (22 August 2021).

Alber, Jan, Stefan Iversen, Henrik Skov Nielsen, and Brian Richardson. "Unnatural Narratives, Unnatural Narratology: Beyond Mimetic Models." *Narrative* 18.2 (2010): 113–136.

Alighieri, Dante. *La Divina Commedia: Inferno*. Ed. Natalino Sapegno. Bologna: La Nuova Italia, 1985.

Alighieri, Dante. *The Divine Comedy of Dante Alighieri*. Vol. 1: *Inferno*. Ed. and transl. Robert M. Durling. Introduction and notes by Ronald L. Martinez and Robert M. Durling. New York and Oxford: Oxford University Press, 1996.

American Psychiatric Association (APA). *Diagnostic and Statistical Manual of Mental Disorders (DSM-5®)*. 5th edition. Arlington, VA: American Psychiatric Association, 2013.

Baixauli, Inmaculada, Carla Colomer, Belén Roselló, and Ana Miranda. "Narratives of Children with High-Functioning Autism Spectrum Disorder: A Meta-Analysis." *Research in Developmental Disabilities* 59 (2016): 234–254.

Baron-Cohen, Simon, Alan M. Leslie, and Uta Frith. "Mechanical, Behavioural and Intentional Understanding of Picture Stories in Autistic Children." *British Journal of Developmental Psychology* 4.2 (1986): 113–125.

Battaglia, Fortunato, Sarah H. Lisanby, and David Freedberg. "Corticomotor Excitability during Observation and Imagination of a Work of Art." *Frontiers in Human Neuroscience* 5 (2011): art. 79.

Bauby, Jean-Dominique. *The Diving Bell and the Butterfly: A Memoir of Life in Death*. Transl. Jeremy Leggatt. London: Fourth Estate, 2019 [1997].

Benussi, Vittorio. *Psychologie der Zeitauffassung*. Amsterdam: Rodopi, 2002 [1913].

Bill, J. Christopher, and Leon W. Teft. "Space-time Relations: Effects of Time on Perceived Visual Extent." *Journal of Experimental Psychology* 81.1 (1969): 196–199.
Bliss, Lynn S., Allyssa McCabe, and A. Elisabeth Miranda. "Narrative Assessment Profile: Discourse Analysis for School-Age Children." *Journal of Communication Disorders* 31.4 (1998): 347–363.
Bonato, Mario, Marco Zorzi, and Carlo Umiltà. "When Time Is Space: Evidence for a Mental Time Line." *Neuroscience & Biobehavioral Reviews* 36.10 (2012): 2257–2273.
Bornstein, Robert, and Thane Pittman, eds. *Perception without Awareness: Cognitive, Clinical, and Social Perspectives*. Conference proceedings of the same titled conference at Gettysburg College, 1–2 March 1991, Gettysburg, PA. New York and London: Guilford Press, 1992.
Boroditsky, Lera. "How Languages Construct Time." *Space, Time and Number in the Brain: Searching for the Foundations of Mathematical Thought*. Eds. Stanislas Dehaene and Elizabeth M. Brannon. Amsterdam: Elsevier, 2011. 333–341.
Bottini, Roberto, Davide Crepaldi, Daniel Casasanto, Virgine Crollen, and Olivier Collignon. "Space and Time in the Sighted and Blind." *Cognition* 141 (2015): 67–72.
Calder, Andrew J., Jill Keane, Facundo Manes, Nagui Antoun, and Andrew W. Young. "Impaired Recognition and Experience of Disgust Following Brain Injury." *Nature Neuroscience* 3.11 (2000): 1077–1078.
Caruana, Fausto, Ahmad Jezzini, Beatrice Sbriscia-Fioretti, Giacomo Rizzolatti, and Vittorio Gallese. "Emotional and Social Behaviors Elicited by Electrical Stimulation of the Insula in the Macaque Monkey." *Current Biology* 21.3 (2011): 195–199.
Caruana, Fausto, Pietro Avanzini, Francesca Gozzo, Stefano Francione, Francesco Cardinale, Giacomo Rizzolatti. "Mirth and Laughter Elicited by Electrical Stimulation of the Human Anterior Cingulate Cortex." *Cortex* 71 (2015): 323–331.
Casasanto, Daniel, and Lera Boroditsky. "Time in the Mind: Using Space to Think about Time." *Cognition* 106.2 (2008): 579–593.
Castelli, Fulvia, Francesca Happé, Uta Frith, and Chris Frith. "Movement and Mind: A Functional Imaging Study of Perception and Interpretation of Complex Intentional Movement Patterns." *Neuroimage* 12.3 (2000): 314–325.
Castelli, Fulvia, Chris Frith, Francesca Happé, and Uta Frith. "Autism, Asperger Syndrome and Brain Mechanisms for the Attribution of Mental States to Animated Shapes." *Brain* 125.8 (2002): 1839–1849.
Chatterjee, Anjan, M. Helen Southwood, and David Basilico. "Verbs, Events and Spatial Representations." *Neuropsychologia* 37.4 (1999): 395–402.
Chen, Yu-Han, Jürgen Dammers, Frank Boers, Susanne Leiberg, J. Christopher Edgar, Timothy P. L. Roberts, and Klaus Mathiak. "The Temporal Dynamics of Insula Activity to Disgust and Happy Facial Expressions: A Magnetoencephalography Study." *Neuroimage* 47.4 (2009): 1921–1928.
Cicchini, Guido M., and Maria Concetta Morrone. "Shifts in Spatial Attention Affect the Perceived Duration of Events." *Journal of Vision* 9.1 (2009): art. 9.
Cohen, John. *Psychological Time in Health and Disease*. Springfield, IL: Charles C. Thomas, 1967.
Cohen, John, Charles Edward Mark Hansel, and J. D. Sylvester. "Interdependence of Temporal and Auditory Judgments." *Nature* 174.4431 (1954): 642–644.
Collyer, C. E. "Discrimination of Spatial and Temporal Intervals Defined by Three Light Flashes: Effects of Spacing on Temporal Judgments and of Timing on Spatial Judgments." *Perception & Psychophysics* 21.4 (1977): 357–364.

Conson, Massimiliano, Fausta Cinque, Anna Maria Barbarulo, Luigi Trojano. "A Common Processing System for Duration, Order and Spatial Information: Evidence from a Time Estimation Task." *Experimental Brain Research* 187.2 (2008): 267–274.

De Jaegher, Hanne, Ezequiel Di Paolo, and Shaun Gallagher. "Can Social Interaction Constitute Social Cognition?" *Trends in Cognitive Sciences* 14.10 (2010): 441–447.

Di Bono, Maria Grazia, Marco Casarotti, Konstantinos Priftis, Lucia Gava, Carlo Umiltà, and Marco Zorzi. "Priming the Mental Time Line." *Journal of Experimental Psychology: Human Perception and Performance* 38.4 (2012): 838–842.

Di Cesare, Giuseppe, Cinzia Di Dio, Massimo Marchi, and Giacomo Rizzolatti. "Expressing Our Internal States and Understanding Those of Others." *Proceedings of the National Academy of Sciences* 112.33 (2015): 10331–10335.

Dolan, R. J., J. S. Morris, and B. de Gelder. "Crossmodal Binding of Fear in Voice and Face." *Proceedings of the National Academy of Sciences* 98.17 (2001): 10006–10010.

Driver, Jon, Gordon C. Baylis, and Robert D. Rafal. "Preserved Figure-Ground Segregation and Symmetry Perception in Visual Neglect." *Nature* 360.6399 (1992): 73–75.

Eikmeier, Verena, Hannes Schröter, Claudia Maienborn, Simone Alex-Ruf, and Rolf Ulrich. "Dimensional Overlap between Time and Space." *Psychonomic Bulletin & Review* 20.6 (2013): 1120–1125.

Filimon, Flavia, Cory Rieth, Martin Sereno, and Garrison Cottrell. "Observed, Executed, and Imagined Action Representations Can Be Decoded from Ventral and Dorsal Areas." *Cerebral Cortex* 25.9 (2015): 3144–3158.

Fludernik, Monika. "How Natural Is 'Unnatural Narratology': Or, What Is Unnatural about Unnatural Narratology?" *Narrative* 20.3 (2012): 357–370.

Fogassi, Leonardo, Pier Francesco Ferrari, Benno Gesierich, Stefano Rozzi, Fabian Chersi, and Giacomo Rizzolatti. "Parietal Lobe: From Action Organization to Intention Understanding." *Science* 308.5722 (2005): 662–667.

Frassinetti, Francesca, Barbara Magnani, and Massimiliano Oliveri. "Prismatic Lenses Shift Time Perception." *Psychological Science* 20.8 (2009): 949–954.

Fuhrman, Orly, and Lera Boroditsky. "Crosscultural Differences in Mental Representations of Time: Evidence from an Implicit Nonlinguistic Task." *Cognitive Science* 34.8 (2010): 1430–1451.

Fusar-Poli, Paolo, Anna Placentino, Francesco Carletti, Paola Landi, Paul Allen, Simon Surguladze, et al. "Functional Atlas of Emotional Faces Processing: A Voxel-Based Meta-Analysis of 105 Functional Magnetic Resonance Imaging Studies." *Journal of Psychiatry & Neuroscience* 34.6 (2009). 418–432.

Gallagher, Shaun. "Understanding Interpersonal Problems in Autism: Interaction Theory as an Alternative to Theory of Mind." *Philosophy, Psychiatry, & Psychology* 11.3 (2004): 199–217.

Gallagher, Shaun. "The Narrative Alternative to Theory of Mind." *Radical Enactivism: Intentionality, Phenomenology, and Narrative: Focus on the Philosophy of Daniel D. Hutto*. Ed. Richard Menary. Amsterdam and Philadelphia: John Benjamins, 2006. 223–229.

Gallagher, Shaun. "Direct Perception in the Intersubjective Context." *Consciousness and Cognition* 17.2 (2008): 535–543.

Gallagher, Shaun, and Daniel Hutto. "Understanding Others through Primary Interaction and Narrative Practice." *The Shared Mind: Perspectives on Intersubjectivity*. Eds. Jordan Zlatev, Timothy P. Racine, Chris Sinha, and Esa Itkonen. Amsterdam and Philadelphia: John Benjamins, 2008. 17–38.

Garcia, Emanuel E. "A Brief Note on 'Jekyll and Hyde' and MPD." *Dissociation: Progress in the Dissociative Disorders* 3.3 (1990): 165–166.

Gibson, William S., Shinho Cho, Osama A. Abulseoud, Krzysztof R. Gorny, Joel P. Felmlee, Kirk M. Welker, et al. "The Impact of Mirth-Inducing Ventral Striatal Deep Brain Stimulation on Functional and Effective Connectivity." *Cerebral Cortex* 27.3 (2017): 2183–2194.

Giummarra, Melita J., and John L. Bradshaw. "Synaesthesia for Pain: Feeling Pain with Another." *Mirror Neuron Systems: The Role of Mirroring Processess in Social Cognition*. Ed. Jamie A. Pineda. New York: Humana Press, 2008. 287–307.

Glenberg, Arthur M., Marc Stato, Luigi Cattaneo, Lucia Riggio, Daniele Palumbo, and Giovanni Buccino. "Processing Abstract Language Modulates Motor System Activity." *Quarterly Journal of Experimental Psychology* 61.6 (2008): 905–919.

Hadjikhani, Nouchine, and Beatrice de Gelder. "Seeing Fearful Body Expressions Activates the Fusiform Cortex and Amygdala." *Current Biology* 13.24 (2003): 2201–2205.

Halligan, Peter W., Gereon Fink, John Marshall, and Giuseppe Vallar. "Spatial Cognition: Evidence from Visual Neglect." *Trends in Cognitive Sciences* 7.3 (2003): 125–133.

Halliwell, Stephen. *The Aesthetics of Mimesis: Ancient Texts and Modern Problems*. Princeton, NJ: Princeton University Press, 2002.

Hardwick, Robert M., Svenia Caspers, Simon Eickhoff, and Stephan Swinnen. "Neural Correlates of Action: Comparing Meta-Analyses of Imagery, Observation, and Execution." *Neuroscience & Biobehavioral Reviews* 94 (2018): 31–44.

Harris, Robert, and Bauke M. de Jong. "Cerebral Activations Related to Audition-Driven Performance Imagery in Professional Musicians." *PloS one* 9.4 (2014): e93681.

Hartmann, Matthias, and Fred W. Mast. "Moving Along the Mental Time Line Influences the Processing of Future Related Words." *Consciousness and Cognition* 21.3 (2012): 1558–1562.

Heider, Fritz, and Marianne Simmel. "An Experimental Study of Apparent Behavior." *The American Journal of Psychology* 57.2 (1944): 243–259.

Helson, Harry, and Samuel M. King. "The Tau Effect: An Example of Psychological Relativity." *Journal of Experimental Psychology* 14.3 (1931): 202–217.

Holck, Pernille, Annika Dahlgren Sandberg, and Ulrika Nettelbladt. "Narrative Ability in Children with Cerebral Palsy." *Research in Developmental Disabilities* 32.1 (2011): 262–270.

Hutto, Daniel D. "Limited Engagements and Narrative Extensions." *International Journal of Philosophical Studies* 16.3 (2008): 419–444.

Hutto, Daniel D. "Folk Psychology as Narrative Practice." *Journal of Consciousness Studies* 16.6–8 (2009): 9–39.

Inman, Cory S., Joseph R. Manns, Kelly R. Bijanki, David I. Bass, Stephan Hamann, Daniel L. Drane, et al. "Direct Electrical Stimulation of the Amygdala Enhances Declarative Memory in Humans." *Proceedings of the National Academy of Sciences* 115.1 (2018): 98–103.

Ishihara, Masami, Peter E. Keller, Yves Rossetti, and Wolfgang Prinz. "Horizontal Spatial Representations of Time: Evidence for the STEARC Effect." *Cortex* 44.4 (2008): 454–461.

Jezzini, Ahmad, Fausto Caruana, Ivilin Stoianov, Vittorio Gallese, and Giacomo Rizzolatti. "Functional Organization of the Insula and Inner Perisylvian Regions." *Proceedings of the National Academy of Sciences* 109.25 (2012): 10077–10082.

Johnson, Mark. *The Body in the Mind: The Bodily Basis of Meaning, Imagination, and Reason*. Chicago: University of Chicago Press, 1987.

Jones, Bill, and Yih Lehr Huang. "Space-Time Dependencies in Psychophysical Judgment of Extent and Duration: Algebraic Models of the Tau and Kappa Effects." *Psychological Bulletin* 91.1 (1982): 128–142.

Keysers, Christian, Jon H. Kaas, and Valeria Gazzola. "Somatosensation in Social Perception." *Nature Reviews Neuroscience* 11.6 (2010): 417–428.

Klin, Ami, and Warren Jones. "Attributing Social and Physical Meaning to Ambiguous Visual Displays in Individuals with Higher-Functioning Autism Spectrum Disorders." *Brain and Cognition* 61.1 (2006): 40–53.

Koch, Sabine C., Stefanie Glawe, and Daniel V. Holt. "Up and Down, Front and Back: Movement and Meaning in the Vertical and Sagittal Axes." *Social Psychology* 42.3 (2011): 214–224.

Kranjec, Alexander, and Laraine McDonough. "The Implicit and Explicit Embodiment of Time." *Journal of Pragmatics* 43.3 (2011): 735–748.

Krolak-Salmon, Pierre, Marie-Anna Hénaff, Jean Isnard, Catherine Tallon-Baudry, Marc Guénot, Alain Vighetto, et al. "An Attention Modulated Response to Disgust in Human Ventral Anterior Insula." *Annals of Neurology: Official Journal of the American Neurological Association and the Child Neurology Society* 53.4 (2003): 446–453.

Lakoff, George, and Mark Johnson. *Metaphors We Live By*. Chicago: University of Chicago Press, 1980.

Lakoff, George, and Mark Johnson. *Philosophy in the Flesh: The Embodied Mind and Its Challenge to Western Thought*. New York: Basic Books, 1999.

Lamarque, Peter. "On Not Expecting Too Much from Narrative." *Mind & Language* 19.4 (2004): 393–408.

Lanteaume, Laura, Stéphanie Khalfa, Jean Régis, Patrick Marquis, Patrick Chauvel, and Fabrice Bartolomei. "Emotion Induction after Direct Intracerebral Stimulations of Human Amygdala." *Cerebral Cortex* 17.6 (2007): 1307–1313.

Maass, Anne, and Aurore Russo. "Directional Bias in the Mental Representation of Spatial Events: Nature or Culture?" *Psychological Science* 14.4 (2003): 296–301.

Marshall, John C., and Peter W. Halligan. "Blindsight and Insight in Visuo-Spatial Neglect." *Nature* 336.6201 (1988): 766–767.

Matthews, William J., Neil Stewart, and John H. Wearden. "Stimulus Intensity and the Perception of Duration." *Journal of Experimental Psychology: Human Perception and Performance* 37.1 (2011): 303–313.

McNally, Kieran. "Schizophrenia as Split Personality/Jekyll and Hyde: The Origins of the Informal Usage in the English Language." *Journal of the History of the Behavioral Sciences* 43.1 (2007): 69–79.

Meletti, Stefano, Laura Tassi, Roberto Mai, Nicola Fini, Carlo Alberto Tassinari, and Giorgio Lo Russo. "Emotions Induced by Intracerebral Electrical Stimulation of the Temporal Lobe." *Italian League Against Epilepsy (LICE): Official Reports 2005–2006 and Proceedings of the 28th National Congress, 18–21 May 2005, Bari, Italy*. Eds. Renzo Guerrini, Maria Paola Canevini, Roberto Michelucci, and Emilio Perucca. Special Issue *Epilepsia* 47.5 (2006): 47–51.

Mighall, Robert. "Diagnosing Jekyll: The Scientific Context to Dr Jekyll's Experiment and Mr Hyde's Embodiment." *The Strange Case of Dr Jekyll and Mr Hyde and Other Tales of Terror*. By Robert Louis Stevenson. Ed. Mighall Robert. London: Penguin Books, 2002. 143–161.

Miles, Lynden K., Katarzyna Karpinska, Joanne Lumsden, and C. Neil Macrae. "The Meandering Mind: Vection and Mental Time Travel." *PLoS One* 5.5 (2010a): e10825.

Miles, Lynden K., Louise K. Nind, and C. Neil Macrae. "Moving Through Time." *Psychological Science* 21.2 (2010b): 222–223.

Morris, J. S., C. D. Frith, D. I. Perrett, D. Rowland, A. W. Young, A. J. Calder, and R. J. Dolan. "A Differential Neural Response in the Human Amygdala to Fearful and Happy Facial Expressions." *Nature* 383.6603 (1996): 812–815.

Murray, Ryan J., Tobias Brosch, and David Sander. "The Functional Profile of the Human Amygdala in Affective Processing: Insights from Intracranial Recordings." *Cortex* 60 (2014): 10–33.

Nedelko, Violetta, Thomas Hassa, Farsin Hamzei, Cornelius Weiller, Ferdinand Binkofski, Mircea Ariel Schoenfeld, et al. "Age-Independent Activation in Areas of the Mirror Neuron System during Action Observation and Action Imagery: A fMRI Study." *Restorative Neurology and Neuroscience* 28.6 (2010): 737–747.

Núñez, Rafael, and Kensy Cooperrider. "The Tangle of Space and Time in Human Cognition." *Trends in Cognitive Sciences* 17.5 (2013): 220–229.

Ogoshi, Sakiko, Yasuhiro Ogoshi, Tomohiro Takezawa, and Yoshinori Mitsuhashi. "An Examination of Gender Differences in Mu Rhythm of the Mirror-Neuron System during the Imagination of Observed Action." *Brain and Health Informatics: Proceedings of the International Conference, BHI 2013, 29–31 October 2013, Maebashi, Japan*. Eds. Kazayuki Imamura, Shiro Usui, Tomoaki Shirao, Takuji Kasamatsu, Lars Schwabe, and Ning Zhong. Cham: Springer, 2013. 489–495.

Ono, Fuminori, and Jun-Ichiro Kawahara. "The Subjective Size of Visual Stimuli Affects the Perceived Duration of Their Presentation." *Perception & Psychophysics* 69.6 (2007): 952–957.

Oosterhof, Nikolaas N., Steven P. Tipper, and Paul E. Downing. "Visuo-Motor Imagery of Specific Manual Actions: A Multi-Variate Pattern Analysis fMRI Study." *Neuroimage* 63.1 (2012): 262–271.

Ouellet, Marc, Julio Santiago, María Jesús Funes, and Juan Lupiánez. "Thinking About the Future Moves Attention to the Right." *Journal of Experimental Psychology: Human Perception and Performance* 36.1 (2010): 17–24.

Papagno, Costanza, Alberto Pisoni, Giulia Mattavelli, Alessandra Casarotti, Alessandro Comi, Francesca Fumagalli, et al. "Specific Disgust Processing in the Left Insula: New Evidence from Direct Electrical Stimulation." *Neuropsychologia* 84 (2016): 29–35.

Pennisi, Paola. *Il linguaggio dell'autismo: Studi sulla comunicazione silenziosa e la pragmatic delle parole*. Bologna: Il Mulino, 2016.

Pennisi, Paola. "Inferential Abilities and Pragmatic Deficits in Subjects with Autism Spectrum Disorders." *Pragmemes and Theories of Language Use*. Eds. Keith Allan, Alessandro Capone, Istvan Kecskes. Cham: Springer, 2017. 749–768.

Pennisi, Paola, and Patrizia Torricelli. "Interazione sociale, cognizione sociale e acquisizione della L2". *Competence, Performance, Enhancement: Corisco X PRIN*. By Sebastiano Nucera and Alessandra Falzone. Rome et al.: corisco, 2018.

Phillips, M. L., A. W. Young, S. K. Scott, A. J. Calder, C. Andrew, V. Giampietro, et al. "Neural Responses to Facial and Vocal Expressions of Fear and Disgust." *Proceedings of the Royal Society of London, Series B: Biological Sciences* 265.1408 (1998): 1809–1817.

Pineda, Jaime A., B. Z. Allison, and A. Vankov. "The Effects of Self-Movement, Observation, and Imagination on/spl mu/rhythms and Readiness Potentials (RP's): Toward a Brain-Computer Interface (BCI)." *IEEE Transactions on Rehabilitation Engineering* 8.2 (2000): 219–222.

Price-Williams, D. R. "The *Kappa* Effect." *Nature* 173.4399 (1954): 363–364.

Ramachandran, Vilayanur S. *The Tell-Tale Brain: A Neuroscientist's Quest for What Makes Us Human*. New York and London: WW Norton & Company, 2011.

Renfrew, Catherine. *Bus Story Test: A Test of Narrative Speech*. 4th edition. Bicester,Oxon: Winslow Press, 1997. Swedish translation and adaptation: Y. Svensson, and A.-M. Tuominen-Eriksson. *Bussagan*. Göteborg, Sweden: Specialpedagogiska Institutet Läromedel, 2002.

Rizzolatti, Giacomo, and Laila Craighero. "The Mirror-Neuron System." *Annual Review of Neuroscience* 27 (2004): 169–192.

Rizzolatti, Giacomo, and Corrado Sinigaglia. *Mirrors in the Brain: How Our Minds Share Actions and Emotions*. Oxford: Oxford University Press, 2008.

Rizzolatti, Giacomo, and Corrado Sinigaglia. *Specchi nel cervello: Come comprendiamo gli altri dall'interno*. Milan: Cortina Raffaello, 2019.

Rochat, Magali J., Fausto Caruana, Ahmad Jezzini, Ludovic Escola, Irakli Intskirveli, Franck Grammont, et al. "Responses of Mirror Neurons in Area F5 to Hand and Tool Grasping observation." *Experimental Brain Research* 204.4 (2010): 605–616.

Rosenbaum, Peter, Nigel Paneth, Alan Leviton, Murray Goldstein, and Martin Bax. "A Report: The Definition and Classification of Cerebral Palsy April 2006." *Developmental Medicine & Child Neurology* 49.suppl. 109 (2007): 8–14.

Rozin, Paul, Jonathan Haidt, and Clark R. McCauley. "Disgust: The Body and Soul Emotion." *Handbook of Cognition and Emotion*. Eds. Tim Dalgleish and Mick J. Power. Chichester et al.: John Wiley & Sons, 1999. 429–445.

Russell, Bertrand. *The Problems of Philosophy*. New York: H. Holt and Company, 1912.

Ryan, Marie-Laure. "Toward a Definition of Narrative." *The Cambridge Companion to Narrative*. Ed. David Herman. Cambridge and New York: Cambridge University Press, 2007. 22–35.

Saarela, Miiamaaria, Yevhen Hlushchuk, Amada Williams, Martin Schürmann, Eija Kalso, and Riitta Hari. "The Compassionate Brain: Humans Detect Intensity of Pain from Another's Face." *Cerebral Cortex* 17.1 (2007): 230–237.

Sacks, Oliver. *The Man Who Mistook His Hife for a Hat and Other Clinical Tales*. New York: Summit books, 1985.

Santiago, Julio, Juan Lupáñez, Elvira Pérez, and María Jesús Funes. "Time (Also) Flies from Left to Right." *Psychonomic Bulletin & Review* 14.3 (2007): 512–516.

Santiago, Julio, Antonio Román, Marc Ouellet, Nieves Rodríguez, and Pilar Pérez-Azor. "In Hindsight, Life Flows from Left to Right." *Psychological Research PRPF* 74.1 (2010): 59–70.

Sapegno, Natalino. "Canto Sesto." *La Divina Commedia: Inferno*. By Dante Alighieri. Ed. Natalino Sapegno. Bologna: La Nuova Italia, 1985.

Sarrazin, Jean-Christophe, Marie-Dominique Giraudo, Jean Pailhous, and Reinoud Bootsma. "Dynamics of Balancing Space and Time in Memory: Tau and Kappa Effects Revisited." *Journal of Experimental Psychology: Human Perception and Performance* 30.3 (2004): 411–430.

Savaki, Helen E., and Vassilis Raos. "Action Perception and Motor Imagery: Mental Practice of Action." *Progress in Neurobiology* 175 (2019): 107–125.

Singh, Shubh M., and Subho Chakrabarti. "A Study in Dualism: The Strange Case of Dr. Jekyll and Mr. Hyde." *Indian Journal of Psychiatry* 50.3 (2008): 221–223.

Stevenson, Robert Louis. *The Strange Case of Dr Jekyll and Mr Hyde and Other Tales of Terror*. Ed. Robert Mighall. London: Penguin Books, 2002 [1886].

Stevenson, Robert Louis. "A Chapter on Dreams" [1888]. *The Strange Case of Dr Jekyll and Mr Hyde and Other Tales of Terror*. By Robert Louis Stevenson. Ed. Robert Mighall. London: Penguin Books, 2002 [1886]. 135–143.

Stiles, Anne. "Robert Louis Stevenson's 'Jekyll and Hyde' and the Double Brain." *Studies in English Literature, 1500–1900* (2006): 879–900.

Sullivan, Jessica L., and Hilary C. Barth. "Active (Not Passive) Spatial Imagery Primes Temporal Judgements." *The Quarterly Journal of Experimental Psychology* 65.6 (2012): 1101–1109.

Szameitat, Diana P., Benjamin Kreifelts, Kai Alter, André J. Szameitat, Annette Sterr, Wolfgang Grodd, and Dirk Wildgruber. "It Is Not Always Tickling: Distinct Cerebral Responses during Perception of Different Laughter Types." *Neuroimage* 53.4 (2010): 1264–1271.

Tammet, Daniel, with Simon Vance. *Born on a Blue Day: Inside the Extraordinary Mind of an Autistic Savant: A Memoir*. United States: Tantor Audio, 2007.

Torralbo, Ana, Julio Santiago, and Juan Lupiáñez. "Flexible Conceptual Projection of Time onto Spatial Frames of Reference." *Cognitive Science* 30.4 (2006): 745–757.

Turner, Mark. *The Literary Mind*. New York, NY: Oxford University Press, 1996.

Ulrich, Rolf, Verena Eikmeier, Irmgard de la Vega, Susana Ruiz Fernández, Simone Alex-Ruf, and Claudia Maienborn. "With the Past Behind and the Future Ahead: Back-to-Front Representation of Past and Future Sentences." *Memory & Cognition* 40.3 (2012): 483–495.

Umiltà, M. A., L. Escola, I. Intskirveli, F. Grammont, M. Rochat, F. Caruana, et al. "When Pliers Become Fingers in the Monkey Motor System." *Proceedings of the National Academy of Sciences* 105.6 (2008): 2209–2213.

Vicario, Carmelo Mario, Patrizia Pecoraro, Patrizia Turriziani, Giacomo Koch, Carlo Caltagirone, and Massimiliano Oliveri. "Relativistic Compression and Expansion of Experiential Time in the Left and Right Space." *PloS one* 3.3 (2008): e1716.

Vicario, Carmelo Mario, Gaetano Rappo, Anna Maria Pepi, and Massimiliano Oliveri. "Timing Flickers across Sensory Modalities." *Perception* 38.8 (2009): 1144–1151.

Walker, Esther J., Benjamin K. Bergen, and Rafael Núñez. "Disentangling Spatial Metaphors for Time Using Non-Spatial Responses and Auditory Stimuli." *Metaphor and Symbol* 29.4 (2014): 316–327.

Weger, Ulrich W., and Jay Pratt. "Time Flies Like an Arrow: Space-Time Compatibility Effects Suggest the Use of a Mental Timeline." *Psychonomic Bulletin & Review* 15.2 (2008): 426–430.

Wicker, Bruno, Christian Keysers, Jane Plailly, Jean-Pierre Royet, Vittorio Gallese, and Giacomo Rizzolatti. "Both of Us Disgusted in My Insula: The Common Neural Basis of Seeing and Feeling Disgust." *Neuron* 40.3 (2003): 655–664.

Xuan, Bin, Daren Zhang, Sheng He, and Xiangchuan Chen. "Larger Stimuli Are Judged to Last Longer." *Journal of Vision* 7.10 (2007): art. 2.

Part IV: Cognitive Approaches to Narrative across Literature and Science

Renata Gambino and Grazia Pulvirenti

Embodied Sublime

A Case-Study from German Romanticism: Heinrich von Kleist's Ekphrasis *Feelings about a Seascape by Friedrich*

Abstract: A controversial and enigmatic author of the Romantic era, Heinrich von Kleist was fascinated by a painting which is now considered one of the most emblematic representations of the Romantic Sublime, *The Monk at the Sea* by Caspar David Friedrich (now at the *Alte Nationalgalerie* in Berlin – NG 9/85). When the picture was first exhibited, in October 1810 at the Academy of Fine Arts in Berlin, together with another painting by Friedrich, *The Abbey in the Oakwood,* it was not positively received by the audience (see Müller 1986; 1995; Begemann 1990; Rühse 2013); it became a controversial case because it radically disrupted the basic principles of the classical landscape (Greiner 1996, 160).

At that time, Heinrich von Kleist, together with Adam Müller, edited a journal, *Berliner Abendblätter,* and they asked the writers Clemens Brentano and Achim von Arnim to write a review of Friedrich's painting. But the text that Kleist published was radically re-written by himself, although he published it using the initials of Brentano's name (cb), giving rise to a large quarrel (Kurz 1988).

What struck Kleist so much to lead him to re-write the text emphasizing its aesthetic power and novelty, which the other authors did not understand as deeply as he did? Why is Kleist's revolutionary ekphrasis of Friedrich's painting so difficult to understand and interpret even now?

The aim of this study is to highlight and explain the innovation in Kleist's use of the traditional rhetorical strategy of *ekphrasis* in relation to the crucial question about the Sublime-debate in the eighteenth century: he attests to an anti-Kantian concept of the Sublime, intending it as a physiological experience of the body, deeply rooted in the senses. In this way he disrupts the transcendental and hegemonic conceptions of the aesthetic experience as related to the abstract intellect and that of the arts as depending on pre-established rules and principles.

We argue in this paper[1] that Kleist's text does not describe an art object, according to the traditional ekphrastic strategy, but performs the aesthetic, emotional, dynamical, and immersive process engaged in by the reader. The immer-

1 This study is the result of a collaborative research carried out by the two co-authors. Sections 1–3 are written by Grazia Pulvirenti, sections 4–5 by Renata Gambino.

https://doi.org/10.1515/9783110782844-012

sion of the viewer-reader is elicited by stimulating embodied simulation, according to recent neuroscientific studies, and leads in Kleist's representation to a new embodied experience of the Sublime.

In this way Kleist takes a position in the eighteenth-century debate about the Sublime: the described experience echoes the concept of the Sublime theorized by Edmund Burke, in opposition to the Kantian one. While for Kant the Sublime results from a mental state of tension between nature and art, provoking an intuition of the supersensible (*Logos*), Burke intended the Sublime as bodily immersive experience, leading to the intuition of the sub-sensible (material essence).

Kleist's writing about the bodily experience of the viewer in front of the picture allows him to render the Sublime by calling on physical immersion and consequently on the emotional and imaginative power of the reader. Reconstructing the interwoven bodily and mental processes of the Sublime experience in front of a natural phenomenon, Kleist successfully links the natural to the artful Sublime in a very modern perspective. The Sublime becomes therefore a peculiar form of aesthetic experience rooted in the senses, an idea forecasting contemporary theories about embodied simulation. Experiencing beauty, according to Gumbrecht (2004) and Zeki (2008), always includes the construction of meaning on one side, and the bodily involvement of the image beholder on the other (see Gallese 2018). In Kleist's opinion, there is almost no difference in the experience of a viewer whether they are before an artifact or a natural landscape, since both experiences are rooted in the body, although the artistic one involves imagination to create the situation models responding to the represented nature, while the experience of nature impacts directly on the body. The major difference consists in the role played by the imagination, which is able to trigger the bodily simulation of nature, thus enhancing the cognitive power of the image beholder or of the reader, who is able to overcome their perceptual and imaginative boundaries.

1 Introduction

This study proposes a transdisciplinary neurohermeneutic approach[2] in literary criticism, linking traditional text-immanent studies to neurocognitive theories

2 On the concept of Neurohermeneutics see Gambino and Pulvirenti (2018a; 2019; 2024). According to the neurohermeneutic approach, the literary text is intended as a complex device that stimulates the reader to construct ever new meanings through the activation of situation models triggered by the peculiarities of the *foregrounding* features of literary devices. The reader is "guided" by different forms of *foregrounding* to experience the text, according to their own experiences rooted in the embodied reality of its perceptual, mnemonic, emotional and cognitive background.

about embodied simulation processes, which are involved in the activation of cognitive simulation-models built during text-reception and the reader's meaning-making process. By applying this approach, we will highlight the revolutionary features in Heinrich von Kleist's text *Feelings about a Seascape by Friedrich* (1810), an innovative *ekphrasis* of Caspar David Friedrich's painting *The Monk at the Sea*. We will interpret this text as a milestone in the crucial German Sublime-debate at the turn from the eighteenth to the nineteenth century, since in our opinion Kleist expresses in this text a sharp anti-Kantian position, featuring the concept of the Sublime represented in Friedrich's painting as a physiological experience of the body, deeply rooted in the senses.

We will demonstrate our hypothesis through the analysis of the referential situation models proposed in Kleist's text, with regard to the embodied simulation process prompted by the text in the reader's mind, and explicating this concept through Burke's idea of an embodied Sublime experience. This will lead us to two different goals:

1) taking into account recent neurocognitive theories about the embodied simulation process activated in the act of reading, it is possible to gain new insight into the relevance of specific textual features activating embodied situation models in order to convey new meanings;
2) applying these theories inherent to the neurohermeneutic approach to Kleist's text, we will gain a new interpretation of it, attesting to his anti-transcendental concept of the Sublime experience.

By doing so, we interpret Kleist's innovation in the use of ekphrasis as consisting in the fact that the author does not directly describe the art object, according to the traditional ekphrastic strategy, but rather the aesthetic, emotional, dynamical, and immersive process engaged in by the viewer looking at the painting. By reading this original text, the imagination of the reader is prompted to create a situation model (see Zwaan 1999), i.e. a mental representation of the viewer's relationship to the pictorial work of art. The situation-model theory implies that the reader is influenced by the peculiar nature of the textual representation, which in this case elicits an embodied simulation process. According to recent neuroscientific studies, this process implies the "possibility that we understand the *meaning* of language by simulating in our minds the experience that the language describes through memories of our own experiences" (Gallese 2005). Therefore, in Kleist's text the reader is guided to experience the Sublime represented in the painting in a new embodied way.

In our opinion, the peculiar embodied dimension proposed in the ekphrasis of Friedrich's painting reveals Kleist's critical position within the debate about the Sublime: the experience referred to in his ekphrasis hints at the concept of the Sublime theorized by Edmund Burke, in opposition to the Kantian one. While for Kant the experience of the Sublime results from a mental state of tension between nature and art, provoking an intuition of the super-sensible (*Logos*), Burke interpreted it as an immersive bodily experience, leading to an intuition of the sub-sensible (material essence). Kleist's stressing the bodily experience engaged in by the reader allows him to render vividly the deictic center of the Sublime and consequently to place the reader within specific spatial, temporal, and psychological dimensions, facilitating the activation of their emotional and imaginative power. Reconstructing the bodily and mental processes of the Sublime experience within a natural space, Kleist successfully links the natural to the artful Sublime in a very modern perspective. The Sublime becomes a peculiar form of aesthetic experience rooted in the senses and not in the power of reason, disrupting the mainstream ideas of that time and anticipating some of the very recent cognitive narratological and neuroaesthetic positions regarding experiencing beauty. So, for instance, according to Gumbrecht (2004), Zeki (2008), and Gallese (2018), beauty and the Sublime always deal with the construction of meaning on one side, and with the bodily involvement of the image beholder on the other (Gallese 2018). In Kleist's ekphrasis there is almost no difference between the experience of a viewer put in front of an artifact and that of a viewer put in front of a natural landscape. Therefore, the main difference between nature and art consists in the role played in response to art by the imagination, which constructs mental representations, enhancing the cognitive power of the image beholder or of the reader. This is obtained by disrupting the limits of the artistic media, and by eliciting a vicarious experience of events, perceptions, and emotions, feeding the never-ending process of constructing meaning, arousing emotions and gaining knowledge through mental imagery.

We will proceed by examining the following issues: a) the features of ekphrasis according to classical rhetoric; b) the concept of embodied simulation according to Gallese's hypothesis and situation-models theory; c) a brief account of the eighteenth-century German debate about the concepts of the Kantian transcendental Sublime vs. Burke's embodied Sublime; d) the analysis of Kleist's ekphrasis as modern representation of an embodied Sublime rooted in the senses and emotions of the human being considered as part of the living *Bios*.

2 Ekphrasis

Ekphrasis is a textual genre that aims to bring a visual subject matter vividly before the eyes (Webb 2009). This definition characterizes the pivotal works from antiquity to the Renaissance: Philostratus' *Εἰκόνες*, Giulio Camillo's *Theatre of Memory*, Vasari's *Lives of the Artists*, Giambattista Marino's *La Galleria* (without forgetting Homer's description of Achilles' shield in the eighteenth book of the *Iliad* and Virgil's description of the decorated cups by Alcimedon in the third eclogue of the *Aeneid*). This genre was rediscovered and much practiced in the eighteenth century, starting with Winckelmann's famous description of the Apollo Belvedere and with Lessing's attempt to challenge the old Horatian paradigm of "ut picture poësis erit." In the aesthetic discourse of this epoch – not only in texts by Johann Joachim Winckelmann (1756), and by Gotthold Ephraim Lessing (1766), but also in those by Wilhelm Heinse (1787) and Johann Wolfgang Goethe (1798) – ekphrasis contributed to the crucial inquiry into the relation between the artwork and its literary description, as well as to experiments of intertwining verbal and visual codes, involving the processes of both reading a text and priming a mental image.

From its origins, ekphrasis was meant to stimulate the imagination (Webb 2009, 1) and to achieve an alteration in the reader/listener's emotional state. This was to be obtained by enhancing through words the "liveliness" of the mental image representing the described object (Henkel 1997, 337), consequently illuminating the quality and essence of the object itself. This technique was considered in Latin culture, and particularly by Quintilian (*Institutio Oratoria*, 6.2.29–30), as a powerful tool giving to the verbal medium new persuasive qualities, pushing and stretching the usual descriptive limits of language and its power of meaningfulness (Boehm 2014, 22). For Latin rhetoricians, ekphrasis was "a technique used to make the audience feel involved in the subject matter, to make them feel as if they were at the scene of a crime, or that they themselves witnessed the achievements for which an emperor is being praised" (Webb 2009, 10). Ekphrasis was best applied to represent not just a scene (i.e. time, place, characters, etc.) but the action itself (i.e. a murder, a battle, etc.).

With regard to the relevance of actions in narrative, recent neuro-cognitive studies have empirically proven that the description of an action triggers a more intense emotional response in the reader/listener than the description of a static scene or of objects, since the somatic components of emotion react more strongly to the activation of the solicited sensory-motor circuitry and, as we shall explain later on, elicit "embodied simulation" ("simulazione incarnate") (Cuccio et al. 2013, 69). Although unaware of these brain processes, ancient writers were well aware of the efficacy of the rhetorical use of the language and aimed to depict subject matter through actions and events, instead of carefully listing the elements in the scene. In fact, the final purpose of ancient ekphrasis was to persuade and emotionally move the audience by creating the fictive and lively illusion of being present at the described scene. At a linguistic level, this was achieved through the quality of *energeia*, i.e. vividness of language, which was theorized not only in ancient times, but also in the Romantic period, mainly by Wilhelm von Humboldt (1830–1835).

In order to uncover the main features ascribed to *energeia*, we have to go back to book VI of Quintilian's *Institutio Oratoria*, in which he describes the production of *energeia* from the orator's point of view and sets out his understanding of the physiological and psychological processes involved in it. As analyzed by Ruth Webb (2009), Quintilian conceived *energeia* as the result of a mental process, pointing out that the force or vividness of a speech relies on the capacity of the orator to transfer his mental images through his words to the listener. This mastery was thought to give great power over the emotions of the audience (Quintilian, *Institutio Oratoria*, 6.2.29–30). Quintilian's claim about *energeia* implies that it is conceived as an active force with a strong impact on the reader's mind. *Energeia* is therefore far more than just a rhetorical strategy: it becomes a way of giving to language the power to transfer the mental images stored in the author's memory to the audience, activating the imagination (Webb 2009, 10) by triggering a situation model corresponding to the peculiarities of the description and simulating the experience of being present bodily at, and involved in the space and time of, the described situation.

Significant changes occurred in the use of ekphrasis during the German Romantic period, mainly due to the general paradigm-shift caused by the scientific, anthropological and aesthetic revolution taking place in the turn from the eighteenth to the nineteenth century. The relevance of a new view of what Jacques Rancière proposes to call the "aesthetic regime of art" (Rancière 2013) determines a deep change in the aesthetic paradigms of the epoch, finding in ekphrasis an ideal device to reflect on and express the radical innovations occurring in the philosophical debate about perception, thought and emotion, as pointed out by many scholars focusing their investigations on the regimes of the visible as related to language (Osterkamp 1991; Boehm and Pfotenhauer 1995; Pfotenhauer 2000; Miller 2002). What is at stake in the ekphrasis of Romanticism is the crisis of the classical aesthetic representation of visible qualities. This led Romantic authors to try to enhance the power of images by using new forms of description developed through an innovative experimental use of language principally conveying actions and inner motion (Gambino and Pulvirenti 2017a). This is a core feature of the new forms of modern ekphrasis arising from the Romantic aesthetic reflection on language as deeply connected to thought and rooted in the body. In fact, a crucial issue of aesthetic discourse during the Romantic period is the commitment to a concept of art and beauty as deeply rooted in all aspects of the *Bios*. In this sense it promoted reflection on some critical issues of human nature, such as consciousness (Berlin 1999, 20) and thought, striving toward a "knowledge of the whole human being" (Schlegel 1977, 241). Moreover, the programmatic project of Romantic aesthetics was to link nature and science to art, and art to nature and science, as pointed out by one of the most active theoreticians of Romanticism, Friedrich Schlegel: "The Romantic imperative demands [that] all nature and science should become art [and] art should become nature and science" (Schlegel 1977, 586).

The relevance of aesthetics in Romantic discourse is in the intention to create a new modeling of scientific, philosophical, social and ethical pursuits according to the forms exemplified in poetry and in the arts, as Novalis' famous imperative of "poeticizing life" claimed. In the Romantic view, poetry aims to reach the absolute essence of life and truth (the Absolute), intended as a whole, an unconditioned totality grounded in the *Bios*, a holistic unity itself open-ended and autonomous.

Therefore, in the Romantic epoch, ekphrasis was regarded as a strategy for enhancing a sensual perception of nature and of art. Moreover, the Romantic author was interested in evoking the subjective act of emotionally relating to an artifact, as in the case of Wordsworth's "tendency to prioritize the act of confronting an artwork over the artwork's appearance," or Keats's "psychological dynamics established between the speaker and the object of his gaze" (Kneisley 2016, 7 and 9).

The Romantics experimented with representing these emotional dynamics by triggering what today we would call a stronger simulation of the embodied experience during the act of reading, in order to convey the moving encounter with an artwork.

3 Embodied simulation and situation models

Embodied simulation is considered in neurocognitive studies as the source of empathy and enaction. By reading literary texts with many sensory descriptions, the bodily "states associated with actions, emotions, and sensations [are] evoked [. . .] as if the reader would be doing a similar action or experiencing a similar emotion or sensation" (Gallese 2007, 216). This internal activity (Varela et al. 1991) relies on an "as-if" process, which implies the empathic and emotional activation of the subject. As Vittorio Gallese claims, when interpreting the actions, emotions and sensations of others, embodied simulation is the core experience of the pre-rational sense-making process (Gallese 2009, 519–536). Freedberg and Gallese draw the following conclusions:

> Activation of the same brain region during first- and third-person experience of actions, emotions and sensations suggests that, as well as explicit cognitive evaluation of social stimuli, there is probably a phylogenetically older mechanism that enables direct experiential understanding of objects and of the inner world of others. (Freedberg and Gallese 2007, 198)

Not only the actions, emotions, and sensations we experience, but also those we read about, activate our internal embodied system of bodily representations "as if" we were directly experiencing similar emotions, feelings or sensations, also evoked by representational forms, colors, lines and words. We argue that the aesthetic and emotional response during literary experience is strongly enhanced by the activation of embodied simulation elicited by the dynamic qualities of literary imagery (Pulvirenti and Gambino 2014). It has already been demonstrated through brain imaging that the activity of motor areas is particularly enhanced by motor imagery triggered both by visual stimuli and by literary texts (Cuccio et al. 2013). There is also scientific evidence of the fact that the neural substrates for visual and haptic processes overlap (see Kim and James 2010; Stilla and Sathian 2008). Information deriving from visual and haptic origins converges in the IPS (intraparietal sulcus), which is not only part of the somatosensory and visual system, but also part of the motor system. This allows the integration of sensory, visual and motor processes. Moreover, the multisensory nature of imagination involves haptic activity bound to movement and visual acts (Goebel et al. 1998). In particular, looking at a painting representing a physical action or reading about a physical action engages part of the motor system, specifically the mirror-neuron system. Discovered by Gallese, Rizzolatti and colleagues (Pellegrino et al. 1992; Gallese et al. 1996; Iacoboni et al. 1999; Rizzolatti and Luigi 2009), mirror neurons fire, both in monkeys and human beings, not only when executing, but also when observing an action, and have also been shown to resonate in humans when observers "infer the intent of an artistic gesture" (Chatterjee 2014; Chatterjee and Vartanian 2014, 370), either explicitly declared in an artwork or implicitly conveyed through colors and forms. The mirror-neuron system offers a scientific explanation for many phenomena relevant to aesthetic experience, such as the empathic processes relying on embodied simulation activated in the sensory-motor system.

The embodied simulation process is relevant in every narrative strategy aiming to strongly engage the reader in the construction of an imaginative world. What neurocognitive scientists call embodied simulation is at the core of any immersive literary experience as well as of the human ability to understand others (Theory of Mind).

In our opinion, embodied simulation is linked to the construction of mental situation models elicited by the words of a text and directly responding to sensory-motor processes. The past 15 years of studies in this field have provided a great deal of evidence suggesting that language comprehension involves sensory-motor representations (Meteyard et al. 2012; Louwerse et al. 2015). According to studies of cognition and text processing referring to the simulation-model theory and the embodied cognition hypothesis, in order to understand the text, the reader's mind creates a situation model representing the narrative world described by the text which goes beyond the "text-base," i.e., the actual words of the text itself (Johnson-Laird 1983; van Dijk and Kintsch 1983). The construction of situation models is an important process for understanding narratives because it allows the reader to connect different aspects and information (time, space, motivation, character and cause-effect) within a coherent and fluid mental representation that can be continually integrated, stabilized, updated and emotionally experienced (Kintsch 1988; Kintsch et al. 1990; van Dijk and Kintsch 1983). In empirical psychology, most studies about reading distinguish three levels of cognitive representations in processing a text: 1) the surface code (wording and syntax); 2) the text-base (explicit text propositions and text-coherence); and 3) the referential situation model (what the text is about). At this last level, readers can take on an active role by placing themselves in the perspective of the characters and their surroundings (i.e., the time frame of the story; the location in which the events are happening; the objects that the characters are carrying, etc.). Unlike the surface structure or the propositional text-base, in which these mental representations rely solely on the information given in the sentences, situation models require the reader's imagination and their interpretation of the information conveyed (Graesser and Zwaan 1995).

As proposed by Anežka Kuzmičová (2014), the reader experiences the actions described in the literary text through their "enactment-imagery," an embodied process that enhances the creation of the situation model and sets the reader virtually within the storyworld (see Alber et al. 2018). In the aesthetic experience of a literary text, the intensity of a reader's emotional response is bound to the intensity of the activation of the neural motor areas, according to Chatterjee and colleagues, who have produced evidence of the interaction among sensory-motor, emotion-valuation and meaning-knowledge circuitry in aesthetic experience (Chatterjee and Vartanian 2014, 370; Shimamura 2012; 2013; Chatterjee 2014). This has been demonstrated in relation to vision, to language, and partially to the process of imagination: even in the absence of any physical movement, the observation of an object involves the activation of the sensory-motor system related to interacting with the object (Freedberg and Gallese 2007, 201). As well, language has been shown to exploit the pre-existing multimodal character of the sensory-motor system (Gallese and Lakoff 2005, 456). Many studies also prove that some parts of the brain activated by seeing are also engaged in visual imagination (imagining that you are seeing) (Farah 2000; Kosslyn 1980; 1983; Kosslyn et al.

2006; Gallese and Lakoff 2005, 463). Finally, as Gabrielle Starr also argues, referring to empirical studies that have shown the implication of the motor system in imagined actions, any kind of imagery, because of its multisensory nature, enhances a bodily kinaesthetic reaction (Starr 2010, 275–291; Starr 2013). Moreover, a high number of references to action and motion within a literary text produces a particularly intense dynamic embodied experience during the construction of the mental simulation in the act of reading (see Zwaan et al. 1995). The concept of mental simulation refers to the hypothesis that the reader captures aspects of the world (Johnson-Laird 1983; van Dijk and Kintsch 1983) represented in the story at an event level (Zwaan and Radvansky 1998). According to such evidence, we claim that Kleist's ekphrasis strongly triggers the embodied simulation of the reader because it enhances the creation of enactment-imagery. This kind of mental simulation (of visual perception and actions) intensifies the activation of the brain's sensory-motor neural areas, evoking an experience of the Sublime rooted in the body. (GP)

4 Transcendental Sublime *vs* Embodied Sublime

The German debate about the Sublime at the turn from the eighteenth to the nineteenth century was characterized by later authors' reflections on Kant's theories, distinguishing the Sublime from the Beautiful, as earlier theoreticians like Burke had already done. Kant described the Sublime as a "negative pleasure" (*Negative Lust*) (Kant 2009 [1790], 76), leading to an inhibition of life's bodily forces; this inhibition is overcome by a transcendental mental elaboration of this experience. The Sublime overwhelms the sensual perception and the imagination, leading to the activation of the rational intellect, which is considered as a "measure of the senses" (*Maßstab der Sinne*) (Kant 2009 [1790], 85). This is particularly enhanced by what Kant defines as the "dynamic Sublime" (*dynamisches Erhabene*) with regard to the "mathematic Sublime" (*mathematisches Erhabene*) (Kant 1764): while the mathematical Sublime is intended as the feeling of a supersensible faculty apprehending the magnitude of nature, the dynamic Sublime is intended as the experience of the dreadful boundless power of nature. The dynamic Sublime evokes feelings of limits and nothingness in the viewer and leads him/her to refuse to identify with their own bodily finiteness and to discover their mental superiority over senses and nature. In this way the experience of the dynamic Sublime leads to the perception of a mental self-experience as empowerment of the rational faculties over nature (Kant 2009 [1790], 102). In both cases, Kant relates the Sublime to the mind's power of surpassing sensory perception, of feeling pleasure in experiencing the superiority of reason over nature, conveying reason to transcendence.

In contrast to this, in his treatise *A Philosophical Inquiry into the Origin of our Ideas of the Sublime and Beautiful* (1757), Edmund Burke rooted his revolutionary concept of the Sublime in the embodied nature of the human being, tracing back the perception of the Sublime to the most natural and elementary feelings of human nature (Meyer-Sickendiek 2005), which regulate survival in the natural and social world: fear and pain. Burke claims that the human body is either in a state of pain or in a state of pleasure, or sometimes in neither of them, i.e. in a condition of indifference. The removal or diminution of pain does not provoke pleasure, but delight (Burke 1887, part 1, sections 3 and 4). Pain and pleasure, or their modifications, like danger, are oriented or finalized to "self-preservation and society" (Burke 1887, 113). This last is intended to mean not only human society, but "general society, which we have with men and with other animals, and which we may in some sort be said to have even with the inanimate world" (Burke 1887, 112). Erasing the big divide between culture and nature, Burke establishes a new conception of natural life, which will be taken up by some authors of the Romantic era, such as Kleist, as in his representation of a cross-species community of humans and animals (Haraway 2008) in *Penthesilea* (Gambino and Pulvirenti 2017b).

Pain and danger are therefore "productive of the strongest emotion which the mind is capable of feeling" and are instantiated by "whatever is in any sort terrible" (Burke 1887, 111). According to Burke, we experience Beauty and the Sublime in art because of our ability to feel sympathy, i.e. a sort of substitution by which "we are put into the place of another man, and affected in a good measure, as he is affected, so that this passion may either partake of the nature of those which regard self-preservation, and turning upon pain, may be a source of the Sublime; or it may turn upon ideas of pleasure [. . .]" (Burke 1887, 117).

In Burke's theories – developed according to the aesthetic principles derived from overcoming Cartesian dualism and empirically confirmed by recent neuroscientific research into emotions and embodied simulation – "sympathy" is at the origin of the immersive processes linked to artistic and literary experience, since it lets us transfuse our passions "from one breast to another, and are often capable of grafting a delight on wretchedness, misery and death itself" (Burke 1887, 117). Burke attests that tragic representations are therefore a source of pleasure, arguing that this pleasure arises not from the reasoning faculty, but from the "mechanical structure of our bodies" (Burke 1887, 118). While in real life passions regarding self-preservation turn on pain and danger, they may also turn on delight "when we have an idea of pain and danger, without being actually in such circumstances. [. . .] Whatever excites this delight, I call *Sublime*" (Burke 1887, 125).

Burke distinguishes two different passions caused by the Sublime: 1) astonishment, i.e. the effect of Sublime in its highest degree, which is related to horror in nature and may cause admiration or reverence or respect; and 2) terror, which is related to fear, considered as "an apprehension of pain or death" (Burke 1887, 131). In nature, such passions are induced by the following features: magnificence; greatness in dimensions; vastness of extent; quantity; obscurity; vacuity; darkness; solitude; silence; and infinity (cause of delightful horror). In the arts, such passions are triggered by the following features: succession of the parts; magnitude of dimension; unfinished and difficult composition which solicits the imagination more strongly; magnificence and disorder.

As Burke points out, since sight may be a source of fear, sublime feelings may be triggered by paintings, in case these last resemble nature, depicting it with dark or confused images, or with features of privation, like "vacuity, darkness, solitude, and silence" (Burke 1887, 146). These features more strongly influence the imagination. Colors are a powerful source of magnificence, since they convey light and darkness, which is exalted through colors overpowering the senses, and producing an effect of grandeur (Burke 1887, part 1, sections 16–18). Therefore, a dark, gloomy, cloudy or night sky is more solemn that a blue daytime sky and this may be obtained by using "sad and fuscous colours, as black, or brown, or deep purple, and the like" (Burke 1887, 159). A further source of fear is words, i.e., the literary texts that convey "the affections of the mind" (Burke 1887, 134). Therefore, words are conceived as the best way to convey passions, giving rise to imagery, which is always of an obscure kind: "So that poetry, with all its obscurity, has a more general, as well as a more powerful dominion over the passions, than the other arts" (Burke 1887, 135). Sounds produce sublime passions, and particularly the cries of animals in pain or danger: "The angry tones of wild beasts are equally capable of causing a great and awful sensation" (Burke 1887, 160), since they express the universal connection among humans, animals, and the inanimate world, without any further need of interpretation, being asemantic (Burke 1887, part 2, section 20). Other sources of the Sublime may be smells and tastes, particularly if they convey modes and degrees of pain, anguish, or torment. Troubles and danger are at the origin of great virtues more than any pleasant experience, since strong unpleasant "affections of the body produce such a distinct emotion of mind" (Burke 1887, 208).

These same feelings are at work when sensuously experiencing exceedingly fearful manifestations of nature: its boundlessness, greatness, infinity obscurity, solitude, silence, etc. This is because, in Burke's opinion, the feelings mentioned above provoke a "violent emotion of the nerves" (Burke 1887, 212) and a "violent pulling of the fibres" (Burke 1887, 245), which are positive feelings, since they enhance the vital forces. In this way, Burke highlights a theory of the Sublime as rooted in the body, forecasting the mechanisms of embodied simulation:

> But I conceive, if we can discover what affections of the mind produce certain emotions of the body; and what distinct feelings and qualities of body shall produce certain determinate passions in the mind, and no others, I fancy a great deal will be done; something not unuseful toward a distinct knowledge of our passions, so far at least as we have them at present under our consideration. (Burke 1887, 298)

For Burke the Sublime does not depend on the intuition of a transcendental realm, but on the embodied mind touching its limits and achieving a feeling of contact with the essence of the material world. He theorizes a corporeal physiological Sublime, to which Kant will later oppose a transcendental concept of the Sublime, focused on the concept of an abstract reason, as Alan Richardson claims:

> Kant's influential theory of the sublime, especially as it has been taken up and rehearsed by later writers, notably leaves the body and its brain behind to posit an idealized Reason that, in the course of the sublime moment, succeeds in touching on the noumenal itself, however obliquely, however indescribably. Kant dismissed Burke's theory of the sublime, after all, as a "physiological" or "merely empirical" exposition of the subject, unqualified to speak to the universality of aesthetic taste revealed by the sublime experience and so required to give place to a transcendental treatment of the subject such as Kant's own. (Richardson 2010, 24)

By contrast, Richardson considers Burke's "neural Sublime" as "distinctly earthly and physical":

> [. . .] the Kantian sublime definitely trends toward valuing the mentalistic over the physiological, the abstract over the empirical. Kant's sublime might itself be described as a cognitive sublime, but a neural sublime would seem to remain, by contrast, tied to the phenomenal, tethered to the body and its limitations – and what kind of a sublime would that be? [. . .] I contend that the neural sublime does not, as in Kant, trigger in the beholder the apprehension of a higher Reason but rather, as in the neuroscientific demonstrations recounted above, yields up a disturbing but compulsive glimpse into the ordinarily secret workings of the brain. (Richardson 2010, 24–25)

A relevant consequence of the embodied concept of the Sublime is its emancipatory implication: the rules of logic and logical thinking are disrupted and the aesthetic taste is seen as a subjective "pluralistic" choice, which is opposed to any possible institution of general principles and *a priori* ideas, which are not deducible from the empirical approach. This freedom related to subjective choice overcomes the principles of aesthetic rules and of transcendental reflection leading to the definition of beauty and taste through rational judgment. The Sublime in Burke's terms, intended as bodily experience of naturalness, is therefore revolutionary, since it emancipates art and aesthetic judgment from rules and principles, creating a new subjective expression of taste and beauty which constituted a revolutionary way of affirming personal subjectivity, without obeying any institution or any normative aesthetics or poetics. In this sense, Burke writes:

> But art can never give the rules that make an art. This is, I believe, the reason why artists in general, and poets principally, have been confined in so narrow a circle; they have been rather imitators of one another than of nature; and this with so faithful a uniformity, and so remote an antiquity, that it is hard to say who gave the first model. [. . .] The true standard of the arts is in every man's power; and an easy observation of the most common, sometimes of the meanest things in nature, will give the truest lights, where the greatest sagacity and industry that slights such observations, must leave us in the dark, or what is worse, amuse and mislead us by false lights. (Burke 1887, 128–129)

Beauty is no longer an exclusive experience for artists or for an instructed audience, but a potentially common experience for any person capable of sensing and feeling beauty in art and nature with their own body and heart.

In Germany, from Moses Mendelssohn's *Rhapsodie* to Kant's theories and Schiller's works *Vom Erhabenen* (1793) and *Über das Erhabene* (1796), the experience of the Sublime was described as provoking a kind of sublimation of the soul with regard to the dangerous and therefore "delightful" experience of the mixed feelings of horror and pleasure, enlarging the human ability to think, to feel one's own self, nature and the world. In opposition to Kant and as a result of the so-called Kantian crisis (Mandelartz 2006), Kleist evokes sublime experiences as powerful ones because they are bound not to a mental empowerment of the self, but to an enhanced bodily perception of nature, already considered by Herder as the core of cognition (1778): the human being is driven to more strongly perceive the material force of nature through the senses, by activating their most archaic emotions, bound not to language but to a natural, we may say animal, reaction to primitive feelings of fear and anguish. This issue is at the core of the text *Feelings about a Seascape by Friedrich*, Kleist's ekphrasis of Friedrich's painting *The Monk at the Sea*,[3] which is now considered one of the most emblematic representations of the Romantic Sublime.

3 It is now exhibited at the *Alte Nationalgalerie* in Berlin – NG 9/85.

5 Kleist's *Feelings about a Seascape by Friedrich* and his embodied concept of the Sublime

When Friedrich's painting was first exhibited in October 1810 at the Academy of Fine Arts in Berlin,[4] it became a controversial case because it radically disrupted the basic principles of the classical landscape (Müller 1986; 1995; Begemann 1990; Rühse 2013). Heinrich von Kleist published in the journal *Berliner Abendblätter*[5] a review of Friedrich's painting by Clemens Brentano and Achim von Arnim, but the text had been re-edited by Kleist himself, who preserved the first part (lines 1–12 of the following text) and added a completely new part (starting at line 12), though he printed the new part using the initials of Brentano's name (cb), giving rise to a major quarrel with his fellow writers.[6] What struck Kleist so powerfully as to lead him to re-write the text, emphasizing in the second totally original part its aesthetic power and novelty, which the other authors did not understand as deeply as he did? Why is Kleist's revolutionary ekphrasis of Friedrich's painting of such great impact and so difficult to interpret even today?

The first issue concerns the fact that the text does not describe the work of art itself, i.e. the pictorial features of Friedrich's painting, but rather the reader/viewer's physical and emotional process of relating to the overwhelming nature depicted; the second issue concerns the physical experience of the Sublime, which is achieved by enhancing the embodied simulation of the reader in relation to the visual experience of the painting. In this way Kleist amplifies Friedrich's intentions of representing the process of seeing and of relating to an overwhelming natural scene, as Friedrich claimed: "It [*The Monk by the Sea*] cannot be defined as a description. It is instead the act of seeing itself" (Friedrich's letter to Johannes Schulze, February 1809; Friedrich 2006, 45).

Before focusing our attention on Kleist's text, we need to point out some revolutionary features of Friedrich's painting, which have been emphasized in Kleist's text. At the end of the eighteenth century, visual perception was deeply modified by innovations in scientific research on light, vision and representation in the field

4 It was exhibited together with another painting by Friedrich, *The Abbey in the Oakwood*.

5 The Journal *Berliner Abendblätter* was edited by Heinrich von Kleist with Adam Müller and published in Berlin by Julius Eduard Hitzig from 1 October 1810 until 30 March 1811.

6 The text was followed by six short dialogues, a sort of ironic report of the comments made by the audience during the exhibition. For the quarrel among Kleist, Brentano and Arnim, see Kurz (1988). For the text-genesis and Kleist's re-writing of Brentano's and Arnim's original contribution, see Begemann (1990), Seeba (2003), and Meyertholen (2013).

Fig. 1: Barker's *London Panorama* (1792).

of physics,[7] as exemplarily shown by the circular representations of *panoramas*, i.e. oversized rounded paintings, like the famous one by Robert Barker shown in London in 1792 (Fig. 1). Panorama exhibitions became a fashion in Europe, as other cases show: in 1808, Karl Friedrich Schinkel exhibited his *Panorama of Palermo* in Berlin, in the round Church of St. Hedwig (Fig. 2). This new way of looking at a wide landscape deprived the observer of the traditional central perspective, compelling him to move around the 360-degree "installation," to change his point of view continuously and to take part in the space of the representation.

Fig. 2: Karl Friedrich Schinkel, *Panorama of Palermo*, Berlin, St. Hedwig (1808).

7 For instance, new theories were developed concerning the principle of polarity, giving rise to studies of electricity (Donndorf 1784); for this, see Breidbach and Burwick (2012). Other studies were devoted to nature in order to identify a common principle to all forms, as in the case of Johann Wilhelm Ritter's (1806) theory of the existence of a proto-unity (*All-Thier*).

This new experience of the gaze disrupted Alberti's central perspective, which was the expression of a rational elaboration of world perception.

Friedrich's painting *The Monk by the Sea* (Fig. 3) interprets this radical shift of paradigms occurring in the pictorial representation of space perception. It subverts almost all the formal rules of landscape painting of that time: Alberti's perspective construction, the use of frame, the relationship between foreground and background, the references to proportion, and the function of *staffage* figures, i.e. humans and animals depicted in a landscape. Moreover, it renders the violence of nature and the loneliness of human endeavors by depicting a mostly schematic scene, horizontally divided into three areas characterized only by colors, the pale ochre of the dune, the black of the sea, the blue-grey of the cloudy sky. Friedrich's representation exceeds the frame of the human gaze and amplifies vision toward boundlessness. This is achieved by applying various strategies: first of all the painting modifies the use of the foreground, introducing a great change in the basic principles of the eighteenth century's "framed gaze" (Stoichita 2013). The importance given to the frame in eighteenth century theory was related to the implicit rational intention of the gaze: this art of exploring the world emphasized precision and sharpness and therefore the need to *cut out* a part from the whole. Isolation and sharpness determined by the framed gaze helped to give a cognitive structure to the elements in paintings. Step by step, the eye was conducted to explore the surface of the canvas, putting single elements together into a sort of complex cluster that gave a sense to the parts and to the whole of the representation. This framed model responded to the taste of the time and to the principles of the educational model called *Anschauungsunterricht* (Gambino 2010), based on observation, recognition and narration, as theorized during the eighteenth century. Friedrich rejected the rules of the framed gaze by opening the sight to immeasurable spaces created by means of color fields composed without any perspective structure (Begemann 1990, 15) and of *staffage* figures disproportionately reduced in dimension.

The presence of just the tiny dark figure of a monk, placed under the line of the horizon, almost completely turned away from the viewer, compels the observer of the painting to physically enter in the depicted natural space, by identifying him/herself with him, who is the only reference for the viewer. Therefore, the viewer is led to imaginatively experience the representation not as a picture but as a real natural space, in which they are part of the scene, perceiving at first not so much the artistic qualities of the painting, but the natural endeavor. In this way Friedrich tries to represent the experience of the subject in front of a boundless and menacing nature. The painting compels the viewer to experience the Sublime as rooted in the bodily sensible perception of overwhelming nature, in

Fig. 3: Caspar David Friedrich, *Der Mönch am Meer*, Berlin, Alte Nationalgalerie (1810).

contrast to the Kantian concept of the Sublime intended as the metaphysical experience of an abstract intellect.

We now quote Kleist's text in a translation by Jeremy Hunt, with numbered lines to make further references clear:

It is magnificent to stand in infinite solitude on the seashore, beneath an overcast sky, and to look on an endless waste of water. Part of this feeling is the fact that one has made life's way there and yet must go back, that one would like to cross over but cannot, that one sees nothing to support life and yet senses the voice of life in the sight of the waves, the murmur of the air, the passing clouds and the lonely cry of birds. Part of this feeling is a claim made by the heart and a rejection, if I may call it that, on the part of nature. But this is impossible in front of the picture, and what I should have found in the picture itself I found only between myself and the picture, namely a claim my heart made on the picture and the picture's rejection of me; and so I myself became the monk, and the picture became the dune, but the sea itself, on which I should have looked out with longing – the sea was absent.

There can be nothing sadder or more desolate in the world than this place: the only spark of life in the broad domain of death, the lonely center in the lonely circle. The picture, with its two or three mysterious subjects, lies there like an apocalypse, as if it were thinking Edward Young's "Night Thoughts" and since it has, in its uniformity and boundlessness, no foreground but the frame, it is as if one's eyelids had been cut off. Yet the painter has undoubtedly broken an entirely new path in the field of his art, and I am convinced that with his spirit, a square mile of the sand of Mark Brandenburg could be represented with a barberry bush, on which a lone crow might sit preening itself, and that such a picture would have an effect that rivalled Ossian or Kosegarten. Why, if the artist painted this landscape using its own chalk and its own water, I believe he would make the foxes and wolves weep: the most powerful praise, without doubt, that could be given to this kind of landscape painting.[8] (Kleist 2013 [1810])

The text is structured in three parts: the first (lines 1–11) describes the magnificent experience of solitude in front of the sea, by physically placing the reader in the depicted landscape "as-if" it were a real landscape, which overwhelms the perceptual limits of the human being, challenging their ability to organize their cognitive experience of nature. In line 7 the reader discovers that the process in which they are engaged is not prompted by a real landscape, but by art. In fact, we find here the first mention of the painting to which the text refers. In the second part (lines 12–16), the perceptual perspective changes and the author shifts to strongly organized rhetorical language, in order to overcome the human limits in the cognitive organization of fundamental coordinates, like space and time, by artistically representing rhetorical images of the sublime experience. Finally, in the third part (lines 16–23) the experience of the Sublime attests itself as a sensual, embodied and pre-linguistic one, accord-

8 "Herrlich ist es, in einer unendlichen Einsamkeit am Meeresufer, unter trübem Himmel, auf eine unbegrenzte Wasserwüste, hinauszuschauen. Dazu gehört gleichwohl, daß man dahin gegangen sei, daß man zurück muß, daß man hinüber möchte, daß man es nicht kann, daß man alles zum Leben vermißt, und die Stimme des Lebens dennoch im Rauschen der Flut, im Wehen der Luft, im Ziehen der Wolken, dem einsamen Geschrei der Vögel, vernimmt. Dazu gehört ein Anspruch, den das Herz macht, und ein Abbruch, um mich so auszudrücken, den einem die Natur tut. Dies aber ist vor dem Bilde unmöglich, und das, was ich in dem Bilde selbst finden sollte, fand ich erst zwischen mir und dem Bilde, nämlich einen Anspruch, den mein Herz an das Bild machte, und einen Abbruch, den mir das Bild tat; und so ward ich selbst der Kapuziner, das Bild ward die Düne, das aber, wo hinaus ich mit Sehnsucht blicken sollte, die See, fehlte ganz. Nichts kann trauriger und unbehaglicher sein, als diese Stellung in der Welt: der einzige Lebensfunke im weiten Reiche des Todes, der einsame Mittelpunkt im einsamen Kreis. Das Bild liegt, mit seinen zwei oder drei geheimnisvollen Gegenständen, wie die Apokalypse da, als ob es Youngs Nachtgedanken hätte, und da es, in seiner Einförmigkeit und Uferlosigkeit, nichts, als den Rahmen, zum Vordergrund hat, so ist es, wenn man es betrachtet, als ob einem die Augenlider weggeschnitten wären. Gleichwohl hat der Maler zweifelsohne eine ganz neue Bahn im Felde seiner Kunst gebrochen; und ich bin überzeugt, daß sich, mit seinem Geiste, eine Quadratmeile märkischen Sandes darstellen ließe, mit einem Berberitzenstrauch, worauf sich eine Krähe einsam plustert, und daß dies Bild eine wahrhaft Ossiansche oder Kosegartensche Wirkung tun müßte. Ja, wenn man diese Landschaft mit ihrer eignen Kreide und mit ihrem eigenen Wasser malte; so, glaube ich, man könnte die Füchse und Wölfe damit zum Heulen bringen: das Stärkste, was man, ohne allen Zweifel, zum Lobe für diese Art von Landschaftsmalerei beibringen kann" (Kleist 1810, 47–48).

ing to an organic vision of nature as a material essence bounding all creatures.

The opening lines are characterized by two issues: on the one hand, the author uses words which seem to refer directly to those used by Burke explaining the principles of the Sublime (in italics in the following quotation); on the other hand, the reader is prompted to imaginatively create a situation model[9] which emotionally transports him/her to the center of the depicted natural scene, thanks to the use of verbs of motion and sight, spatial deixis and locative expressions (in bold in the following quotation):

> It is *magnificent* **to stand** in *infinite solitude* **on** the seashore, **beneath** an *overcast sky*, and **to look on** an *endless waste* of water. Part of this feeling is the fact that **one has made his life's way there** and yet must **go back**, that one would like **to cross over** but cannot, that one **sees** nothing to support life and yet **senses** the voice of life in the **sight** of the waves, the *murmur* of the air, the **passing** clouds and the *lonely cry* of birds. (lines 1–5)

The embodied-simulation process of the reader activates a situation model combining the ideas derived from the text with the background knowledge and experience of the reader into a coherent mental representation of the described situation (Zwaan 2005, 534). In this case the reader is plunged into the "infinite solitude on the seashore, beneath an overcast sky" and cast in front of "an endless waste of water" (lines 1, 2). The embodied-simulation process is elicited by the use of verbs setting the body in space (e.g. "to stand", line 1), which solicits the sensation of a stable position, and expresses possible movements: "life's way there" (line 2), "yet must go back" (line 3), and "to cross over" (line 3). The use of such verbs of movement elicits a mental representation strongly bound to space and motion, two of the fundamental information parameters on which the situation model is constructed, allowing the reader to situate him/herself in the represented space of nature.

Kleist triggers the reader's sensory-motor system within the mental representation of natural space, guiding him/her into an "as-if" process by using verbs of bodily motion (see lines 2, 3), and soliciting the reader to establish a dynamical perceptual relationship with the space inside the picture, by using verbs of perception, like "to look" (line 1–2), "to see" (line 3), "to sense" (line 4). In the empathic experience triggered by the first part of the text, the "as-if" simulation pro-

9 We refer here for our analysis to the Immersed Experiencer Framework based on discourse processing framework models introduced by Zwaan et al. (1995), later integrated by several research findings attesting that during the comprehension of a text, brain regions are activated that are very close to or even overlap with brain regions that are active during the perception or the action to which the words refer (Kuzmičová 2014).

cess, instantiated by verbs of motion and sensory perception and by the deictic shift of the prepositions, solicits the reader's mind within a "mental simulation" (Zwaan 2016) that reproduces the somatic states seen in or implied by the description of the action of getting inside the space, "as if" the body were present.

The use of verbs related to the act of perceiving real material nature triggers the sensory-motor system, engaging the reader in an embodied simulation process involving the visual, auditory and proprioceptive systems, which elicits the creation of a stable and clear situation model. The action of standing on the seashore, of moving in this endeavor and of going back, activates with particular efficacy the process of bodily simulation and of empathic identification. As Vittorio Gallese claims, embodied simulation is instantiated not only in the direct subjective experience of the world, but also when observing the actions and emotions of others or when reading about them (Gallese and Cuccio 2015, 11). The author further triggers the reader to perceive the landscape, suggesting the double identification of the reader, not only with the viewer in front of the painting but also with the monk represented in the painting (in bold in the following quotation):

> [. . .] what I should have found in the picture itself **I found only between myself and the picture**, namely a **claim** my heart made on the picture and the picture's **rejection** of me; and so **I myself became the monk, and the picture became the dune**, but the sea itself, on which I should have looked out with longing – the sea was absent. (lines 7–11)

The embodied simulation is provoked by the spatial situation model through verbs of movement, driving the experiencing self into a disproportional landscape, which they can perceive through the senses but not cognitively understand. The focus of vision in the painting – the sea – (lines 10–11) is revealed to be a vacuum, with no references to cognitive coordinates or visual boundaries. The landscape rejects the viewer of the painting and the reader of the ekphrasis since the space references and the cause-effect references are shifted to an immaterial and boundless panorama.

In the second part (lines 12–16), the situation model of being within the natural landscape is shifted to a symbolic and meta-representational level. The failed challenge of the cognitive process in front of nature is featured again by rejecting the experience of nature and by gaining a relation to the artwork.

In order to convey in verbal code the aesthetic experience of the viewer in front of the painting, Kleist shifts to a rhetorical language, intertwining figures like hyperbole ("There can be nothing sadder or more desolate in the world than this place", line 12), prosopopoeia ("the lonely center in the lonely circle", line 13), metaphor ("the only spark of life in the broad domain of death", lines 12–13), and simile ("like an apocalypse", line 14), which determines the climax of this second part. The boundlessness of the vision represented in the painting leads to a "failure" of the gaze, which is no longer able to frame and stabilize the image, leaving the observers/readers with their "eyelids [. . .] cut off" (line 16).

The double failure, of sight to frame the vision of the sublime natural landscape, and of the imagination to construct a corresponding spatial situation model through rhetorical language, attests in Kleist's view to the absolute power of Friedrich's painting to surpass cognitive limits and enter into a boundless emotional experience that brings the subject to feel their own being as deeply related to the very essence of nature. The sublime experience does not take form as a transcendental intuition over the body, but as a bodily retrieval of the material, sensual dimension.

In the last part (lines 16–23) Kleist expresses his own opinion about the painting in the first person. Writing as an art critic, he attests to the deep relationship existing between nature and art, underlying the fact that the "artful" power of Friedrich's picture consists in its ability to create a bodily perception of the material components of the landscape, pushing the subject to regain their animal core. In Kleist's opinion, the material and perceptual experience of the viewer/reader is achieved by the painting because it has dismissed every symbolic and allegorical reference, presenting instead the physical elements in their pure materiality ("chalk" and "water"). The experience of nature becomes a pre-linguistic one, so that finally the true human emotional response cannot be expressed through semantic language but rather through something similar to the natural cries of animals, "the weep of foxes and wolves" (line 21):

> If the artist painted this landscape, using **its own chalk and its own water**, I believe **he would make the foxes and wolves weep**: the most powerful praise, without doubt, that could be given to this kind of landscape painting. (lines 20–23).

The cries of animals in front of nature were a significant issue in the bodily conception of the sublime experience and had already been considered by Burke as examples of a powerful expression of the embodied Sublime. Kleist's representation of the Sublime through the comparison between the natural Sublime and the artful one is relevant in German aesthetic discourse because the two forms of experiencing the Sublime were not regarded as equally effective, since art and nature were thought to affect the human mind in different ways. Burke, on the contrary, considered both equally powerful in affecting the body and the emotions.

Kleist's text is an expression of a bodily conception of the Sublime. The most evident peculiarity of Kleist's text is that it is not focused on the description of the painting, but rather on the process engaged in by the viewer looking at the picture. Kleist apparently denies the possibility of experiencing the powerful pictorial image by relocating the viewer inside the depicted natural scene. Through the rhetorical construction of the second part of the text, the reader is able to imagine not only the real natural scene, but also the powerful effect of the language presenting the painting as medium of the natural landscape elaborated by the painter in terms of the embodied Sublime, which is stressed in the last part of the text by abolishing every intermedial experience and by relocating the reader in the materiality of the natural world.

Kleist was not so much a theorist in aesthetics and poetics, but rather an aphoristic declaimer of his ideas and experimented with them in his works, essays and letters. In this perspective, we do not find the enunciation of a theory of the Sublime, but many passages that are representations of sublime experiences, where nature and culture do not appear as opposites. A letter from Leipzig about his stay in Dresden, sent to Wilhelmine von Zenge on May 21st, 1801 (Kleist, *Briefe*, letter 45), is the best demonstration of this issue: the author describes the masterworks of the *Gemäldegalerie* in Dresden as natural landscapes evoking the experience of the Sublime, which is again and again expressed in relation to the perceptions of the body ("senses and heart" / "auf Sinn u. Herz") and "not of the intellect" ("den Verstand nicht braucht"). The landscape around Dresden and the natural environment experienced during his wanderings is also described in terms of a sensual Sublime, with a handful of expressions concerning the grandeur of the sights, the irregular movement of the river, and the immensity of the landscape, which is compared to a "turbulent sea of earth" ("ein bewegtes Meer von Erde"). The complementation – and not opposition – of nature to art is achieved in the description of the mountains which look "like colossal Pyramids [. . .] made by the hand of Angels playing in the sand" ("wie colossalische Pyramiden, in den schönsten Linien geformt, als hätten die Engel im Sande gespielt"). In this simile everything conduces to the Sublime emotion: nature, art, and the evocation of a superior power in terms of the bodily action of the angel's hand. This comparison between art and nature as source of the Sublime again evokes Burke's theories and their revolutionary implications for a new conception of the human being and human cognitive and imaginative faculties:

> [. . .] the images in paintings are exactly similar to those in nature; and in nature dark, confused, uncertain images have a greater power on the fancy to form grand passions [. . .]. (Burke 1887, 136)

For Kleist, verbal as well as visual arts are able to give rise to the Sublime since they both concern the bodily experience of Nature. Stretching bodily perception to its limits, Kleist represents man not as a being placed above Nature, able to transcend it thanks to reason, but as a living part of Nature, experiencing it with body and affects.

The experience of the Sublime leads to a stronger perception of a nature common to humans and animals, whereby the power of bodily perception is conveyed by the animal "reduction" of experience. This issue appears again in the short story *St. Cecilia, or the power of music* (published in November 1810). Here four brothers are deeply struck by an artistic-mystic Sublime experience of the music played in the Cathedral for Corpus Christi; they become crazy and are kept in the city's house for the insane. After listening to the composition, probably mysteriously conducted by Saint Cecilia, they begin praying in front of a crucifix and singing the *Gloria in excelsis Deo* at midnight (Greiner 1996). The relevant point is the way in which their singing is described, more similar to animals' cries than to human song:

> [. . .] they began to sing the *Gloria in excelsis* in the most hideous and horrible voice. The sound of leopards and wolves, when on an icy winter's night they roar at the sky, may be something like it. The pillars of the house [. . .] were shaken, and the window-panes smitten by the visible breath from their lungs, rattled and threatened to fall in, as if handfuls of heavy sand were dashed against their surface. (Kleist 1844, 303)

Their "fearful and revolting howls" are the expression of an experience that is defined not by the rational mind, overcoming the fear and horror of the Sublime, as in Kant's theory, but by the mind surrending to the body's most archaic affects. According to Burke, the Sublime is a product of feelings of fear and pain, instantiated by the perception of a bodily danger, which is not felt as real, since it is mediated by the artistic experience, as explained by the embodied simulation hypothesis. By giving poetic evidence to this idea of the Sublime, Kleist disrupts the transcendental and hegemonic conceptions of aesthetic experience as related to the abstract intellect and of the arts as depending on pre-established rules and principles, opening new modern insights into the humanly embodied concept of the mind, as well as into the concept of narration as an experiment of the mind's adventures (Gambino and Pulvirenti 2018b) beyond the known, in a never-ending process of constructing new meanings for human emotional survival in a world menacing our fragile and ephemeral nature. (RG)

Reference list

Alber, Jan, Marco Caracciolo, and Irina Marchesini. "Mimesis: The Unnatural between Situation Models and Interpretive Strategies." *Poetics Today* 39.3 (2018): 447–471.

Begemann, Christian. "Brentano und Kleist vor Friedrichs Mönch am Meer: Aspekte eines Umbruchs in der Geschichte der Wahrnehmung." *Deutsche Vierteljahrsschrift für Literaturwissenschaft und Geistesgeschichte* 64 (1990): 89–145.

Berlin, Isaiah. *The Roots of Romanticism*. Princeton: Princeton University Press, 1999.

Boehm, Gottfried. "Bildbeschreibung: Über die Grenzen von Bild und Sprache." *Grenzen der Bildinterpretation*. Eds. Michael R. Müller et al. Wiesbaden: Springer, 2014. 15–37.

Boehm, Gottfried, and Helmut Pfotenhauer. *Beschreibungskunst, Kunstbeschreibung: Ekphrasis von der Antike bis zur Gegenwart*. Munich: W. Fink, 1995.

Breidbach, Olaf, and Roswitha Burwick, eds. *Physik um 1800: Kunst, Wissenschaft oder Philosophie?* Munich: Wilhelm Fink, 2012.

Burke, Edmund. *A Philosophical Inquiry into the Origin of Our Ideas of the Sublime and Beautiful. The Works of The Right Honourable Edmund Burke*. Vol 1. Ed. John C. Nimmo. London, 1887. 67–263.

Burke, Edmund. *A Philosophical Inquiry into the Origin of Our Ideas of the Sublime and Beautiful and Other Pre-revolutionary Writings*. London: Penguin, 1998 [1757].

Chatterjee, Anjan. *The Aesthetic Brain: How we Evolved to Desire Beauty and Enjoy Art*. New York: Oxford University Press, 2014.

Chatterjee, Anjan, and Oshin Vartanian. "Neuroaesthetics." *Trends in Cognitive Sciences* 18.7 (2014): 370–375.

Cuccio, Valentina, Marco Carapezza, and Vittorio Gallese. "Metafore che risuonano. Linguaggio e corpo tra filosofia e neuroscienze." *E/C: Rivista dell'Associazione Italiana Studi Semiotici* 7.17 (2013): 69–74.

Donndorf, Johann August. *Die Lehre von der Elektricität theoretisch und practisch auseinander gesetzt*. 2 vols. Erfurt, 1784.

Farah, Martha. J. "The Neural Bases of Mental Imagery." *The Cognitive Neurosciences*. Ed. Michael Gazzaniga. Cambridge, MA: MIT Press, 2000.

Freedberg, David, and Vittorio Gallese. "Motion, Emotion and Empathy in Aesthetic Experience." *Trends in Cognitive Sciences* 11.5 (2007): 197–202.

Friedrich, Caspar David. *Die Briefe*. Hamburg: ConferencePoint Verlag, 2006.

Gallese, Vittorio. "Embodied Simulation: From Neurons to Phenomenal Experience." *Phenomenology and the Cognitive Sciences* 4 (2005): 23–48.

Gallese, Vittorio. "Embodied Simulation: From Mirror Neuron Systems to Interpersonal Relations." *Novartis Found Symposium* (2007): 216–221.

Gallese, Vittorio. "Mirror Neurons, Embodied Simulation, and the Neural Basis of Social Identification." *Psychoanalytic Dialogues* 19 (2009): 519–536.

Gallese, Vittorio. "The Problem of Images: A View from the Brain-Body." *Phenomenology and Mind* 14 (2018): 70–79.

Gallese, Vittorio, and Valentina Cuccio. "The Paradigmatic Body: Embodied Simulation, Intersubjectivity, the Bodily Self, and Language." *Open MIND*. Eds. Thomas Metzinger and Jennifer M. Windt. Frankfurt a. M.: MIND Group, 2015. Ch. 14.

Gallese, Vittorio, and George Lakoff. "The Brain's Concepts: The Role of the Sensory-motor System in Conceptual Knowledge." *Cognitive Neuropsychology* 22.3/4 (2005): 455–479.

Gallese, Vittorio, Luciano Fadiga, Leonardo Fogassi, and Giacomo Rizzolatti. "Action Recognition in the Premotor Cortex." *Brain* 119 (1996): 593–609.

Gambino, Renata. "Auf dem Weg zum *Mittelpunkt des Schönen*: Das pädagogische Konzept von Karl Philipp Moritz' italienischer Reisebeschreibung." *Karl Philipp Moritz: Signaturen des Denkens*. Ed. Anthony Krupp. Amsterdam and New York: Rodopi, 2010. 254–236.

Gambino, Renata, and Grazia Pulvirenti. "The Paradox of Ekphrasis: Metacritic Discourse, Perception and Imagination in Art Description." *Metacritic Journal for Comparative Studies and Theory* 3.1 (2017a): 151–179.

Gambino, Renata, and Grazia Pulvirenti. "Vom Zweikampf zur Hetzjagd: Die politische Bedeutung der Tiere in Heinrich von Kleists 'Penthesilea'." *Weimarer Beiträge* 3.63 (2017b): 367–388.

Gambino, Renata, and Grazia Pulvirenti. *Storie, menti, mondi: Approccio neuroermeneutico alla letteratura*. Milan: Mimesis, 2018a.

Gambino, Renata, and Grazia Pulvirenti. *La mente narrativa di Heinrich von Kleist*. Milan: Mimesis, 2018b.

Gambino, Renata, and Grazia Pulvirenti. "Neurohermeneutics: A Transdisciplinary Approach to Literature." *Gestalt Theory* 41.2 (2019): 185–200.

Gambino, Renata, and Grazia Pulvirenti. Mind the Text! Neurohermeneutics for Suspicious Readers. Cambridge: Scholars Publishing, 2024.

Goebel, Rainer, Darius Khorram-Sefat, Lars Muckli, Hans Hacker, and Wolf Singer. "The Constructive Nature of Vision: Direct Evidence from Functional Magnetic Resonance Imaging Studies of Apparent Motion and Motion Imagery." *European Journal of Neuroscience* 10.5 (1998): 1563–1573.

Goethe, Johann Wolfgang. "Über Laokoon" (1798). *Sämtliche Werke: Münchner Ausgabe*. Eds. Karl Richter et al. Vol. 4.2: *Wirkungen der französischen Revolution 1791–1797*. Eds. Karl Kiefer et al. Munich: Hanser, 2006. 73–88.

Graesser, Arthur C., and Rolf A. Zwaan. "Inference Generation and the Construction of Situation Models." *Discourse Comprehension: Essays in Honour of Walter Kintsch*. Eds. Charles A. Weaver, Suzanne Mannes, and Charles R. Fletcher. New York and London: Routledge, 1995. 117–140.

Greiner, Bernhard. "Das ganze Schrecken der Tonkunst: Die heilige Cäcilie oder Die Gewalt der Musik: Kleists erzählender Entwurf des Erhabenen." *Zeitschrift für deutsche Philologie* 115 (1996): 501–520.

Gumbrecht, Hans U. *Production of Presence: What Meaning Cannot Convey*. Stanford, CA: Stanford University Press, 2004.

Haraway, Donna. *When Species Meet*. Minneapolis and London: University of Minnesota Press, 2008.

Heinse, Wilhelm. *Ardinghello und die glückseligen Inseln*. Lemgo: Meyersche Buchhandlung, 1787.

Henkel, Nikolaus. "Descriptio." *Reallexikon der deutschen Literaturwissenschaft*. Vol. 1. Eds. Georg Braungart et al. Berlin: De Gruyter, 1997. 337–339.

Herder, Johann Gottfried. "On the Cognition and Sensation of the Human Soul" (1778). *Philosophical Writings*. Ed. Michael N. Forster. Cambridge: Cambridge University Press, 2002. 187–246.

Humboldt, Wilhelm von. "Ueber die Verschiedenheit des menschlichen Sprachbaues und ihren Einfluss auf die geistige Entwicklung des Menschengeschlechts" (1830–1835). *Schriften zur Sprachphilosophie*. Darmstadt: Wissenschaftliche Buchgesellschaft, 1972. 368–375.

Iacoboni, Marco, Roger P. Woods, Marcel Brass, Harold Bekkering, John C. Mazziotta, and Giacomo Rizzolatti. "Cortical Mechanisms of Human Imitation." *Science* 286 (1999): 2526–2528.

Johnson-Laird, Philip. *Mental Models*. Cambridge: Cambridge University Press, 1983.

Kant, Immanuel. *Beobachtungen über das Gefühl des Schönen und Erhabenen*. Königsberg: Kanter, 1764.

Kant, Immanuel. “Analytik des Erhabenen.” *Kritik der Urteilskraft*. Ed. Heiner F. Klemme. Hamburg: Meiner, 2009 [1790]. 105–153.

Kim, Sunah, and Thomas W. James. “Enhanced Effectiveness in Visuo-haptic Object-Selective Brain Regions with Increasing Stimulus Salience.” *Hum. Brain Mapp*. 31.5 (2010): 678–693.

Kintsch, Walter. “The Role of Knowledge in Discourse Comprehension: A Construction-Integration Model.” *Psychological Review* 95.2 (1988): 163–182.

Kintsch, Walter, David M. Welsch, Franz Schmalhofer, and Susan Zimny. “Sentence Memory: A Theoretical Analysis.” *Journal of Memory and Language* 29 (1990): 133–159.

Kleist, Heinrich von. *Briefe*. *KLEIST.digital*. Ed. Günter Dunz-Wolff. http://kleist-digital.de/briefe (21 August 2021).

Kleist, Heinrich von. “Empfindungen vor Friedrichs Seelandschaft.” *Berliner Abendblätter* 12 (1810): 47–48.

Kleist, Heinrich von. “Saint Cecilia or the Power of Music.” *Tales from the German, Comprising Specimens form the Most Celebrated Authors*. Trans. John Oxenford and C. A. Feiling. London: Chapman and Hall, 1844. 298–305.

Kleist, Heinrich von, Clemens Bretano, and Achim von Arnim. “Various Expressions Experienced before a Seascape with a Monk by Caspar David Friedrich, 1810.” Trans. Jeremy Hunt. *Art in Fiction: Anthology of Art in Literature*. 11 July 2013. https://artinfiction.wordpress.com/2013/07/11/heinrich-von-kleist-clemens-bretano-achim-von-arnim-various-expressions-experienced-before-a-seascape-with-a-monk-by-caspar-david-friedrich-1810/ (22 October 2019).

Kneisley, John Anderson. “‘A Shadow of Magnitude’: The Condition of Artworks in John Keats’s Ekphrastic Poetry.” *Dickinson College Honors Theses*. Paper 254. Dickinson College, 2016. https://scholar.dickinson.edu/student_honors/254 (21 August 2021).

Kosslyn, Stephen M. *Image and Mind*. Cambridge, MA: Harvard University Press, 1980.

Kosslyn, Stephen M. *Ghosts in the Mind’s Machine*. New York: W.W. Norton, 1983.

Kosslyn, Stephen M., William L. Thompson, and Giorgio Ganis. *The Case for Mental Imagery*. Oxford: Oxford University Press, 2006.

Kurz, Gerhard. “Vor einem Bild: Zu Clemens Brentanos Verschiedene Empfindungen vor einer Seelandschaft von Friedrich, worauf ein Kapuziner.” *Jahrbuch des Freien Deutschen Hochstifts* (1988): 128–140.

Kuzmičová, Anežka. “Literary Narrative and Mental Imagery: A View from Embodied Cognition.” *Style* 48.3 (2014): 275–293.

Lessing, Gotthold Ephraim. “Laokoon oder über die Grenzen der Mahlerey und Poesie” (1766). *Lessings Werke*. 7 vols. Ed. Georg Witkowski. Vol. 4. Leipzig and Wien: Bibliographisches Institut, 1900.

Louwerse, Max M., Hutchinson Sterling, Tillman Richard, and Recchia Gabriel. “Effect Size Matters: The Role of Language Statistics and Perceptual Simulation in Conceptual Processing.” *Language, Cognition and Neuroscience* 30 (2015): 430–447.

Mandelartz, Michael. “Von der Tugendlehre zur Lasterschule: Die sogenannte ‘Kantkrise’ und Fichtes ‘Wissenschaftslehre’.” *Kleist-Jahrbuch* (2006):120–136.

Meteyard, Lotte, Sara Rodriguez-Cuadrado, Bahador Bahrami, and Gabriella Vigliocco. “Coming of Age: A Review of Embodiment and the Neuroscience of Semantics.” *Cortex* 48.7 (2012): 788–804.

Meyer-Sickendiek, Burkhard. *Affektpoetik: Eine Kulturgeschichte literarischer Emotionen*. Würzburg: Königshausen & Neumann, 2005.

Meyertholen, Andrea. “Apocalypse Now: On Heinrich von Kleist, Caspar David Friedrich, and the Emergence of Abstract Art.” *The German Quarterly* 86.44 (2013): 404–420.

Miller, Norbert. *Von Nachtstücken und anderen erzählen Bildern*. Munich: Hanser, 2002.

Müller, Gernot. "Kleists Rhetorik der Innerlichkeit." *Studia Neophilologica* 58.2 (1986): 231–242.
Müller, Gernot. *Kleist und die bildende Kunst*. Tuebingen: Francke, 1995.
Osterkamp, Ernst. *Im Buchstabenbilder: Studien zum Verfahren goethischer Bildbeschreibungen*. Stuttgart: Metzler, 1991.
Pellegrino, G. di, L. Fadiga, L. Fogassi, V. Gallese, and G. Rizzolatti. "Understanding Motor Events: A Neurophysiological Study." *Experimental Brain Research* 91 (1992): 176–180.
Pfotenhauer, Helmut. "Kleists Rede über Bilder und in Bildern: Breife, Bildkommentare und literarische Werke." *Sprachbilder: Untersuchungen zur Literatur seit dem achtzehnten Jahrhundert*. By Helmut Pfotenhauer. Würzburg: Königshausen und Neumann, 2000. 85–108.
Pulvirenti, Grazia, and Renata Gambino. "Immaginazione come poetica della cognizione: Faust nel Regno delle Madri." *Linguaggio, letteratura e scienze neurocognitive*. Eds. Stefano Calabrese and Stefano Ballerio. Milano: LEDIpublishing, 2014. 128–168.
Rancière, Jacques. *Aisthesis: Scenes from the Aesthetic Regime of Art*. London: Verso, 2013.
Richardson, Alan. *The Neural Sublime: Cognitive Theories and Romantic Texts*. Baltimore: The Johns Hopkins University Press, 2010.
Ritter, Johann Wilhelm. *Physisch-chemische Abhandlungen in chronologischer Folge*. 3 vols. Leipzig 1806.
Rizzolatti, Giacomo, and Cattaneo Luigi. "The Mirror Neuron System." *Archives of Neurology* 66 (2009): 557–560.
Rühse, Viola. "Dies wunderbare Gemählde: Ästhetische und kunstpolitische Aspekte in Texten von Clemens Brentano, Achim von Arnim und Heinrich von Kleist zu Caspar David Friedrichs Landschaftsgemälde Mönch am Meer." *Kleist-Jahrbuch 2013*. Ed. Günter Blamberger. Stuttgart: J.B. Metzler, 2013. 238–255.
Schlegel, Friedrich. "Fragments on Literature and Poesy." *Theory as Practice: A Critical Anthology of Early German Romantic Writings*. Eds. Jochen Schulte-Sasse et al. Minneapolis: University of Minnesota Press, 1977.
Schlegel, Friedrich. *Fragmente zur Poesie und Literatur I. Kritische Ausgabe seiner Werke*. Ed. E. Behler. Vol. 16. Paderborn: Schöningh, 1981.
Seeba, Hinrich C. "The Eye of the Beholder: Kleist's Visual Poetics of Knowledge." *A Companion to the Works of Heinrich von Kleist*. Ed. Bernd Fisher. New York: Camden House, 2003. 103–122.
Shimamura, Arthur P. "Towards a Science of Aesthetics." *Aesthetic Science: Connecting Minds, Brains and Experience*. Eds. Arthur P. Shimamura and Stephen E. Palmer. New York: Oxford University Press, 2012. 3–28.
Shimamura, Arthur P. *Experiencing Art: In the Brain of the Beholder*. New York: Oxford University Press, 2013.
Starr, Gabrielle. "Multisensory Imagery." *Introduction to Cognitive Cultural Studies*. Ed. Lisa Zunshine. Baltimore: The Johns Hopkins University Press, 2010. 275–291.
Starr, Gabrielle. *Feeling Beauty: The Neuroscience of Aesthetic Experience*. Cambridge, MA: MIT Press, 2013.
Stilla, Randian, and Krish Sathian. "Selective Visuo-haptic Processing of Shape and Texture." *Hum. Brain Mapp*. 29.10 (2008): 1123–1138.
Stoichita, Victor. *L'invenzione del quadro*. Milan: Il Saggiatore, 2013.
Van Dijk, Teun A., and Walter Kintsch. *Strategies of Discourse Comprehension*. New York and London: Academic Press, 1983.
Varela, Francisco J., Evan Thompson, and Eleanor Rosch, eds. *The Embodied Mind: Cognitive Science and Human Experience*. Cambridge, MA: MIT Press, 1991.
Webb, Ruth. *Ekphrasis, Imagination and Persuasion in Ancient Rhetorical Theory and Practice*. Farnham: Ashgate, 2009.

Winckelmann, Johann Joachim. *Gedanken über die Nachahmung der griechischen Werke in der Malerey und Bildhauerkunst.* Dresden and Leipzig: Walther, 1756.

Zeki, Semir. *Splendors and Miseries of the Brain: Love, Creativity, and the Quest for Human Happiness.* Chichester: Wiley-Blackwell, 2008.

Zwaan, Rolf A. "Situation Models: The Mental Leap into Imagined Worlds." *Current Directions in Psychological Science* 8.1 (1999): 15–18.

Zwaan, Rolf A. "Situation Model." *The Routledge Encyclopedia of Narrative Theory.* Eds. David Herman et al. London: Routledge, 2005. 534–535.

Zwaan, Rolf A. "Situation Models, Mental Simulations, and Abstract Concepts in Discourse Comprehension." *Psychonomic Bulletin & Review* (2016): 1028–34. doi:10.3758/s13423-015-0864-x.

Zwaan, Rolf A., and Gabriel A. Radvansky. "Situation Models in Language Comprehension and Memory." *Psychological Bulletin* 123 (1998): 162–185.

Zwaan, Rolf A., Mark C. Langston, and Arthur C. Graesser. "The Construction of Situation Models in Narrative Comprehension: An Event-Indexing Model." *Psychological Science* 6.5 (1995): 292–297.

Michael Sinding

Energy Models of Causation in Mind, Culture and Science

Electro/Magnetic Thinking in Blake, Galvani and Mesmer

Abstract: To explain how we imagine coherent storyworlds as we experience narratives, literary semantics needs new conceptual principles. Sensorimotor schemas (*image schemas, force dynamics, fictive motion*) are conceptual structures that are central to storyworld structure, as they are central to how we construe persons, objects, settings and events, and to narrative metaphors. People often see themselves as invisible soul-*substances* in body-*containers*, *moved* by emotions that can be *stirred, build up*, and *burst forth*.

Yet to understand how these schemas work in literary thought, we need to develop them into richer notions that can approximate embodied experience. To pursue this aim, I draw on "intuitive ontology," our non-scientific conceptual models of our everyday worlds. I propose an *energy principle* to characterize how the whole palette of intuitive-physics concepts (*space, substance, force, motion*) interact in our modeling of physical structure and causation. Basic causation is seen as *transfer of energy between moving objects via contact*; and energy is seen as a *force*-bearing *substance*. Energy-patterns of movement are interpreted in terms of intuitive psychological causes (beliefs, desires, goals, fears) to produce basic human narrativity. Other storyworld energies (light, heat etc.) are often metaphorical in literary narratives.

I compare forms of "electro/magnetic" energy-thinking across the domains of literature, science and pseudoscience, focusing on key eighteenth-century examples: Blake, Galvani and Mesmer. The comparison helps us to characterize both the distinctions and the interactions of those domains, in terms of their modeling of local physical causes vs. mental-spiritual worldviews, use of repeated experiments in order to refine models, and attitudes to metaphor.

1 Introduction: Energy thinking

Consider a very simple text that relies on scenarios of simple physical concepts to express a vision of persons and worldview. The 1950s song "Something's Gotta Give" opens by framing the start of a relationship in terms of physics:

https://doi.org/10.1515/9783110782844-013

When an irresistible force such as you,
Meets an old immovable object like me,
You can bet just as sure as you live,
Something's gotta give,
Something's gotta give,
Something's gotta give.[1]

We know immediately that on one level, this *refers to* the physical story of a force striking an object and causing a reaction, and that on another level, it *implies* a much larger and richer human story of a relationship between one vibrant, presumably emotionally warm person, and one "solid," presumably emotionally cool person. The lyrics allude to laws of physics, but even if we overlook that, we understand the metaphorical story because we understand how the pattern of motion, contact and energy-transfer among objects evokes in capsule form a complex system of physical, emotional and social relationships among people.

This chapter examines how such inter-related concepts of energy, from the intuitive and "natural" kinds to the culturally enriched and elaborated kinds, underpin meaning-making in everyday life, art and science. First, I sketch a big picture of cognition and meaning with remarks on cognitive semantics, specifically about how simple concepts of the physical world are combined into a more complex concept of *energy*, which is used to structure narratives and metaphors to create complex conceptualizations of the psychosocial world. Second, I elaborate on the energy principle and how it is used in cultural semiotics by analyzing examples from various art forms. Third, I use the energy principle as a basis for comparing thinking across literature, science, and a grey zone in between them, pseudoscience.[2] I concentrate on the early period of what I'll call electro/magnetic science, which recognizes some connection between electricity and magnetism, but precedes the concept and term "electromagnetism," which states the unity of the two forces. I take writings by William Blake, Luigi Galvani and Franz Anton Mesmer as exemplars of these domains, and examine the different ways they use energy-thinking in narrative and metaphor to model the electro/magnetic concepts that define their causal concerns. I offer a few conclusions about how these divergent manifestations of energy-thinking help differentiate literature, science

1 By Johnny Mercer, written in 1954 for Fred Astaire in *Daddy Long Legs* (1955).

2 I am aware that "pseudoscience" is a controversial label, because of how difficult it can be to demarcate the boundaries between science and non-science (as we shall see below). However, after considering the theoretical issues, and the particular case of Mesmer's relation to the science of his time, I am convinced that the term usefully identifies thinking that claims to be scientific but isn't, and that Mesmer is a clear case of this. See e.g. Hanen et al. (1980) on the demarcation problem, and Weyant (1980) on Mesmer in particular.

and pseudoscience, explain the cognitive appeal of pseudoscience, and tease out the interplay of nature and culture in cognition.

2 Cognitive semantics in everyday language and belief

To explain how meaning works in the arts, in cultural worldviews and in science, we need to delve into how narrative and metaphor are central to thinking. Cognitive semantics concerns the basic *conceptual* structures that underlie meaning. Cognitive linguists show how the semantics of ordinary language is grounded in intuitive ontology, yet also full of fictive and metaphorical ontology. Narrative and metaphor share certain conceptual structural elements and forms of operation: They typically involve human-scale scenarios as units, and they link such concrete scenarios to more abstract conceptualization. They do so by projecting the conceptual structures that cognitive linguists call *image schemas, fictive motion,* and *force dynamics* (I refer to these as *sensorimotor schemas*). For example, in the conventional metaphor of Life-as-Journey, an important part of the Journey concept is the Path; it's a schematic image, but one that is grounded in commonplace experience. You start from one point, move towards an endpoint, and pass through all the points in between. These schematic concepts are very important, but much-debated; and recent work urges us to go beyond skeletal and decontextualized views of them, to get at the "kinesthetic feel" of "full bodied experiences that have textures and a felt sense of three-dimensional depth," as Ray Gibbs Jr. says (2005, 119). Leonard Talmy has gone some way in this direction. He points out how ordinary language is full of "fictive motion": The "fictive emanations of linguistic construals" as in "The tree *threw* its shadow *across* the valley," which is fictive and metaphorical in that trees can't throw, and shadows can't be thrown, "are conceptualised (and perceived?) as being invisible and intangible, [. . .] emitted by some entity" and perhaps hitting some distant one (Talmy 1996, 233–235). Such ordinary language concepts remarkably echo concepts of spirits and ghosts, and other cultural forms of "imagined energy," such as "mana, power, fields of life force, or magical influence" and the "folk iconography" of comics and movies. Everyday language assumes that the world is full of invisible virtual paths, and everyday utterances use these paths to frame its descriptions, metaphors, narrations, explanations and arguments. Literature builds on these deep-running patterns to build its resonant symbolic worlds.

2.1 Forms of metaphor: Words, scenarios, models, blends

To better understand how this works, we can look to current research on metaphor, which views metaphor as existing in several forms: *words*; *scenarios*; *models*; and *blends*.[3] In quoted text I will bold key metaphorical words, and underline words related to them.[4]

To explain, consider the example of an accurate (though hostile) sketch of the idea of animal magnetism, from Benjamin Franklin's 1785 *Report* on it:

> The object of this system was a **fluid** extremely subtle, [. . .] which was pretended to be **diffused** through the whole space occupied by the material **creation**, to animate the system of nature, to **penetrate** all substances, and to be the **vehicle** to animated bodies in general, and their several regions in particular, of certain forces of attraction and repulsion, by means of which they explained the phenomena of nature.
>
> [. . .] they flattered themselves that they were able [. . .] to render themselves masters of this fluid, and to direct its operations. [. . .] they affirmed that this fluid, in which they admitted a species of **flux and reflux**, exerted an important degree of action upon the nerves and had a grand analogy with the **vital principle**; [. . .] that it was possible to **impregnate** with it, either certain powders, [. . .] or fluids, or different parts of the bodies of animals; that this agent was like **light reflected by mirrors**, [. . .]. (Franklin 1785, 2–3; emphasis by MS)
>
> The action and the virtue of the animal magnetism are capable of being **communicated** from one body to another, animated or inanimate; [. . .] this action is [. . .] **propagated and augmented** by sound; and the virtue itself is capable of being **accumulated, concentrated and transferred**. (Franklin 1785, 20; emphasis by MS)

Metaphorical *words* are used in non-standard senses to create comparisons. "Achilles is a lion" is a classic example. But words may also be ambiguous between literal and metaphorical: here, "impregnate" and "communicate" are conventional metaphors (at the time) for a transfer of substance from A to B; there is likely no deep comparison, but rather a partial mapping of conceptual structure. Such words are often tied to metaphorical models, or concepts in flux.

Metaphor *scenarios* are more specific, visualizable multi-word chunks of text, and are often explicit comparisons; for example, "this agent was like light reflected by mirrors".

3 For metaphor in words, see Steen (2007). For metaphor scenarios, see Musolff (2006). For metaphor as conceptual mapping, see Lakoff and Johnson (2003 [1980]; 1999), and Lakoff and Turner (1989). For blends, see Fauconnier and Turner (2003).

4 This is an imperfect method, used for purposes of clarification. Words are often ambiguous between metaphorical and literal, and scenarios, models and blends can not always be clearly linked to specific words.

Metaphorical *models* use one concept to think about another by a general and abstract mapping of structural properties, conceived in sensorimotor-schema terms, typically from the more perceptually familiar concept (the "source") to the less (the "target"). Cognitively, this is the most important level. Seeing magnetism as *fluid* motivates the use of words "diffused" and "flux and reflux," and the capacity for "being accumulated, concentrated and transferred."

Metaphorical *blends* are complex constructions that involve mappings from multiple concepts, and further elaboration after the mapping. Blending analyses speak of "input spaces" mapping to a "blend space," rather than source-target mapping. With inputs of magnetism, light, sound and the vital principle, the theory is a metaphorical blend rather than a simple mapping.

2.2 Intuitive energy in perception and popular belief

I suggest we can go further in the directions Gibbs Jr. and Talmy signpost by adapting "intuitive ontology" research to better specify and elaborate sensorimotor schemas. Intuitive ontology refers to those "naïve" or "folk" theories we use to understand and navigate the everyday world of objects, plants, animals, people. They're unscientific, simplified, informal; but they work well enough and we use them constantly. Hence they go deep into our habits of thought and our ways of making meaning, including in the semantics of texts, literature and worldview. Intuitive *physics*, Alan Leslie says, represents "the arrangement of cohesive, solid, three-dimensional objects embedded in a system of mechanical relations, such as pushing, blocking, and support" (1994, 124). A major part of this is "find[ing] the 'sources of energy' that produce [object] motions [. . .] or, more generally, the distribution or dynamics of forces in a scenario" (Leslie 1994, 125). The key here is that *physical causation* is seen as *transfer of energy* between objects that come into contact (like billiard balls). In turn, *energy* is seen as *substance*, and objects are seen as *containers*. This kind of causation is *perceived*, not inferred or deduced. As Michotte's studies showed, people can't *not* see billiard-ball type interactions as contact causation, even when only watching 2-D shapes on film. The intuitive physics that Leslie and others describe is rather like medieval physics, where the substance transferred was called "impetus": when you throw something, the impetus passes from your arm into the thing, carries it through the air, and drains out as the thrown object falls (cf. McCloskey 1983). Yet despite the deadness of medieval physics, even today people (including physicists!) struggle to get over the idea that energies like electricity, light and heat are substances (cf. Reiner et al. 2000).

This intuitive physics can be seen as grounding many other kinds of thinking. It feeds into intuitive biology and psychology. We recognize animals because they move in certain ways (unlike objects, they start, stop, and change direction by themselves). And we read the minds of people (and things!) through the energy-relations revealed in what they do, and how. Actions can appear happy or sad, angry or fearful, just by their manner (see Stern 2010 on "vitality forms"). This, we might say, is a kind of intuitive *psycho*physics, or perception of *narrative* causality. In the classic "Heider-Simmel Illusion," interacting geometric shapes irresistibly evoke an anthropocentric narrative, of (approximately) two small shapes (triangle and circle) teaming up to escape the threat of a larger shape (triangle), which then destroys the square room they were in, in a fit of rage (cf. Heider and Simmel 1944). This kind of reading of movement as story is related to the fact that humans are "natural dualists" – meaning that we tend to accept that body and self, soul and/or spirit are different *kinds* of substances, with soul suffusing body, feeling through its senses and controlling its movements (cf. Bloom 2004).

People develop corresponding beliefs about life and death. People see the self as "initially contained in bodily mass," "exiting or taking temporary leave of the body at some point after [its] expiration"; "The *soul* is [. . .] the conscious personality of the decedent and the once animating force of the now inert physical form" (Bering 2006, 453). As the language here reveals, the soul still behaves in bodily ways: it is an animating force, it is contained, it exits. This motivates the traditional paradox that spirit is an "immaterial substance," which infuriated Thomas Hobbes and others. Yet it can be traditional only because we have, curiously, no trouble imagining and representing it. More generally, according to cognitive anthropologists, conceptualizations of beings like gods, spirits, ghosts and saints rely on "ghost physics." The behavior of these beings departs from intuitive physics in a few key ways that make them memorable (cf. Boyer 2001). For example, saints and their relics are imbued with a "*virtus*," a force from God that "infused their souls, their bodily remains, and objects that may have come into close physical contact with them." The force "extended to the clothes saints wore, to their tomb, to the soil that abutted their tomb, to liquids that [. . .] oozed from beneath it" (Scott 2006, 215). Relics "were treated as if surrounded by a force field of spirituality that could be transmitted to anyone who came close to it" (Scott 2006, 216). Spirits and ghosts are typically quasi-substantial and quasi-visible: they can pass through walls and other barriers, but they can make their presence known by making sounds, moving objects, partial appearances, ectoplasmic effusions. Sometimes people just *feel their energy*.

2.3 Generalization: The energy-dynamics schema

Given these examples, we can generalize about what energy is and how it works. *Energy* blends concepts of force (termed "umph" in the literature) with substance (termed "stuff"). It is umph-stuff, force-imbued substance. In a narrative, argument or description that models causation in the world, energy can be:

- contained within bodies and other entities
- accumulated, stored and drained
- moved around within an entity
- expended to exert force (e.g. movement)

Through movement, it can

- be transmitted (given/taken, by contact of some kind)
- change condition (volume, intensity, shape, density, speed)
- create relations between entities (individuals or groups) (attract/repel, connect/conflict)

3 Cultural semiotics: Energy in stories and storyworlds – high, low and in-between

Everyday language and belief are important, but to find *imaginative models of experience* presented in their full depth of detail, we must turn to cultural semiotics – to meaning-making in the arts and sciences.

3.1 Literary cosmology and artistic convention: Religious epic

Interaction patterns of core sensorimotor schemas seem to be key to the traditional conventions that support artistic imagination. Northrop Frye describes a general "literary cosmology", a complex but highly coherent diagrammatic "schematism of correspondences" growing out of religious mythology, which supplies a framework for artistic symbolism, even after scientific and mythological world-pictures diverge (1971, 160–161). Such cosmologies are clearest in major epics. Dante's *Divine Comedy* describes an allegorical storyworld in which God is a "river of living light," and the divine energy is distributed throughout the universe as the soul is distributed throughout human bodies, supplying its substance, holding it together and keeping it moving, and organizing its parts into a spatial setting. This defines how the inhabitants *relate* to the world, and to each other:

> [. . .] and as the soul within your living **dust**
> **diffuses** through your body's different parts,
> adapted to its various faculties,
> just so does this Intelligence **unfold**
> its **bounty** which the stars have multiplied [. . .]. (Alighieri 1984, 22; emphasis by MS)

In this metaphorical order, all things have a place or station. God is uppermost and outermost (paradoxically also central), Satan lowest and inmost; humans are in the middle, drawn both upwards and downwards. The vertical scale maps to other spatial schemas, and other physical schemas and quantities. Dante's descent through Inferno reveals increasing degrees of evil in circles that become ever-narrower, darker, colder and more fixed. His ascent up Purgatory and to Paradise gets ever-easier, swifter, brighter and warmer.[5]

3.2 Supernatural plots and storyworlds: Pop culture

Having touched on the highest of high-culture texts, some cultural slumming will help us demonstrate the ubiquity of the specific pattern of personal energy inhabiting and controlling bodies, and transferring between them. Consider how it can govern whole *plots*, including those that develop their own idiosyncratic cosmologies. Think of all the films and comics in which some form of "magical" or "spiritual" or pseudo-scientific quasi-physical energy like a stream of light is transferred from one character to another – often via hands, eyes and mouth. The transfer can be experienced as good or bad. Good guys may give positive energy or absorb bad energy. Bad guys infect with bad energy, and/or drain good energy. This is a basic underlying structure, on which is built plenty of variation. Energy can inhabit organisms, objects, places, even the whole universe (as in the Force, in *Star Wars*). Battles between energies are common, as in the magic worlds of Tolkien and *Harry Potter*. Indeed the pattern is fundamental to "archetypal" plots of myths and legends. For example:

- Possession: An evil spirit enters and takes control of an entity (person, animal, object, etc.), and may be driven out by a good or holy figure
- Scapegoating: The sins of a community are projected onto a figure (animal or human), who is driven out and punished/sacrificed, in order to purify the community

5 For a fuller analysis along these lines, see Sinding (2014, ch. 2, 69–90).

These have parallels in fantasy plots about monsters like vampires and werewolves, and superheroes who acquire powers from extraordinary energies: Spiderman's radioactive spider, The Hulk's gamma rays, the "X-gene" producing mutant X-Men, the cosmic rays that create the Fantastic Four, Superman's alien sensitivity to different lights (yellow sunlight activates his powers, red sunlight deactivates them, green "kryptonite" radiation incapacitates him).[6]

Let's consider in detail an example that shows that energy-plots pre-date the CGI technology that makes energy imagery so familiar in movies today. The 1972 film *Horror Express* (with Peter Cushing, Christopher Lee, and Telly Savalas, directed by Eugenio Martin) is something of a B-movie classic (and it helped crystallize these ideas for me). It's a sci-fi horror story of what might be called "evolutionary astro-neurobiology," because it connects evolution and the missing link, brains and knowledge, alien species and space travel, possession and zombies. To sketch the plot: Professor Saxton has found an ape-like creature he believes to be the "missing link" frozen in the Himalayas. He's taking it back to Europe by train. Inevitably, it awakens and starts killing people. When the creature attacks, its eyes glow red, the victim's eyes bleed, then turn white and blank, and suddenly the creature knows what the victim used to know. Saxton and his rival Dr. Wells conduct an autopsy on a victim, find the brain strangely smooth, and conclude that it has been somehow "drained" of all knowledge and memory. Later, Wells has an "aha!" moment when he sees a boiled fish with the same white eye on a dinner plate: the creature must be absorbing mental contents through the eyes. Not too long afterwards, the monster is killed, and another autopsy reveals "images embedded in liquid inside the eye," images of a prehistoric earth, *and* a view of the earth from space. They deduce that the creature is really an "alien [. . .] inhabiting the body" of the missing link, and it must have "transferred itself to someone else" on the train. The alien mind/soul goes on a body-hopping rampage, planning to combine the knowledge it absorbs to build a spacecraft. Eventually it reveals that it was left behind by its fellows millions of years ago; it is indeed "a formless energy" and "has survived [. . .] by transferring itself to each form of life that has evolved"; hence "it possesses all of the knowledge that Saxton so desper-

---

6 Some of these details of superhero powers have been taken from the relevant pages on Wikipedia: "Spider-Man." *Wikipedia*. https://en.wikipedia.org/wiki/Spider-Man (10 August 2022); "Hulk." *Wikipedia*. https://en.wikipedia.org/wiki/Hulk (10 August 2022); "X-Men." *Wikipedia*. https://en.wikipedia.org/wiki/X-Men (10 August 2022); "Fantastic Four." *Wikipedia*. https://en.wikipedia.org/wiki/Fantastic_Four (10 August 2022); "Superman." *Wikipedia*. https://en.wikipedia.org/wiki/Superman (10 August 2022).

ately wants." It tries to bargain with him, offering to help end disease and hunger, but at last dies a fiery death when the train crashes.[7]

We can step back and notice that these actions of heat-attack, brain-drain, and body-transfer are crucial to the plot; and because they are intensely emotional and dramatic scenes, they also help generate the pervasive mood of fear. As so often happens, the biopsychosocial causality here is metaphorical, but has an intuitive coherence (however ludicrous). But that coherence reveals something important about how we process the story. The hokum goes down without a hitch because it has the right kind of energy-structure. The creature is essentially a unified umph-stuff that can enter nearby bodies and take them over, and then depart when they die. Causation works by *energy-transfer*, which requires (at least virtual) *contact*. It attacks by making eye-contact; this involves other physical *quantities* – it turns up the heat in a kind of ray, by its glowing red eyes – and has effects on *substances*, effects that are at once physical and mental. The optical energy-contact causes bleeding, and boils the eyeballs and brain white, which allows mental *draining*, which *smooths* the brain back to a kind of *tabula rasa*, and *draws* knowledge, reified as (contained in?) a kind of *immaterial fluid*, into its brain, to be *imprinted* on its own eye-fluid.

It's notable that while this is not a very traditional monster, its causality works along the lines of traditional monster-models. Like a demon, it possesses bodies; like a vampire, it infects and drains life; like a gorgon or basilisk, it kills by a look; like a dragon, it cooks its victims; like zombies, it ingests brain-material, which gives it a kind of telepathy.

3.3 Realistic storyworlds and worldview argument: *News from Nowhere*

Non-supernatural narratives are often full of mixes of metaphors that convey clear meanings because they create quite coherent underlying energy scenarios and structures. Yet even narratives with highly realistic storyworlds, quotidian topics, and plain non-figurative styles tend to turn to energy metaphors and stories in their most important moments. An interesting example is William Morris's 1891 "utopian romance" *News from Nowhere*. Despite the fact that the story's premise of a man's visit to an ideal future society is fictional overall, the story-

7 Some of this account is adapted from the IMDb pages for the film: "Synopsis" of "Horror-Express (1972)." *IMDb*. https://www.imdb.com/title/tt0068713/plotsummary?ref_=tt_stry_pl#synopsis (10 August 2022).

world of that future society, and most of the individual scenes within it, are realistic, and the language is predominantly literal and straightforward in style (in accordance with Morris's values of nature, common-sense reason, and solid craftsmanship). There is hardly a metaphor to be found, except of the most incidental kind. That is, while the fictional society is organized according to Morris's radical socialist-communist values, the persons and objects and events encountered by the protagonist (Edward Guest) are of the average-everyday variety, and Morris's language is predominantly non-figurative. Guest meets people, has conversations and meals, observes bodies and clothes and architecture and gardens and forests and rivers and food and behavior. The most "unrealistic" facts of the storyworld are that the inhabitants appear so comparatively healthy, young and attractive – due to their living by principles of nature, equality and freedom, having escaped from the oppressive economics, politics, and physical grime of the late nineteenth century.

For a flavor of this, consider a passage in which Guest's main interlocutor, Hammond, explains how this society essentially does without government:

> It is true that we have to make some arrangements about our affairs, [. . .]; and it is also true that everybody does not always agree with the details of these arrangements; but, further, it is true that a man no more needs an elaborate system of government, with its army, navy, and police, to force him to **give way** to the will of the majority of his *equals*, than he wants a similar **machinery** to make him understand that his head and a stone wall cannot occupy the same space at the same moment. (Morris 2003, 122; italics in original, all other emphasis by MS)

There is the metaphor of "machinery," and that comparing the simple point about social cooperation with the simple point about heads cooperating with walls. Lakoff-and-Johnsonists would point to arguably metaphorical expressions "force" and "give way," and the arguably metonymical "will of the majority." One could also argue that "arrangements" refers primarily to a more basic sense of physical arrangements (e.g. of objects in space) and secondarily to a metaphorical sense of political arrangements. But for the most part with Morris, when abstract words are useful, they are used ("government," "arrangements," "affairs," "system," "majority"), and typically not modified with metaphors or other figures.

However, tellingly, Morris turns to metaphorical energy-thinking in the key passage that recounts how the system of the pre-utopian world worked, in order to explain how the happy revolution came about.[8] This account is partly a (gen-

8 Chapter XV (136–143), titled "On the Lack of Incentive to Labour in a Communist Society" because this is the assumption that prompts the question of how it came to be that all work is rewarding because pleasurable.

eral) physical description of actions like producing, selling and buying wares, but it is blended with metaphorical psychosocial analysis to explain how the former system was perpetuated despite its villainous consequences. I suggest that Morris needs to blend physical with psychosocial causality, in order to compose a coherent, compelling, dynamic image of an out-of-control socio-politico-economic system – and this need leads him to reach for metaphor, to blend local metaphors into local scenarios, and to blend local scenarios into an overall narrative, all of which hold together because of underlying energy-patterns. Hammond blames the ills of the past age on a monstrous "World-Market" that personifies an image of the idiom "vicious circle":

> [. . .] in the last age of civilization men had got into a **vicious circle** in the matter of production of wares. [. . .] they had gradually created (or allowed to **grow**, rather) a most elaborate **system** of buying and selling, which has been called the World-Market; and that World-Market, once **set a-going, forced** them to go on making more and more of these wares, whether they needed them or not. So that while [. . .] they could not **free** themselves from the toil of making real necessaries, they created in a never-ending series sham or artificial necessaries, which became, under the **iron rule** of the aforesaid World-Market, of equal importance to them with the real necessaries which supported life. By all this they **burdened** themselves with a **prodigious mass** of work merely for the sake of keeping their wretched **system going**. [. . .] [S]ince they had **forced themselves to stagger along under this horrible burden** of unnecessary production, it became impossible for them to **look upon** labour and its results from any other point of view than one – to wit, the ceaseless endeavour to expend the least possible amount of labour on any article made, and yet at the same time to make as many articles as possible. To this 'cheapening of production,' as it was called, everything was **sacrificed**: [. . .] [the workman's life] did not **weigh a grain of sand in the balance** against this dire necessity [. . .]. [E]ven rich and powerful men, the masters of the poor **devils** aforesaid, **submitted** [. . .], in order that their riches might **bolster up** this supreme folly. The whole community, in fact, was **cast into the jaws of this ravening monster,** 'the cheap production' forced upon it by the World-Market. (Morris 2003, 138–139; emphasis by MS)

There is a comparison with the dung beetle getting used to living in dung, then the idea of "labour-saving machines" is mockingly rejected, and the global expansion of markets is savaged:

> The **appetite** of the World-Market **grew** with what it **fed on**: the countries **within the ring** of 'civilization' (that is, organized misery) were **glutted** with the **abortions** of the market, and force and fraud were used unsparingly to '**open up** ' countries ***outside*** that **pale**. (Morris 2003, 139–140; italics in original, all other emphasis by MS)

There is some ambiguity as to whether a term like "system" is metaphorical here (it could be a mechanical system, or an abstract one, but it seems organic because it "grows"). But it is clear from the set of word choices that there is an underlying metaphor scenario, of the World-Market as a man-eating monster (both machine-like and demonic) that, having been created by society (largely unwittingly) then enslaves that society: the monster forces society mercilessly to work in its service – to feed it, carry it (as a burden), and help it to expand into other countries.

The causal logic of the metaphor story is that of a "vicious circle," as noted, but with additions. Simply by coming into being, this particular "system" sets up a feedback loop where it must make as much product as possible for the lowest cost possible; therefore it always needs more input, and it must always expand. Moreover, because the creators are part of the system, not external to it, the creature/system becomes an ever-increasing burden on them and takes away their ability to choose another path or system, so that they become *its* creatures or subjects. A feedback loop is not in itself a bad thing, but clearly there are emotional, aesthetic and ethical evaluations projected on to the causal logic of *this* feedback loop (because of the specific details of the participants in this logic, and the consequences for their experiences). The image of this monster "glutted with abortions" of the market conveys Morris's severe ethical and aesthetic judgment of this economic system in vivid, memorable, forceful terms.

The key point here is that an argument about a quite real socioeconomic situation (one that is necessarily both concrete and abstract, due to its complex linking of physical production to legal institutions, politics, and economic "laws" of supply and demand) can be conveyed using metaphors that mix together to construct a fictional and fantastic multi-scene narrative frame with strong causal coherence. That is, although the scenes and story have multiple complexly interacting dimensions – physical, psychic, and social – the metaphors for those dimensions mix and fuse to create a powerfully unified energy pattern.

We could develop these analyses in more formal and detailed terms, to describe how the discourse unfolds as a sequence of target concepts that are metaphorically framed by source concepts, themselves structured by sensorimotor schemas. The overlapping target domains: production of wares > economic system > World-Market > globalization. The overlapping source domains: feeding > growing > burden > iron rule > ravening monster. But the less formal analysis establishes the point that the causal logic of this discourse, which integrates narrative, argument and evaluation, is constituted by energy-thinking – the *action* and the *description* of the scene cohere because of the *spatially organized structure of energy-transfer*.

This account of Morris's highly polemical narrative fiction could easily lead into consideration of how worldviews are built up in forms of writing that are primarily discursive arguments about facts, rather than either narrative or fiction. Schematic metaphorical thinking can feed into pundits' clichés, but it is also central to major canonical worldview texts.[9]

4 Scientific concepts of energy

Our ideas about energy are now strongly shaped by science. Beyond our informal understanding of "energy," primarily relating to 1. a sense of bodily vigor, and 2. whatever it is that makes our machines run, we also know that "energy" has multifarious scientific meanings. Indeed energy is one of the coins of the scientific realm, one of a handful of concepts with very broad application and very long standing. It cuts across a range of physical qualities and their sciences, with divergent definitions and mathematizations. There is mechanical energy, perhaps the most basic kind – the push and pull of physical objects moved by forces. There are electrical energy, magnetic energy, electromagnetic energy, thermal energy, chemical energy, and more. All of these energy types can be converted into one another. They can be analyzed in terms of different, more abstract, divisions of energy: potential energy (stored energy waiting to do work in a system), kinetic energy (energy in motion as the system operates) and thermal energy (exhaust energy due to resistance to force). Energy is the basis of some of the most fundamental physical laws. The Law of Conservation of Energy: Energy can neither be created nor destroyed, only transformed (more formally, the total amount of energy in an isolated system remains constant). Within particular energies, laws state mathematical relationships of factors (often conceived metaphorically). For electrical circuits, Ohm's law states that voltage (force) = current (rate of flow) x resistance (friction). Further laws build on these: "power" = voltage x current; "work" = power x time. Einstein's famous formula $E = mc^2$ tells us that energy equals mass times the speed of light squared: matter and energy are two sides of the same coin, mass-energy.

Hence science in general shares commonalities of intuitive ontology with culture while developing it in distinct ways. Scientists are concerned with energy flows at several levels of purpose: understanding the abstract nature of energy (theoretical), examining how it behaves (experimental), and evaluating system energy to system purposes (practical). These types of thinking can be deeply entwined with one another, and with aesthetic, emotional and moral cognition. Consider how theoretical and social hopes and dreams drive enormous feats of engi-

9 See Sinding (2016), on the origins of conservative and liberal worldviews in texts by Edmund Burke and Thomas Paine.

neering like CERN's Large Hadron Collider. As with cultural energy, scientific purposes involve concern with knowing sources and destinations of energy, patterns of flow, and how energy increases, decreases and transforms along flow-paths. Yet scientific energy-concepts are much more fully theorized, strictly-defined and mathematized, and the ontological and epistemological targets remain different, centred on physical structures and quantities rather than human life-experience and worldview. True, scientific energy-thinking is highly imaginative, and it becomes more cosmological as it becomes more theoretical (e.g. electromagnetism as one of the four fundamental forces of the universe). But even cosmology only ever has indirect implications for human life and worldview, because "ought" doesn't follow from "is." When theories in the human sciences are relatively explicitly focused on energy, as with Freud's hydraulic-like model of psychic drives and blocks, or J. G. Frazer's electricity-like anthropological model of energy shared across nature, communities and kings, any formalizations (semantic and/or mathematic) never approach scientific exactness.

With this sketch of what's shared and distinct in patterns and purposes of scientific vs. cultural energy-thinking, we may turn to examine in depth a decisive phase in the evolution of electro/magnetic energy-thinking – when it was not yet so far from its wellspring in intuitive ontology, and its future division into separate streams of intellectual/professional domains was accordingly less clear, more fraught with unknowns, and more traversable.

4.1 Electro/magnetic thought in the late eighteenth century

A snapshot of early electro/magnetic science, and its offshoots in culture, will contextualize the following textual analyses. In the eighteenth century, new sciences of electricity and magnetism fed into enthusiastic speculation and applications connecting the physical with the mental – in medicine, psychology, politics, religion, the arts. Galvani's studies of "animal electricity" and Mesmer's claims about "animal magnetism" were both inspired by biological discoveries about how fish like the torpedo and the electric eel controlled electrical charges for attack and defense, and by physical discoveries about how electricity could be collected in Leyden jars, and then discharged.

Some suggested that magnetism and/or electricity were the life-force, and the substance of spirit. William Gilbert's *De Magnete* (1600) had proposed magnetism "as the binding agent of the universe," a view that was only gradually displaced by Newton's theory of gravitation (1687) (cf. Fara 1995, 135). Galvanism and Mesmerism were taken to have profound moral and political implications (because moral laws should parallel natural laws), especially in connection with the French

Revolution; and they inspired many artists, including Coleridge, Shelley, Balzac and Poe.

Both theories were closely akin to, and supported by, theories of the "Aether," the concept, rooted in ancient mythology and science, of an air-like substance filling the realm of the gods, and/or the celestial spheres. In its more materialistic versions, it was often regarded as a "subtle" or perhaps "immaterial" fluid-like substance pervading all space. Among the most important developments was Newton's (1687) proposal that an aethereal medium could explain the action of gravitation and light. Indeed, the hypothesis of a "luminiferous aether" was only finally disproven by the Michelson-Morley experiment in the 1880s.

In the end, Galvani was largely vindicated, and became part of the history of science, while Mesmer was debunked, and became part of the history of pseudoscience. But at the time they seemed to occupy the same spectrum of open conceptual possibilities. We can scrutinize this spectrum to find where and how boundary-drawing between science, pseudoscience and literature becomes puzzling. Cultural historian Robert Darnton, aiming to capture "the mentality of literate Frenchmen on the eve of the Revolution" focuses on "the hottest topic – science in general, mesmerism in particular" because his subjects "found that mesmerism offered a serious explanation of Nature, of her wonderful, invisible forces, and even, in some cases, of the forces governing society and politics" (Darnton 1968, vii).[10] Mesmerism easily posed and passed as science:

> [Mesmerism] corresponded perfectly to the interests of literate Frenchmen in the 1780's. Science had captivated Mesmer's contemporaries by revealing to them that they were surrounded by wonderful, invisible forces: Newton's gravity, made intelligible by Voltaire; Franklin's electricity, popularized by a fad for lightning rods and by demonstrations in the fashionable lyceums and museums of Paris; and the miraculous gases of the Charlières and Montgolfières that astonished Europe by lifting man into the air for the first time in 1783. Mesmer's invisible fluid seemed no more miraculous, and who could say that it was less real than the phlogiston that Lavoisier was attempting to banish from the universe, or the caloric he was apparently substituting for it, or the ether, the "animal heat," the "inner mold," the "organic molecules," the fire soul, and the other fictitious powers that one meets like ghosts inhabiting the dead treatises of such respectable eighteenth-century scientists as Bailly, Buffon, Euler, La Place, and Macquer. Frenchmen could read descriptions of fluids very like Mesmer's under the articles "fire" and "electricity" in the *Encyclopédie*. If they de-

10 Although it "originally had no relevance whatsoever to politics, it became, in the hands of radical mesmerists like Nicolas Bergasse and Jacques-Pierre Brissot, a camouflaged political theory very much like Rousseau's. The mesmerist movement therefore serves as an example of the way in which, on a vulgar level, politics became enmeshed with fads, providing radical writers with a cause that would hold their readers' attention without awakening that of the censors" (Darnton 1968, 3).

> sired inspiration from a still greater authority, they could read Newton's description of the "most subtle spirit which pervades and lies hid in all gross bodies" in the fantastic last paragraph of his *Principia* (1713 edition) or in the later queries of his *Opticks*. (Darnton 1968, 10–11)

Moreover, scientific thought was not then well-distinguished, in topics and methods, from philosophy and religion (cf. Darnton 1968, 11–12). Mesmerism seemed to have a noble lineage, an honourable vocation and a respectable circle:

> [M]esmerism had most in common with the vitalistic theories that had multiplied since the time of Paracelsus. [. . .] his system descended directly from those of Paracelsus, J. B. van Helmont, Robert Fludd, and William Maxwell, who presented health as a state of harmony between the individual microcosm and the celestial macrocosm, involving fluids, human magnets, and occult influences of all sorts. Mesmer's theory, however, also seemed related to the cosmologies of respectable writers who sponsored a variety of fluids, which they sent swirling through the universe under familiar names like gravity, light, fire, and electricity. (Darnton 1968, 14)

Mesmer's imaginings echoed electrical terminology, experiments, spectacles, fantasies, and even equipment:

> Galvani was experimenting with "animal electricity" in Italy at the same time that Mesmer used animal magnetism to cure hundreds of persons in France. Meanwhile, [. . .] others had discovered miraculous powers in the universal electric fluid. [. . .] [It] made plants grow faster and [. . .] electric eels cured gout. [. . .] Some detected a note of charlatanism in Mesmer's treatments, but his apparatus resembled the all-popular Leyden jar and the machines illustrated in standard works on electricity [. . .]. These amateurs often sent electrical charges through "chains" of persons like Mesmer's and often regarded electricity as a magic potion that would conquer disease and even [. . .] help to create life. (Darnton 1968, 14–15)

Darnton concludes that "mesmerism seems to have occupied a place somewhere near the middle of the spectrum in which science shaded off into pseudoscience and occultism" (Darnton 1968, 36–38).[11] This positioning of Mesmer's vision within the developing world-models of the time indicates how easily (naturally?) the imaginations of scientists, artists and the public fixed on numerous overlapping visions of invisible forces operating in different domains (physical/biological/psychological/social), but in very similar ways – that is, as systems of energy or umph-stuff.

---

11 Fara cautions that mesmerism was not as enthusiastically received in England as it was in France, given its discrediting by a prestigious committee, and its use by satirists to mock opponents in various debates (1995, 130–131).

5 Electro/magnetic energy-thinking across domains: Literature/science/pseudoscience

We now consider how each text uses narrative and metaphor to frame its electro/magnetic thinking in terms of our energy model of causality: *Force-imbued substance transfers from one container-object to another via contact.* By comparing their specific uses of this complex conceptual structure, we gain a sense of how these authors adapt an arguably natural and universal pattern of causal thinking in ways that 1. serve their specific purposes in grappling with the particular kinds of causality that mattered most to them, and 2. reveal the forms of causal cognition that characterize their domains of literature, science and pseudoscience.

5.1 Literature: William Blake's *Milton*

Blake disliked certain aspects of science. He repeatedly attacks Bacon, Locke and Newton for the narrow "single vision" that limits poetic imagination. On the other hand, he absorbed the ambiguously materialist and mystical ideas of Paracelsus and Emanuel Swedenborg – both scientists and mystics, and sometimes linked with animal magnetism and electricity. Blake may well have known the variant of Mesmerism associated with the most eminent British magnetizer, John de Mainauduc. De Mainauduc propounded a "theory of an energetic divinity operating on 'the nervous system'" (Rix 2002, par. 19, citing De Mainauduc 1798, 42): "all beings [. . .] are made up of 'pores,' which 'allow the passage [. . .]' of [an] invisible 'fluid'" (Rix 2002, par. 19, citing De Mainauduc 1798, 19–21). "His key diagnosis of sickness was that the fluidity and circulation of the atoms [. . .] was blocked" (Rix 2002, par. 21). For treatment, the magnetizer "uses his spiritual power to [. . .] [break] down the impediment" (Rix 2002, par. 21).

Electro/magnetic ideas inform a number of Blake's works. In general, they are associated with positive revolutionary human energy – poetic inspiration, and charismatic power – and folded into an allegorical mythology, often of spiritual sickness and cure. Again, the magnetizer worked through eyes, hands and breath. Rix (2002) discusses magnetic "religion" and "cure" in aspects of *The Marriage of Heaven and Hell, America, The Book of Urizen, The Book of Los, Europe,*

The Four Zoas, *Milton* and *Jerusalem*; and later (Rix 2010) "The French Revolution."[12]

Blake's most magnetic poem is arguably *Milton* (cf. Rix 2002, par. 27). In this prophecy, Milton's spirit visits Blake, delivering a vision of his quest to learn and correct his errors – chiefly his Puritanical elevation of Reason and undervaluing of the feminine. Milton's self is divided into various aspects distributed through the cosmos; this drives his quest for reunification by self-correction. The poem opens by recasting the traditional epic invocation to the Muse in very bodily and indeed neurobiological terms: "Come into my hand, / By your mild power descending down the Nerves of my right arm / From out the portals of my Brain" (Blake 1993, 111, ll. 5–7). Similarly, when Milton appears to Blake and enters his spirit, it is in a way that is oddly physically specific: "Then first I saw him in the Zenith as a falling star. / Descending perpendicular, swift as the swallow or swift: / And on my left foot falling on the tarsus, enterd there" (Blake 1993, 141, ll. 47–49). Milton later descends to inspire the rest of the world, seemingly via electrical energy that translates into physical symptoms of agitating dreams: "Now Albions **sleeping** Humanity began to turn upon his Couch; / Feeling the **electric flame** of Miltons awful precipitate **descent**" (Blake 1993, 152, ll. 25–26; emphasis by MS).

Milton is a primarily narrative text, and as an allegorical epic poem, its narrative cries out for interpretation of non-literal meanings. Allegories refer (at least implicitly, relatively continuously, and typically by metaphor) to a parallel set of ideas and/or events, so the reader is pressed to ask, what do these conspicuously curious details mean? The poem is also autobiographical in a sense – it analyzes the tribulations of Blake's imagination under pressures from his patron Hayley during a sojourn at Hayley's cottage – and that context gives some clues. Blake's wife Catherine underwent a cure by what Blake called "electrical magic" for rheumatism in her foot. Her doctor, John Birch, composed an essay on medical electricity, which opens describing a star-shaped appliance (cf. Schott 2003). Such biographical fact must be framed in the literary context: why this allusion now? Fara shows how Blake's contemporaries were updating ancient ideas about inspiration. The actor David Garrick celebrated how "sympathetic imagination" enabled him to "take on other personalities" and how "a keen sensibility should shoot through the spectators' bodies like electric fire," echoing Plato's *Ion*'s account of "how artists are divinely inspired through a magnetic chain linking them

12 See Schuchard (1989) and Schott (2003) for broader pictures of these cultural backgrounds to Blake's work, Fulford (2004) on Mesmerism in the politics and poetics of the 1790s, and Fara (1995) on animal magnetism in eighteenth-century England.

to God" (Fara 1995, 147). Blake similarly connects electric with traditional magnetic chains of inspiration. In the poem (and Blake's illustrations), the image of inspiration by electrical contact with the foot is linked with comets and lightning. Symbolically, the falling star echoes Milton's image of Satan in *Paradise Lost*, and Blake saw Satan as the true hero of that poem. Here, Milton is the corresponding Satanic hero, the foot connects Blake's body with the earth, and the star turns that "Vegetable World" of life and death into a sandal he can wear (an artful/biblical form) so that he can journey in Milton's vision:

> And all this Vegetable World appeard on my left Foot
> As a bright sandal formd immortal of Precious stones & gold:
> I stooped down & bound it on to walk forward thro' Eternity. (Blake 1993, 155, ll. 12–14)

The tarsus presumably alludes to the apostle Paul's revelation and conversion on the road to Damascus, since Paul began as Saul of Tarsus. The poem attacks Paul's militant and anti-sexual errors, inherited by the church and Milton, but it closes with Blake floored by a "Revelation."

Overall, the electro/magnetic references in the poem are few, but they are highly significant, as they play the role of the forces of inspiration, and artistic imagination, which provide spiritual "cure." The electrical descent of Milton's spirit is central to the plot; it connects the main characters to one another, to the cosmos, and to humanity in a vivid image, which blends with more traditional images, of light, star, fire, and birdsong.

More broadly, the new discoveries align with Blake's insistence on the unity of body and soul (his poems treat body parts and mental faculties as characters). The thought of electricity traveling via nerves to affect sensation and muscles, drive biological life and cause healing may have bolstered his vision of matter as subject to mind, and of consciously metaphorical perception as the means to redeem a material world. In Blake's work, the world of objects and mechanism is a kind of enemy (cold, hard, heavy, closed, "opake"), and it is opposed by a world of living, moving energies (fiery, warm, flowing, light). That's not to say that mind can change matter directly and mechanically, as in Mesmer. Rather, our perception and experience of the physical world change according to our imaginative vision, so the metaphors and stories of art that feed imagination are essential to navigating the world.

5.2 Science: Luigi Galvani's *Commentary*

The famous discoveries recounted in Galvani's *Commentary on the Effect of Electricity on Muscular Motion* (1791) began with an accident. He placed a dissected frog near an "electrical machine"; then, "When by chance one of those who were assisting me gently touched the point of a scalpel to the medial crural nerves [. . .] of this frog, immediately all the muscles of the limbs seemed to be so contracted that they appeared to have fallen into violent tonic convulsions" (Galvani 1953, 24). Galvani seeks to explain this rudimentary cause-effect link by using extensive experimentation to fill in all details – both to identify the full range of possible causal entities and processes, and to narrow them down to the smallest possible range of plausibly actual causal chains. He describes a series of many similar experiments with minor variations, in connection with his thought processes as to how observations relate to possible causation (e.g. regarding the "location" of electrical fluid: nerves or muscles).

The Commentary is divided into four sections, "The Effects of Artificial Electricity on Muscular Motion" (Galvani 1953, 24–35), "The Effects of Atmospheric Electricity on Muscular Motion" (Galvani 1953, 36–39), "The Effects of Animal Electricity on Muscular Motion" (Galvani 1953, 40–59), and "Conjectures and Some Conclusions" (Galvani 1953, 60–81). This reflects Galvani's carefully systematic thinking: three different kinds of electricity being studied extensively in relation to the same phenomenon of electrically-induced muscular motion. We follow Galvani seeking to account for the phenomenon by constantly proposing hypotheses about the causal relations between the electrical charge and the muscle twitch, and constantly posing challenging questions to his own hypotheses and testing those hypotheses in the lab.

To demonstrate with a few examples. In section 1, Galvani first wonders if the motion is due to the contact of the point of the scalpel with the nerve, or the spark induced, or both (Galvani 1953, 24). He then considers and tests conductile vs. non-conductile materials linking the electric current to the frog (Galvani 1953, 25). His observations tell him that the motions require not only applying a conducting body to the nerves, but also that the body must be of a certain magnitude and extension (Galvani 1953, 27). He then investigates and discovers that the phenomenon can be observed in an animal wired to a distant (150 m) machine (Galvani 1953, 27). He wonders and tests if the effect diffuses in all directions, and in a circle (Galvani 1953, 28). He inquires which more powerfully gets the effect, attaching the conductor to the muscle or to the nerve (Galvani 1953, 28). He tests if the effect occurs when the flow through the conductor is intercepted by another body, or completely interrupted (Galvani 1953, 28–29). He tests whether the (apparent) electricity pervades the substance or only the surface of the conductor (Galvani 1953, 29).

At this point he records generalizations about how particular variables affect the contractions:

> Muscular contractions of this sort, then, seemed to us, within certain limits, to vary directly with the strength of the spark and of the animal and especially the extent of the nerve-conductors, but inversely with the distance from the conductor of the machine. Likewise these contractions generally appeared to us greater when the animal was placed on the same table as the machine and the table was covered with oily pigment, or when the animal, removed from the table, was laid on an insulating, rather than on a conducting, substance. (Galvani 1953, 29)

He notices that contractions are greater from the conductor of the machine, so supposes that greater contractions would result from "the electric flame which breaks out when the circuit is overloaded," but finds that "the thing turned out altogether differently," and no motions ensued (Galvani 1953, 30). Galvani's habitual practice of correcting expectations derived from hypothesis in light of results derived from experiment is largely alien to Mesmer and his followers.

Galvani goes on to do tests for negative electricity similar to those done for positive electricity (Galvani 1953, 30–31), using a Leyden jar and an electrophor ("that we might omit no kind of electricity exhibiting a spark") (Galvani 1953, 31) for different purposes. An electrometer is used to further confirm electricity in animal conductors (Galvani 1953, 31). He takes pains to isolate the animal in a glass jar, to ensure that the electric machine cannot be acting on both the animal and the conductors (Galvani 1953, 32). He compares contractions in living vs. dead animals (Galvani 1953, 33–34), and warm-blooded vs. cold-blooded animals (Galvani 1953, 34–35). This illustrates the nature of Galvani's energy-thinking and how it is expressed in narrative. As in the second and third sections, the narrative recounts a long series of fine-grained experiments that systematically vary and test every factor imagined to be potentially causally relevant, accompanied by insistent probing hypothesizing that shifts according to results. The fourth section summarizes results and proposes tentative explanations for his observations.

As for the role of metaphor in this energy-thinking, a few additional passages are worth noting. Addressing the key question of the *location* of electricity, Galvani develops a method on the basis of analogy:

> But since nothing seemed to be more suitable for discovering so obscure and difficult a thing, namely the location of each electricity, than to increase and apply the electricity, I therefore began to meditate sedulously concerning the method of accomplishing this, and, following analogy, this method presented itself first, that I should cover the nerves in which electricity seemed to prevail, and whose nature we had ascertained, with some metal foil,

> preferably of tin, no less than the physicists are accustomed to accomplish in their magic square[13] and Leyden jar [. . .]. (Galvani 1953, 47)

Galvani turns to metaphor not only for particular theoretical problems, but also for methods for solving practical problems, in a way that bridges fields (biology and physics) and guides actions. Analogizing physics equipment, a known conductive "stuff" (tinfoil) is inserted into a key node of the energy-flow in order to increase "umph." The method leads to further speculations on how electricity is increased, and how it works differently when increased.

Metaphors are also central in the final section, "Conjectures and Some Conclusions." Galvani's final conclusions about the location and distribution of electricity in bodies are developed through arguments that rely directly and heavily on metaphors consciously and explicitly used to make highly specific statements about the structure and operations of this animal electricity. Certain key steps in these conclusions reveal how metaphorical thinking helps Galvani change or reconsider main points. First, consider how he arrives at his view of the nerve as a conductor, and the muscle as the "seat" of the electric nerve fluid:

> Wherefore, if the great number of contractions obtained in the prepared animal be considered, for which surely the small quantity of electricity contained in the small part of the nerve remaining in the prepared muscles is adequate; if, moreover, the many arguments be considered which are sought from the animal functions, which openly declare that the nerve **fluid**, already demonstrated by us is electric and **flows** freely and swiftly through the nerves; if finally the obvious and simple explanation of the phenomena from each electricity be sought residing in the muscle itself: it will not seem beside the point, [. . .] that the muscle should be the proper seat of the electricity investigated by us, but that the nerve performs the function of a conductor. (Galvani 1953, 61; emphasis by MS)

His summary of this situation is an analogy, hedged about heavily to highlight that it is hypothetical and tentative, and immediately modified to increase its accuracy:

> These things being admitted, it would perhaps be a not inept hypothesis and conjecture, nor altogether deviating from the truth, which should compare a muscle fibre to a small Leyden jar, or other similar electric body, charged with two opposite kinds of electricity; but should liken the nerve to the conductor, and therefore compare the whole muscle with an assemblage of Leyden jars. (Galvani 1953, 61; emphasis by MS)

13 A magic square (or *quadratum magicum*) is one early version of an electrical condenser or capacitor, in the form of a pane of glass with metal foil on top and bottom. It was invented by Benjamin Franklin to show that electricity could be collected without a bottle.

Moreover, having set out the analogy, his next steps involve describing the physical structure of muscle-nerve connections in as much detail as possible, stressing complexity beneath apparent simplicity (Galvani 1953, 61).

The metaphor of the nerve as conductor soon leads Galvani to face metaphoric inferences that seem contradictory, within his model of the nerve-conductor as a single substance:

> But indeed, although this hypothesis and comparison present no slight **appearance** of truth, nevertheless there are some things which seem not slightly to **oppose** them. For either nerves are of an insulative nature, [. . .] and cannot then perform the function of conductors; or they are conductive: and how then could it be that the electric **fluid** should be **contained** within them and not be permitted to **escape** and **diffuse** to **neighboring** parts, not without great detriment surely of muscular contractions? (Galvani 1953, 64; emphasis by MS)

He resolves the contradictions by adjusting his model of the nerve, and he does so by using schematic spatial reasoning about containers, surfaces, and contact between substances: in short, umph-stuff and its paths. He proposes two possibilities: nerves are hollow; or they are a more complex structure of at least two kinds of substance, organized such that an inner surface of one substance promotes conduction, and an outer surface of another substance ensures insulation:

> But this inconvenience and difficulty will easily be **met** by him who imagines the nerves so constituted that they are hollow within, or composed of some material suitable for conveying electric **fluid**, but externally they are either oily or are fused with some other substance which prevents the **effusion** and **dissipation** of the said electric **fluid running** through them. (Galvani 1953, 64; emphasis by MS)

The electrical-fluid metaphor motivates the view that nerves are hollow within, like tubes, which would allow the fluid to flow. The alternative is that some other material could allow such passage. Thus the "fluid" metaphor often seems to be taken tentatively, as it was already known that electricity could pass through certain materials that were impassable to familiar fluids like water – it was a "subtle" fluid and not an ordinary one.

These are small adjustments to his model of animal electricity, but they are essential to analyzing the phenomenon with maximal detail and accuracy. And they are complex inferences combining several cognitive operations at once: observation (e.g. electricity passes from the conductor through the limb; nerves are tube-shaped); general knowledge (e.g. materials conduct or insulate electricity); and schematic spatial reasoning (e.g. some insulating surface must prevent the electrical fluid from dissipating along the nerve's path).

A little later Galvani turns to speculating about specific possible causes of the main phenomenon (electrical excursions). Here again explicitly hypothetical explanations are proposed tentatively, with variant reasonings about causation based on variant metaphors:

> But in what way an **excursion** of this kind of electricity causes and induces contractions, whether, for example, by some mechanical **irritation** and stimulus **striking** either the nerves or the muscle fibre and, as they say, exciting the irritability thereof, or, by the custom and character of common electric **vapor** exciting a peculiar and violent attraction between the particles composing, for example, a muscle fibre, by its swift **excursion** through it, so that, as they mutually approach one another, the fibre becomes shorter; or whether, as is easier to believe, it is acting for some other reason not yet understood, is a very difficult thing to be known, and very much involved in **darkness**. More and more experiments under taken on this subject may perhaps bring some **light** some time. (Galvani 1953, 68; emphasis by MS)

The passage sets out alternate metaphorical models as options for the true causality, but actually casts doubt on both in favor of the as-yet-unknown.

His caution was justified. The explanation of electrical causality in nerves and muscles had to wait for a new metaphorical model to arise to displace the "fluid" model.[14] As Piccolino explains, in the controversy between Galvani and Volta, the "greatest conceptual difficulty in understanding the mechanism of electric signalling in excitable cells" was "the difference between ordinary electric conduction and 'active' propagation along nerve fibres" (1997, 445). Most importantly, "To overcome the attenuation inherent in passive electric conduction, the electric nerve signal needs to be regenerated during its propagation, and an electric energy must be present locally along the entire extension of the nerve fibre ready to be discharged." The "gunpowder" metaphor captured the process of "active" propagation, foregrounding the inferences that each grain of powder would set off the next, hence a distant spark could produce a greater result than a nearby one. This replaced inferences derived from "fluid" and "cable" models, that the signal should gradually attenuate, so closer signals should have stronger results. Galvani's notion of "irritability" implied that "the energy for the reaction is accumulated in its inside, and is not directly related to the external energy" (Piccolino 1997, 447), but the gunpowder image clinched it. Note how this shift in explanations is essentially a deep reconfiguration of substance-force-space-movement relations in the causal-chain model. It seems that gradual shifts in the

14 Galvani "believed that the electricity flowed from the muscles to the nerve. We now know that the stimulation is produced by small currents applied to the nerve: the electricity in Galvani's experiment was produced by chemical reaction from acids present on the frog's skin. His work led, however, to the modern understanding of the electrical basis of neural activity" ("Galvanic Stimulation" 1987, 283–284).

sensorimotor model of the event (narrative thinking) get solidified when a new image captures the essential details of that event model (metaphorical thinking).

Galvani's thinking is occasionally reminiscent of Mesmer. For example, he speculates on how electricity relates to health and sickness, proposing that positive electricity could be administered to reduce "rheumatic contractures of the muscles": "forsooth the stagnating humors may be attenuated by the mechanical and repellent force of electricity, and then more easily removed from the affected part by increased impulse" (Galvani 1953, 78–79). This recalls the Mesmeric idea of directing magnetic flow to break down blockages to fluid circulation. Galvani also evokes the traditional idea of a direct link between the cosmos and the human condition when he speaks of "causes of agreement between the vicissitudes of atmospheric electricity and of our own health, and between some diseases and the sudden changes of the former" (Galvani 1953, 80). Yet Galvani's speculations along these lines are clearly marked off in a section labeled "Conjectures." And since he had shown that electricity has a central role in the functioning of nerves and muscles, it was fair to conjecture about how electricity might affect body, mind and health more generally. By contrast, Mesmer's leaping speculations about his all-encompassing and all-powerful single principle rest on very thin justifications of loose analogies posited (but not tested) between multiple domains of the physical world.

Galvani's style is personal and literary in places, but the careful experimentation and discussion sound like modern science as we know it. Yet it's framed almost entirely in terms of stories about metaphorical electric fluid. This makes it an excellent window into the processes whereby intuitive energy-thinking can be developed in scientific ways. Galvani elaborates complex hypotheses about umph-stuff structures and flows, tests them by experiments that map those hypotheses to real-world set-ups, and reconfigures structures and flows in light of results.

5.3 Pseudoscience: Franz Anton Mesmer's *Mémoire*

Mesmer's *Mémoire sur la découverte du magnétisme animal* (1779)[15] consists of narratives of cure supplemented by metaphorical statements of theory. It is structured as follows: an "Avis" to the public and the reader (Mesmer 1779, iii–vi); an

15 There is an obstacle to analyzing Mesmer's texts for cognition in the fact that "Mesmer's own cures, published with elaborate testimonials, spoke more eloquently for his system than his brief and cryptic publications" (Darnton 1968, 14). But Darnton advises that Mesmer's *Mémoire* (1779) and De Veaumore's *Aphorismes de M. Mesmer dictés à l'assemblée de ses élèves [. . .]* (1785) are

extended narrative of Mesmer's discovery, application and development of his principle of animal magnetism (Mesmer 1779, 1–73); a list of 27 Propositions that outline the essentials of his system (Mesmer 1779, 74–83); and a brief conclusion (Mesmer 1779, 83–85). In some important ways, Mesmer's text types are strikingly different from Galvani's. Most notably, the *Mémoire* contains no narration of painstaking experiment, but rather narrations of Mesmer's "discovery," of a series of "cures" (like case studies), and of his struggles with his "adversaries." The propositions are *exposition* text types, and simply stipulate, rather than argue for, Mesmer's view of the cosmos, with no hedging at all. A few examples illustrate the canonical building-blocks of the system, in the canonical form in which they are framed:

1. There exists a mutual influence between the Celestial Bodies, the Earth, and Animate Bodies.
2. A **fluid** universally **diffused**, and **continuous** in such a manner as to permit no vacuum, of which the subtlety permits no comparison, and which, by its nature, is susceptible of **receiving, propagating and communicating** all impressions of movement, is the means of this influence. (Mesmer 1779, 74; emphasis by MS)

Strikingly, proposition-sequences often leap across topics and lump them together analogically:

15. It is augmented and reflected by mirrors, like light.
16. It is communicated, propagated and augmented by sound.

[. . .]

21. This system furnishes new insights into the nature of Fire and Light, as well as into the theory of Attraction, of Flux and Reflux, of Magnets and Electricity. (Mesmer 1779, 78–80; emphasis by MS)

Similarly, Caullet de Veaumore's *Aphorisms* give us numbered statements that expound magnetic cure, by narrating actions, results and causes. Aphorism X states "You **strengthen** the Magnetic action by **multiplying** the medicines on the patient. [. . .] For the same reason we find the utility in ropes, trees, irons, and chains" (De Veaumore 1785, 5–6; emphasis by MS). The first sentence presents treatment advice in the form of general actions; the latter elaborates the advice about multiplying medicines by specifying further means of doing so. Yet these are not narratives of particular events, and the narrative qualities serve if-then type statements of general laws of cause and effect, which serve the propositional goals of unconditional assertion. Again, they explain without arguing, as if it is

some of the best of the many pamphlets explaining the theory and practice of mesmerism (Darnton 1968, 10, fn. 2). Translations of Mesmer's *Mémoire* are my own.

settled fact that merely needs conveying, as in a textbook. The absence of any kind of self-questioning in these texts tends to suppress any mental questioning or exploration in the reader. The textual atmosphere is the opposite of Galvani's relentless probing and revising.

Still, a significant portion of Mesmer's memoir narrates particular events, and here the relation of textual patterns to cognition and to scientific status is more complex. The narrative section is a kind of intellectual autobiography centered on Mesmer's system and his practice of it, mixed with reflections on the attendant professional and personal experiences. It begins with his dissertation in Vienna in 1766 and ends with the writing of the *Mémoire* to explain and defend his system in 1779. The body of the book moves from theory to practice, from addressing doctors by setting out the system proposed in his dissertation to improve medicine, to 12 years of undertaking and observing medical cases, after the traditional-minded reject him for the "crime" of his "propensity to leave the ordinary route of medicine" (Mesmer 1779, 8). There is little evidence that Mesmer was consciously hoaxing the public, or that he did not believe his system to be true. Quite the opposite: addressing accusations of deception is central to his story of injured merit. Mesmer includes as evidence multiple case studies of cures, and often comments on what he regarded as unfair, confused, malicious and disloyal attacks on him and his system, sometimes by other scholars and doctors, sometimes by the families of patients. (There are implications that the attackers are driven by jealousy and self-interest, seeking to steal his ideas or displace them by their own.) He even accuses a "cabal" of plotting to label his work as delusion and deception. He is quite open and detailed about exactly how he has been criticized and rejected, and by whom; and he often recurs to "facts" and "truth" in his own defence. He is eager to discuss how he demonstrated cures to prominent persons, even when those persons later betray him, to his astonishment, by disavowing publicly what Mesmer says they witnessed and believed in private.

The introductory narrative (as with the *Aphorisms*) serves to soften up readers by setting theory and advice in the context of the narrator's personal experience. The personal account of how the writer developed/learned the system, and why he is publishing this text, naturally produces some emotional and intellectual identification with him, deepening the reader's understanding of his reasons for thinking and acting as he does.

Much of the persuasive power of the memoir, however, comes from Mesmer's accounts of cures performed on young ladies. These present a curious mixture of details, with some fairly technical accounts of specific medical conditions, but notably lacking corresponding details of just how his cures were accomplished (and details offered are often unconnected with animal magnetism). These stories of extended treatment, observation, testing and cure might be compared with those of Galvani, in terms of scientific ideals of repeated experimentation and testing of hypotheses. But of course Mesmer and his human subjects are far more susceptible than are Galvani and his animals and machines to all the perception-distorting influences of ego and emotion tangled in knotty psychological and social relationships.

As for Mesmer's presentation of his system, the early parts of the memoir (1779, 6–8) and the 27 Propositions indicate that the system is a blend of analogies from a broad swathe of elementary contemporary ideas about physics. Mesmer's 1766 Dissertation argued,

> according to the known principles of universal attraction [. . .] that these spheres [planets, sun and moon] also **exert a direct action** on all the parts constitutive of animated bodies, particularly on the *nervous system*, by means of a **fluid** that **penetrates** all: I established this **action** by THE INTENSION AND REMISSION of properties of *matter and organized bodies*, such as *gravity, cohesion, elasticity, irritability, electricity*. [. . .] I claimed that, just as the alternating effects, in the case of gravity, produced in the sea the sensible phenomenon that we call flux and reflux, THE INTENSION AND REMISSION of the said properties, being **subject to the action of the same principle**, **occasioned**, in animated bodies, **alternating effects** analogous to those which the sea experiences. By these considerations, I established that the animal body, being **subject to the same action**, also experiences a sort of **flux & reflux**. I applied this theory to various examples of periodic revolutions. I named the property of the animal body, which makes it **susceptible** to the action of celestial bodies and the earth, ANIMAL **MAGNETISM**; I explained by this **magnetism**, the periodical revolutions that we remark in sex, and generally those that physicians of all times and countries have observed in maladies. (Mesmer 1779, 6–8; italics and capital letters in original, all other emphasis by MS)

Mesmer's thinking here is consciously analogical, though the analogies serve the development of a *literal* belief in a single comprehensive system in which magnetism is the force ultimately behind all other forces, and the root of all human health and illness. The *Mémoire* indicates in interesting detail how Mesmer's multiple analogies were constructed and blended. The most important mapping is from gravity (as a universal force of attraction among heavenly bodies) to magnetism (as a force of attraction and repulsion among mineral bodies, especially metals). Mesmer extends the general mapping in key ways that spell out implications for the human body. The known effects of gravity on the tides, and assumed effects on human bodies (e.g. lunar cycles taken to cause menstrual cycles, waves of illness and recovery) supports the speculative extension of further effects to bodies. The human ability to artificially magnetize metals like iron, so that compass needles align with the earth's magnetic field, supports the extension of magnetic powers to animate bodies (which contain iron), and hence the projection, to the medical aim of *cure*, of magnetizing bodies to re-align them with the cosmic magnetic flow. The analogies from fire, light and heat project how they can be created, destroyed, propagated and concentrated. The analogy from electricity projects how it can be collected and stored, then released and directed; and how it acts on nerves to cause muscular movements.

How did this new force work on people, and how could it be used for medical purposes? The *Aphorisms* provide extensive instruction. Aphorism I. in "General Notions on Cures Performed by Magnetism" asserts that health is "the perfect **harmony** of all our organs and their functions, ... A malady is only the **aberration** from this **harmony**: the cure thereof consists to re-establish **disturbed harmony**. The general remedy is, the applying a magnetic power by the indicated means" (De Veaumore 1785, 16; emphasis by MS). Aphorism I. of "On the Crisis of a Disorder" asserts that

> No disorder can be cured without being brought to a crisis, which is an **effort** of nature, against the malady by an increase of motion, tone and intension, together with the action of the **magnetic fluid**, to **dissipate** the **obstructions** that prevent **circulations** to **dissolve and evacuate the particles** that formed them, and re-establish **harmony** and an **equilibrium** in every part of the body. (De Veaumore 1785, 24; emphasis by MS)

As for the mesmerist's specific healing practices, the instructions in the *Aphorisms* are highly revealing of certain patterns of analogical energy-thinking. Aphorism XX. tells us that to increase the magnetic flow, ropes should be tied from a tree to patients:

> To give it a proper action, some ropes must be tied, at a certain height, to the trunk and the principal branches of the tree, the number and length of them according to the quantity of persons who are for the operation: the patient will twist them round the affected parts, sitting on chairs or straw, with their faces to the tree, all in a circular or oval form, as at the great tub[16] – the more they turn the ropes round them, the **more effect** they will find, and yet not very violent. The greater quantity of patients, the **greater power in the magnetism**, for thereby the **mediums multiply**, and the **fluids gain strength** by the **union** and the **contacts**; if the wind shakes the branches of the tree it will **add force to the action**; the same will happen at a rivulet or cascade, if you are so happy as to meet with it: if many trees should stand near, they may be **all magnetized**, and by the means of ropes **made to communicate with one another**. (De Veaumore 1785, 12–13; emphasis by MS)

Without explicit comparison, the rather magical logic of augmenting the magnetic flow maps structure across several concepts, abstracting from simple physical experience. That is, just about *any kind* of physical force can supply magnetic power: wind, water, light, music, trees, people, etc. To *augment* it, increase *quanti-*

---

16 The "great tub" refers to a common instrument of magnetic cure, a tub (*baquet*) that was used to gather magnetic fluid to be distributed to patients.

ties, and *proximity or contact* among them. These mappings suggest that the practitioners are perceiving the world via metaphors believed and felt to be real.

Mesmer's shift from using instruments to using purely mental powers comes during 1773–1774. He reports repeated experiments on a Lady OEsterline to relieve convulsions, first using custom-made magnets (1779, 12–16), then using the magnetic fluid directly (Mesmer 1779, 22–28). He writes that "observations of these effects, combined with ideas about the general system [. . .] taught me that another principle caused the magnet to act, incapable by itself of this action on the nerves" (Mesmer 1779, 15–16). That is, as his later rebuttals to his critics put it, magnets are mere useful but imperfect conductors of the fluid, and animal magnetism and its properties are not to be confounded with mineral magnetism (Mesmer 1779, 34–35). (He ceases to use magnets and electricity after 1776.)

Considering the paramount importance of this distinction, and of the claim to heal by directly manipulating a universal fluid, Mesmer's silence here on how this principle and its techniques were discovered and developed is deafening. We get very limited description of these treatments, despite the extended period of interactions, and despite some detailed descriptions of Lady OEsterline's conditions. Hence Mesmer's first account of directly manipulating the animal-magnetic fluid requires close scrutiny. He wishes to demonstrate the fluid's effects to a M. Ingenhousze, a member of the Royal Academy of London who had disparaged Mesmer (1779, 21–26). Mesmer engages the assistance of Ingenhousze and a young doctor colleague in communicating the fluid to Lady OEsterline, whose earlier fainting and convulsions have returned. Mesmer has Ingenhousze approach and touch the patient, with no result. He then purports to "communicate the animal magnetism to him, by taking him by the hands" (Mesmer 1779, 23), and has him repeat the touch, while Mesmer remains at a distance. The patient convulses. Mesmer has him repeat this experiment with his fingertip, varying the direction of the finger each time. Convulsions continue to result, to Ingenhousze's astonishment. Mesmer offers a second proof. They distance themselves from the patient, such that she would not perceive them even if she were conscious. He asks Ingenhousze to choose one of six porcelain cups into which Mesmer should "communicate the magnetic virtue" (Mesmer 1779, 24). Ingenhousze then applies the six cups successively to the hand of the patient; when the chosen cup arrives, she moves and shows signs of pain. They repeat this experiment, but with Mesmer touching Ingenhousze's hand to use him as a medium to communicate the fluid to a chosen cup, instead of Mesmer touching the cups directly. They conduct a third experiment, to demonstrate the fluid's action at a distance, and its penetrating virtue (Mesmer 1779, 25). Mesmer directs his finger toward the patient from eight paces; her body convulses, rising on her bed with signs of pain. Continuing in the same position, he places M. Ingenhousze between them, and she experiences the same sensations.

The story is highly revealing. The circulation of such anecdotal narratives of dramatic personal experience were central engines in the rage of mesmerism as a popular belief. We can identify the interconnected narrative qualities that make this kind of story so compelling:

- The plot focuses on exciting events that push at the frontiers of the known and the believable, and thematically foregrounds crucial human concerns of sickness and cure, and cosmic connection
- The protagonist becomes a heroic wise man and healer, having recently discovered a universal force that can cure all ills; indeed his power over the mysterious fluid, and thereby human bodies, gives an aura of magical, superheroic or divine power
- The causality reflects the deep-seated mechanical-magical logic of energy-thinking: an invisible magnetic fluid controls bodies by virtual contact from fingers to nerves

Alas, the marvelous story is undercut when, two days later, Ingenhousze publicly contradicts what Mesmer says he then believed, dismisses the cures performed, and uses magnets to unmask Mesmer's practice as ridiculous deception (1779, 26–27). Mesmer is astonished, and blames this reversal for Lady OEsterline's relapse (1779, 27–28).

Mesmer's practice in this anecdote relies on some of the key qualities that recur in literary energy stories and that were prominent in standard magnetic treatments: the energy is fluid-like in that it can be simply poured into and gathered in cups; it is tied in to human senses of touch, muscle control and gesture in that it is "communicated" by hand contact and "directed" by finger pointing (magnetizers often also used wands). Thus such stories get repeated not because the events they depict are replicable in a controlled situation, but because the stories have a certain pattern of dramatic heroic plot, dreams-coming-true theme, and mysterious but resonant causality and cosmology. As for what actually happened in that situation, it is impossible to know at this distance, but the demonstration sounds like a parlor-trick, and invites explanation by all the cognitive biases, and unconscious giving and taking of cues, that can distort uncontrolled psychological experiments, along the lines of Clever Hans.[17]

17 Dennett explains the special importance of control in psychological experiment: "The history of *folie à deux* and Clever Hans phenomena suggests that quite unwittingly the experimenter and the subject may reinforce each other into artifactual mutual beliefs that evaporate when prop-

6 Conclusions: Following the energy: Imagination, discourse, domains and history

Following the energy can sharpen our understanding of the relations of science, literature and pseudoscience, and help lead us through questions beyond those relations.

6.1 Follow the energy I: Intuitive grounding of scientific imagination

Following the energy can enrich our understanding of scientific imagination. Atran's "cognitive anthropology of science" addresses "how cognition constrains culture in producing science" (1998, 547). One of his conclusions is:

> Only by laborious cultural strategies like those involved in science can minds accumulate the knowledge to transcend the bounds of their phenomenal world and grasp nature's subtleties. But this requires continued access to the intuitive categories that anchor speculation and allow more sophisticated knowledge to emerge, much as the universal intuition of solid bodies and contingent movement has anchored scientific speculation about mass, matter, and motion. (Atran 1998, 567)

Hence a better understanding of scientific imagination, and its products in history, requires a better understanding of intuitive physics, and how it interacts with intuitive biology, psychology and sociology. Developmental psychology supplies a rich picture of this intuitive ontology core, and humanists can explore how it operates in meaning-making of all kinds.

Einstein famously remarked on the importance for his thinking of an "'associative play' of 'more or less clear images' of a visual and muscular type" (Witelson, quoted in Abraham 2002, 307). James Clerk Maxwell similarly, but more vividly, described how some people think by "project[ing] their whole physical energies into the scene which they conjure up [. . .]. They calculate the forces with which the heavenly bodies pull at one another, and they feel their own muscles straining with the effort" (Maxwell, quoted in Cat 2001, 407). The energy principle could connect these intriguing remarks more fully with research in thought and language in the arts and sciences. They indicate a potentially deep

erly probed," so "standard experimental practice" is to erect "barriers and foils that keep subjects from too intimate an appreciation of what the experimenters have in mind" (Dennett 2003, 11).

and persistent connection between the cognitive semantics of language and literature and creative scientific theory-building.

6.2 Follow the energy II: Model-building by narrative and metaphor, across domains

Following the energy can help us grasp *how texts create models,* including storyworlds and worldviews, because it gives us a more complete picture of the underlying conceptual-structuring process. This helps us map domain-specific thinking to domain-specific discourse, and thus to see better how to carve the joints between them. Differences in energy-thinking across our domains are to some extent reflected in the ways narrative and metaphor are used by the writers.

In terms of narrative model-building and the energy-tracing that defines its causality, Blake and Mesmer are really proposing comprehensive worldviews, while Galvani is trying to explain a very small, very particular slice of the physical world. That is, Mesmer is akin to Galvani in his focus on physical causal factors and explanations. But he is relatively uninterested in enacting experiments to test causal "stories." In fact, he is closer to Blake in his focus on cosmology, human values and the links between them. Mesmer even shares some of Blake's prophetic ambitions, with the vast scope and psychological/spiritual implications of his curative purpose, with its echoes of "inspiration," and easy transfer into religious registers. Fueling these expansive ambitions is a similarly literary style of thinking, a free-wheeling mental leaping that sees analogies and posits connections, testing them for resonance rather than factuality. But Blake's leaps and connections are all contained in the context of his "vision," an imaginative construct that frames reality indirectly; while Mesmer's are simply claimed to be real, theoretical propositions based on metaphor that get illustrated in stories of cure. Mesmer borrows and blends the epistemological and ontological thinking of both science and literature, yet adheres to neither's epistemological and ontological standards. This may help explain why Mesmer's enterprise was a short-term fad but a long-term fallacy.

We find differences in the relative intensity of metaphors, in the writers' attitudes towards them, and in how they use them. Literary artists tend to maximize the coherence of their metaphors' sensorimotor structures. As with Morris's "vicious circle" of capitalism, a sequence of metaphors blends into an overall complex metaphorical world-model. Blake uses metaphor expansively, creating allegorical worlds peopled by personified mental forces, and advocating for conscious metaphorical perception. Likewise, Mesmer's sequence of metaphors cascade into one vast metaphorical world-model. He maps the qualities of gravity, fluids, light, magnetism, sound, electricity, and other physical forces to the new universal force he proposes. Mesmerists get drawn in to maximizing the presence, power and detail of their force by metaphorically transferring to it the literal quantities of connecting objects (e.g. size and strength of trees, length and strength and windings of ropes) and intensities of feeling. In the process, they seem to deny or forget that they're thinking metaphorically. Galvani avoids metaphors in his accounts of experimentation as much as possible, but allows them freer play in his theoretical conjectures. When he does use them, he is explicit and cautious about proposing his metaphors *as* metaphors, and he avoids mixing them. Thus he does not, like Blake and Mesmer, build up one global world-model, but focuses on trying out variant possible metaphors as explanatory models for a local physical phenomenon. Variant models of muscles, nerves and electrical flow inherit (some of) the metaphoricity of the fluid model of electricity. These variant models drive the reasoning that seeks to map the causality of animal electricity by varying space-force-movement schemas to develop and revise multiple candidate models. His search for the right model is scientific in that it includes ongoing testing of models by experiment – and even accepting that all known models may be flawed and the truth may lie hidden in the future.

6.3 Follow the energy III: Cognitive cultural historicizing

By following the energy, we can also combine cognitive analyses with historicizing ones, to sharpen the contours of the *interplay between the natural and the cultural* in artistic and scientific meaning-making. Historicizing analyses can detail the complex influences of sources. But the local historical influences are suffused by more general, more persistent forms of thinking, and historical analysis should address this too. Complementing Atran, we can consider how culture constrains natural cognition, by trying to subtract or filter out the effects of each from the other. Humanists can trace how our (apparently natural and universal) intuitive ontology feeds into its cultural elaborations – though because *imagined* energies are *grounded* in real ones, this effort may be as delicate as extracting a plant from its soil.

Broadly speaking, there are natural-cognitive reasons for the spontaneous creation of, and enthusiastic belief in, systems of invisible fluids: those systems are based partly on the mind's intuitive disposition to map the world of causal forces in terms of umph-stuff. And there are natural-evolutionary reasons for the existence of those cognitive intuitions. Humans have evolved to seek and model the causal structure of the natural world, and that structure really is based on energy – it's just that our energy intuitions are only locally and partially accurate, so it's fairly easy to come up with *possible* energy-maps of the world, but very difficult to fine-tune those energy-maps in completely correct ways. But our evolved minds tend to be content with what works well enough to keep us alive, rather than demanding scientific proof for our everyday mental models.

Cognition is also suffused by emotion, and another aspect of the task of mapping the fuzzy-bordered terrain of pseudoscience is understanding the *cognitive appeal* that it draws from art, science and the intuitive ontology underlying both. What exactly are the cognitive qualities that made animal magnetism so magnetic for so many? They are partly aesthetic, partly logical and empirical, and partly intuitive. The system is mechanistic in a quite straightforward way (its mechanics are much simpler than those of Galvani's phenomena), but it is addressed to the imagination. As Richardson says, "given how the mind-brain seems to work, some thoughts are easier to think than others" (2010, 8). Mesmerism was a little too easy to think, we might say, *because* it was a simple model for deep human concerns about personal connection with and control of the universe, and health of body and spirit, *and because* it is built from the nuts and bolts of energy-thinking. Consider how it fulfills cognitive-emotional ideals. It was:

- Naturally familiar, because grounded in embodied cognitive experience of physical and mental energies, and their interplay
- Comprehensive, because cosmological from the start, covering "celestial bodies, the earth, and animal bodies"
- Coherent, because unified by its own intuitive logic; it is easy to understand and visualize
- Existential, because satisfying deep desires for control of and harmony with nature, by directly linking the dynamics of the human body with the universe
- Culturally resonant, because it fits deep cultural narratives from magical, religious and mystical traditions

Indeed, Franklin shrewdly observed natural groundings for these supernatural ideas, the "influence of imagination upon the animal frame" (Franklin 1785, 2) as he put it:

> There is so intimate a connection, whatever be the vehicle of that connection, between the volitions of the soul and the motions of the body, that it is not easy to prescribe limits to the influence of attention [. . .]. When we recollect that the arm is moved by the will as it pleases, how can we be certain, that the attention being fixed upon some interior part of our frame, may not [. . .] so modify its actual situation as to produce in it new sensations? (Franklin 1785, 40)

As for the power of the magnetists' eyes, he proposed the "very simple" reason that "it is the eyes that convey the most energetic expressions of passion, it is in them that is developed all that the human character has of the commanding or the attractive. [. . .] but this power consists merely in the aptitude they possess of moving the imagination" (Franklin 1785, 77). It is an intriguing sociocognitive question just *why* this should be so.

Following Franklin, we can make more detailed educated guesses on how other core tenets are grounded in embodied energy experience:

– *There is an omnipresent invisible magnetic fluid*:

Underlying Mesmer's system is *intuitive fluid mechanics*, based on perception of causality in tides of lakes and oceans, the flow of rivers and streams, and the everyday experience of using water and similar fluids, including bodily fluids. Fluids move and interact with spaces and objects in certain ways. We perceive similar causation, and potential human control over forces, in *invisible* substances and forces. We have everyday experience of gases like air (human breath; wind moving trees, clouds, water; birds flying), and of the partially-visible radiant forces of heat and light, from fire and the sun. Even unconsciously, our bodies continually process the invisible contact-less causation (action at a distance) of gravity. It not such a stretch to imagine an invisible fluid permeating the world and its bodies.

– *The invisible fluid controls bodies, and bodies can control it*:

Everyday experience of using water (and also air, fire, light and heat) for practical purposes would motivate the inference that the key problem in fluid systems is control of flow, including regulating pressure and direction. Blockage would lead to under/over-pressure, under/over-flow, damage to channels etc. Such experience also grounds the inference that humans could control the magnetic flow, in the sense of starting and stopping its motion, adjusting its strength, and directing its path. We can move fluids, and use them to pressure and move other things (filling and emptying containers, including bodies, using channels to direct force, e.g. to break down blockages).

– *The fluid is directed by eyes, hands and breath*:

We also make *virtual* contact with distant objects by focusing senses and attention on them. That mental focus creates virtual paths, which we can move, and which we actualize as we execute plans of interaction (tracing a virtual path with our eyes as we browse a bookshelf, walking another virtual path to the shelf, reaching along another virtual path to grasp the book). And via gaze and gesture and speech, we can direct *others'* mental paths to guide emotions and actions ("Listen! Look! Lunch!").

Existing medical knowledge would reinforce experiential fluid mechanics: bodily-fluid "humours" must be balanced, and imbalances (excesses, coagulations) could be cured by various means of draining and heating and disrupting. Recent discoveries and widespread experiments on electricity would reinforce the prospect of control over pervasive invisible forces. Galvani's work supplied the missing link of nerves and muscles as the specific electro/magnetic channel for influence on bodily movement. This allowed Mesmer to tap into an even deeper level of intuitive appeal for his magnetic manipulation: the constant experience of, and dependence on, contact-less causation in muscular control of bodily actions. We don't *see* muscular energy as we move our limbs and feel modulating intensities of "umph" (effort and resistance) as we interact with the world's "stuff," but we know that supply of force must be getting from A to B somehow. If my will can control my body's muscular forces by directing forceful fluids via nerves, perhaps my will can also control the universe's fluid powers.

In a fascinating study, Weyant concludes that Mesmer's system is pseudoscience because it is both self-contradictory and unfalsifiable. But he argues that Mesmer's real mistake was using metaphors that set his genuine cures against the wrong epistemological standards:

> Mesmer took real phenomena, functional cures of hysterical disorders, which had traditionally been considered a part of the hermetic tradition of renaissance natural magic and attempted, through the use of metaphors, to make those phenomena acceptable to the dominant mechanical philosophy. The metaphors were taken literally, the metaphoric explanations of the phenomena were tested within the framework of mechanical causes and effects, and were rejected. (Weyant 1980, 104–105)

Treatments such as hypnotherapy were inspired by Mesmer, and are now recognized by medical authorities as effective for relieving stress, anxiety and pain, and changing behaviors.[18] The cognitive appeal of Mesmer's metaphors and stories may help explain why hypnotherapy works.

The final word is "Follow the Energy" consciously because it's what we do in all our causal thinking: what we do intuitively as we track causation in everyday life; what the natural and human sciences do as they struggle to separate hypothesized from real causation; and what the arts do as they model such causation in imagined experience, to assign emotional and moral value in larger visions and interpretations of the human world. As for a more complete mapping of the overlapping terrains of artistic, scientific and pseudoscientific thinking, we can edge along these lines to investigate both what Franklin called the "influence of imagination upon the animal frame," and its Möbius-strip mirror-image, the influence of the animal frame upon imagination.

18 See the article on "hypnosis" at the Mayo Clinic: https://www.mayoclinic.org/tests-procedures/hypnosis/about/pac-20394405 (10 August 2022). The article also warns that hypnosis for psychotherapy may lead to strong emotions and the creation of false memories.

Reference list

Abraham, Carolyn. *Possessing Genius: The True Account of the Bizarre Odyssey of Einstein's Brain.* Toronto: Penguin, 2002.

Alighieri, Dante. *The Divine Comedy*. Vol. III: *Paradise*. Trans. Mark Musa. New York: Penguin, 1984.

Atran, Scott. "Folk Biology and the Anthropology of Science: Cognitive Universals and Cultural Particulars." *Behavioral and Brain Sciences* 21 (1998): 547–609.

Bering, Jesse. "The Folk Psychology of Souls." *Behavioral and Brain Sciences* 29 (2006): 453–498.

Blake, William. *Milton: A Poem, and the Final Illuminated Works*. Eds. Robert N. Essick and Joseph Viscomi. Princeton: The William Blake Trust/Princeton University Press, 1993.

Bloom, Paul. *Descartes' Baby*. New York: Perseus-Basic, 2004.

Boyer, Pascal. *Religion Explained*. London: William Heinnemann, 2001.

Cat, Jordi. "On Understanding: Maxwell on the Methods of Illustration and Scientific Metaphor." *Studies in History and Philosophy of Modern Physics*32.3 (2001): 395–441.

Darnton, Robert. *Mesmerism and the End of the Enlightenment in France*. Cambridge, MA: Harvard University Press, 1968.

De Mainauduc, John Boniot. *The Lectures of J. B. de Mainauduc, M.D.* London: printed for the Executrix, 1798.

De Veaumore, Caullet. *Mesmer's Aphorisms and Instructions*. London: Glass Warehouse, 1785.

Dennett, Daniel. "Who's on First? Heterophenomenology Explained." *Journal of Consciousness Studies* 10.9/10 (2003): 19–30.

Fara, Patricia. "An Attractive Theory – Animal Magnetism in Eighteenth-Century England." *History and Philosophy of Science* 33 (1995): 127–177.

Fauconnier, Gilles, and Mark Turner. *The Way We Think*. New York: Basic, 2003.

[Franklin Commission]: *Report of Dr. Benjamin Franklin and Other Commissioners Charged by The King of France with the Examination of the Animal Magnetism as Now Practised at Paris*. Translated from the French with an historical introduction. London: J. Johnson, 1785. https://www.woodlibrary museum.org/wp-content/uploads/rare-books/ACLL.pdf (4 August 2022).

Frye, Northrop. *Anatomy of Criticism: Four Essays*. Princeton University Press, 1971 [1957].

Fulford, Tim. "Conducting the Vital Fluid: The Politics and Poetics of Mesmerism in the 1790s." *Studies in Romanticism* 43.1 (Romanticism and the Sciences of Life) (2004): 57–78.

Galvani, Luigi. *Galvani on Electricity: A Translation of Luigi Galvani's* De Viribus Electricitatis in Motu Musculari Commentarius*: Commentary on the Effect of Electricity on Muscular Motion*. Trans. Robert Montraville Green. Baltimore: Waverley, 1953 [1791].

"Galvanic Stimulation." *The Oxford Companion to the Mind*. Ed. Richard L. Gregory. Oxford: Oxford University Press, 1987. 283–284.

Gibbs, Ray, Jr. "The Psychological Status of Image Schemas." *From Perception to Meaning*. Eds. Beate Hampe and Joseph Grady. Berlin: Mouton de Gruyter, 2005. 113–135.

Hanen, Marsha P., Margaret J. Osler, and Robert G. Weyant, eds. *Science, Pseudoscience and Society*. Waterloo and Ontario: Wilfrid Laurier University Press, 1980.

Heider, Fritz, and Mary-Ann Simmel. "An Experimental Study of Apparent Behavior." *American Journal of Psychology* 13 (1944): 243–59.

Lakoff, George, and Mark Johnson. *Philosophy in the Flesh*. New York: Basic Books, 1999.

Lakoff, George, and Mark Johnson. *Metaphors We Live By*. 2nd edition. Chicago: University Chicago Press, 2003 [1980].

Lakoff, George, and Mark Turner. *More Than Cool Reason: A Field Guide to Poetic Metaphor*. Chicago: University Chicago Press, 1989.

Leslie, Alan M. "ToMM, ToBy, and Agency: Core Architecture and Domain Specificity." *Mapping the Mind: Domain Specificity in Cognition and Culture*. Eds. Lawrence A. Hirschfeld and Susan A. Gelman. Cambridge: Cambridge University Press, 1994. 119–148.

Martin, Eugenio, dir. *Horror Express*. Written by Arnaud d'Usseau and Julian Zimet (Julian Halevy). Granada Films/Benmar Productions, 1972.

McCloskey, Michael. "Intuitive Physics." *Scientific American* 248.4 (1983): 122–130.

Mesmer, Franz Anton. *Mémoire sur la découverte du magnétisme animal*. Geneve: P. Fr. Didot le jeune, 1779. www.loc.gov/item/50047308/ (4 August 2022).

Mesmer, Franz Anton. *Aphorismes de M. Mesmer dictés à l'assemblée de ses élèves et dans lesquels on trouve ses principes, sa théorie et les moyens de magnétiser; [. . .] Ouvrage mis au jour par M. C. de V.* Paris, 1785. http://ark.bnf.fr/ark:/12148/bpt6k754717 (10 August 2022).

Michotte, Albert. *The Perception of Causality*. Andover, MA: Methuen, 1963.

Morris, William. *News from Nowhere*. Ed. Stephen Arata. Peterborough: Broadview Press, 2003 [1891].

Musolff, Andreas. "Metaphor Scenarios in Public Discourse." *Metaphor & Symbol* 21.1 (2006): 23–38.

Piccolino, Marco. "Luigi Galvani and Animal Electricity: Two Centuries after the Foundation of Electrophysiology." *Trends in Neurosciences* 20.10 (1997): 443–448.

Reiner, Miriam, James D. Slotta, Micheline T. H. Chi, and Lauren B. Resnick. "Naive Physics Reasoning: A Commitment to Substance-Based Conceptions." *Cognition and Instruction* 18.1 (2000): 1–34.

Richardson, Alan. *The Neural Sublime: Cognitive Theories and Romantic Texts*. Baltimore: The Johns Hopkins University Press, 2010.

Rix, Robert W. "Healing the Spirit: William Blake and Magnetic Religion." *Romanticism and Victorianism on the Net* 25 (2002). https://www.erudit.org/en/journals/ron/2002-n25-ron437/006011ar/ (4 August 2022).

Rix, Robert W. "Magnetic Cure in William Blake's THE FRENCH REVOLUTION." *The Explicator* 68.3 (2010): 167–171.

Schott, G. D. "William Blake's *Milton*, John Birch's 'Electrical Magic' and the 'Falling Star.'" *The Lancet* 362.9401 (2003): 2114–2116.

Schuchard, Marsha Keith. "Blake's Healing Trio: Magnetism, Medicine, and Mania." *Blake: An Illustrated Quarterly* 23 (1989): 20–32.

Scott, Robert A. "Making Relics Work." *The Artful Mind*. Ed. Mark Turner. Oxford: Oxford University Press, 2006. 211–223.

Sinding, Michael. *Body of Vision: Northrop Frye and the Poetics of Mind*. Toronto: University of Toronto Press, 2014.

Sinding, Michael. "From World to Worldview: An Energy Principle of Psychosocial Dynamics." *The Cognitive Humanities: Embodied Mind in Literature and Culture*. Ed. Peter Garratt. London: Palgrave Macmillan, 2016. 133–150.

Steen, Gerard. *Finding Metaphor in Grammar and Usage*. Amsterdam: John Benjamins, 2007.

Stern, Daniel N. *Forms of Vitality*. Oxford: Oxford University Press, 2010.

Talmy, Leonard. "Force Dynamics in Language and Cognition." *Cognitive Science* 12.1 (1988): 49–100.

Talmy, Leonard. "Fictive Motion in Language and 'Ception.'" *Language and Space*. Eds. Paul Bloom, Mary A. Peterson, Lynn Nadel, and Merrill F. Garrett. Cambridge, MA: The MIT Press, 1996. 211–76.

Weyant, Robert G. "Protoscience, Pseudoscience, Metaphors and Animal Magnetism." *Science, Pseudoscience and Society*. Eds. Marsha P. Hanen, Margaret J. Osler, and Robert G. Weyant. Waterloo and Ontario: Wilfrid Laurier University Press, 1980. 77–114.

Brook Miller

Reading Narratives

Abstract: This chapter considers the cognitive processing that occurs during the reading of narratives. It surveys recent research on narrative as well as studies of the links between experiences of the real world and imagined fictional worlds (Gallese, Iser, Armstrong, Gerrig, Speer, Zwaan, Caracciolo, Jacobs, et. al.). It builds on Fludernik's concept of "experientiality" to describe the abstraction processes of both ordinary experiences and narrative reading in terms of the role the felt quality of experiences plays in information processing, rather than vice versa. This leads to a taxonomy of emergent structures in reading and ordinary experiences, including emergent qualities such as relational salience and dynamic valuation. The essay concludes by describing the bottom-up aspects of reading narrative from these experiential building blocks.

1 Introduction

Humans have the remarkable capacity to have experiences while reading narratives. Even before accounting for how this works, understanding the information processing involved in reading narratives is challenging. Reading narratives is a complex informational task. To read a narrative text, a reader confronts written words and forms concepts that, through the reading process, are developed and related to other concepts into a complete story, about which the skilled reader can derive summary information. The extraction and synthesis of information draws upon the reader's repositories of social, natural and aesthetic knowledge, and as such may be thought of as a process of translation.

Anyone who has seen a filmic adaptation of a beloved novel and has been disconcerted by the casting and acting of a key character has confronted this. When a protagonist seems not as you imagined her, the fact of your translation of words into concepts that are not precisely indicated in the text becomes clear. The making and synthesis of concepts from words involves significant cognitive processing. Given the nature of the concepts and the significance of time to their relations, the reader must coordinate discrete concepts with the whole of a narrative, as well as new concepts that emerge because of the relations of concepts that become apparent as the narrative is read. Both the working through of a reading of a text and subsequent reflection involve, then, complex cognitive work. To un-

https://doi.org/10.1515/9783110782844-014

derstand reading narratives, we must closely consider how readers translate texts into time-dependent informational/conceptual networks.

It would seem even more daunting to understand how these concepts generate experiences as they are formed and coordinated. As one prominent narrative theorist puts the matter, a fundamental feature of narratives is that they convey qualia.[1] On first consideration, this would seem to make narratives a kind of supercapacity. We generate and inter-relate concepts by translating words, and as we generate them we have a phenomenologically rich experience of them.

This chapter argues that, while aesthetic experiences engendered by reading narratives are indeed extraordinary, the processes they are grounded in are mostly not. As a number of scholars have pointed out (see below), the processing model just described puts the cart before the horse: so-called qualia are not subsequent to information processing, but integral to it. This chapter contributes to this conversation by offering a set of gap-filling ideas to further develop models that foreground the phenomenology of reading narrative. Specifically, it argues for isomorphic processes in reading and ordinary experiences. Subsequently, it offers a taxonomic description of conceptual processing in reading narrative that distinguishes between high-level processes distinctive to reading and low-level processes common to all experience, and it describes their interrelations. This foregrounds processes necessary for understanding and experiencing narrative

1 Narratologist David Herman defines narrative as "a basic human strategy for coming to terms with time, process, and change" (Herman 2009, 2). Narrative representations include several elements, including "a specific discourse context or occasion for telling [. . .] [a process that] cues interpreters to draw inferences about a structured time-course of particularized events [. . .] [that] [. . .] introduce some sort of disruption or disequilibrium into a storyworld involving human or human-like agents [while] also convey[ing] the *experience* of living through this storyworld-in-flux, highlighting the pressure of events on real or imagined consciousnesses affected by the occurrences at issue" (Herman 2009, 14). Herman concludes from this description that the phenomenology of narrative is crucial and connects it to the nature of cognition: "narrative is centrally concerned with *qualia*, a term used by philosophers of mind to refer to the sense of 'what it is like' for someone or something to have a particular experience [. . .] recent research on narrative bears importantly on debates concerning the nature of consciousness itself" (Herman 2009, 14).

Herman thus identifies four parameters for narrative representation. The inclusion of qualia raises questions about how a post hoc representation can include 'going through' the experience represented. That is, can processing ink on a page, for example, have the phenomenology of an event depicted in a narrative – say, the traumatic experience of a shipwreck. In Herman's view, it is "conveyed," meaning it is representation that elicits a subjective feeling that resonates with what the subject imagines being-a-bat, or some other qualia, might be like. What's attractive, though, about narrative, is that it seems to give us the experience of "living-through" a storyworld via the "pressure" of events on consciousness. In other words, qualia are stimulated in the reading process (within the reader) that simulate the qualia represented.

that are also present in some non-reading experiences. In sum, reading narrative draws upon basic capacities that generate the phenomenology of experience and builds upon them. Moreover, in light of recent research blurring traditional lines between perception and conceptualization, the phenomenology and informational features of translating language likely combine in generating the imaginary simulations produced in reading narratives.

2 Current ideas about reading and reading narrative

Reading is a late developing, complex activity, and reading narrative engages emotional, analytical, spatial, temporal, and other capacities. There are debates about whether reading is primarily a "bottom-up" or "top-down" activity, though standard models acknowledge the need to accommodate both perspectives (Rayner et al. 2012, 22).[2] Theories of literacy have articulated a variety of models of the engagement of readers, evolving from views of readers as passive recipients to makers of meaning to, in recent decades, contingently positioned within social networks of meaning (Hruby et al. 2016).[3]

Recent research in cognitive neuroscience finds that reading "recycles" particular brain areas used for other activities, though there is persistent debate about the degree of localization of reading (Dehaene 2014). Nevertheless, these findings support the idea that reading narrative engages ordinary cognitive processes in ways that can make reading "feel" like non-reading experiences.

Vittorio Gallese has claimed, for example, that "solid empirical evidence shows that the neurobiological mechanisms interfacing us with the 'real world' largely overlap with those activated when we imagine fictional worlds both through images and words [. . .] the world of literature is 'felt' not too differently from how we feel the more prosaic world of our daily life" (Gallese et al. 2018).

Phenomenological critics have historically sought to understand how the reader interacts with the text to produce meaning, describing a "virtuality" of the work and the reader's creative participation in building meaning through processing that shares deep features with Husserl's phenomenological model of how we experience time (Iser 1972, 280–282).

2 Rayner et al. reflect that "our model of the reading process can best be described as a bottom-up model in which the reader gets some help from top-down processes" (2012, 22). Rayner et al. (2012, Ch. 1) provide a strong overview of the history of reading theories.

3 Hruby et al. (2016) provide a history of theories of literacy that variously model readers.

Paul Armstrong, citing Gabrielle Starr and others, argues that reading narratives is an embodied form of "skilled coping" that recruits emotional, cognitive, and other capacities. Armstrong argues that reading engages the "hermeneutic circle" by which readers' developing understandings of parts of and anticipations of the wholes of narratives mutually inform one another (the whole is hypothesized during ongoing reading) (Armstrong 2013, see ch. 3 in particular).

Richard Gerrig, Nicole Speer, Rolf Zwaan, and others have authored pioneering explorations of the cognitive processing required for rendering narratives meaningful, the various kinds of readerly experiences stimulated by narratives, and the linkages between them. "Schemas" are called upon in ongoing sensemaking of narrative texts.[4] Zwaan and colleagues propose an "Event Indexing Model" of narrative comprehension in which "comprehenders construct situation models from text by extracting events from the text and integrating these events on five different situational dimensions: time, space, causation, motivation, and protagonist" (Zwaan et al. 2001, 72). Speer and colleagues more recently propose that readers spontaneously segment narratives into events as part of ongoing situation modeling (Speer et.al. 2007).[5]

The literature on schema and situation modelling in narrative comprehension emphasizes the importance of "participation," or ongoing hypothesis and inference building, as well as emotional "transport" (Gerrig 1993; 2001). Burke builds upon the hedonic elements of this theory to hypothesize the existence of "disportation"[6] and "oceanic cognition" in the experience of reading stories, which reflect "distributed blends and combinations of basic multidirectional neural processes" (Burke 2015, 5). This "bushy" or distributed model of cognition applies to reading and non-reading activities alike.

In describing participation, transport, coping, and model building, these researchers propose that reading is simultaneously experiential (that is, reading is a kind of experience) and representational. Literary theorist Monika Fludernik (2010; 2018) has explored this relation by distinguishing between experience and

4 Paul Armstrong notes a change in Gerrig's model from schemas to "fluid[,] [. . .] idiosyncratic" processing of narrative texts through reference to "general knowledge" (Armstrong 2013, 196). See also Gerrig and Egidi (2003) and Gerrig (2010).

5 Zwaan hypothesizes that event processing precedes language acquisition: "Our general hypothesis is that the mechanisms involved in situation-model construction from language are derived from the mechanisms of situation-model construction in the real world. Children first learn to understand basic events, such as the movement of a person or an object in the environment, before they learn to understand the phrases that describe these events" (Zwaan et al. 2001, 73).

6 Refers to a heightened affective state that creates a felt sense of being in motion. Burke argues literary reading stimulates disportation, and that the existence of disportation points to the interrelatedness of affect and cognition (see Burke 2015, 4–5).

“experientiality.” Experience is constituted through ordinary cognitive processes. Experientiality, by contrast, is representation or evocation of experience within artifactual narrative. But what is the cognitive work that goes into reading narratives? That is, how do we experience experientiality?

Marco Caracciolo explains experientiality as “tension between the linguistic structures [readers] engage with and their own past experiences” (2014, 47). He argues that two central mechanisms are responsible for this engagement: “‘consciousness-attribution,’ [i.e.,] the simple ‘recognition’ that a particular event referred to by the story is experiential [. . .] [and] ‘consciousness-enactment’ [which] [. . .] consists in empathizing with or mentally simulating the experience that we attribute to a fictional character” (Caracciolo 2014, 41).

These character-oriented processes may have larger implications for the relations between narrative and ordinary cognition. Mar (2011) has found that “consciousness-enactment,” as Caracciolo puts it, may indeed be integral to reading narratives: Theory of Mind activities in general and Narrative Processing share significant overlaps in neural networks (Mar 2011).[7]

Various accounts of the experiential dimension of reading narrative may seem difficult to reconcile. Some involve intensive, and perhaps intentional, participation (as top-down activities), while others happen bottom-up and automatically. Some focus on emotion as integral to cognition, while others take an information processing view. Jacobs (2015a) and colleagues offer an “eclectic, comprehensive” Neurocognitive Poetics Model (NCPM), which describes different forms of processing for “foreground” and “background” aspects of narratives,[8] located in different brain hemispheres (left and right). The model also accounts for differential aspects of readers, texts, and contexts as determinants of the reading process.

The model is of particular interest because it describes reading language-based narratives simultaneously in terms that comport with ordinary cognitive processing and in terms that emphasize processing language for specific features. Background generally maps to “immersive processes” and foreground to “aesthetic” experiences (and in both cases, to accompanying emotional systems). Yet

7 Mar notes several important caveats and grounds for future investigation, including: a lack of deep understanding of how TOM and narrative comprehension are linked; the implication of the same neural network(s) in other activities, like future thinking and autobiographical memory; and a need to understand our tendency to anthropomorphize as part of understanding (Mar 2011, 125). For an introduction to how literary studies has approached Theory of Mind, see Zunshine (2006).

8 Jacobs adopts a pragmatic approach to identifying foreground and background features, emphasizing different effects. Whereas foreground elements are textual sites of significant focus and attention, background elements facilitate a sense of immersion (Jacobs 2015a, 7).

these mappings are partial. Jacobs' model suggests that experiences of narratives draw upon ordinary forms of processing by employing the same systems that generate foreground/background distinctions in non-reading situations.

Reading text may draw upon these systems via processes of simulation. Caracciolo argues that "engaging with stories can trigger imaginative responses that simulate more basic models of interaction with the environment, thus digging below mental representations and higher order cognition" (2014, 11). For Caracciolo, "our experiential responses to non-actual events and existents depend on a simulative mechanism whereby we use our cognitive resources in an 'off-line' mode [. . .], reacting to them as if they were actual" (2014, 32).

There are different systems that contribute to phenomenological feelings in ordinary experience, and potentially also to the simulative aspect of reading narrative. Emotion, in Jacobs' model, is integral to language processing, rather than subsequent to it. Emotional systems mediate the felt sense of experiences being embedded in a background "field that is variously and continuously shaped by the biological make-up, culture, social positioning, and personal experience of every individual" (Carraciolo 2014, 57). Other theorists, notably Lawrence Barsalou, have argued for the identity of perceptual and cognitive processes ("cognition is inherently perceptual"), and linked the "feel" of experiencing to non-linguistic perception (Barsalou 1999, 577). In this model, perceptions are stored in a manner that facilitates re-simulation, rather than categorical, taxonomic archiving of their properties and contents.[9]

The temporal aspects of having an experience and simulating one while reading narrative may provide a key for understanding simulation. Reading a fictional text, similarly to some forms of ordinary experience, can be described as accumulative sensemaking:[10] that is, sensemaking emerges and accrues via the temporal "working-through" of experience. Generally, this occurs through two forms of emotionally laden work associated with reading narrative: event segmentation and anticipation. First, reading a narrative involves the experience of meanings that depend on changing relations through time, which often reflect emotional

9 Some scholars have articulated models that fold together the linguistic and non-linguistic. For a perspective on how linguistic and embodied information contribute to word meanings, see Speed et al. (2015). Other scholars have theorized that emotion is crucial to perception. See, for example, Pessoa (2013).

10 Sensemaking is a term used in a variety of contexts, but here I am referring to the version associated with Ezequial Di Paolo (2009). This version places the individual at the center of dynamic couplings that engender meaning. However, the dominant strand of theorizations of sensemaking, associated with Weick (1995), focuses on sensemaking's deep and inherent sociality.

dynamism and are structured in some relation to a conventional pattern.[11] Second, reading a narrative involves anticipatory feelings that grow from the prior experiences of the reader. As a reader, one begins (usually) as an unknowing subject who encounters narrative features and makes predictions about them. The effort and selection involved in these predictions creates a structure of anticipation that is inseparable from developing a sense of care for particular features of the narrative, usually particular agents or groups of agents. Taken together, the readerly experience of temporal extension and of an anticipatory, empathic economy operate in processing the representations of temporal extension and agentive care (i.e., the cares of agents within the story). This processing creates an experience in reading roughly parallel to the experience represented, helping to simulate qualia.[12]

Consider the emotional economy activated when we read, for example, a climactic moment in the fairy tale "Little Red Riding Hood." When the girl approaches the wolf in bed dressed in her grandmother's clothes, our foreknowledge of the wolf's plans generates an anticipatory terror that we may or may not associate with the girl herself as she considers the wolf's features. This state in the reader is the result of a) understanding that the wolf is a wolf, and knowing the threat wolves pose; and b) prior knowledge of the wolf's machinations and consumption of the grandmother. The moment of the girl's observations – "What big eyes you have," etc. – is shaped by terror at her vulnerability and anticipation of the conclusion of this event in the wolf's attack. The heightening of our emotions runs parallel to the girl's increased wonder at the non-grandmotherly features of the figure in the bed. Our evaluation of the girl in this moment, which may have earlier involved censure at her careless behavior during her journey, now modulates into fear for her safety grounded in an evaluation of her credulity. This double dynamism – at once experiencing the textual present in terms of what is to come and simultaneously experiencing an agent in a manner embedded in the moment (terror for her safety, in contrast to our prior evaluation) – makes the reading experience a working-through with a dynamic phenomenology.

11 The time spent reading is also experienced by readers. In reading, readers respond affectively to the duration of reading, and to this duration's relation to the representations drawn from the text.

12 I believe the processing described here and Caracciolo's 'consciousness attribution' and 'consciousness enactment' are mutually constitutive.

3 Emergent meaning in reading and life

Time is a key element of the "working through" of narrative.[13] As they read, readers make provisional recognitions and associations between elements of their texts, which remain available for further synthesis. As they accumulate these recognitions, associations and syntheses contribute to sensemaking – answering the who, what, why, when, and how of a particular moment in the narrative by linking it to the past moments of the narrative as well as by projecting future developments.

Thus sensemaking pervades the experiential aspects of simulation in narrative reading in a manner similar to ordinary experience. In ordinary experience, the subject-in-experience processes the contacts of world and subject, (re)marking the familiar and unfamiliar, about which there are already well-formed beliefs, and about newness, both of new pieces of experience and of new dynamisms within and between the familiar.

What elements would characterize accumulative sensemaking in reading narrative? Katherine Nelson's describes "narrative dynamics" as "the structural glue that ties together the who, what, where, when, and why" (Nelson 2003, 28). Nelson views processing narrative as an active "gluing" activity involving "three essential components of narrative": "temporal perspective, the mental as well as physical perspective of self and of different others, and essential cultural knowledge of the unexperienced world" (Nelson 2003, 28). These components are emergent within reading experiences.

4 Cognizing artifactual narratives

The emergent nature of meaning via these three components reflects a number of underlying, and quite ordinary, cognitive processes. Before articulating these, it is worthwhile to consider the qualities of narrative that readers are attracted to, and around which they build meanings. Meir Sternberg (2003) has identified several universal emotions – "narrative appetites" – associated with narrative processing: curiosity, suspense, and surprise. Curiosity is stimulated by past events, which prompt care about what is currently happening. Suspense is antici-

13 The felt sense of working-through or working towards understanding is common to non-narrative reading as well. However, the nature of narratives contributes to a conceptual integration of the working-through into the meaning of the whole, whereas in most other reading working-through is quickly forgotten as the reader integrates concepts into a summative memory of the reading.

patory – we predict and care for what will happen. Surprise reflects a breakdown in our expectations and prompts the emergence of new predictions.

These appetites drive and are satiated by several kinds of sensemaking activities that mutually influence one another: sequencing, selection, and evaluation for salience. To be clear, while narrative texts reflect such activities on the part of an author during composition, this section focuses on how characteristics of narrative stimulate sensemaking work in readers, comparing this readerly work to sensemaking in non-reading activity.

5 Sequencing

First, reading narrative involves active, ongoing attribution of meaning to unfolding sequences.[14] The text offers events and exposition in a particular order, and the reader actively processes each as s/he reads, making sense of the new in light of what has already been presented as well as in light of what's expected.

Sequencing may be linear, as in the following narrative:

> Bored with his circumstances and feeling unloved, George hoped something big would happen to him on his birthday. He threw a penny into a wishing well and made his wish. The next day was his birthday. He arrived home after a dull day at work. When he flipped on the light, a tremendous shout of "Surprise!" rose from his living room. He found his friends and a large chocolate cake in his apartment. His lover had thrown him a birthday party. George wept with joy. He felt lucky for the first time in years.

In an artifactual narrative, the sequencing may also be more complex: say, for instance, the above might be retold by beginning with George's tears, and then jumping back in time to the previous day and continuing the narrative back to the present.[15] In this case the order of presentation and the chronology of events differ. However, in both versions sequencing is important. This narrative, let's call it "George's Story," reflects chronological sequences which may be preserved in George's memories and reflections after the sequence is complete. Other sequences also exist, particularly the "ordering" of the narrative as it is presented. Non-chronological sequences cue cognitive processing for meaning, as the reader ponders why the events are presented as they are. We see extreme versions of

14 By meaningful sequencing, I mean that reading a narrative requires coordinating parts by linking them in meaningful ways.

15 See Zwaan and Radavansky (1998) for an analysis of the particulars of linear ("iconic") sequencing and the achronological temporal structures often found in literary texts.

this in novels like Martin Amis's *Time's Arrow* (1991) and the film *Memento* (2000), in which important sequences (in the novel's case, subsuming the entirety of the artifact) are presented in reverse chronological order.

While chronological succession is characteristic of some sequences, it is typically not the only operant logic. Meaning-attribution depends on a reader's prior experiences with different narrative genres. Readers use these past experiences to make sense of what Vladimir Propp called the "morphemes" of narratives, and to anticipate how particular units likely presage others in a familiar narrative grammar. Meaning depends upon how individual elements fulfill expectations regarding particular narrative functions, as well as upon how these elements are related through time.

Some sequencing is (in David Herman's phrasing) "causal-sequential." For example, when George weeps, it appears to be caused by the presence of friends and/or the realization that he is not alone. If George's weeping is "caused" by the presence of friends (in a behaviorist manner), it is immediate causal sequencing; if the weeping is caused by a realization that he is not alone, then it is an unrepresented cognitive process that has been anticipated earlier in the story. What seems most likely is that the weeping involves an amalgam of the experiential quality of his friends' presence (i.e., feeling their loving gazes and smiles, etc.) and an unrepresented cognitive process (George realizes "oh, I am not alone after all").[16]

However, readers attribute meaning to other forms of narrative sequencing as well. For example, what if the last three sentences of George's story were replaced with "George felt a sudden convulsion in his chest and fell to the floor. He shook and gasped for air. As the dark closed around him, George thought he was sorry to miss the party."

What does the effect of such a change show? As we explored above, reading a narrative is participatory, rather than passive. In this case, the earlier portion of the story creates anticipation: will George's loneliness be cured? In the first version, the answer is an emphatic "yes." In the second, however, the answer is mixed – a "yes, sort of, but the shock induces a heart attack, and so George's death mitigates the rec-

16 Apprehending causal history is crucial, not only to following the plot, but also for extracting the kinds of meanings Currie emphasizes. That is, in processing (hearing, viewing, reading) narrative, this causal history is crucial for higher order evaluation (who is morally culpable, for example?). However, experience is inseparable from understanding causal history. And the way causal history is understood arguably involves 3 components: 1) selection of components (delimiting the when, the where, the what (occurs), and the who (the agents and receivers of action, not necessarily human)). (Importantly, because experience is rich, this is selective – much (most?) is excluded); 2) baseline evaluations of each component; 3) evaluations of dynamics. That is to say, we process for causal history through interleaved processes of selection and evaluation. This interleaving occurs within interleaved acts of perception and higher order cognition.

ognition of community." In this case, the feeling is more complex, and may stimulate reflection. By grappling with the dissonance between emotional anticipation, partial fulfillment and sudden catastrophe, the reader may reflect on competing values, which can lead to more complex meanings: some things, like life, are more important than recognition from friends. Or, we have a psychological portrait of George, however hazily sketched, as valuing the wrong things; or, we have an implied statement about the desperate need for human community, couched in irony. In terms of Sternberg's (2003) narrative universal "surprise," this juxtaposition of versions reveals that surprise reflects re-evaluation of particular elements of a narrative, and also (potentially) revaluation of the values readers deploy in processing the narrative. To put this simply, while both versions involve dynamism regarding George's loneliness, in the second version loneliness itself turns out to have been overvalued.

The key to appreciating the narrative in this case is to grapple with how emotional anticipation and fulfillment or disappointment are forms of cognitive processing associated with reading. This processing potentially leads to meanings. The reading experience might (and frequently does) involve empathic responses: the reader's grasping of the story as meaningful involves feeling along with the focalized consciousness. This partly explains why narratives in which the focalized consciousness is an unlikely subject for empathy (say, an evildoer) can be disturbing – processing the story involves resolving tension between a judgment and an empathic reading strategy.

External contexts that supervene as a narrative unfolds can cause retroactive revaluations of earlier events and expository statements. For example, what if the last four sentences of the original story were replaced with the following: "He found his friends, bound and gagged, and a chocolate cake with detonator wires strung to the front door. The agent from the enemy's spy service, masked but for a pair of twinkling eyes, gave him a wave as he rappelled away from the open window. George should never have allowed anyone to get close to him; it was the first rule of spycraft. The blast wave felt strangely warm, then unbearable, and then everything stopped." In this version of George's story, the last four sentences provide a shock via juxtaposition: the context of George's occupation and the grisly ending function as internal indexes of the final meaning of the story, making the opening emotional references resonate far differently.[17] (Admittedly, different readers will have different reactions to the story – most probably find it emotionally over-

17 Paul Ricœur (1984) describes the ongoing importance of "configuration" in grasping narrative. Such juxtapositions point to how ongoing configuration can involve dramatic swings in a reader's understanding of a text.

wrought. However, regardless of the reader's reaction, these new four sentences force a kind of re-thinking: things were not what they appeared.)

The effect of this sequence is relatively complex. On the one hand, the frisson one might feel in confronting this ending involves a radical reframing. You read the first bit with the belief that you understood the context in which the story occurred sufficiently. However, the narrative turn reveals that you have been left in the dark – that a key context has been withheld. This sort of thrill – perhaps a singular pleasure of narrative – results from a disruption and re-establishment of a normative contextual framework through which the narrative is processed. Context-based reframing exposes the ongoing situational processing that governs sequencing relations during reading.

6 Selection

Underlying sequencing processing are ongoing processes of selection. These would seem to be distinctively part of pre-formed narrative artifacts. Novels, for example, contain a finite number of words, and arguably a finite amount of information. Details come pre-selected – they are simply part of the text. As such, selection appears to be entirely part of authorship. Take our example, "George's Story." In composing the story, our author sat at her writing desk, processing bits of her observational experience, her research, and her imaginative creations during an exhausting process of synthesizing, reimagining, researching (through introspection or through life experiences or texts), writing, and revising. It's all after the fact that she composes "George's Story," even if the genesis of imaginative details is mysterious.

Pre-formed narratives offer additional layers of selection as features of the text. Suppose George himself "tells the story" that comprises the major plot of the piece, and "George's Story" is actually something he has ruminated upon in his mind as a memory of the year he turned 30. Novels with this kind of structure are quite common, including works featuring the character Marlow as a narrator in some of Joseph Conrad's fictions, and in more recent work like Julian Barnes' *The Sense of an Ending* (2011). Surely creative, post-experiential processing has occurred, as George has "configured" the story for us. These structures are distinctive to narratives that are read or told, rather than being aligned with ordinary elements of experience.[18]

---

18 Some literary theorists might argue that these structures are homologous to Theory of Mind, which is common to both narrative texts and ordinary cognitive experience. However, I believe the structure described is distinctive, and since this argument is reconcilable with the general position of TOM theorists, it does not seem worth pursuing at length.

However, attention to the selective design of a text should not obscure the role of selection in the work of reading. For example, George's emotional response to the surprise party as it is experienced probably reflects how elements of the surprise party (the good will reflected in his friends' and loved ones' smiles, for example) are selected for the ongoing story of his loneliness. As readers, part of our cognitive work is to detect the selections that have been made, to select among them, and to draw inferences about them. For example, how do George's emphases in telling the story reflect his view of himself? How does his presentation of his younger self reflect an older George? What details does he focus upon, and what does this reveal?

As we read, in other words, we do not simply receive selected information and extract pre-given meanings from it. Instead, reading involves intensive cognitive processing of selections (or selections of selections, to be precise) for meaning (see Jacobs and commentary on foreground and background, pages 146–53). The "givens" of the artifact exist, but they are continually prioritized, associated, and revisited.[19] In this sense, reading an artifactual narrative involves complexity – we make sense of its representations by building an internal representation of them. Just as we noted in the discussion of sequencing, in selection there is something like an "end point" – after reading is complete, or perhaps after delay occasioned by forgetfulness, discussion, or other factors, we more or less settle on an understanding of a work that we might communicate if asked what we thought of it. However, the selection processing we undertake during reading is continual, and it has heuristic value, contributing to anticipation and emotional response. So, we are continuously engaged in selection within a growing set of selections in order to anticipate the form of the final set.

7 Salience and relationality

So, if selection underlies the work of sequencing, on what basis are selections made? This points to a previously mentioned feature of readerly cognition, salience. To select a particular detail is to value it above others. Salience can be related to primary organismal needs, like hunger, or to threats, like a snake emerging from the wall above your bed. These can suppress or overshadow other features of concern. Certainly if you woke to a snake slithering from a crack in the corner of

19 To illustrate this, consider the experience of reading a mystery novel. As you proceed through the work, you are weighting particular information and constructing theories that render the crime meaningful in terms of the social agents who exist within the framework of the plot.

your bedroom wall you probably wouldn't think as much about the ceiling needing repainting! Salience can also be related to beliefs, such as George's belief that he is isolated and his life is dull. As readers, we might take George's circumstances as evidence for or against this belief. Thus, something quite unremarkable becomes remarkable. There's an important distinction here – in the snake example, salience is more or less a characteristic of an individual element of experience (the snake). This is how salience is generally treated in the scientific literature, along with a focus on salience related to personal well-being. In the "beliefs about George" example, by contrast, salience emerges through the relation of otherwise unremarkable details to a key concern of the reader about George.

Because salience in individual elements is linked to overarching meanings, a feedback mechanism emerges in the reading process: meaning emerges from the entailment by salient elements of otherwise unremarkable elements, but most forms of salience are fundamentally guided by meaning. When a particular detail is freighted with implications for survival or well-being (a snake coming from the wall), comports with immediate goals, or has significant cultural meaning (such as a religious symbol), salience is experienced as "given." But in other cases, how does this work? In reading, threats and goals are generally not immediately pertinent. First, consider a device that is employed in narrative, in music, in virtually any art form: repetition. When repetition occurs, we attend to the element repeated. Indeed, if we recognize no other significance to it, we actively seek it. Thus, repeated references to George's habit of brushing his hair from his eyes might take on additional significance if we see it as revealing of his inner psychological character. If this represents a "depth-finding," or "vertical" form of significance seeking, another form is "horizontal," in the sense that salience is relational. Salience emerges through relations between particular features of texts. In literary studies, terms such as "antagonist" and "foil" describe characters whose salience exists primarily in relation to other characters.

Relationality is thus a crucial building block of understanding in the reading process. Two forms of relationality, at minimum, are processed more or less continuously during reading: first, condition changes for salient elements; second, relations between these parsed elements. The first of these we might term "dynamic persistence." We follow a particular feature of a narrative while processing its changes, creating a new analysis of the feature that can supervene upon the salience assigned to it in its relational role within the narrative. Processing for dynamic persistence occurs when the reader compares different instances of the feature, either creating a static composite or deriving meaning from both the identity and difference in the feature as it changes through time. This occurs most visibly when we track the changes that occur in a particular character or place. The second kind of relationality involves relation to other elements in the narrative. Some examples of this are straightforward, as in a character who functions as a foil to another, but others entail processing the salient feature in relation to condition changes within the narrative. For example, there are many condition changes in "George's Story," including: temporal (from one day to another); spatial (from work to home); character-field (from only George to George and friends and loved ones); and also, condition changes that reveal dynamic persistence, such as the character-psychological change from George's despair to George's joy.

There are a few insights we might immediately derive from describing relationality as a key focus of cognitive work during reading narratives. First, of the many relations between parsed elements that have significance, some follow a semantic, rather than a causal, circuit. For example, George's wish, early in the narrative, is related to his surprise party at the end. The reader is unlikely to infer that George's surprise party is caused by his wish, but it is a fulfillment of the wish nevertheless. That is to say, individual elements are implicitly related via associative or causal-sequential bonds. Second, relationality is not limited to human agency. Objects, changes, and features can have inchoate relational and emotional features. An object, such as a bridge the devil is rumored to have appeared on in a story, may become imbued with a felt sense of empathic terror in the reader for the child wandering across it late at night.

More ordinary objects may derive their value from implicit relationality. For example, the same home (clean, unoccupied except by the perceiver, nice, in a safe location, and unattended by any unusual circumstances) as experienced in the switch from work to home differs from home as a place one seldom is able to leave. Some texts, including famously *The Yellow Wallpaper* and *The Room*, offer features related to a single domestic space that shift with the passage of time to dramatize psychological changes in a primary character. Time spans can have relational properties as well. For example, the same day (filled with the same activities and interactions) takes its value from its relation to others. This is dramatized in the film *Groundhog Day*, in which the repetition of a single day creates changing significance for the central character, and for the viewer.

Relationality exists between elements in a narrative or experience, but also between elements and themselves, elements and concepts, and elements and readers. One of the common structures of cognitive processing in reading artifactual narratives, as several examples above demonstrate, is derived from the reader's care for a particular central agent (in literary studies, usually designated the protagonist). Processing for relationality often centers around this agent in a multitude of ways. Tab. 1 offers a taxonomy of typical forms of relationality centered around a protagonist (see below).

Arguably, reading involves negotiating any number of elements and tracking multiple relationalities for each. In viewing reading narrative as such, I am sup-

Tab. 1: A Taxonomy of Relationalities.

1. to reader	How does the element relate to the reader's sense of self and to prior experiences?
2. to prior manifestation	How does a particular incarnation of the element relate to an earlier incarnation?
3. to another element	How does the presence of another valued element condition the valuation of the element in question?
4. to future manifestation	What change is anticipated in the element, and how does that shape present evaluation of the element?
5. to the element as a concept	How does this manifestation of the element relate to a broader concept of this element (e.g., "home," or "car" vs. a dilapidated structure or luxury vehicle)

A taxonomy of relationalities: common relationships parsed in reading narrative
Note: a particular element can be engaged in multiple relationships

plementing Armstrong's argument that reading involves ongoing processing of a "polyphonic harmony of value qualities" (2013, 24). In the examples I've given of spaces and time spans, readerly cognitive processing involves a doubled analysis of meaning – in assessing relational differences in time and space, readers (or viewers) assess the dynamics of a particular character or characters. However, relational processing is ongoing whether or not a particular object, time, or place reveals the inner psychological makeup of a character. Rather, it is part of what narratologists call "world-building."

The point is that the meanings that emerge through narrative features are constructed in part through the sequencing, selection, salience and forms of relationality attendant upon the sensemaking activity of reading. Simultaneously, inchoate meanings (we might term them "meaning-theories") guide processing for relationality, salience, selection and sequencing. But our valuations of individual textual elements may seem to us natural, even "given" rather than constructed. Music provides a useful analogue: a middle-C on the piano can sound quite different if it is the tonic (the main note in a scale) rather than the seventh. Our auditory cognition, in what might be understood as a regular attribution error, parses the note relationally but attributes its value to the note itself. Moreover, this slippage is conventionally associated with values associated with cultural codes. For example, the final C# in a western classical piano sonata in the key of C#-minor creates a sense of finality or closure. This sense involves hearing the note differently from earlier iterations of the C# (for example, during an interlude in which the key shifts), but the valuation is associated with the note. Other features of the piece as a whole – its tempo, for instance – will similarly shape the valuation of the note.[20]

8 Valuation

The lowest level of our charting of cognitive processes engaged in reading artifacts involves "dynamic valuation." Valuing, as I'm using the term here, means assigning characteristics to particular features of a narrative artifact that may cue additional processing. Value is a "folded" concept, synthesizing characteristics of the object (which may be social, as in the price of a car, or situational, as in the temperature of a skillet on the stovetop) with relevance to the subject (as a source of pleasure, as a source of danger).[21]

That is to say, the objective and subjective values of a particular object are relative to other values (both like and apparently unlike), and they are subject to change. A subjective experience of narrativity thus begins with coordinated value processing through time, and develops in relation to higher order values, primarily future prediction.

Valuing begins with basic features that relate to readers' intuitions about the need for this additional processing, which may manifest in at least three forms: first, valuing elements-in-themselves as stimulants of predictions. Famously, a loaded gun on a wall in the beginning of a story may generate the prediction of it being fired. Predictions extend widely through the reading process, and can be understood as the basis of the characterization of individual elements as well as the sociocultural milieu of an artifact. This processing can even be directly cued by the text, as in Austen's famous opening to *Pride and Prejudice*: "It is a truth universally acknowledged that a single man in possession of a good fortune must be in want of a wife." In this sentence, the reader is cued to make predictions about the marriages of single men and women in the novel as we encounter them.

Second, readerly affective responses (involving the first type of relationship described in Tab. 1) to particular characters trigger valuations that are available for future relational processing. These affective responses reflect sociocultural conditioning, but also the reader's processing of the rest of the text. For example,

20 Armstrong notes deeps similarities between music and literary reading in the sense that both involve working through experience to produce emergent meanings: "melody is not an objective entity but a developing temporal construct" (2013, 96).

21 This folding is useful – it entails enfolded subjective and objective processes. Simply, "valuing" is both objective and subjective.

an angry, antisocial character in *Pride and Prejudice* might trigger valuations of him as a threat to the social order; however, an angrier, even more rebellious character in a film like *Rebel Without a Cause* or *Cool Hand Luke* would likely trigger quite different affective responses. As the reader (or viewer) works through the narrative, these affective valuations may themselves become dynamic – in *Pride and Prejudice*, the angry character may turn out to be angry for good reason, or misunderstood, or in need of sympathy.[22]

Third, valuation in cognitive processing involves a subtle dynamic element, undergirding the predictive and affective dimensions. Certain features of a narrative may by nature of their inherent value as concepts (events such as marriages, objects such as diamond rings) attract more attention and focus than others. Other features, in virtue of greater focus (manifested in long, detailed descriptions, for instance) may be established as high value. In addition, working through a narrative involves dynamism regarding the nature and amount of attention devoted to a particular feature. Tracking value dynamism is a key building block for the salience, selection, and sequencing functions described above. Consider Henry James' novel *The Golden Bowl* (1904), for example. As the novel progresses, the titular bowl changes from being valued as an aesthetic artifact, the presence of which has implications for how we understand the people who interact with and through it. Later, it becomes something quite different (it becomes an indicator of undisclosed interpersonal relations), but in this dynamism demands much deeper attention than we might otherwise accord it. This focus itself is cued by focus from a particular character on fine details of the bowl's flawed structure. In any event, for our purposes the bowl receives increasing attention as its role in the text shifts. That is, referring to the previous section, its salience increases as the register in which it is valued shifts.

Through these dynamic valuations (of agents, objects, settings, and the like), readers derive structures of understanding the narrative artifact that take shape around the continual work of heuristic prediction and revision, a changing affective economy, and through changing attentional commitments. While narrative artifacts inevitably contain external indexicals (i.e., references that point to particular known features of the real world), valuation dynamics develop meanings contingent upon relationships through time. To give a simple example, a novel might begin with an evil billionaire firing his secretary to blame her for what he calls an "accounting error" in his portfolio. The secretary, who supports a cancer-ridden mother with her paltry salary, is put into an untenable situation. By the end, the billionaire's cor-

22 The point isn't that the character is good or bad – the point is that the reader assigns a value (this manipulability is a signature of the role of processing).

ruption is exposed, and he is left in much reduced, though still wealthy, circumstances. Meanwhile, the secretary has recovered from grief at her mother's passing and is starting out as a small business owner. Conceivably, the reader's experience of such a narrative could involve feeling the secretary has triumphed and the billionaire has fallen, although in terms of the most straightforward marker of value identified here – money – the billionaire still has considerably more. One way this works is that both characters experience change through the novel, with the billionaire on a downward trend and the secretary rising upwards. The comparison of such trends provides a sense of satisfying closure to the crisis initiated by the billionaire's selfish malevolence and the secretary's vulnerability.[23]

Dynamic valuation, then, provides the basis upon which the higher levels of cognitive processing of narrative artifacts occur. Note that I am not arguing that narrative artifacts don't contain characters, or scenes, or other narrative features – instead, I am focused on the cognitive processing readers undertake in encountering and making sense of them. As I tried to establish at the beginning of this chapter, understanding a definition of narrative, or narrativity, that depends upon the interaction of readers with artifacts to describe what's special about narrative requires that we look carefully at the cognitive processing that readers undertake when they read.

9 Emergent structures in reading and other ordinary experiences

Consider the typology of processing activities employed in reading and ordinary cognition in Tab. 2.[24] As described in earlier sections, the cognitive work involved in reading narrative mostly has analogues in ordinary cognition. No reading experiences are the same, of course, but there is a universal: as they read, subjects have experiences. What Gallese and others have noted is that this experiencing involves

23 One opportunity of process-oriented rather than object-oriented models of cognition is that it may lead to re-examination of valuation and salience. Rather than the abstractions *good, bad, beautiful, ugly, advantageous, threatening* and the like, they highlight values based on imagined (sometimes fantasized) actions related to identity concerns.

24 By exploring these dynamics, we complement the top-down approach (where the seminal example of a narrative is a finished, communicable verbal or written text) with a bottom-up approach (where acts of narrative comprehension take center stage as the reader synthesizes and evaluates features of the narrative, ultimately developing an emergent sense of the meanings of parts or the whole). This bottom-up perspective may shed a different light upon what is at the top (that is, the nature of finished, communicable verbal or written texts).

Tab. 2: Processing Structures in Narrative Reading and Ordinary Experiences.

	Who	Emergent processing structures	Emergent temporal structures
I	Readers	Configured wholes	Time remembered
II	Readers	Proto-configurations (theorized parts and arcs based upon parsed events)	Whole-directed, anticipatory temporal figuration
III	Readers + Ordinary Subjects	Events	Processing time into "temporality" via boundary processing & selection
IV	Readers + Ordinary Subjects	Structuration via context, causality, and association	Time-based associative networks
V	Readers + Ordinary Subjects	Dynamic valuations	Time primitives (i.e., proto-events)

simulation, and this simulation draws upon the same neural and cognitive resources as processing ordinary experiences. Simulation is ongoing – that is, it is not exclusively a retrospective "making sense" of collected elements of an experience.[25]

Let's begin with the top level. This is associated with reading narrative artifacts, and generally not with other experiences. "Configured wholes" refers to Paul Ricœur's (1984) theory that *configuration* is a crucial cognitive process in understanding narrative. At this level, connections between major plot arcs, themes, contexts, and the like might be processed into a synthesis of something like the "overall message" of the artifact. It is here that we often locate the inferential work of understanding the "intention" embedded into the artifact. While this kind of processing does not occur exclusively after a reader finishes an artifact, this is prototypically the moment when the reader makes sense of the ending by recalling earlier parts.

In artifacts, some temporal markers are obvious – for example, in Virginia Woolf's *To the Lighthouse*, the book begins with a scene and ends with a parallel scene ten years later. However, some might be intuitive – in interpreting a scene, we might not have a definite idea of how long it took, but we categorize it as a scene and have some intuitive range within which it falls (e.g., less than a day, more than a second; perhaps even something as specific as more than 30 minutes,

25 To characterize the simulation in reading as "closed-form" would be to repeat the errors associated with artifactual views of narrative in earlier cognition.

but less than an hour). These units, in more or less fuzzy/defined form, become the material for top-level configuration.

The top level of configured wholes involves readerly work in which those explicit and implicit temporal units (formally, scenes, chapters, and other structural units, but also uncodified parts) are synthesized into a notion of overall structure. If you have a poor memory, this configuration may be quite fuzzy, and may favor recent and highly dramatic or emotional points in the narrative over others. High-level configuration may also be associated with disportation (see page 404). Nevertheless, in the course of reading, we frequently feel we've grasped the book, including the meaningfulness of its temporal structures.

Partly this has to do with the second level, again associated primarily with reading. This level involves *proto-configurations*. As the reader progresses through the artifact, meta-concepts related to theme, tone, and plot and character arcs, informed by known conventions, emerge both consciously and intuitively. This configuring process is ongoing and dynamic, rather than summative (as in the first level). In addition to the reader not having yet reached the end of the artifact, a key difference is that configuration occurs as part of making sense of particular passages in the artifact. Linking earlier passages to the present can serve two functions simultaneously – it is sensemaking of the new, providing a localized form of context; in addition, it may be an anticipatory configuring activity, like a hypothesis.

Armstrong argues that reading creates an ongoing, spiraling "hermeneutic circle," a concept initially popularized by reader response critics. For Armstrong, this circle is strongly related to the multi-dimensional temporality of Edmund Husserl, Martin Heidegger, and others (2013, 56). Since this notion of temporality is not exclusive to reading, it is important to point to a particularity – the *spiraling* involved in literary reading operates via an expectation of narrative closure. In ordinary experiences, schemas and scripts may condition such an expectation, albeit loosely. What's distinctive about the working-through experience of reading a work of literature (as opposed to ordinary experience), however, seems to be anticipation of a retrospective evaluative position regarding the structure of the work as a whole.[26]

26 This might not be entirely true. Consider a rather ordinary example: say you've gotten tickets with a friend to a concert for a band you love, but you are not a "night person." The experience of the concert might be one of mixed positive and negative feelings (enjoying the music vs. fighting off sleepiness and grouchiness). In addition, though, you might experience feelings of processing the experience with an eye towards how you will recount it later, including your evaluation. This may subtly alter your experience of the concert.

This anticipation manifests most obviously when a reader (or a viewer, in a film) feels an emergent sense of following the intention of the author or director.[27]

In artifactual narratives featuring a configuring narrator, the reader experiences temporal units that have been prefigured, as in the sentence: "three hours later, the couple were married in a brief service." Here the time marker from the previous scene is clear. However, at this level there is implicit cognitive work happening – the reader has to imagine the brief service, which in an Anglo-American context would probably mean fewer than 30 minutes, but more than two. In addition, the reader engages in some conscious and some intuitive cognitive work to make sense of implied temporal units, as in: "Later, she would feel the marriage had been a mistake."[28]

While the text may offer clear definitions of the temporal parameters and structures of the narrative, often these are left to the reader. For example, consider the following sentence: "After the tragedy, Amaya came to understand Nicholson better." The exact timeframe of Amaya's developing understanding is unclear, though the sequence (understanding occurring after the tragedy) is clear. In rendering this sequence meaningful, the reader is likely to do some additional cognitive work to form a fuzzy idea of the time parameters of Amaya's understanding, and perhaps indeed of the understanding as a process that unfolds in time.

In addition to grappling with the presence of a more or less explicit narrator, reading narratives often involves the configuration of multiple perspectives. Point of view shifts, in particular, are a striking feature of artifactual narratives. While ordinary experiences might include "perspective taking" through theory of mind activities, the dramatic juxtaposition of different fully articulated perspectives requires unique cognitive work for the reader.[29] David Miall and Don Kuiken describe the work of perspective management and synthesis into a "felt

---

27 One certainly might argue that this kind of anticipation occurs in other circumstances. However, what's distinctive about consuming artifacts is how the working-through experience is permeated by a sense of anticipating this retrospective evaluative moment, which (again) can manifest as a feeling of tracking authorial intention.

28 Armstrong describes "gap-filling" as a key form of cognitive processing that distinguishes reading. See Armstrong (2013, 84).

29 There is a potential critique of this position. Say that you believe that juxtaposed perspectives in narratives have correlatives in ordinary experiences in the retrospective gathering of perspectives on a particular situation or event (the juxtaposition of perspectives in artifactual narratives sometimes works like this). Then, *in situ* ordinary experiences might include a mentalized juxtaposition of imagined perspectives in the mind of the subject. However, this example seems overblown: the complexity of processing a particular experience would seem to preclude this kind of mental gymnastics.

perspective, often a personal one, which inheres in a literary text's expressively rich aesthetic components" (2001, 292).

There is an additional kind of temporal experience that conditions the reader. As the reading experience progresses, the reader, additionally, "goes through" the temporality – the literal time it takes to read has a relationship to the temporal units represented within the work. To make sense of this, consider a famous technique from the French New Wave movement in film. In Jean-Luc Godard's *Weekend* (1967), a traffic jam is depicted as the camera slowly pans along a line of stationary automobiles. The film forces the viewer to endure a boringly long duration that may induce an experience, or at least emotions, sympathetic to that of actually being stuck in a traffic jam. Earlier, I argued that readerly qualia are stimulated through the substance and form of a particular representation of experience (the experiential component of narrative), and that this combination of representations of experiences and stimulated experiences for the reader together embody what I think Herman means about how qualia are "conveyed" in narrative artifacts. The time and work of reading, in other words, conditions the reader's appraisal of the artifact's representations, including appraisal of the temporality of narrative events.

These top two levels, then, seem largely confined to the experiences of readers engaging artifacts. The next two levels, however, clearly are parts of the ordinary cognitive processing of subjects in experience. Reading engages these processes too, but in ways that reflect the particular conditions of this activity. Just as proto-configurations, in level two, can be considered an important anticipatory architecture for configured wholes, in level one, so levels three and four represent underlying frameworks through which the top levels might be developed. However, whereas proto-configurations mostly precede the retrospective configuration of the whole, the bottom three levels are simultaneous and interleaved. Moreover, the bottom three levels are each constituent of the level just above them.

Before we consider the lower three levels, let's consider the role of language in simulation. Simulation is complex: it is representation typically focused on perceptual faculties, with language playing a lesser or facilitating role. Earlier, we considered the idea, introduced above through Vittorio Gallese and Marco Caracciolo, that reading literature involves simulation that draws upon the same neural areas that help us navigate ordinary experiences. One route to understanding how this works is to consider how ordinary perception is translated into information that is designed to be employed in simulations, and how memory processes facilitate simulation.

Some scholars have described non-linguistic "perceptual symbol systems" that support, and indeed require, re-enactment of an earlier state for the retrieval of memory. In the model from Lawrence Barsalou (adapted by Speed et al. 2015), amodal and perceptual symbol systems are contrasted. Whereas in amodal systems perceptual information is translated into non-representational, taxonomic information, in perceptual symbol systems perceptual information is preserved in an association area, causing neural simulation of the original information upon re-activation. Speed, Vinson and Vigliocco (2015) propose a semantic architecture that combines linguistic and non-linguistic components, and that would thus help account for how narrative reading involves a combination of simulation and difference from ordinary experience.

Without the grammar afforded by language, then, what is the grammar of the simulations afforded by perceptual symbol systems? In part, ongoing simulations predict and respond to the coordination of dynamic relational salience (level 3 in our diagram). This means, as explored earlier, identification of what matters based on its relation to or association with other features of experience that matter. A common example of this involves attention to agents in experience. If a particular agent has strong attentional focus, and that agent is herself strongly focused on another agent, then that other agent may grow in salience – in relative importance for the future provision of attentional resources.

Relational salience does not need to be about agents, however. A context might make one feature of experience particularly important, and a causal chain might, in Rube Goldberg fashion, make an ordinarily unimportant feature crucial. As such, relational salience plays an important role in the ongoing work of sense-making, by familiarizing the unfamiliar through its relations with the familiar.[30]

Relational salience draws upon the reader's affective and analytic responses as well. As Gerrig's pioneering work on participation demonstrates, responses to texts are not passive – instead, readers bring particular biases to the text, and the text's internal cues about affective identification (such as the use of first person perspective and different perspectives) bias how events, agents, and other elements are evaluated (Gerrig 2001, esp. 310–323).

As the reader builds networks of dynamic relational salience, textual time begins to take on meaningful shape. Via the influence of context, or shifts in relational networks away from and back to states of relative equilibrium, the subject's experience is marked with temporally anchored changes in relational salience that provide memorable future reference points, and provide a framework about which proto-configurations occur.

The third processing level, focused on defining events, occurs for both the reader and the non-reading subject in experience. Event segmentation and processing creates units out of the continuous flow of a read text or an experience.

---

30 Armstrong describes familiar/unfamiliar "grafting," while many scholars, notably Gerrig, emphasize the dynamic, ongoing deployment of schemas (Armstrong 2013, 72).

However, reading for events leverages textual cues not available in ordinary life. Whereas for the subject in experience relational salience and temporal processing are emergent, growing out of processing in relation to prior experience, for the reader the artifact is more or less strewn with expository, formatting, and other cues embedded in features as subtle as parataxis and other grammatical forms that suggest particular relational and temporal dynamics, and thus give shape to events.

These textual features frequently create a distinctive experience of temporality by creating surprising juxtapositions of events. Zwaan and colleagues describe this property as "displacement":[31] "In real life, we experience events as a continuous flow. Because of its design feature of displacement, language allows speakers and writers to jump forward or backward in time. This creates discrepancies between how we experience events in real life and how we experience them vicariously through language" (Zwaan et al. 2001, 71–72). Nevertheless, the experience engendered in reading often contributes to a form of active processing that depends on the displacement we experience in mediating present activities via memory and anticipation. Consider, to illustrate this idea, the well-known scene at the end of the film *The Usual Suspects* (1995) in which the heretofore unwitting police detective realizes that the story he's been told is in fact an assemblage of details taken from articles and paraphernalia the suspect had observed in the detective's office. The realization offers an example of radical reconfiguration of the detective's understanding, and we witness his remembrance of moments that revealed the sources of the suspect's narrative. The scene is recounted as a form of active, real-time processing (the detective drops a cup of coffee, which falls in slow motion as he recalls details from the witness's testimony and connects them to items on his office walls). While most examples of displacement do not have this immediate revaluative role, typically displacement is deployed for the continued construction of the narrative's dynamic relational salience network, which in turn helps revalue the diegetic present. The point is that texts have particularly complex tools for depicting the emergence and evolution of our understanding of narrative salience networks.

This form of processing is crucial to understanding how the simulations engendered by reading relate to the ongoing sensemaking we ordinarily associate with experience. This level suggests an immediately fruitful type of processing to

---

31 The displacement creates cognitive work that results in a proto-configuration, although one might want to call the displacement itself the proto-configuration. Here the point is to assess the materials from which protoconfigurations are drawn.

explore if we'd like to understand what the work of reading narratives might tell us about cognition more generally.

The second level, or *structuration*, involves the primitive forms of the sequencing, selection and relational valuation dynamics described in the previous section. Note that sequencing, selection and relational valuation occur throughout the upper levels of the processing chart as well, but that these dynamics are emergent in primitive forms at this level.

Finally, the most basic level involves *dynamic valuations* (also dealt with in the previous section – see pages 424–6). This means, simply, that portions of experience are assigned values, and that those values change (i.e., the elements of experience are tracked through time and assigned new values as they change). Consider a simple example of this: you see your child looking at a pencil you've been using to write the grocery list. You might notice features of the pencil – its age and condition, its sharpness – and value the pencil accordingly in the same moment that you recognize it. If your child picks up the pencil, its value might change – perhaps you see the pencil as an instrument of your child's curiosity, or as a container of dangerous lead he might ingest. If your child brings the pencil to show you and the sharp tip gets close to your eye, you might revalue the pencil as a threat. These valuation changes, both within a particular set of predictions about the future and in terms of a break from one set of predictions in favor of another, as when the pencil becomes a threat to your well-being, are quite ordinary.

Further, these values and changes in them are crucial networks from which meanings emerge. This kaleidoscopic level of dynamic valuation not only precedes time-sense, but is the stuff out of which primitive time units are constructed. While dynamic valuations are constructed via cognitive processing, they are experienced as "givens" by the subject-in-experience, as they happen automatically, and can only be constructed consciously through dedicated attention. For readers, these dynamic valuations can be taken as signs of intention, as a subtle form of perspective even when a narrative's narrator is relatively transparent. They are built into the language of the text, and the reader receives them in the course of simply processing the language into meaningful concepts. Jacobs and colleagues have noted the importance of stylistic devices in literature for "foregrounding" particular ideas (see pages 405–6). Certainly, alliteration, metaphor and other devices highlight particular textual details, bringing them into the "slow" evaluative reading streams Jacobs hypothesizes.[32] Additionally, textual perspective is revealed through valuative claims embedded in texts.

---

32 Foregrounding may have important effects with regard to the reader's self-perspective as well: Miall and Kuiken (2001) offer a "defamiliarization-reconceptualization cycle" theory, which states

For a hyperbolic example of these phenomena, consider the opening lines of *101 Dalmatians*:

> A large car was coming towards the Dearlys. It drew up at a big house just ahead of them, and a tall woman came out onto the front-door steps. She was wearing a tight-fitting emerald satin dress, several ropes of rubies, and an absolutely simple white mink cloak. Her hair was parted severely down the middle and one half of it was black and the other white – rather unusual.
>
> "Why, that's Cruella de Vil," said Mrs. Dearly. "We were at school together. She was expelled for drinking ink."
>
> "Come in and meet my husband," said Cruella. The absolutely simple white mink cloak slipped from her shoulders to the floor. Mr. Dearly picked it up.
>
> "What a beautiful cloak," he said. "But you'll find it too warm for this evening."
>
> "I never find anything too warm," said Cruella. "I wear furs all the year round. I sleep between ermine sheets. I worship furs, I live for furs! That's why I married a furrier."
>
> Then Mr. De Vil came in. Cruella introduced him and then said, "Where are those two delightful dogs?"
>
> Pongo and Missis came forward politely.
>
> "Wouldn't they make enchanting fur coats?" Said Cruella to her husband.
>
> Pongo gave a sharp, menacing bark. (Smith 1984, 1)

The passage is suffused with aptronyms, including the Dearlys' goodness and naivete, and Cruella's cruelty. There are narratorial evaluations, indicated in phrases such as "absolutely simple" and "rather unusual." There are social codes built in, such as the connotation of wealth associated with the de Vils through the "big house," "ropes of rubies" and the like. The characters offer valuative observations of one another, such as Mrs. Dearly's report of the reason for Cruella's expulsion from school. Finally, while there is little dynamism in the passage because it is introductory, there is value dynamism in Pongo's change from civility to assertion.

This feature of textuality – the insertion of dynamic valuation claims at a basic semantic level – is a fundamental differentiator between reading narrative and other experiences. Values associated with particular word associations evoke particular value systems intuitively, thus making the experience of experientiality overdetermined in a manner that seems unique to narrative. However, the central idea here is that valuation is a multimodal, ongoing form of basic processing that enriches ongoing experience, and by so doing creates the conditions for the higher levels of processing. While such valuations seem "built in," processing a text is a continuous form of cognitive work in which the reader coordinates

---

that when a reader encounters stylistically foregrounded language in a text, his/her schemata may be inadequate for comprehension. As a result, the feelings evoked by the text facilitate alternative perspectives which can direct and motivate readers to search for new understandings.

them, reacts to them, and employs them to develop networks of dynamic relationality, as well as meaningful units of time.

In sum, these levels all describe forms of cognitive processing, some of which seem more particular to acts of extended reading, and others which occur in both reading and in other forms of ordinary cognition, with differences regarding the activity and passivity of parts of the processing.

One of the key elements that, for many theorists, distinguishes narrative reading experiences from ordinary experiences is the inclusion of narrative *perspectives* within the artifacts themselves. The model I've offered here suggests additional complexity to an already complex notion – perspectives may manifest at lower and higher levels of cognitive processing of the artifact. This is because levels two, three, and four are simultaneous and interleaved. Proto-configurations are provisionally based on relational salience, which is provisionally based on provisional dynamic valuations in a daunting complexity.

These bottom three levels thus describe reading, but they also describe the processing that occurs in ordinary cognition. While the subject-in-experience may seldom, or never, achieve the summative activity suggested in level one, negotiating the world involves continuous valuation, assessments of multiple forms of salience and relationality, and redoubts of proto-configurations that contribute to the sensemaking of ongoing experience. This all occurs in the course of parsing time into an increasingly semantic temporality.

This section has presented a taxonomy of processes engaged in reading and in other ordinary forms of cognition, and drawn linkages between reading and experience at the level of structuring elements of experience, whether perceptual or representational, into networks of dynamic valuation that help define the meaningful temporal units of narratives.

10 Conclusion

Reading narratives is experiential, and so it combines cognitive processes associated with reading and dynamics associated with narrative. Prior scholarship addressing the experiences created by narrative reading through concepts like experientiality can be developed by considering how narrative reading is ordinary and how it is particular. The experiential aspects of narrative reading are integral to, rather than the effect of, sensemaking processes at lower levels of comprehension.

These views have a number of implications for studies of reading, cognition and narrative around the relationships between representation and feeling. Briefly, an orthodox view is that narratives are felicitous information structures, offering a flexibility for recording and extracting information from heterogenous experience. In this view, narrative is a representational structure. An alternative view is that narrative is comprised of associative and temporal relationships that bridge representations and feelings through complex, dynamic forms of valuation. As a (third) supplement to this view, these relationships are often inextricable from, and constitutive of, meaning, including the meaning attributed to particular temporal extensions and dynamics.

If the first view is orthodox, the second is progressive. It can accommodate the first view because information structures are relational – what's new is the inclusion of multimodal, non-representational valuation mechanisms, such as emotion. The third, supplementary view is potentially radical. If valuation is primary, relationality and informational structuring exist downstream of the most basic form of sensemaking.

Reference list

Armstrong, Paul. *How Literature Plays with the Brain: The Neuroscience of Reading and Art*. Baltimore: Johns Hopkins University Press, 2013.

Barsalou, Lawrence W. "Perceptual Symbol Systems." *Behavioral and Brain Sciences* 22.4 (1999): 577–660. doi:10.1017/S0140525X99002149.

Barsalou, Lawrence W., W. Kyle Simmons, Aron K. Barbey, and Christine D. Wilson. "Grounding Conceptual Knowledge in Modality-Specific Systems." *Trends in Cognitive Sciences* 7.2 (2003): 84–91.

Burke, Michael. "The Neuroaesthetics of Prose Fiction: Pitfalls, Parameters and Prospects." *Frontiers in Human Neuroscience* 9 (2015): art. 442. doi:10.3389/fnhum.2015.00442.

Caracciolo, Marco. *The Experientiality of Narrative: An Enactivist Approach*. Berlin and Boston: De Gruyter, 2014. doi:10.1515/9783110365658.

Dehaene, Stanislas. "*Reading in the Brain* Revised and Extended: Response to Comments." *Mind & Language* 29.3 (2014): 320–335. doi:10.1111/mila.12053.

Di Paolo, Ezequial. "Extended Life." *Topoi* 28 (2009): 9–21. doi:10.1007/s11245-008-9042-3.

Fludernik, Monika. "Narratology in the Twenty-First Century: The Cognitive Approach to Narrative." *PMLA* 125.4 (2010): 924–930. http://www.jstor.org/stable/41058291 (22 August 2021).

Fludernik, Monika. "Towards a 'Natural' Narratology Twenty Years After." *Partial Answers: Journal of Literature and the History of Ideas* 16.2 (2018): 329–347. doi:10.1353/pan.2018.0023.

Gallese, Vittorio, and Hannah Chapelle Wojciehowski, with Patrick Colm Hogan. Interview. "Embodiment and Universals." *Literary Universals Project*. University of Connecticut, 25 September 2018. https://literary-universals.uconn.edu/2018/09/25/embodiment-and-universals/# (22 August 2021).

Gerrig, Richard J. *Experiencing Narrative Worlds: On the Psychological Activities of Reading*. New Haven and London: Yale University Press, 1993.

Gerrig, Richard J. "Perspective as Participation." *New Perspectives on Narrative Perspective*. Eds. Willie van Peer and Seymour Chatman. Albany: State University of New York Press, 2001. 303–323.

Gerrig, Richard J. "Readers' Experiences of Narrative Gaps." *Storyworlds: A Journal of Narrative Studies* 2.1 (2010): 19–37.

Gerrig, Richard J., and Giovanna Egidi. "Cognitive Physiological Foundations of Narrative Experiences." *Narrative Theory and the Cognitive Sciences*. Ed. David Herman. Stanford, CA: CSLI, 2003. 33–55.

Herman, David. *Basic Elements of Narrative*. Malden, MA et al.: Wiley-Blackwell, 2009.

Hruby, George G., et al. "The Metatheoretical Assumptions of Literacy Engagement: A Preliminary Centennial History." *Review of Research in Education* 40.1 (2016): 588–643.

Iser, Wolfgang. "The Reading Process: A Phenomenological Approach." *New Literary History* 3.2 (1972): 279–299. doi:10.2307/468316.

Jacobs, Arthur M. "Neurokognitive Poetik: Elemente eines Modells des literarischen Lesens." *Gehirn und Gedicht: Wie wir unsere Wirklichkeiten konstruieren*. By Raoul Schrott with Arthur M. Jacobs. Munich: Carl Hanser Verlag, 2011. 492–520.

Jacobs, Arthur M. "Affektive und ästhetische Prozesse beim Lesen: Anfänge einer neurokognitiven Poetik." *Sprachen der Emotion: Kultur, Kunst, Gesellschaft*. Eds. Gunter Gebauer and Markus Edler. Frankfurt a. M. and New York: Campus, 2014. 134–154.

Jacobs, Arthur M. "Neurocognitive Poetics: Methods and Models for Investigating the Neuronal and Cognitive-Affective Bases of Literature Reception." *Frontiers in Human Neuroscience* 9 (2015a): art. 186. doi:10.3389/fnhum.2015.00186.

Jacobs, Arthur M. "Towards a Neurocognitive Poetics Model of Literary Reading." *Towards a Cognitive Neuroscience of Natural Language Use*. Ed. Roel M. Willems. Cambridge: Cambridge University Press, 2015b. 135–159.

Mar, Raymond A. "The Neural Bases of Social Cognition and Story Comprehension." *Annual Review of Psychology* 62.1 (2011): 103–134. doi:10.1146/annurev-psych-120709-145406.

Miall, David S., and Don Kuiken. "Shifting Perspectives: Readers' Feelings and Literary Response." *New Perspectives on Narrative Perspective*. Eds. Willie van Peer and Seymour Chatman. Albany: State University of New York Press, 2001. 289–301.

Nelson, Katherine. "Narrative and the Emergence of a Consciousness of Self." *Narrative and Consciousness: Literature, Psychology and the Brain*. Eds. Gary D. Fireman, Ted E. McVay, and Owen J. Flanagan. Oxford: Oxford University Press, 2003. 17–36. doi:10.1093/acprof:oso/9780195140057.003.0002.

Pessoa, Luiz. *The Cognitive-Emotional Brain: From Interactions to Integration*. Cambridge: MIT Press, 2013.

Rayner, Keith, Alexander Pollatsek, Jane Ashby, and Charles Clifton. *Psychology of Reading*. 2nd edition. New York and London: Psychology Press, 2012.

Ricœur, Paul. *Time and Narrative*. Vol. 1. Trans. Kathleen McLaughlin and David Pellauer. Chicago and London: The University of Chicago Press, 1984.

Smith, D. *The 101 Dalmatians*. Puffin Books, 1984.

Speed, Laura J., David P. Vinson, and Gabriella Vigliocco. "Representing Meaning." *Handbook of Cognitive Linguistics*. Eds. Ewa Dąbrowska and Dagmar Divjak. Berlin and Boston: Mouton de Gruyter, 2015. 190–211.

Speer, Nicole K., Jeffrey M. Zacks, and Jeremy R. Reynolds. "Human Brain Activity Time-Locked to Narrative Event Boundaries." *Psychological Science* 18.5 (2007): 449–455. doi:10.1111/j.1467-9280.2007.01920.x.

Sternberg, Meir. "Universals of Narrative and Their Cognitivist Fortunes (I)." *Poetics Today* 24.2 (2003): 297–395.

Weick, Karl E. *Sensemaking in Organizations*. Thousand Oaks et al.: Sage Publications, 1995.

Zwaan, Rolf A., and Gabriel A. Radvansky. "Situation Models in Language Comprehension and Memory." *Psychological Bulletin* 123.2 (1998): 162–185. doi:10.1037/0033-2909.123.2.162.

Zwaan, Rolf A., Carol J. Madden, and Robert A. Stanfield. "Time in Narrative Comprehension: A Cognitive Perspective." *The Psychology and Sociology of Literature: In Honor of Elrud Ibsch*. Eds. Dick H. Schram and Gerard Steen. Amsterdam and Philidelphia: John Benjamins, 2001. 71–86.

Zunshine, Lisa. *Why We Read Fiction: Theory of Mind and the Novel*. Columbus: Ohio State University Press, 2006.

Dirk Vanderbeke

The Comforts of Constancy in Literature and Science

> The laws of memory are subject to the more general laws of habit. Habit is a composite effected between the individual and his own organic eccentricities, the guarantee of a dull inviolability, the lightning conductor of his existence. (Beckett, *Proust*, 19)

> In the arts as in the sciences, the things that work are likely to be those that preserve the successes of the past. (Weinberg, "The Craft of Science and the Craft of Art", 199)

Abstract: It is a cherished assumption that humans love novelty, crave new experiences and thus prefer variety in their appreciation of artistic products over repetition or sameness. Studies in evolutionary criticism postulate a human *neophilia* and, in consequence, the necessity for literature and the arts to change continuously in order to fulfil the demands of a public that is easily bored and thus requires new stimuli to tickle their tastes and cognitive faculties. Similarly, in the sciences we intuitively suppose that innovative ideas and original new approaches which challenge existing beliefs and paradigms ought to be met with interest and appreciation. This is not always the case, and in this paper, I want to explore the pleasures of repetition in our artistic preferences and their possible analogies in scientific research. I argue that the erroneous assumption results from a focus on a small segment of the arts and the sciences, the great works and momentous scientific discoveries that have lasting effects and are, in consequence, canonized in the histories of their respective fields. Below these masterpieces and revolutionary scientific achievements, the vast mass of popular literature is published and consumed by immense audiences, and the scientific day to day work proceeds and has to find grace in the eyes of reviewers and readers who favor the confirmation of existing concepts over transgression and the breaking of cherished scientific glassware. Indeed, I want to argue that we as human beings enjoy the security of the familiar, the well-known narrative patterns and the certainties of prevailing paradigms, even though, of course, there are always those who press forward and who, will, ultimately be remembered and praised – in case their creative minds bring forth innovations that we may eventually learn to enjoy or to use for productive research and not be relegated into the blind alleys of artistic or scientific oblivion.

https://doi.org/10.1515/9783110782844-015

1 Literature and the pleasures of repetition

Charles Darwin famously ended *The Origin of Species* on a celebratory note:

> There is grandeur in this view of life, with its several powers, having been originally breathed into a few forms or into one; and that, whilst this planet has gone cycling on according to the fixed law of gravity, from so simple a beginning endless forms most beautiful and most wonderful have been, and are being, evolved. (Darwin 2011, 490)

It is easy to imagine that Darwin, having just destroyed all human hopes for a happy hereafter, felt the need to offer something more consoling and uplifting than the gruesome vision of a nature "red in tooth and claw" (Tennyson 1994, Canto 56, 315). Since then, evolutionary theory has become the accepted scientific view on natural history, at least among the rational segment of our species, even though, of course, research and occasionally controversial discussions about specific aspects and processes are still ongoing. In some studies, however, the desire to present a positive perspective on evolutionary processes still seems to persist, especially when our own kind is under scrutiny. In the face of all arguments that evolution has no purpose or goals, when human nature is discussed the direction frequently seems to point upward to a present state that is cherished as the result of a continuous improvement of our mental and social skills. In particular, in the application of evolutionary concepts to human culture, artistic achievements and aesthetic appreciation, we seem to have been guided by adaptive processes, occasionally attributed to sexual selection, which allegedly favored a development toward self-domestication, pro-sociality and creativity (e.g. Dutton 2009, 135–163; G. Miller 1998, 115–117), and culminated in the pleasure in "truth, beauty and goodness" that marks our preferences in art, literature and other cultural products.

I fully agree with the theoretical tenets of evolutionary studies of human behavior and culture, but in this paper I want to challenge some of the assumptions and conclusions promoted in research into our cultural preferences, arguing that the endearing image of humanity reveling in great works of art and literature may well be dampened by the obvious fact that less glorious products in print and on the screen are also highly successful and may even reach larger audiences than canonical works that are celebrated by the cultural elite. In *The Origins of Creativity*, Edward O. Wilson writes unreservedly that "we [. . .] value Proust" (2017, 39), but do "we," or, who exactly is the "we" that values Proust? If an admiration of Proust did count as a defining feature of humanity, approximately 99.9% of our species would not pass the test. A few pages earlier Wilson writes:

"A reader knows he is in the presence of greatness when he reads the following opening lines [of Lolita] by Nabokov" (Wilson 2017, 36), and he compares those lines favorably with the beginning of Jonathan Franzen's *The Corrections* which he describes as "a painfully ostentatious display of learning by a Harvard Sophomore with a great deal of promise" (Wilson 2017, 38). Again, who exactly is "a reader," and does that category exclude all those readers who prefer Franzen to Nabokov – quite a few I would assume. Similarly, Denis Dutton in *The Art Instinct* celebrates "the masterpieces that have withstood Hume's Test of Time and show every indication of maintaining their hold on the human imagination" (2009, 236) but bemoans "the decline of great art in cynical, ironic ages, such as our own" (Dutton 2009, 240), raising the questions of whether all those people who do not appreciate Shakespeare, Beethoven, Hokusai or Wagner (Dutton 2009, 240) actually partake in "human imagination," and whether our age is regressing to a prehuman condition.

Wilson, in the chapter that addresses Nabokov, Franzen, and "our" evaluation of art, also discusses creative innovation, i.e. we judge art "by its innovation of style and metaphor, by its aesthetic surprise, by the lasting pleasure it gives" (2017, 37). Moreover, he suggests a kind of bodily response to creative success that then triggers the desire for more of it: "breakthroughs will race the heart of the innovators, stiffen their spine, encourage their search for breakthroughs" (Wilson 2017, 40). Great artists, then, are driven by an energy that expresses itself in innovative works, and the audience – we as human beings – craves novelty, the lack of which would render our experiences stale and unsatisfying. Geoffrey Miller, indeed, argued that the pleasure in novelty, i.e. *neophilia*, is an inherited trait in humans that results from evolutionary processes:

> Of course, in modem society, human neophilia is the foundation of the art, music, television, film, publishing, drug, travel, pornography, fashion, and research industries, which account for a substantial proportion of the global economy. Before such entertainment industries amused us, we had to amuse each other on the African savanna, and our neophilia may have demanded ever more creative displays from our mates. This hypothesis can explain the mysterious cultural capacities that are universally and uniquely developed in humans, such as language, music, dance, art, humor, intellectual creativity, and innovative sexual play. These are all highly valued during mate choice and highly useful during courtship. Such displays all use a uniquely human trick: the creative recombination of learned semantic elements (e.g., words, notes, movements, visual symbols) to produce novel arrangements with new emergent meanings (e.g., stories, melodies, dances, paintings). (G. Miller 1998, 116)

Miller's list of areas in human culture that are influenced by neophilia is long, including not only the usual cultural products but also scientific research, and some of them are surprising – pornography, in particular, does not seem to have involved much novelty in the course of its history and may well be regarded as the most repetitive and unsurprising genre. But there seems to exist an almost unanimous agreement that cultural products must include some kind of originality to attract their audiences. Colin Martindale in "The laws governing the history of poetry" argues that "Whatever else they must be, works of art and literature must be novel" (2008, 230). This is regarded as a fundamental requirement, because "[i]f a series of artists kept producing the same or very similar works of art, liking for their productions would decrease over time" (Martindale 2008, 232). In consequence, "the pressure for novelty is unremitting, constant, and consistent" (Martindale 2008, 233). In "Aesthetic Evolution" he suggests that "[a] work of art – or any stimulus for that matter – gradually loses its arousal potential with repetition" (Martindale 1986, 450), and in consequence "the more an audience is exposed to a type of art, the faster the art should change" (Martindale 1986, 454). Nancy Easterlin claims that "if we consider literary canons in the past two to three hundred years – since the rise of literacy and the proliferation of print material – originality and creativity seem to have been consistently valued" (2012, 53), and David S. Miall and Don Kuiken state boldly that "one reason we do not become weary of novelty is because it provides us with a window on the truth" (1994, 391).

Easterlin writes about canonical literature, the studies by Miall and Kuiken are based on student responses to stories by Katherine Mansfield, Virginia Woolf and Sean O'Faolain, and Colin Martindale is chiefly concerned with the history and evolution of poetry. Indeed, when we look at the history of canonical literature, creative innovations are a persistent phenomenon, if only because innovative works will more easily become canonical than works that walk on well-trodden paths. It is also noticeable, that, as Easterlin implies, momentous innovations have sped up over the last 200 years and that in previous eras the development of literature and the arts was paced considerably more slowly. For these periods, Ernst Robert Curtius diagnosed a kind of rhythm between Classicism and Mannerism. Classicism appears as a kind of plateau, a tradition and norm that defines the apt formal und aesthetic features of artistic works. As such, Classicism is not a singular era in cultural history, instead there are several Classicisms over time. Similarly, Mannerism is not a unique phenomenon within one specific era, and for his argument Curtius writes that he "must free the word from all art-historical connotations and broaden its meaning until it represents simply the common denominator for all literary tendencies which are opposed to Classicism, whether they be pre-classical, post-classical, or contemporary with any Classicism" (2013, 273). This "artistic 'manner,' which can find expression in the most various forms, overruns the classic norm. [. . .] Mannerism is a constant in European literature. It is the complementary phenomenon of the Classicism of all periods" (Curtius 2013, 273). This suggests that the Classicist literature of all times, the works usually regarded as canonical, adheres to the dominant literary norms and rules, while innovation would chiefly appear in Mannerist works which disrupt or overturn prevailing literary styles and standards.

Discussing innovation and the avant-garde, James S. Ackerman suggests that the perspective on them has changed over time and that the favorable view is a fairly recent phenomenon: "In a closed and stable society, innovation is likely to be rather feared than welcomed; in the seventeenth and 18th centuries, it was sometimes used as a synonym for 'revolution'; in an open society like our own, innovation is considered a virtue" (Ackerman 1969, 371). Indeed, since the end of the nineteenth century, the speed of artistic change has greatly increased, and innovation for innovation's sake can be diagnosed for some developments, e.g. the rapid succession of styles and radical manifestoes that was an inherent element of early twentieth century art and literature. At the same time, a recalcitrance in the literary work was proposed as a hallmark of artistic greatness by Victor Shklovsky, who argued that if aesthetic appreciation offers pleasures, an artificially prolonged reading process would extend those pleasures and thus add to the value of the work in question: "The technique of art is to make objects 'unfamiliar,' to make forms difficult, to increase the difficulty and length of perception because the process of perception is an aesthetic end in itself and must be prolonged" (Shklovsky 1988, 20). Defamiliarization appears as the artistic countermeasure to habituation, and thus art needs to find ever newer and more unfamiliar forms of expression to fulfil its function. Twentieth century literary and aesthetic theory is then loaded with concepts that celebrate artistic innovation and experiment. But then it is worthwhile to take a closer look at human cultural pleasures and ask in how far these are really based in a desire for novelty.

Among the usual methods to establish human universals are research into child behavior and comparative anthropology. As every parent knows, children crave repetition, and the usual response to any pleasant experience is the exclamation "Again!". In particular, children love to hear the same stories over and over again and will not permit even minimal changes to the familiar tales.

If we look at the cultural significance of narratives outside the Western world and the developments of canonized cultural products within the last few hundred years, we can notice that continuity is of far greater importance than novelty. In cultural environments that are not prone to rapid change, i.e. under the conditions that humans lived in for most of their natural history, stability is more important than innovation, and as social information is retained better than factual information (cf. Mesoudi et al. 2006), stories are the storehouses of cultural knowledge in communities that rely on oral transmission and memory. The wise men and women of indigenous cultures were, and still are, famed not for their ingenious inventions of ever new stories but as repositories of knowledge encoded in traditional tales that were persistently re-told to happy listeners and so passed on from generation to generation. In addition, conversations and arguments in native cultures frequently take recourse to received knowledge in the form of micro-narrations like proverbs, which Achebe describes as "the palm oil with which words are eaten" (1986, 5). And, finally, in the face of Miller's suggestion that song and dance produce constantly new forms and arrangements, they are actually rather part of rituals and thus of repetitions that offer stability rather than novelty. In Western Europe the unchanged rituals of the church service were for centuries one of the most important cultural experiences for most of the population.

But even if we look at our own recent cultural sphere, the question of whether we really prefer novelty over repetition in narratives may lead to surprising answers, once we turn away from highbrow literature and focus on popular genres instead. Research into the production and commercial success of books indicates that classics amount only to a tiny percentage of the fiction sold – 3% in 1985 according to Peter Swirski (1999, 4). Those data may be a bit dusted, but then Catherine Salmon calculated in 2012 that romance novels alone "account for 13.2% of mass market paperback sales in the United States" (2012, 152). Genre literature, of course, tends to employ specific narrative patterns – occasionally the term formula fiction is used to describe works that are based on recurrent narrative structures. For the readers of such works, Nancy K. Miller diagnoses a "pleasure that comes from genre satisfaction" (2007, 541), and Janice A. Radway argues that contemporary "romance novels obviously provide a reading experience enjoyable enough for large numbers of women so that they wish to repeat that experience whenever they can" (1991, 44). Similarly, Thomas Disch writes about genre literature as such:

> Genre fiction may be distinguished from other kinds of writing in being shaped by the (presumed) demands of its audience rather than by the creative will of its writers. The writers accommodate their talents to the genre's established formulae. These formulae exist in order to guarantee readers the repetition of pleasures fondly remembered. (Disch 2005, 6)

In consequence, commercially profitable works spawn sequels and prequels, imitations and adaptations, and occasionally one particularly successful work may even lead to the establishment of a whole new sub-genre repeating the formula endlessly – e.g. J.R.R. Tolkien's *The Lord of the Rings* and so-called High Fantasy, or *Superman* and the genre of superhero comics. As a matter of fact, comics are an especially interesting case in point. Comic strips and the funnies have a particularly long shelf life, and some of them have been around for over a century, repeating the same stories and motifs with minimal variation on a daily basis for their obviously faithful audiences. Charles Schultz's *Peanuts*, one of the most successful comic strips since 1950, has actually been described as "repetition and monotony as art form" (Fresnault-Deruelle 2014, 124).

But then, some narrative patterns have been around for millennia, and the quest of the wandering hero forms the basis of myths, fairy tales, Arthurian romances, the Western and a large segment of science fiction, while the romance plot has similarly survived since the earliest love stories have been recorded. Of course, these patterns can be filled with diverse stories and include surprising twists and new elements, but then many narratives are highly repetitive with only minimal changes and departures from successful predecessors. And it does not seem as if the audiences lose interest in the familiar tales of heroic fights and romantic closure. But as popular genres are far less frequently in the center of research than the canon and mainstream literature, their features and the pleasures they offer do not register quite as strongly in literary criticism and aesthetic theory, while those scholars who actually focus on popular literatures attest to their artistic complexity and subtle nuances in order to raise their status and so legitimize the respective studies.

But is the delight in repetition only characteristic of popular narratives and their readership or can we also detect it in the literary preferences of the so-called cultural elites? If the connoisseurs of canonical works really craved novelty and new stimuli, they should be very varied in their selection of texts. Being wholly enamored of a book, they should dismiss further works by the same author and rather turn to something completely different in search of a truly novel experience. But that is not the common behavior of readers, who usually have their favorite authors, epochs and styles. Moreover, the books we love are not only devoured once but repeatedly, and while it can, of course, be argued that every reading is a new reading that allows us to discover fresh aspects and new details, the incentive for re-readings also results from the expectation of familiarity and the repetition of previous pleasures. Similarly, we watch our favorite films decidedly more than once and listen to our favourite tunes and songs over and over again. Actually, the core repertory of operatic music consists of hardly more than a few dozen works that are re-staged continuously all over the world, admittedly with changes and alterations in the individual productions, but nevertheless also to be enjoyed by educated audiences which take delight in "the same again, with a twist," while more radical experiments in staging and performance draw protest from the huge segment of traditional-minded aficionados.

In addition, the revolutionary developments in literature and the arts in the 19th and early 20th centuries were not always unanimously embraced by enthusiastic readers and spectators, not even by the cultural elites and the feuilletons, but frequently rejected, sometimes violently so. Famously, the first exhibitions of modern art met with scorn and scandal, or the suspicion that the artists collectively suffered from one or the other form of eye disease (cf. Howe 1895). Innovative authors had to struggle to get their work published, and some of them, e.g. Herman Melville, did not live to see the fame of their previously dismissed texts. Innovative and transgressive works – Curtius's mannerist experiments – are frequently celebrated and promoted by only a minority of likeminded artists, authors, and critics in an uphill fight against the normative requirements of good taste and tradition. Ackerman states that: "Reviewing the critical fate of artists of the nineteenth and earlier twentieth centuries, we find that the innovations that proved productive were vigorously rejected by the prevailing taste and were assimilated by that taste only after further innovations based on them produced a new shock and drew off the fire" (Ackerman 1969, 371–372). Indeed, the relationship between innovative artists or authors and the public may well be seen as a kind of conflict. The "Manifesto: The Revolution of the Word," published by Eugene Jolas in *transition* 16–17 in June 1929, ends with the blunt statement: "The plain reader be damned" (Ellmann 1983, 588) and Renato Poggioli claims that the "disdain of the avant-garde artist is [. . .] directed exactly against that part of the public which claims best to represent the civilization of its epoch" (1968, 51).

Only with time and growing familiarity will those works be admitted into the halls of high culture. As early as 1965, Robert B. Zajonc suggested that "mere repeated exposure of the individual to a stimulus is a sufficient condition for the enhancement of his attitude toward it" (1968, 1), obviously in direct contrast to the assumption that repeated stimuli will quickly lead to a decrease in response and ultimately to boredom. James Cutting then writes about our aesthetic experience of pictures: "We like the ones we have seen before and, particularly, those we may have seen many times," and he argues that this is "part of our human nature, built on an evolutionary substrate that makes very good sense. It helps ensure steadiness in culture more generally and relative constancy in artistic canons more particularly" (Cutting 2003, 335).

These findings tie in with neurological research into our aesthetic experiences according to which "repeated exposure leads to the subjective feeling of perceptual fluency, which in turn influences liking" (Reber et al. 1998, 45), and while this research is once more chiefly based on the perception of pictures, the authors point out that: "Mere exposure effects have been obtained with a variety of stimuli (faces, ideographs, words, melodies) and a variety of measures (judgments of preference, behavioral choices, physiological responses), indicating the robustness of the phenomenon" (Reber et al. 2004, 370). Of course, these results challenge the tenets of the previously mentioned aesthetic theories that consider defamiliarization and intricate innovation a prerequisite to aesthetic appreciation. In the light of neurological research, but also of popular products, it may be time to reconsider some cherished assumptions and beliefs about our preferences and pleasures in the arts and literature.

2 Science and the adherence to established knowledge

These observations are not necessarily restricted to our response to art and literature but can possibly be extended to the field of scientific research, i.e. the complementary way in which we encounter, negotiate, and construct external and internal realities. In the sciences, even more than in the various arts, we also intuitively assume that the march of progress is the ultimate objective, driven by unbiased creative minds and innovative theories. However, reservations about such an optimistic view of the sciences and their practitioners have been voiced for quite some time. In 1949, Max Planck suggested in his "Scientific Autobiography" that: "a new scientific truth does not triumph by convincing its opponents and making them see the light, but rather because its opponents eventually die, and a new generation grows up that is familiar with it" (Planck 1950, 33–34), which would indicate that death is one of the most important driving forces in scientific change.

Similarly, if not quite as drastically, Charles Darwin wrote in his "Conclusion" to *The Origin of Species*:

> Although I am fully convinced of the truth of the views given in this volume under the form of an abstract, I by no means expect to convince experienced naturalists whose minds are stocked with a multitude of facts all viewed, during a long course of years, from a point of view directly opposite to mine. [. . .] A few naturalists, endowed with much flexibility of mind, and who have already begun to doubt the immutability of species, may be influenced by this volume; but I look with confidence to the future, to young and rising naturalists, who will be able to view both sides of the question with impartiality. (Darwin 2011, 481–482)

These skeptical statements from both Planck and Darwin are, unsurprisingly, also quoted in Thomas S. Kuhn's *The Structure of Scientific Revolutions* (1996, 151), and the famed alternation of paradigmatic science and theoretical upheaval followed by the rise of a new paradigm shares some features with Curtius's pattern of Classicism and Mannerism. In a 1969 special issue of *Comparative Studies in Society and History* on cultural innovation, E.M. Hafner, drawing on Kuhn's concepts, suggested a close similarity between the developments in the sciences and the arts:

> Whether we speak of science or art, we recognize the essential role played by revolution. Every epoch is marked by its current paradigms, coherent traditions of observation and interpretation which set the stage for normal activity. But every epoch ends in revolution, after which the old paradigms give way to the new. (Hafner 1969, 389)

If this view already relegates the search for truly novel ideas to the phases of scientific revolutions when an anomaly in the experimental results forces the scientists to abandon their cherished beliefs in a "proliferation of versions of a theory [which] is a very usual symptom of crisis" (Kuhn 1996, 71), at least some of the revolutions can also be broken down into smaller steps and so become almost invisible to those who actually take part in them. In his last lecture, "The Trouble with the Historical Philosophy of Science," Kuhn outlined the story of scientific change and concluded: "By the end of the narrative those changes maybe considerable, but they have occurred in small increments, each stage historically situated in a climate somewhat different from that of the one before" (Kuhn 1992, 11).

Of course, Kuhn's seminal book and the special issue of *Comparative Studies in Society and History* are no longer cutting-edge research, but the topic has not dropped from the discussion and is addressed with some regularity and in a variety of garbs. One of them is the question of scientific "schools" and their resistance to conflicting research. Daniel J. Kevles describes this phenomenon in the case of cancer research in the mid-twentieth century. After the search for cancer-causing microorganisms had failed in the early twentieth century, there was an almost unanimous agreement that viruses could be dismissed as influencing agents in the growth of cancerous cells. John J. Bittner, one of the scientists who still found evidence for a viral impact, felt the need to evade any reference to the discredited concepts and used euphemisms instead; later he wrote: "If I had called it a virus, my grant applications would automatically have been put into the category of 'unrespectable proposals'. As long as I used the word 'factor', it was respectable genetics" (quoted from Kevles 1995, 81). This seems, as Kevles points out, to be an exceptional case "where almost every one of the key pioneers encountered pointed resistance from his community of peers," and the resistance was not motivated by any political or religious ideology but "derived from the skepticism of a professional community of biomedical scientists whose beliefs were grounded in laboratory evidence" (Kevles 1995, 106). In consequence, he celebrates not only the "professional courage, imagination, and persistence" (Kevles 1995, 106) of the innovative scientists, but also the "tolerance and pluralism of the basic biomedical research system – the tolerance of deviant ideas and the pluralism that provides niches" (Kevles 1995, 106–107), a claim that rings a bit hollow in the face of the tricks that had to be used to find funding. And if we look at the history of human knowledge, the resistance, occasionally even violent resistance, to new ideas is a constant topic. The march of scientific progress is not a downhill slide but an uphill fight.

The ambivalence between a resistance to change and the ultimate acceptance of conflicting arguments is also present in recent research on the cognitive phenomenon of confirmation bias or, as Hugo Mercier and Dan Sperber call it, myside bias in the sciences. Once more, a hypothetical neophilia would suggest that humans embrace alternative views and unexpected perspectives, but then it seems as if we feel safe and at home when we encounter ideas and concepts that confirm those we already hold. As pointed out before, we crave stability and familiarity as a source of security within our cultural environment. In Mercier and Sperber's book *The Enigma of Reason*, the ambivalence is openly displayed in the titles of the subchapters: "Scientists Are Biased, Too" (Mercier and Sperber 2017, 317–318) and "Yet Scientists Are Sensitive to Good Arguments" (Mercier and Sperber 2017, 318–319). In the second of those chapters, the authors suggest that Planck's law (see above) is far too pessimistic and that, of course, scientists can be converted to new views and beliefs if confronted with sound facts and arguments – Einstein's life-long rejection of the bewildering principles of quantum physics would appear as a particularly stubborn form of resistance rather than as a behavior common among his peers. But then experimental data have to be replicated to find acceptance, theoretical frameworks need to be established and tested against subsequent hypotheses, and in many cases it still takes a long time before the majority of the relevant scientific community has been won over. One might claim that new theories or models are accepted once they are no longer in conflict with the expectations of the community but have yielded further hypotheses that are then confirmed by experimental results – this would be the scientific equivalent of mere exposure. In general, the scientific narrative underlying research is famously not the exploration of unknown territories, but the confirmation of a hypothesis formulated at the beginning of a paper.

The existence of a confirmation bias is, of course, in itself a strong argument against a human desire for innovation, and, as pointed out above, it makes evolutionary sense when we consider the world of our ancestors, i.e. a world that did not undergo constant changes but remained stable over long times. Group coherence, communal values and collective memory and identity were supported by stable traditions and the transmission of received knowledge in the form of well-known narratives. New knowledge was in all probability accumulated slowly and had to be tested and confirmed repeatedly before it was integrated into these tales and traditions over time to enter the repository of useful information.

3 Conclusion

The image of humans who are conservative rather than inventive, who cling to the safety of long cherished beliefs rather than venture out into the unknown, and who prefer the same old tales rather than enjoy a firework of novel and experimental ideas is not quite as endearing as the one suggested by the proponents of a human neophilia. But then human nature is not as we would like it to be, and in the face of scholars who claim that our evolution and sexual selection have domesticated us to be generous, pro-social, creative and charming, we have to admit that we are also a nasty, selfish, greedy and violent lot. It may be useful to keep our flaws and imperfections in mind rather than to indulge in dreams of a humanity that happily embraces improvement and ever new artistic and scientific visions.

4 Postscript

Elisabeth Haefs' contribution to this volume discusses the persistence of specific narrative patterns in the popular transmission of scientific knowledge. In an environment in which science meets a non-academic audience, it needs to attract the attention and hold the interest of the readers in order to achieve its objective. I want to add a short account about a specific motif that has resurfaced in narratives over the centuries or rather millennia and recently made its appearance in a scientific context. It would be fallacious to generalize this observation and to draw far reaching conclusions from it, so I offer it as an anecdote rather than a basis for extensive theoretical considerations.

In the context of a project that employs phylogenetic tools, measuring the degree of formal relatedness between variants to reconstruct the evolution and geographical distribution of folktales and myths, Julian D'Huy suggests that the story of Polyphemus that we now mostly know from the *Odyssey* is one of the oldest myths of mankind. He defines the Polyphemus type as "a tale in which a person gets into the homestead of a master of animals or of a monstrous shepherd; the host wants to kill the hero, but the hero escapes by holding on to the fleece or fur of an animal who is going out, concealing himself under an animal's skin or with a living animal" (D'Huy 2015, 50). Variants can be found world-wide, and D'Huy writes: "Comparative evidence supports the probability that the Polyphemus tale was current in some form in the Paleolithic era, and its *longue durée* is a relevant indicator that it held social interest and relevance" (D'Huy 2015, 55). To explain how this and similar motifs could have evolved and survived for so many millennia, D'Huy once more draws on a biological model, the punctuated equilibrium as developed by Stephen Jay Gould and Niles Eldredge (1977) – and I would like to point out that Gould in *The Structure of Evolutionary Theory* claimed that he was strongly influenced by Thomas S. Kuhn's similarly "punctuational theory" (Gould 2002, 967). D'Huy suggests for his research that "[i]f an analogy may be drawn, newly mythological sister versions would tend to diverge rapidly, which would be followed by extended periods of stability with little net evolutionary change" (2015, 59).

Monsters that trap wandering heroes in caves to feast on them are a rarity in recent science, but there is a phenomenon that is regularly described in similar terms, the black hole. A search on Google for Black Hole + Monster produced 454,000 hits, + Cannibal 418,000 hits, + Devour 410,000 hits, and + Escape 462,000 hits. Here are a few examples of fairly typical, but also sensationalistic accounts of black holes:

> **Scientists Catch A Black Hole Munching On A Giant Planet**
> Like virtually every other galaxy in the universe, the Galaxy NGC 4845 has a central black hole. For the past few decades, though, that black hole has lain dormant, without giving much sign it was there. That changed recently, however, when scientists caught the black hole the in process of eating a little snack. "Little" in this case being a planet larger than Jupiter. [. . .] The scientists were able to examine the event and determine that the mass of the planet being devoured by the black hole was somewhere between 14–30 times that of Jupiter. (Knapp 2013)
>
> **Space Cannibal: Ginormous Black Hole Caught Eating Another**
> A monstrous black hole at the heart of one galaxy is being devoured by a still larger black hole in another, scientists say. (Choi 2011)
>
> **Evidence of Galactic Cannibalism Revealed**
> Some neighbors you just don't want, and that applies if you are a galaxy as well. NGC 1316 looks to be one of those. It's a deep space Hannibal, based on evidence from the European Southern Observatory (ESO) that reveals traces of its past consumption of a spiral galaxy three billion years ago.
> Galactic cannibalism is not rare. (Luntz 2014)

Black holes are, of course, one of the most fascinating astronomical phenomena for the general public and for scientists alike; moreover, as Stephen Hawking has pointed out at the very beginning of his 1988 Hitchcock lecture on "Black Holes and Baby Universes" in Berkeley, "Falling into a black hole has become one of the horrors of science fiction" (1994, 104). But then, of course, quite a few movies have been made, e.g. *Event Horizon* and *Interstellar*, in which travelers survive the passage into a black hole, even if the physics involved is usually rather inventive and certainly not compatible with the laws of nature that govern our universe. Hawking in 1988 made it quite clear that there is no escape from a black hole and that the imaginary transportation schemes of science fiction would run into a snag as "an astronaut who fell into a black hole would come to a sticky end. He would be torn apart by the difference between the gravitational force on his head and feet" (Hawking 1994, 110).

But that was in 1988, and since then things have changed and new concepts arisen. In a lecture in 2008, Hawking suggested that black holes "are not the eternal prisons they were once thought [. . .] If you feel you are trapped in a black hole, don't give up. There is a way out" (2008). The situation addressed here is, however, not the actual escape of a human or a spaceship approaching or passing an event horizon but rather the more abstract problem of the "black hole information paradox" and the necessity "to determine whether information really was lost in black holes, or whether in principle, it could be recovered" (Hawking 2008).

In 1971, Remo Ruffini and John Wheeler had coined the peculiar metaphor that "A black hole has no hair" (1971, 52), indicating that black holes do not have unique properties or complex features. In 2016, however, Stephen Hawking, Malcolm Perry and Andrew Strominger countered the "no-hair theorem" and argued that indeed "Black holes have a lush head of 'soft hair'" (2016, 2). The precise mathematical and physical features of the phenomena termed "soft hair" are of no relevance here, but the metaphor is. And it seems as if information that was previously thought to be irretrievably lost in the black hole could be at least partly stored in this "hair" that forms a kind of halo around the black hole.

We have returned to one of the oldest mythical motifs, and it may be possible to escape the devouring monster by clinging to soft hair.

Reference list

Achebe, Chinua. *Things Fall Apart*. Oxford et al.: Heinemann, 1986 [1958].

Ackerman, James S. "The Demise of the Avant Garde: Notes on the Sociology of Recent American Art." *Comparative Studies in Society and History* 11.4 (1969): 371–438.

Beckett, Samuel. *Proust and Three Dialogues with Georges Duthuit*. London: John Calder, 1999.

Buss, David. "Evolutionary Personality Psychology." *Annual Review of Psychology* 42.1 (1991): 459–491.

Choi, Charles Q. "Space Cannibal: Ginormous Black Hole Caught Eating Another." *Space.com*. 31 August 2011. http://www.space.com/12790-monster-black-hole-cannibal-galaxy-collisions.html (6 December 2018).

Curtius, Ernst Robert. *European Literature and the Latin Middle Ages*. Trans. Willard R. Task. Princeton and London: Princeton University Press, 2013 [ger 1948].

Cutting, James E. "Gustave Caillebotte, French Impressionism, and Mere Exposure." *Psychonomic Bulletin & Review* 10.2 (2003): 319–343.

Darwin, Charles. *The Annotated Origin: A Facsimile of the First Edition of* The Origin of Species. Annotated by James T. Costa. Cambridge, MA and London: The Belknap Press of Harvard University Press, 2011.

D'Huy, Julien. "Polyphemus: A Palaeolithic Tale?" *The Retrospective Methods Network Newsletter* 9 (2015): 43–64.
Disch, Thomas M. "The Embarrassments of Science Fiction" (1978). *On SF*. By Thomas M. Disch. Ann Arbor: The University of Michigan Press, 2005. 3–15.
Dutton, Denis. *The Art Instinct: Beauty, Pleasure, & Human Evolution*. London: Bloomsbury Press, 2009.
Easterlin, Nancy. *A Biocultural Approach to Literary Theory and Interpretation*. Baltimore: The Johns Hopkins University Press, 2012.
Ellmann, Richard. *James Joyce*. Oxford et al.: Oxford University Press, 1983 [1959].
Fresnault-Deruelle, Pierre. "From Linear to Tabular." *The French Comics Theory Reader*. Eds. Ann Miller and Bart Beaty. Leuven: Leuven University Press, 2014. 121–138.
Gould, Stephen Jay. *The Structure of Evolutionary Theory*. Cambridge, MA and London: The Belknap Press of Harvard University Press, 2002.
Gould, Stephen Jay, and Niles Eldredge. "Punctuated Equilibria: The Tempo and Mode of Evolution Reconsidered." *Paleobiology* 3.2 (1977): 115–151.
Hafner, E. M. "The New Reality in Art and Science." *Comparative Studies in Society and History* 11.4 (1969): 385–397.
Hawking, Stephen. "Black Holes and Baby Universes." *Black Holes and Universes and Other Essays*. Toronot et al.: Bantam Books, 1994. 104–114.
Hawking, Stephen. "Into a Black Hole" (2008). *Hawking.org.uk*. https://www.hawking.org.uk/in-words/lectures/into-a-black-hole (23 August 2021).
Hawking, Stephen W., Malcolm J. Perry, and Andrew Strominger. "Soft Hair on Black Holes." *Physical Review Letters* 116.23 (2016): art. 231301. doi:10.1103/PhysRevLett.116.231301.
Howe, Lucien. "Art and Eyesight." *Popular Science Monthly* 47 (1895): 458–471. https://en.wikisource.org/wiki/Popular_Science_Monthly/Volume_47/August_1895/Art_and_Eyesight (23 August 2021).
Kevles, Daniel J. "Pursuing the Unpopular: A History of Courage, Viruses and Cancer." *Hidden Histories of Science*. Ed. Robert B. Silvers. New York: A New York Review Book, 1995. 69–112.
Knapp, Alex. "Scientists Catch A Black Hole Munching On A Giant Planet." *Forbes*. 3 April 2013. https://www.forbes.com/sites/alexknapp/2013/04/03/scientist-catch-a-black-hole-munching-on-a-giant-planet/#d5427ca5967d (6 December 2018).
Kuhn, Thomas S. "The Trouble with the Historical Philosophy of Science." Robert and Maurine Rothschild Distinguished Lecture, 19 November 1991. Cambridge, MA: Harvard University Press, 1992.
Kuhn, Thomas S. *The Structure of Scientific Revolutions*. Chicago: The University of Chicago Press, 1996 [1962].
Luntz, Stephen. "Evidence of Galactic Cannibalism Revealed." *Iflscience*. 4 April 2014. http://www.iflscience.com/space/evidence-galactic-cannibalism-revealed/ (6 December 2018).
Martindale, Colin. "Aesthetic Evolution." *Poetics* 15 (1986): 439–473.
Martindale, Colin. "The Laws Governing the History of Poetry." *Directions in Empirical Literary Studies: In Honor of Willie van Peer*. Eds. Sonia Zyngier, Marisa Bortolussi, Anna Chesnokova, and Jan Auracher. Amsterdam and Philadelphia: John Benjamins, 2008. 229–242.
Mercier, Hugo, and Dan Sperber. *The Enigma of Reason*. Cambridge, MA: Harvard University Press, 2017.
Mesoudi, Alex, Andrew Whiten, and Robin Dunbar. "A Bias for Social Information in Human Cultural Transmission." *British Journal of Psychology* 97 (2006): 405–423.
Miall, David S., and Don Kuiken. "Foregrounding, Defamiliarization, and Affect: Response to Literary Stories." *Poetics* 22 (1994): 389–407.

Miller, Geoffrey F. "How Mate Choice Shaped Human Nature: A Review of Sexual Selection and Human Evolution." *Handbook of Evolutionary Psychology Ideas, Issues, and Applications*. Eds. Charles Crawford and Dennis L. Krebs. Mahwah, NJ: Lawrence Erlbaum Associates, 1998. 87–129.

Miller, Nancy K. "The Entangled Self: Genre Bondage in the Age of the Memoir." *PMLA* 122.2 (2007): 537–548.

Planck, Max. "A Scientific Autobiography." *A Scientific Autobiography and Other Papers*. Trans. Frank Gaynor. London: Williams and Norgate, 1950. 13–51.

Poggioli, Renato. *The Theory of the Avant-Garde*. Trans. Gerald Fitzgerald. Cambridge, MA and London: The Belknap Press of Harvard University Press, 1968.

Radway, Janice A. *Reading the Romance: Women, Patriarchy, and Popular Literature*. Chapel Hill and London: The University of North Carolina Press, 1991 [1984].

Reber, Rolf, Piotr Winkielmann, and Norbert Schwarz. "Effects of Perceptual Fluency on Affective Judgements." *Psychological Science* 9.1 (1998): 45–48.

Reber, Rolf, Norbert Schwarz, and Piotr Winkielman. "Processing Fluency and Aesthetic Pleasure: Is Beauty in the Perceiver's Processing Experience?" *Personality and Social Psychology Review* 8.4 (2004): 364–382.

Ruffini, Remo, and John Wheeler. "Relativistic Cosmology and Space Platform." *Significance of Space Research for Fundamental Physics: Proceedings of an ESRO Colloqium Held at Interlaken, Switzerland, 4 September 1969*. Ed. European Space Research Organization. ESRO Special Publication *SP-52*. Neuilly-sur-Seine: European Space Research Organization, 1971. 45–174.

Salmon, Catherine. "The Pop Culture of Sex: An Evolutionary Window on the Worlds of Pornography and Romance." *Review of General Psychology* 16.2 (2012). 152–160.

Shklovsky, Victor. "Art as Technique" (1917). *Modern Criticism and Theory: A Reader*. Ed. David Lodge. London and New York: Longman, 1988. 16–30.

Swirski, Peter. "Popular and Highbrow Literature: A Comparative View." *CLCWeb: Comparative Literature and Culture* 1.4 (1999): Art. 4. doi:10.7771/1481-4374.1053.

Tabbi, Joseph. "Cognitive Science." *The Routledge Companion to Literature and Science*. Eds. Bruce Clarke and Manuela Rossini. London and New York: Routledge, 2011. 77–88.

Tennyson, Alfred Lord. "In Memoriam A.H.H." *The Collected Poems of Alfred Lord Tennyson*. Were, Hertfordshire: Wordsworth Editions, 1994. 285–365.

Weinberg, Steven. "The Craft of Science and the Craft of Art." *Third Thoughts*. Cambridge, MA, and London: The Belknap Press of Harvard University Press, 2018. 194–210.

Wilson, Edward O. *The Origins of Creativity*. New York and London: Liveright Publishing Corporation, 2017.

Zajonc, Robert B. "Attitudinal Effects of Mere Exposure." *Journal of Personality and Social Psychology* 9.2 (1968): 1–27.

Contributors

Víctor Bermúdez
Literary Theory and Comparative Literature
University of Salamanca
victorbermudez@usal.es

Renate Brosch
English Studies
University of Stuttgart
renate.brosch@ilw.uni-stuttgart.de

Jeffrey K. Bye
Educational Psychology
University of Minnesota
jbye@umn.edu

Marco Caracciolo
English Studies
Ghent University
marco.caracciolo@ugent.be

Federico Corni
Faculty of Education
Free University of Bozen-Bolzano
federico.corni@unibz.it

Elisabeth Dumont
Institute of Applied Mathematics and Physics
Zurich University of Applied Sciences at Winterthur
elisabeth.dumont@zhaw.ch

Anouschka Foltz
English Linguistics
University of Graz
anouschka.foltz@uni-graz.at

Rudolf Freiburg
English Studies
Friedrich-Alexander-Universität Erlangen-Nürnberg (retired)
rudolf.freiburg@fau.de

Hans U. Fuchs
Institute of Applied Mathematics and Physics
Zurich University of Applied Sciences at Winterthur (retired)
hansufuchs@me.com

Ali J. Fulsher
Educational Psychology
University of Minnesota

Renata Gambino
German Studies
University of Catania
renatagambino@gmail.com

Monica Gonzalez-Marquez
Intercultural Studies
Jülich Forschungszentrum
m.gonzalez-marquez@fz-juelich.de

Elisabeth Haefs
German Committee Future Earth
Helmholtz-Zentrum Hereon
elisabeth.haefs@hereon.de

Aura Heydenreich
German Studies and Comparative Literature
Friedrich-Alexander-Universität Erlangen-Nürnberg
aura.heydenreich@fau.de

Sébastien Lemerle
Centre de recherches sociologiques et politiques de Paris
Université Paris Nanterre
slemerle@parisnanterre.fr

Klaus Mecke
Theoretical Physics
Friedrich-Alexander-Universität Erlangen-Nürnberg
klaus.mecke@fau.de

https://doi.org/10.1515/9783110782844-016

Brook Miller
Centre for Humanities and Digital Research
University of Central Florida
brook.miller@ucf.edu

Paola Pennisi
Department of Adult and Childhood Human Pathology "Gaetano Barresi"
University of Messina
ppennisi@unime.it

Grazia Pulvirenti
Humanities
University of Catania
pulvigra@unict.it; grapulvir@gmail.com

Michael Sinding
Literary Studies
Independent Scholar
mjsinding@gmail.com

Dirk Vanderbeke
English Studies
Friedrich-Schiller-Universität Jena (retired)
Palacký University Olomouc
vanderbeke@t-online.de

Matthias Wilde
Biology didactics (zoology/human biology)
Bielefeld University
matthias.wilde@uni-bielefeld.de

Index

https://doi.org/10.1515/9783110782844-017